Mechanics of Materials

Anthony Bedford • Kenneth M. Liechti

Mechanics of Materials

Second Edition

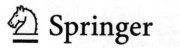 Springer

Anthony Bedford
University of Texas
Austin, TX, USA

Kenneth M. Liechti
University of Texas
Austin, TX, USA

ISBN 978-3-030-22084-6   ISBN 978-3-030-22082-2   (eBook)
https://doi.org/10.1007/978-3-030-22082-2

This Springer imprint is published by the registered company Springer Nature Switzerland AG
The registered company address is: Gewerbestrasse 11, 6330 Cham, Switzerland

Preface

Mechanics of materials is the study and the analysis of the internal forces within materials and the deformations that result from those forces. This book appears in a time of transition for education in the mechanics of materials. The traditional course in "strength of materials" that long formed an important part of the engineering curriculum had as one of its primary goals, acquainting the students with the details of many analytical and empirical solutions that could be applied to structural design. This reliance on a catalog of results has lessened as the finite element method has become commonly available for stress analysis. Another important development is that current research in mechanics of materials is beginning to bring to reality the dream of the merger of continuum solid mechanics and material science into a unified field. For these reasons, the emphasis in the first course in mechanics of materials is becoming oriented more toward helping students understand the theoretical foundations, especially the concepts of stress and strain, the stress-strain relations including the meaning of isotropy, and the criteria for failure and fracture.

In Chap. 1, we provide an extensive review of statics, with problems, that the instructor may choose to cover or simply have students read. In reviewing distributed loads, we lay the groundwork for our definitions in Chap. 2 of the normal and shear stresses. Chapter 2 also introduces the longitudinal and shear strains in terms of changes in infinitesimal material elements. Chapters 3 and 4 cover bars subjected to axial and torsional loads and introduce the definitions of the elastic and shear moduli. In Chaps. 5 and 6, we discuss the internal forces and moments and the states of stress in beams. With the examples in Chaps. 2, 3, 4, 5, and 6 as motivation, in Chaps. 7 and 8, we discuss the general states of stress and strain and their transformations. Chapters 9 and 10 cover deformations of beams and the buckling of columns. Energy methods are introduced in Chap. 11, and in Chap. 12, we discuss failure criteria for general states of stress and introduce modern fracture mechanics, which we believe should become an integral part of the first course in mechanics of materials. Most of the topics in Chaps. 1, 2, 3, 4, 5, 6, 7, 8, 9, and 10 and selected topics from Chapters 11 to 12 can be covered in a typical one-semester course.

The first course in mechanics of materials prepares students for subsequent courses in structural analysis, structural dynamics, and advanced mechanics of deformable media. In comparison with the courses in statics and dynamics, it is an interesting challenge for instructors and students. Engineering students begin the study of statics and dynamics having had some prior experience with the basic concepts in high school and college physics courses. In contrast, the core concepts in mechanics of materials are new to most students. It is therefore essential that a textbook introduce and explain these concepts with great care and reinforce them with many examples. This has been our objective in writing this book. Our approach is to present the material as we do in the classroom, emphasizing understanding of the basic principles of mechanics of materials and demonstrating them with examples drawn from contemporary engineering applications and design.

Features

- **Examples That Teach** We continue reinforcing the problem-solving skills students have learned in their introductory mechanics courses. Separate *strategy* sections that precede most examples and selected problems teach students how to approach and solve problems in engineering: What principles apply? What must be determined? And in what order? Many examples conclude with *discussion* sections that discuss ways of checking and interpreting answers, point out interesting features of the solution, or suggest alternative methods of solution.
- **Free-Body Diagrams** Correct and consistent use of free-body diagrams is the most essential skill that students of mechanics must acquire. We review the steps involved in drawing free-body diagrams in Chap. 1 and emphasize their use throughout the book. Many of our figures are designed to teach how free-body diagrams are chosen and drawn.

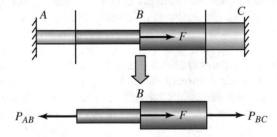

- **Design** The Accreditation Board for Engineering and Technology (ABET) and many practicing engineers strongly encourage the introduction of design throughout the engineering curriculum. By expressing many of our examples and problems in terms of design, we demonstrate the use of mechanics of materials within the larger context of engineering practice. Chapters 3, 4, 6, and 7 contain sections explicitly addressing the design of particular structural elements, and in Chap. 12, we discuss failure and fracture criteria used in structural design.

Austin, TX, USA   Anthony Bedford
                                                                    Kenneth M. Liechti

Contents

Unit Conversion Factors

LENGTH

$1\,\text{m} = 100\,\text{cm} = 3.281\,\text{ft} = 39.37\,\text{in}$

$1\,\text{in} = 0.08333\,\text{ft} = 0.02540\,\text{m}$

$1\,\text{ft} = 12\,\text{in} = 0.3048\,\text{m}$

ANGLE

$1\,\text{rad} = 180/\pi\,\text{deg} = 57.30\,\text{deg}$

$1\,\text{deg} = \pi/180\,\text{rad} = 0.01745\,\text{rad}$

$1\,\text{revolution} = 2\pi\,\text{rad} = 360\,\text{deg}$

$1\,\text{rev}/\text{min}\ (\text{rpm}) = 0.1047\,\text{rad}/\text{s}$

AREA

$1\,\text{mm}^2 = 1.550\text{E-}3\,\text{in}^2 = 1.076\text{E-}5\,\text{ft}^2$

$1\,\text{m}^2 = 10.76\,\text{ft}^2$

$1\,\text{in}^2 = 645.2\,\text{mm}^2$

$1\,\text{ft}^2 = 144\,\text{in}^2 = 0.0929\,\text{m}^2$

VOLUME

$1 \text{ mm}^3 = 6.102\text{E-5 in}^3 = 3.531\text{E-8 ft}^3$

$1 \text{ m}^3 = 6.102\text{E4 in}^3 = 35.31 \text{ ft}^3$

$1 \text{ in}^3 = 1.639\text{E4 mm}^3 = 1.639\text{E-5 m}^3$

$1 \text{ ft}^3 = 0.02832 \text{ m}^3$

MASS

$1 \text{ kg} = 0.0685 \text{ slug}$

$1 \text{ slug} = 14.59 \text{ kg}$

$1 \text{ t (metric tonne)} = 1000 \text{ kg} = 68.5 \text{ slug}$

FORCE

$1 \text{ N} = 0.2248 \text{ lb}$

$1 \text{ lb} = 4.448 \text{ N}$

$1 \text{ kip} = 1000 \text{ lb} = 4448 \text{ N}$

$1 \text{ ton} = 2000 \text{ lb} = 8896 \text{ N}$

WORK AND ENERGY

$1 \text{ J} = 1 \text{ N-m} = 0.7376 \text{ ft-lb}$

$1 \text{ ft-lb} = 1.356 \text{ J}$

POWER

$1 \text{ W} = 1 \text{ N-m/s} = 0.7376 \text{ ft-lb/s} = 1.340\text{E-3 hp}$

$1 \text{ ft-lb/s} = 1.356 \text{ W}$

$1 \text{ hp} = 550 \text{ ft-lb/s} = 746 \text{ W}$

PRESSURE, STRESS

$1 \text{ Pa} = 1 \text{ N/m}^2 = 0.0209 \text{ psf} = 1.451\text{E-4 psi}$

$1 \text{ bar} = 100{,}000 \text{ Pa}$

1 psi $\left(\text{lb/in}^2\right) = 144$ psf $= 6891$ Pa

1 ksi $= 1\,\text{kip/in}^2 = 1000$ psi

1 psf $\left(\text{lb/ft}^2\right) = 6.944\text{E-3}$ lb/in$^2 = 47.85$ Pa

1 ksf $= 1\,\text{kip/ft}^2 = 1000$ psf

STRESS INTENSITY FACTOR

1 Pa-m$^{1/2} = 9.10\text{E-4}$ psi-in$^{1/2} = 0.0379$ psf-ft$^{1/2}$

1 psi-in$^{1/2} = 41.57$ psf-ft$^{1/2} = 1100$ Pa-m$^{1/2}$

1 psf-ft$^{1/2} = 0.02406$ psi-in$^{1/2} = 26.4$ Pa-m$^{1/2}$

About the Authors

Anthony Bedford is a Professor Emeritus of Aerospace Engineering and Engineering Mechanics at the University of Texas at Austin. He has been on the faculty since 1968. He also has industrial experience at Douglas Missiles and Space Systems Division, Sandia National Laboratories, and at TRW, where he worked on the Apollo program. His main professional activity has been education and research in engineering mechanics. He is the coauthor of *Engineering Mechanics: Statics and Dynamics* (with Wallace T. Fowler) and *Introduction to Elastic Wave Propagation* (with Douglas S. Drumheller).

Kenneth M. Liechti is a Professor of Aerospace Engineering and Engineering Mechanics at the University of Texas at Austin and holds the Zarrow Centennial Professorship in Engineering. He received his B.Sc. degree in Aeronautical Engineering at Glasgow University and M.S. and Ph.D. degrees in Aeronautics at the California Institute of Technology. He gained industrial experience at General Dynamics Fort Worth Division prior to joining the faculty of the University of Texas at Austin in 1982.

Dr. Liechti's main areas of teaching and research are in the mechanics of materials and fracture mechanics. He is the author or coauthor of papers on interfacial fracture, mechanics of contact, adhesion and friction at smaller and smaller scales, and the nonlinear behavior of polymers. He has consulted on fracture mechanics problems with several companies.

He is a fellow of the Adhesion Society, the Society of Experimental Mechanics, the American Society of Mechanical Engineers, and the American Academy of Mechanics, and an associate fellow of the American Institute of Aeronautics and Astronautics. He the joint editor of the *Journal of the Mechanics of Time-Dependent Materials*.

Chapter 1
Introduction

Mechanics of materials is one of the sciences underlying the design of any device that must support loads, from the simplest machines and tools to complex vehicles and structures. Here we discuss the central questions addressed in mechanics of materials and review some background needed to begin studying this subject.

1.1   Engineering and Mechanics of Materials

In the study of statics, we were concerned with the analysis of forces and couples acting on objects in equilibrium. It was tacitly assumed that objects could support the forces and couples to which they were subjected without collapsing. Furthermore, deformations, or changes in the shapes of objects due to external loads, were disregarded. It was assumed that objects were effectively rigid, meaning that their deformations were small enough to be neglected. In mechanics of materials, we develop the concepts and tools of analysis needed to examine the ability of objects to support loads and determine their resulting deformations. We examine three fundamental questions:

1. ***Will an object or structure support the loads acting on it?*** This is the principal question faced by structural design engineers. Will the parts of a machine perform their functions without breaking? Will the frame of a building support the weight of the building itself, weights of the building's contents and occupants, and environmental loads without collapsing? Will an airplane's wing support the gravitational and aerodynamic forces to which it will be subjected? To show how these questions are answered, we introduce the state of stress, which describes the forces within a material. We show how the states of stress within an object determine whether it will support specified external loads. Although the examples we present are relatively simple, the underlying procedure applies to all

© Springer Nature Switzerland AG 2020
A. Bedford, K. M. Liechti, *Mechanics of Materials*,
https://doi.org/10.1007/978-3-030-22082-2_1

structural designs. The state of stress must be determined throughout a structure to ensure that it does not exceed the capacities of the materials used.

2. *What is the change in shape, or deformation, of an object subjected to loads?* Pulling on a rubber band causes it to stretch. An airliner's wings can be observed to flex when they are subjected to loads by turbulent air. Any object deforms when it is subjected to loads. We introduce the concept of strain, which describes the state of deformation in the neighborhood of a given point of an object, and show how it is used in determining the deformations of simple objects. For the model of material behavior called linear elasticity, we present the relationships between the state of stress at a point and the state of strain. Design engineers must be concerned with how objects change shape due to the loads acting on them—for example, the members of a linkage must not be allowed to deform so much that the functioning of the linkage is affected—but engineers also determine deformations for another important reason. Doing so enables them to solve statically indeterminate problems.

3. *What can be done if the external loads on an object cannot be determined by using the equilibrium equations?* When the number of unknown reactions on the free-body diagram of an object exceeds the number of independent equilibrium equations that are available, the object is said to be statically indeterminate. Such problems, which are very common in engineering, cannot be solved using the methods of statics alone. We show that by supplementing the equilibrium equations with the relationships between the loads acting on an object and its deformation, the unknown reactions can be determined.

Before we begin answering these three questions in Chap. 2, we need to discuss units and review essential concepts from statics.

1.2   Units

We use both the International System of Units, called SI units, and the US customary system of units in examples and problems.

1.2.1   SI Units

Length is measured in meters (abbreviated m), mass in kilograms (kg), and time in seconds (s). These are the *base units* of the SI system. The unit of force is a derived unit, the newton (N), which is the force required to give an object of 1 kilogram mass an acceleration of 1 meter per second squared. The relationship between an object's mass m in kilograms and its weight W at sea level in newtons is $W = mg$, where $g = 9.81$ m/s^2 is the acceleration due to gravity at sea level.

Prefix	Abbreviation	Multiple
nano-	n	10^{-9}
micro-	μ	10^{-6}
milli-	m	10^{-3}
kilo-	k	10^{3}
mega-	M	10^{6}
giga-	G	10^{9}

Table 1.1 Common prefixes used in SI units and the multiples they represent

Pressures and stresses are usually expressed in the SI system in newtons per square meter (N/m^2), which are called pascals (Pa). Occasionally pressures and stresses are expressed in bars (1 bar $= 10^5$ Pa).

To express quantities in SI units by numbers of convenient size, multiples of units are indicated by prefixes. The most common prefixes, their abbreviations, and the multiples they represent are shown in Table 1.1. For example, 1 GPa (gigapascal) is 1×10^9, or 1E9 Pa.

1.2.2 US Customary Units

Length is measured in feet (ft), force in pounds (lb), and time in seconds (s). These are the base units of the US customary system. In this system the unit of mass is a derived unit, the slug, which is the mass of material accelerated at 1 foot per second squared by a force of 1 pound. The relationship between an object's mass m in slugs and its weight W at sea level in pounds is $W = mg$, where $g = 32.2$ ft/s^2. We frequently use the inch (12 in $= 1$ ft) and kilopound (1 kip $= 1000$ lb).

Pressures and stresses are expressed in US customary units in terms of pounds per square foot (lb/ft^2 or psf) or pounds per square inch (lb/in^2 or psi). We also use thousands of pounds per square foot (kip/ft^2 or ksf) and thousands of pounds per square inch (kip/in^2 or ksi).

1.3 Review of Statics

Many of the fundamental concepts of mechanics of materials that we discuss in this book—including the states of stress and strain in materials—are applicable to both static and dynamic situations. However, for simplicity in this introductory treatment, we consider only materials and objects that are in equilibrium. In this section we review concepts from statics that are needed to understand the developments in subsequent chapters.

1.3.1   Free-Body Diagrams

A *free-body diagram* is simply a drawing of a particular object that shows the external forces and couples acting on it. The term *free body* means that the drawing shows the object (body) isolated from its surroundings, with no other material object included. A good rule of thumb to remember is that a free-body diagram can't be attached to anything or it isn't free. The purpose of a free-body diagram is to identify unambiguously all the external forces and couples acting on a given object. *External* forces and couples mean those exerted on the object by everything that is not included in the free-body diagram. For example, suppose we want to draw the free-body diagram of the standing person in Fig. 1.1(a). We begin by making a

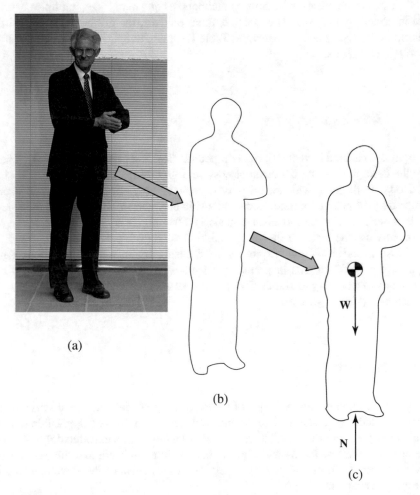

(a)

(b)

(c)

Fig. 1.1 Drawing the free-body diagram of a standing person: (**a**) The person. (**b**) Sketch of the isolated person. (**c**) Completed free-body diagram

drawing of the person in which he is isolated from his surroundings, meaning that we imagine removing everything but the person (Fig. 1.1(b)). Then we add to the drawing the forces and couples exerted by everything that was removed. His weight, exerted by the earth, is an external force, and the forces exerted on his feet by the sidewalk are external (Fig. 1.1(c)). Thinking about what was removed in order to isolate the person helps us recognize the external forces. In Fig. 1.1(a), the person is pressing his hands together. His hands exert forces on each other, but in a free-body diagram of the person, those forces are internal, not external, and are therefore not shown in a free-body diagram of the person.

Drawing a free-body involves three steps, which are illustrated in Fig. 1.2 with a typical example from statics:

*Step 1. **Identify the object to be isolated**. The choice is often dictated by unknown forces and couples that are to be determined. The L-shaped bar in Fig. 1.2 is

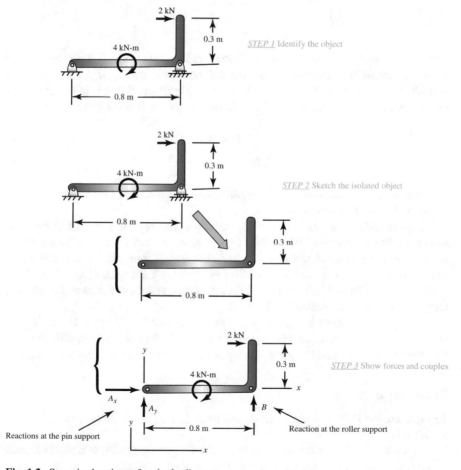

Fig. 1.2 Steps in drawing a free-body diagram

subjected to two loads—a 2-kN force and a 4-kN-m couple—and has pin and roller supports. Our objective is to draw a free-body diagram that would allow us to determine the reactions exerted on the bar by the two supports.

Step 2. Draw a sketch of the object isolated from its surroundings. We carry out this step by drawing a sketch of the bar isolated from its supports.

Step 3. Indicate and label the external forces and couples that act on the isolated object. We complete the free-body diagram of the bar by adding the applied loads and indicating and labeling the reactions exerted by the pin and roller supports. Common supports and the reactions they can exert are shown in Table 1.2.

Notice in Fig. 1.2 that the weight of the L-shaped bar is not included in its free-body diagram. The reason is that the bar's weight is regarded as negligible in comparison to the external loads acting on it.

1.3.2   Equilibrium

We say that an object is in *equilibrium* during an interval of time if it is stationary or in steady translation relative to an inertial reference frame. If an object is in equilibrium, the sum of the external forces acting on it and the sum of the moments about any point due to the external forces and couples acting on it are zero:

$$\Sigma \mathbf{F} = \mathbf{0}, \tag{1.1}$$

$$\Sigma \mathbf{M}_{\text{any point}} = \mathbf{0}. \tag{1.2}$$

In some situations we can use the equilibrium equations to determine unknown forces and couples acting on objects.

A *system of forces and moments* is simply some particular set of forces and moments due to couples. We define two systems of forces and moments to be *equivalent* if the sums of the forces in the two systems are equal and the sums of the moments about any point due to the two systems are equal. Notice that if the equilibrium equations (1.1) and (1.2) hold for a given system of forces and moments, they also hold for any equivalent system.

A *two-force member* is defined to be an object that is subjected to two forces acting at different points and no couples. If a two-force member is in equilibrium, the two forces must be equal in magnitude, opposite in direction, and have the same line of action (Fig. 1.3).

Study Questions

1. What are the units of length, mass, and force in the SI system of units?
2. What is a pascal (Pa)?
3. In drawing the free-body diagram of an object, you do not show internal forces. How do you decide whether a force is external or internal?
4. If two systems of forces and moments are equivalent, what do you know about them?

Supports	**Reactions**
Rope or Cable Spring	A Collinear Force T
Contact with a Smooth Surface	A Force Normal to the Supporting Surface A
Contact with a Rough Surface	Two Force Components y x A_x A_y
Pin Support	Two Force Components y x A_x A_y
Roller Support Equivalents	A Force Normal to the Supporting Surface A
Constrained Pin or Slider	A Normal Force A
Fixed (Built-in) Support	Two Force Components and a Couple M_A y A_x A_y x

Table 1.2

Fig. 1.3 A two-force
member

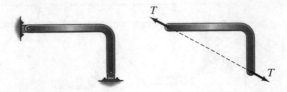

Example 1.1 Determining Reactions at a Support
The beam in Fig. 1.4 supports a 100-kN force and a 600-kN-m couple.
Determine the reactions at the fixed support A.

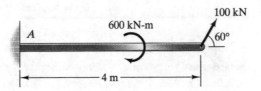

Fig. 1.4

Strategy
We will draw a free-body diagram of the beam by isolating it from the fixed
support and then apply the equilibrium equations to determine the reactions.

Solution
We draw the free-body diagram of the beam in Fig. (a). The terms A_x and A_y
are the components of the force, and M_A is the couple exerted on the beam by
the support.

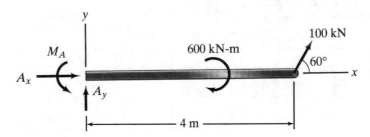

(a) Free-body diagram of the beam

From the equilibrium equations

$$\Sigma F_x = A_x + (100 \text{ kN}) \cos 60^\circ = 0,$$
$$\Sigma F_y = A_y + (100 \text{ kN}) \sin 60^\circ = 0,$$
$$\Sigma M_{\text{point } A} = M_A - 600 \text{ kN-m} + (4 \text{ m})\left[(100 \text{ kN}) \sin 60^\circ\right] = 0,$$

we obtain the reactions (to three significant digits) $A_x = -50$ kN,
$A_y = -86.6$ kN and $M_A = 254$ kN−m.

Discussion

Suppose that the beam in this example has a roller support at the right end instead of the 100-kN external force (Fig. (b)). The beam is now statically indeterminate. There are four unknown reactions (Fig. (c)), but we can write only three independent equilibrium equations:

$$\Sigma F_x = A_x = 0,$$
$$\Sigma F_y = A_y + B = 0,$$
$$\Sigma M_{\text{point } A} = M_A - 600 \text{ kN-m} + (4 \text{ m})B = 0.$$

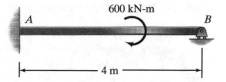

(b) Placing a roller support at the right end of the beam

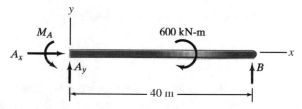

(c) There are four unknown reactions

We cannot determine the reactions A_y, M_A, and B from these equations. In Chap. 9, problems such as these are solved by using the equilibrium equations together with equations that relate the loads acting on the beam to its lateral deformation.

1.3.3 Structures

Here we consider structures composed of interconnected parts or *members*. A structure is called a *truss* if it consists entirely of two-force members. A typical truss is made up of straight bars connected at their ends by pins, it is loaded by forces at joints where the members are connected, and it is supported at joints (Fig. 1.5). Because the members of a truss are two-force members, they are subjected only to axial forces. In Example 1.2, we review methods for determining the axial forces in the members of a statically determinate truss.

A structure of interconnected members that does not satisfy the definition of a truss is called a *frame* if it is designed to remain stationary and support loads, and it is called a *machine* if it is designed to move and exert loads. Such structures can be analyzed to determine the reactions at the supports of the structure and at the connections between the members by drawing free-body diagrams of the individual members. We review this process for a statically determinate frame in Example 1.3.

Fig. 1.5 Howe bridge truss.
Notice that it is loaded and
supported at its joints

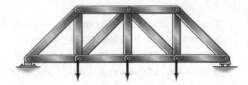

Example 1.2 Axial Force in a Truss Member
Determine the axial force in member *BH* of the Howe truss in Fig. 1.6 and
indicate whether the member is in tension or compression.

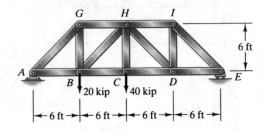

Fig. 1.6

Strategy
We will apply what is called the method of joints, using equilibrium to analyze
carefully chosen joints until we have determined the axial force in
member *BH*.

Solution
We first draw the free-body diagram of the entire truss (Fig. (a)). From the
equilibrium equations for this free-body diagram, we determine that the
reactions at the supports are $A_x = 0$, $A_y = 35$ kip and $E = 25$ kip.

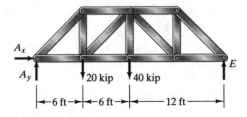

(a) Free-body diagram of the entire truss

Applying the method of joints, we first draw the free-body diagram of joint
A (Fig. (b)), because it is subjected to a known force, the 35-kip reaction at *A*,
and only two unknown forces, the axial forces in members *AB* and *AG*. From
the equilibrium equations

$$\Sigma F_x = P_{AB} + P_{AG}\cos 45^\circ = 0,$$
$$\Sigma F_y = 35 \text{ kip} + P_{AG}\sin 45^\circ = 0,$$

we obtain $P_{AB} = 35$ kip and $P_{AG} = -49.5$ kip. (The signs of these answers indicate that member AB is in tension and member AG is in compression.) Now that we know P_{AG}, the free-body diagram of joint G has only two unknown forces (Fig. (c)). From the equilibrium equations

$$\Sigma F_x = P_{GH} - P_{AG}\sin 45^\circ = 0,$$
$$\Sigma F_y = -P_{BG} - P_{AG}\cos 45^\circ = 0,$$

we obtain $P_{BG} = 35$ kip and $P_{GH} = -35$ kip. We can now use the free-body diagram of joint B to determine the axial force in member BH (Fig. (d)). From the equation

$$\Sigma F_y = P_{BG} + P_{BH}\sin 45^\circ - 20 \text{ kip} = 0,$$

we obtain $P_{BH} = -21.2$ kip. Member BH is subjected to a *compressive* axial force of 21.2 kip.

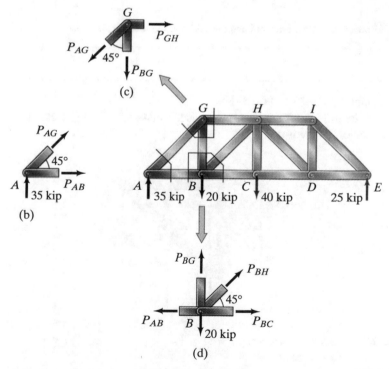

Free-body diagrams of (b) joint A; (c) joint G; (d) joint B

Discussion

We can also determine the axial force in member BH by applying what is called the method of sections. In Figure (e), we isolate a section of the truss by "cutting" members BC, BH, and GH. From the equilibrium equation

$$\Sigma F_y = 35 \text{ kip} - 20 \text{ kip} + P_{BH} \sin 45° = 0,$$

we obtain $P_{BH} = -21.2$ kip. The method of sections can sometimes provide the information that is needed much more easily than the method of joints, but it isn't always possible to find a section that works.

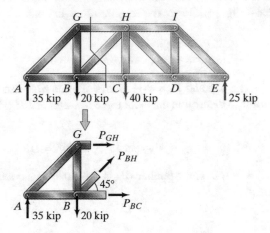

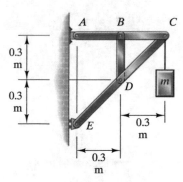

(e) A section suitable for determining the axial force in member *BH*

Example 1.3 Analysis of a Frame
The frame in Fig. 1.7 supports a suspended mass $m = 20$ kg. Determine the reactions on its members.

Fig. 1.7

Strategy
We will draw the free-body diagrams of the individual members and apply the equilibrium equations to each member to determine the reactions. Notice that, although members *ABC* and *CDE* are not two-force members, member *BD* is. We can take advantage of this observation to simplify the free-body diagrams of members *ABC* and *CDE*—member *BD* exerts equal and opposite forces at

B and D that are parallel to member BD—and we do not need to write equilibrium equations for member BD.

Solution
In Fig. (a) we draw the free-body diagrams of members ABC and CDE. We denote the equal and opposite forces exerted by member BD by P. Also, we have assumed that the force exerted by the weight mg acts on member ABC. We could have assumed instead that it acted on member CDE.

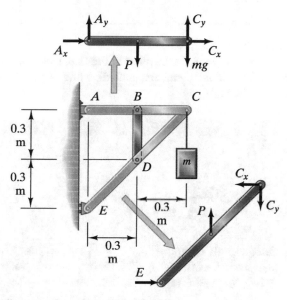

(a) Free-body diagrams of members ABC and CDE

We can write three independent equilibrium equations for each free-body diagram, obtaining six equations in terms of the unknown reactions A_x, A_y, P, C_x, C_y and E. The equilibrium equations are
Member ABC:

$$\Sigma F_x = A_x + C_x = 0,$$
$$\Sigma F_y = A_y + C_y - P - mg = 0,$$
$$\Sigma M_{\text{point } A} = -(0.3\text{ m})P + (0.6\text{ m})C_y - (0.6\text{ m})mg = 0.$$

Member CDE:

$$\Sigma F_x = E - C_x = 0,$$
$$\Sigma F_y = P - C_y = 0,$$
$$\Sigma M_{\text{point } E} = (0.3\text{ m})P + (0.6\text{ m})C_x - (0.6\text{ m})C_y = 0.$$

Setting $m = 20\,\text{kg}$ and $g = 9.81\,\text{m/s}^2$ and solving yields $A_x = -196\,\text{N}$, $A_y = 196\,\text{N}$, $P = 392\,\text{N}$, $C_x = 196\,\text{N}$, $C_y = 392\,\text{N}$, and $E = 196\,\text{N}$.

Discussion
Why is it useful to determine the forces acting on the members of the frame? To determine whether a structure will safely support the loads to which it is subjected, the first step is to determine all of the forces and couples acting on its members, as we do in this example. The next step is to determine whether the individual members will support those forces and couples. We develop the concepts and methods required to do this in the following chapters.

Example 1.4 Analysis of a Machine
The person exerts 40-N forces on the pliers in Fig. 1.8. Determine the resulting reactions on the individual members, including the forces exerted on the members by the object at *A*.

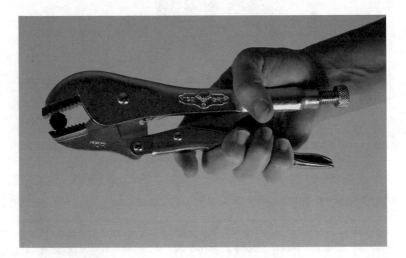

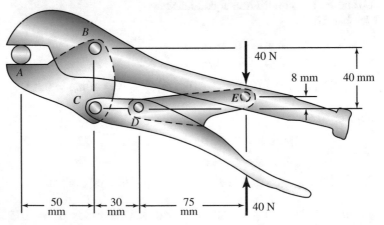

Fig. 1.8

Strategy

By drawing free-body diagrams of the individual members and applying the equilibrium equations, we can determine the forces on the members.

Solution

In Figs. (a), (b), and (c), we draw the free-body diagrams of the individual members of the pliers. Notice that in free-body diagrams (a) and (c) we use the fact that DE is a two-force member. The angle $\alpha = \arctan(8/75) = 6.09°$. Next, we write the equilibrium equations for each free-body diagram.

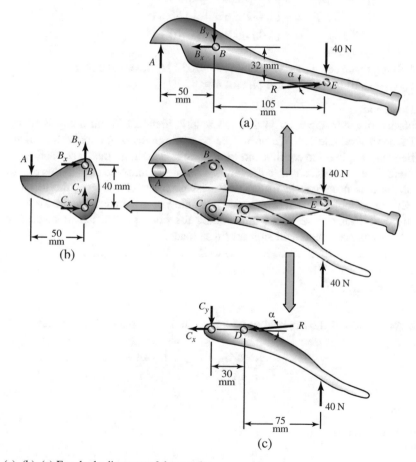

(a), (b), (c) Free-body diagrams of the members

Free-body diagram (a):

$$\Sigma F_x = -B_x + R\cos\alpha = 0,$$
$$\Sigma F_y = A - B_y + R\sin\alpha - 40\,\text{N} = 0,$$
$$\Sigma M_{\text{point } B} = -(0.05\,\text{m})A + (0.105\,\text{m})R\sin\alpha$$
$$+ (0.032\,\text{m})R\cos\alpha - (0.105\,\text{m})(40\,\text{N}) = 0.$$

Free-body diagram (b):

$$\Sigma F_x = B_x + C_x = 0,$$
$$\Sigma F_y = -A + B_y + C_y = 0,$$
$$\Sigma M_{\text{point } B} = (0.05 \text{ m})A + (0.04 \text{ m})C_x = 0.$$

Free-body diagram (c):

$$\Sigma F_x = -C_x - R\cos\alpha = 0,$$
$$\Sigma F_y = -C_y - R\sin\alpha + 40 \text{ N} = 0,$$
$$\Sigma M_{\text{point } C} = -(0.03 \text{ m})R\sin\alpha + (0.105 \text{ m})(40 \text{ N}) = 0.$$

Solving these equations, we obtain $A = 1050$ N, $B_x = 1310$ N, $B_y = 1150$ N, $C_x = -1310$ N, $C_y = -100$ N, and $R = 1320$ N.

Discussion
Notice that our approach in this example is identical to our analysis of the frame in Example 1.3. We drew free-body diagrams of the individual members and applied the equilibrium equations to determine the forces acting on them. The motivation is also the same. To design a machine such as this pair of pliers, it is necessary to insure that *each member* will support the loads to which it is subjected, and the first step in doing so is to determine the loads. In the next chapter, we begin introducing the concepts needed to determine whether the members will support those loads.

1.3.4   Centroids

The coordinates of the centroid of an area A in the x–y plane (Fig. 1.9) are

$$\bar{x} = \frac{\int_A x\, dA}{\int_A dA}, \qquad \bar{y} = \frac{\int_A y\, dA}{\int_A dA}.$$

Fig. 1.9 Coordinates of the centroid of A

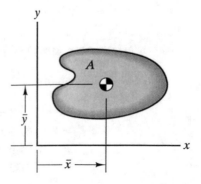

Fig. 1.10 Composite area
showing the coordinates of
the centroids of the parts

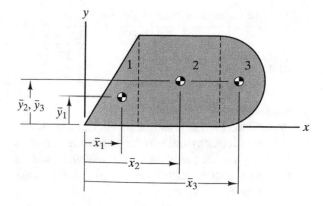

The centroid of a composite area consisting of parts 1, 2, ... whose centroid
locations are known (Fig. 1.10) can be determined from the relations

$$\bar{x} = \frac{\bar{x}_1 A_1 + \bar{x}_2 A_2 + \cdots}{A_1 + A_2 + \cdots}, \qquad \bar{y} = \frac{\bar{y}_1 A_1 + \bar{y}_2 A_2 + \cdots}{A_1 + A_2 + \cdots}.$$

A "hole" or cutout can be treated as a negative area (see Example 1.5).

Example 1.5 Centroid of a Composite Area
Determine the x coordinate of the centroid of the area in Fig. 1.11.

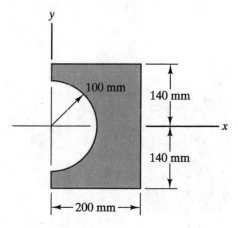

Fig. 1.11

Strategy
We can determine the x coordinate of the centroid by treating the area as a
composite area consisting of the rectangle without the semicircular cutout and
the (negative) area of the cutout.

Table 1.3 Information for determining $\bar{x}$

	$\bar{x}_i$ (mm)	A_i (mm^2)	$\bar{x}_i A_i$ (mm^3)
Part 1 (rectangle)	100	$(200)(280)$	$(100)[(200)(280)]$
Part 2 (cutout)	$\frac{(4)(100)}{3\pi}$	$-\frac{1}{2}\pi(100)^2$	$-\frac{(4)(100)}{3\pi}\left[\frac{1}{2}\pi(100)^2\right]$

Solution

We will treat the area as a composite area consisting of the rectangle without the semicircular cutout and the area of the cutout, which we call parts 1 and 2 (Fig. (a)). Let the areas of the rectangle without the semicircular cutout and the cutout be called parts 1 and 2 (Fig. (a)). From Appendix D, the x coordinate of the centroid of the cutout is

$$\bar{x}_2 = \frac{4R}{3\pi} = \frac{4(100 \text{ mm})}{3\pi}.$$

The information for determining the x coordinate of the centroid is summarized in Table 1.3. (Notice that the cutout is treated as a negative area.) The x coordinate of the centroid is

$$\bar{x} = \frac{\bar{x}_1 A_1 + \bar{x}_2 A_2}{A_1 + A_2}$$

$$= \frac{(100 \text{ mm})[(200 \text{ mm})(280 \text{ mm})] - [(4)(100 \text{ mm})/3\pi]\left[\frac{1}{2}\pi(100 \text{ mm})^2\right]}{(200 \text{ mm})(280 \text{ mm}) - \frac{1}{2}\pi(100 \text{ mm})^2}$$

$$= 122 \text{ mm}.$$

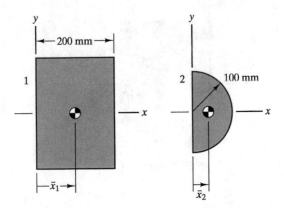

(a) Rectangle and semicircular cutout

Discussion

If a thin plate of homogeneous material with uniform thickness has the shape of the area in this example, the location of the centroid of the area coincides with the center of mass of the plate. That is one of the motivations for determining the location of the centroid of an area. In mechanics of materials, there are other reasons for determining centroids of areas. We show in Chap. 6 that in order to determine whether a beam will support the loads acting on it, the location of the centroid of its cross section must be known.

1.3.5   Distributed Forces

To describe a force that is distributed along a line, we define a function w, called the *loading curve*, such that the *downward* force exerted on each infinitesimal element dx of the line is $w\,dx$ (Fig. 1.12). For example, the force exerted by a floor of a building on one of the horizontal beams supporting the floor is distributed along the length of the beam. This type of distributed load can also be used to model a large number of discrete loads acting along a line, such as the forces exerted by the wheels of the traffic on a bridge.

The total downward force exerted by the distributed load on a portion L of the x axis is

Fig. 1.12 (a) Force distributed along the x axis. (b) The force on each element dx is $w\,dx$

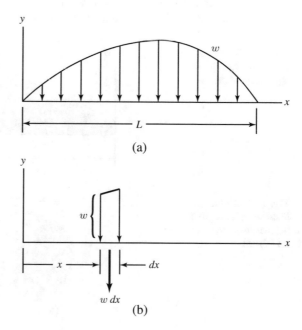

$$F = \int_L w \, dx,$$

and the total clockwise moment about the origin is

$$M = \int_L xw \, dx.$$

We can calculate the area A between the x axis and the loading curve in Fig. 1.12(a) by integrating the infinitesimal element of area $dA = w \, dx$ shown in Fig. 1.12(b):

$$A = \int_A dA = \int_L w \, dx = F.$$

The force F is equal to the area A (Fig. 1.13). The x coordinate of the centroid of A is

$$\bar{x} = \frac{\int_A x \, dA}{\int_A dA} = \frac{\int_L xw \, dx}{\int_L w \, dx} = \frac{M}{F}.$$

This equation shows that if the force F is placed at the centroid of A (Fig. 1.13), the moment about the origin due to F is equal to the total moment about the origin due to the distributed load. The force F is then equivalent to the distributed load. We review the use of this equivalence in Example 1.6.

The force exerted by pressure on an object submerged in a stationary liquid is distributed over the surface of the object (Fig. 1.14(a)). The pressure p is defined such that the normal force exerted on an infinitesimal element dA of the object's

Fig. 1.13 Representing a distributed load by an equivalent force

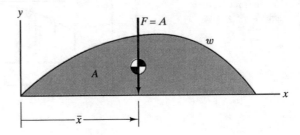

Fig. 1.14 (a) Pressure is a distributed load. (b) The normal force on an element dA is $p \, dA$

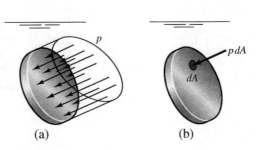

Table 1.4 Distributed loads

Type of distributed load	Domain	Force on an element
Distributed load on a beam	Line	$w\,dx$
Pressure	Area	$p\,dA$
Weight	Volume	$\gamma\,dV$

surface is $p\,dA$ (Fig. 1.14(b)). The pressure increases with depth, so it is usually necessary to integrate to determine the total force exerted on a finite part of an object's surface. The total normal force on a plane surface of area A is

$$F = \int_A p\,dA.$$

An object's weight is another example of a distributed force. The *density ρ* of a material is defined such that the mass of an infinitesimal element of the material of volume dV is $\rho\,dV$. The *weight density γ* of a material is defined such that the weight of an infinitesimal element of the material of volume dV is $\gamma\,dV$. The weight density at sea level and the density are related by $\gamma = \rho g$. For example, the mass density of tungsten is $\rho = 19,300$ kg/m³, so its weight density is

$$\gamma = \left(19,300 \text{ kg/m}^3\right)\left(9.81 \text{ m/s}^2\right) = 189 \text{ kN/m}^3.$$

An object's total weight W is obtained by integrating its weight density over its volume

$$W = \int_V \gamma\,dV,$$

and if the object is homogeneous ($\gamma = $ constant), its total weight is $W = \gamma V$.

We have discussed loads distributed over lines, areas, and volumes (Table 1.4). In each case the load is characterized by a function that determines the force acting on an infinitesimal element of its domain of definition. In Chap. 2 we show that the stresses in materials are described in terms of distributed loads that are defined in exactly the same way as these familiar examples.

Example 1.6 Beam with a Distributed Load
The beam in Fig. 1.15 is subjected to a triangular distributed load whose value at the right end of the beam is 120 lb/ft. What are the reactions at A and B?

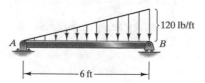

Fig. 1.15

Strategy
We will draw the free-body diagram of the beam by isolating it from its supports. Then by representing the distributed load by an equivalent force and applying the equilibrium equations, we can determine the reactions at the supports A and B.

Solution
We draw the free-body diagram of the beam in Fig. (a), showing the reactions at the pin support at A and the roller support at B. To represent the distributed force by an equivalent force, we determine the magnitude of the force by calculating the triangular area under the loading curve:

$$F = \frac{1}{2}(120 \text{ lb/ft})(6 \text{ ft}) = 360 \text{ lb},$$

and we place the force at the centroid of the triangular area, $\bar{x} = \frac{2}{3}(6 \text{ ft}) = 4 \text{ ft}$ (Fig. (b)). From the equilibrium equations

$$\Sigma F_x = A_x = 0,$$
$$\Sigma F_y = A_y + B - F = 0,$$
$$\Sigma M_{\text{point } A} = (6 \text{ ft})B - \bar{x}F = 0,$$

we obtain $A_x = 0$, $A_y = 120$ lb and $B = 240$ lb.

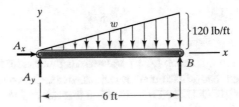

(a) Free-body diagram of the beam

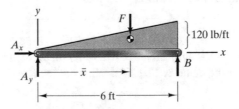

(b) Representing the distributed load by an equivalent force

Discussion

This example demonstrates that representing a distributed load by an equivalent force can be much simpler than using integration to determine the force
and moment exerted by the distributed load. However, that is true only when
the distributed load is sufficiently simple that the area under the loading curve
and the position of the centroid of the area are easy to determine, as they are in
this example.

Review Problems

*Answers to even-numbered problems are given in Appendix G. Numbers in the
answers, and in examples, are usually expressed to three significant digits. Problems
that are more difficult and/or lengthy are marked with an asterisk.*

1.1 The beam is subjected to a force and a couple. (**a**) Draw the free-body diagram
of the beam. (**b**) Determine the reactions at A and B.

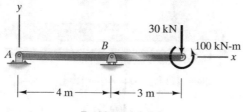

Problems 1.1–1.2

1.2 Determine the reactions at A and B if the 100-kN-m couple is moved to point A.
Compare your answers to the answers to Problem 1.1.

1.3 (**a**) Draw the free-body diagram of the bar. (**b**) Determine the reactions at the
fixed support A.

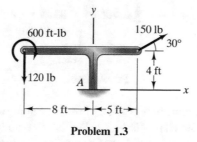

Problem 1.3

1.4 The floor exerts vertical forces on the drill press at A and B. What are they?

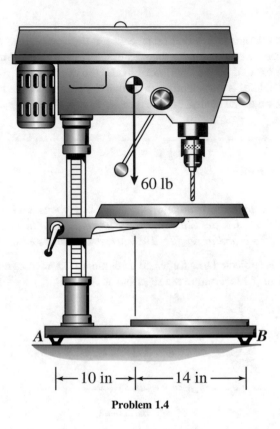

60 lb

|← 10 in →|← 14 in →|

Problem 1.4

1.5 Determine the reactions exerted on the L-shaped bar by its supports at *A* and *B*.

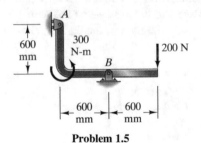

Problem 1.5

1.6 The free-body diagram of a portion of one of Fallingwater's decks is shown.
The mass of the isolated part is 14,700 kg. The free-body diagram is acted upon
by its weight *mg*, forces *P* and *V*, and a couple *M*. Determine the values of *P*, *V*,
and *M*.

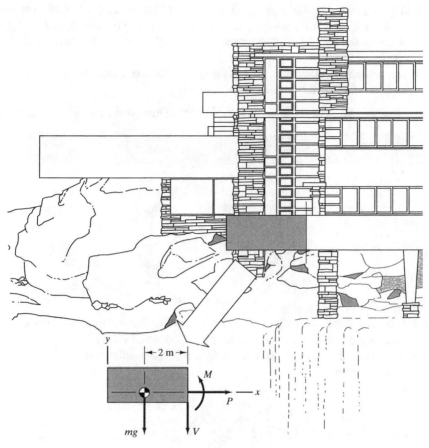

Problem 1.6

1.7 The two systems of forces and moments (a) and (b) are equivalent. Determine the magnitude of the force F, the angle α, and the magnitude of the clockwise couple C.

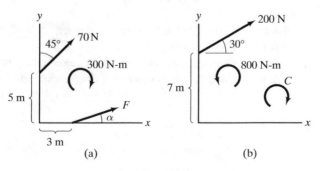

Problems 1.7–1.8

1.8 If you represent the equivalent systems of forces and moments (a) and (b) by a new equivalent system consisting of a single force **R** acting at the origin of the coordinate system and a couple **M**, what are the force **R** and the couple **M**?

1.9 The suspended mass $m = 20$ kg. Determine the axial force in the bar AB and indicate whether it is in tension or compression.

Strategy: The bar is a two-force member. Draw the free-body diagram of joint B.

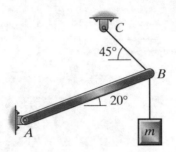

Problems 1.9–1.10

1.10 If the bar AB will safely support a compressive force no greater than 400 N, what is the largest mass m that can be suspended as shown?

1.11 Write the equilibrium equations for the entire truss and determine the reactions at the supports at A and D.

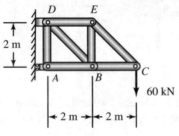

Problems 1.11–1.12

1.12 Determine the magnitude of the axial force in member BD of the truss and indicate whether it is in tension (T) or compression (C).

1.13 The force $F = 3$ kip. Determine the axial forces in members AC and CD of the truss and indicate whether they are in tension (T) or compression (C).

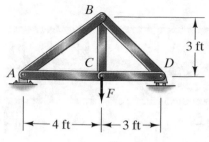

Problems 1.13–1.14

1.14 The members of the truss will safely support a tensile axial force of 4 kip and a compressive axial force of 2 kip. Based on these criteria, what is the largest safe value of F?

1.15 The compressive axial force in member CD due to the suspended mass m is 4 kN. Determine the axial forces in members BC and BD, and indicate whether they are in tension (T) or compression (C).

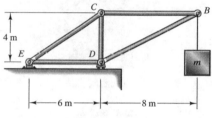

Problems 1.15–16

1.16 The suspended mass $m = 200$ kg. Determine the magnitudes of the axial forces in members CE and DE, and indicate whether they are in tension (T) or compression (C).

1.17 The loads $F_B = 40$ kN, $F_C = 60$ kN and $F_D = 20$ kN. Determine the magnitudes of the axial forces in members BC, CG, and GH, and indicate whether they are in tension (T) or compression (C).

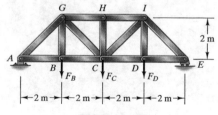

Problem 1.17

1.18 The Warren truss supporting the walkway is designed to support 50-kN vertical loads at B, D, F, and H. If the truss is subjected to these loads, what are the resulting axial forces in members AC, BC, and BD? Indicate whether the axial forces are tensile (T) or compressive (C).

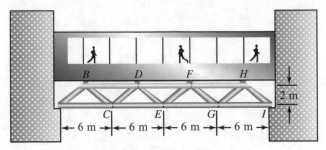

Problems 1.18–1.19

1.19 The Warren truss supporting the walkway is designed to support 50-kN vertical loads at *B*, *D*, *F*, and *H*. If the truss is subjected to these loads, what are the resulting axial forces in members *CE*, *DE*, and *DF*? Indicate whether the axial forces are tensile (T) or compressive (C).

1.20 Determine the reactions on member *CDE* of the frame.

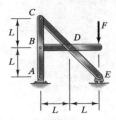

Problem 1.20

1.21 The bucket of the front-end loader is supported by a pin support at *C* and the hydraulic actuator *AB*. If the mass of the bucket is 180 kg and the system is stationary, what is the axial load in the hydraulic actuator?

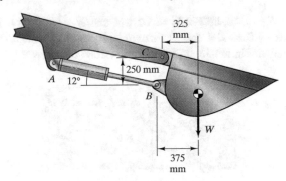

Problem 1.21

1.22 The suspended crate weighs 2000 lb and the angle $\alpha = 40°$. If you neglect the weight of the crane's boom *ACD*, what is the axial force in the hydraulic cylinder *BC*?

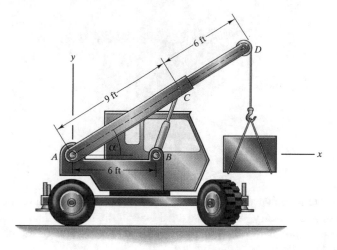

Problems 1.22–1.23

1.23 The suspended crate weighs 2000 lb and the angle $\alpha = 40°$. If you neglect the weight of the crane's boom *ACD*, what are the *x* and *y* components of the reaction exerted on the crane's boom by the pin support *A*?

1.24 Determine the axial force in member *BE* of the frame.

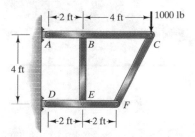

Problems 1.24–1.25

1.25 Determine the reactions at the frame's supports *A* and *D*.

1.26 The system shown supports half of the weight of the 680-kg excavator. If the system is stationary, what is the axial load in member *AB*?

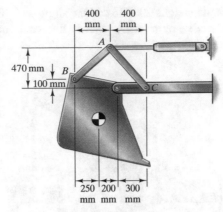

Problems 1.26–1.27

1.27 The system shown supports half of the weight of the 680-kg excavator. If the system is stationary, what is the axial load in member *AC*?

1.28 A person applies the 150-N forces shown to the handles of the pliers. Determine the axial force in the link *AB*.

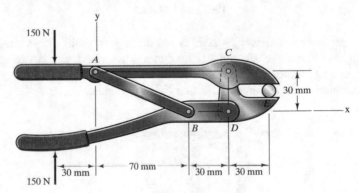

Problems 1.28–1.29

1.29 A person applies the 150-N forces shown to the handles of the pliers. Determine the *x* and *y* components of the reaction exerted on the upper handle *AC* of the pliers at *C*.

1.30 Member *ABC* of the frame has a pin at *C* that is supported by a smooth horizontal slot in member *DC*. Determine the reactions at the pin supports at *A* and *E*.

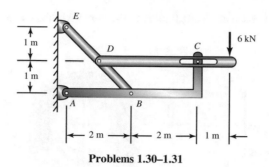

Problems 1.30–1.31

1.31 Member *ABC* of the frame has a pin at *C* that is supported by a smooth horizontal slot in member *DC*. Determine the magnitude of the force exerted on member *BDE* by the pin at *B*.

1.32 Determine the coordinates of the centroid of the area.

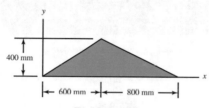

Problem 1.32

1.33 The area in Fig. (a) is the cross section of an L203 × 152 × 25.4 rolled steel shape (see Appendix E). Approximate the area of the cross section by ignoring the curved corners as shown in Fig. (b) and determine the coordinates of the centroid of the cross section. Compare your results with those given in Appendix E.

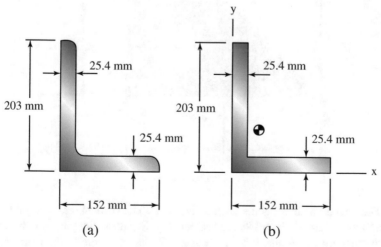

(a)   (b)

Problem 1.33

1.34 Determine the x and y coordinates of the centroid of the area.

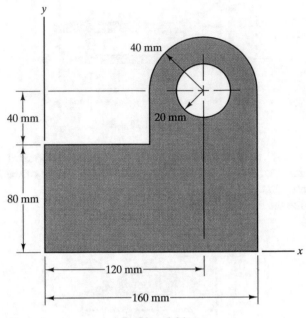

Problem 1.34

1.35 The steel plate is homogeneous, of uniform thickness, and weighs 10 lb. **(a)** Determine the x coordinate of the plate's center of mass. **(b)** What are the reactions at A and B?

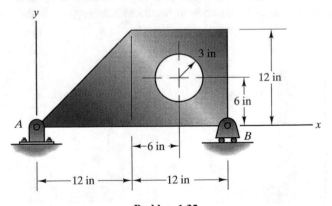

Problem 1.35

1.36 The area of the homogeneous plate is 10 ft^2. The vertical reactions on the plate at A and B are 80 lb and 86 lb, respectively. Suppose that you want to equalize the reactions at A and B by drilling a 1-ft diameter hole in the plate. At what horizontal distance from A should the center of the hole be placed?

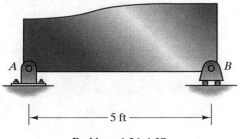

Problems 1.36–1.37

1.37∗ In Problem 1.36, what are the reactions at A and B after the hole is drilled in the plate?

1.38 The beam has a circular cross section with a diameter of 100 mm and consists of aluminum alloy with mass density $\rho = 2900$ kg/m³. **(a)** Determine the reactions at A and B. **(b)** If you represent the bar's weight by a uniformly distributed load w, what is the value of w?

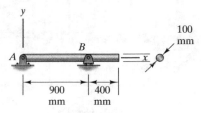

Problem 1.38

1.39 Determine the reactions at A and B.

> *Strategy*: Treat the distributed load as two triangular distributed loads, and represent each distributed load by an equivalent force.

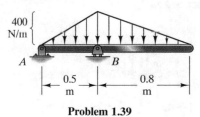

Problem 1.39

1.40 The beam is subjected to a distributed load given by the equation shown. Determine the reactions at A and B.

> *Strategy*: Integrate to determine the total force and moment exerted by the distributed load.

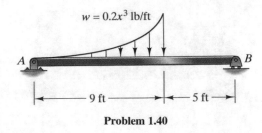

Problem 1.40

1.41 The aerodynamic lift of the wing is described by the distributed load $w = -300(1 - 0.04x^2)^{1/2}$ N/m. Determine the magnitudes of the force and the moment about R exerted by the wing's lift.

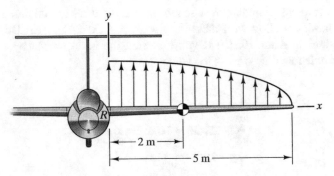

Problems 1.41–1.42

1.42 The mass of the wing in Problem 1.41 is 27 kg, and its center of mass is located 2 m to the right of the wing root R. Including the effects of the wing's lift, determine the reactions exerted on the wing at R where it is attached to the fuselage.

1.43 Determine the reactions at A and C.

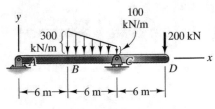

Problem 1.43

1.44∗ A plane surface of area A in the x-y plane is subjected to a uniform pressure p_0. Show that this distributed load can be represented by an equivalent force of magnitude p_0A whose line of action passes through the centroid of A.

Strategy: Write integral expressions for the total force and the total moment about the origin due to the uniform pressure, and use the definitions of the centroid of an area.

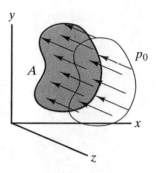

Problem 1.44

1.45 The beam is subjected to a distributed couple c, defined such that each infinitesimal element dx of the beam is acted upon by a counterclockwise couple $c\,dx$. If $c = c_0 = $ constant from $x = 0$ to $x = L$, determine the reactions at A and B.

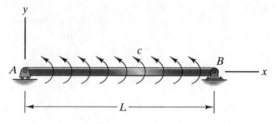

Problems 1.45–1.46

1.46 The distributed couple acting on the beam is given by the equation $c = (x/L)c_0$, where c_0 is a constant. Determine the reactions at A and B.

1.47 A pile is being slowly pushed into the ground by a vertical force F. The friction of the ground exerts a distributed axial force q, defined such that each infinitesimal element dx of the pile is acted upon by an axial force $q\,dx$. If $q = 400(1 - 0.4x^2/L^2)$ lb/ft and $L = 30$ ft, determine the force F at the instant shown. (The pile's weight and the force exerted by the ground at the bottom of the pile are negligible in comparison to the frictional force exerted along the pile's length.)

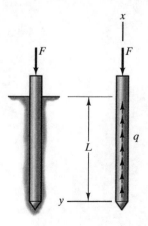

Problem 1.47

Chapter 2
Measures of Stress and Strain

We now define stresses and strains, which are measures used to describe the internal forces and deformations in materials. We need these measures to assess the capability of structural elements such as a bridge's cables and arches to support loads and to determine the deformations that result from those loads.

2.1 Stresses

Suppose that a volume V of some solid material, such as iron, is subjected to the system of forces shown in Fig. 2.1. We assume that the material is in equilibrium and that its weight is negligible in comparison to the forces shown, so that $\mathbf{F}_1 + \mathbf{F}_2 + \mathbf{F}_3 = \mathbf{0}$. Let us cut the volume by an imaginary plane $\mathcal{P}$, dividing it into volumes V' and V'', and isolate volume V' (Fig. 2.2). From Fig. 2.2(b), we can see that volume V' cannot be in equilibrium if it is subjected only to the forces $\mathbf{F}_1$ and $\mathbf{F}_2$. Clearly, volume V'' must exert forces on volume V' at the area A where they are joined.

What are those forces? They are literally the forces that hold the material together. Iron in its solid state consists of atoms "connected" to neighboring atoms by chemical bonds, electromagnetic forces that can be modeled as springs. The external forces acting on the volume V alter the distances between atoms, changing the forces exerted by the bonds. It is those bond forces, exerted on volume V' by volume V'' at the area A, that keep volume V' in equilibrium (Fig. 2.3).

As we will demonstrate in subsequent chapters, these internal forces within materials are an essential factor in determining whether objects will support the external loads to which they are subjected. The question is how can we describe and analyze the internal forces? To do so, *we will represent them as a distributed load on the area A.*

© Springer Nature Switzerland AG 2020
A. Bedford, K. M. Liechti, *Mechanics of Materials*,
https://doi.org/10.1007/978-3-030-22082-2_2

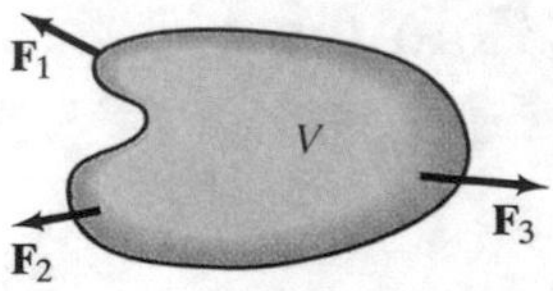

Fig. 2.1 A volume of
material subjected to forces

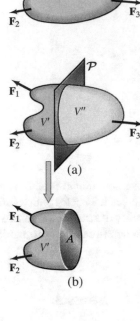

Fig. 2.2 (**a**) The plane $\mathcal{P}$
divides the volume
into parts V' and V''.
(**b**) Isolating part V'

Fig. 2.3 Bond forces acting
on volume V' at the area A

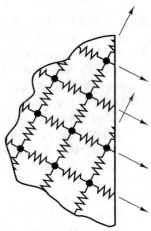

2.1.1 *Traction, Normal Stress, and Shear Stress*

Let a vector-valued function **t**, the *traction*, be defined such that the force exerted on
each infinitesimal element dA of the area A is $\mathbf{t}\,dA$ (Fig. 2.4). Notice that the
definition of the traction **t** is very similar to the definition of the pressure p in a

Fig. 2.4 (a) Representing
the forces on the area
A by a distributed load.
(b) The force on an element
dA is **t** dA

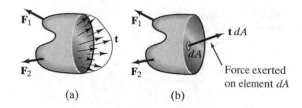

(a) (b)

Fig. 2.5 The components
of **t** normal and tangential to
A are the normal and shear
stresses

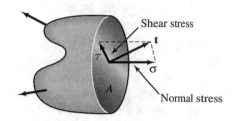

liquid (Fig. 1.14). However, there is an important difference. The force exerted on an element of area by pressure can be described by the scalar function p because the force is perpendicular to the surface. The direction does not need to be specified. But a vector-valued function **t** is needed to describe the force exerted on an element of area by the internal forces in a solid material, because the internal forces can act in any direction relative to the surface.

Because the traction **t** is a vector, we can express it in terms of components parallel and normal to A (Fig. 2.5). The normal component σ is called the *normal stress*, and the tangential component τ is called the *shear stress*. The normal stress σ is defined to be positive if the traction points outward or away from the material, and the normal stress is said to be *tensile*. If the traction points inward, the normal stress σ is defined to be negative and is said to be *compressive*.

The force exerted on an element of area dA is $\sigma\,dA$ in the normal direction and $\tau\,dA$ in the tangential direction. Because the products **t** dA, $\sigma\,dA$, and $\tau\,dA$ are forces, the dimensions of **t**, σ, and τ are (force)/(area). In SI units, the traction and the normal and shear stresses are expressed in pascals (Pa), which are newtons per square meter. In US customary units, they are usually expressed in pounds per square foot (psf) or pounds per square inch (psi).

We have used the example of a volume of iron to motivate the definitions of the normal and shear stresses, but these definitions are used in mechanics of materials and in fluid mechanics to describe the internal forces in other media, and they need not be in equilibrium. Depending on the medium, the internal forces represented may be very different in nature from the bond forces between the atoms in crystalline iron. For example, the internal forces in a gas arise from impacts between the molecules of the gas resulting from their thermal motions. Nevertheless, the forces on a surface are represented by distributions of normal and shear stress.

Fig. 2.6 The normal and tangential scalar components of the average traction are the average normal and shear stresses

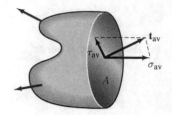

2.1.2 Average Stresses

Structural design generally requires that the stresses be determined throughout the members of a structure. This determination is called *stress analysis*. In the following chapters, we discuss the stress distributions in simple structural members subjected to external loads. In this section, our goal is less ambitious. We want to show how the *average* stresses on a plane can be determined when an object is in equilibrium.

The average value of the traction **t** on the area A in Fig. 2.5 is defined by

$$\mathbf{t}_{\mathrm{av}} = \frac{1}{A} \int_A \mathbf{t} \, dA, \tag{2.1}$$

where the subscript A on the integral sign signifies that integration is carried out over the entire area A. The scalar components of the average traction $\mathbf{t}_{\mathrm{av}}$ normal and tangential to A are the average normal stress σ_{av} and average shear stress τ_{av}, respectively (Fig. 2.6). The normal and tangential forces exerted on the area A are therefore $\sigma_{\mathrm{av}}A$ and $\tau_{\mathrm{av}}A$. When the external forces acting on an object in equilibrium are known, we can use equilibrium to determine the average normal and shear stresses acting on an arbitrary plane $\mathcal{P}$. Two particular examples are important in applications.

2.1.2.1 Average Normal Stress in an Axially Loaded Bar

A straight bar whose cross section is uniform throughout its length is said to be *prismatic*. (The familiar triangular glass prism is a prismatic bar.) Figure 2.7(a) shows a prismatic bar subjected to axial forces parallel to its axis. In Fig. 2.7(b), we pass a plane $\mathcal{P}$ perpendicular to the bar's axis and draw the free-body diagram of the part of the bar to the left of the plane. We show the force exerted by the average normal stress at $\mathcal{P}$, where A is the bar's cross-sectional area. (*When we refer to the cross-sectional area of a bar, we will mean the area obtained by passing a plane perpendicular to the bar's axis unless we specify otherwise.*) We know that the average shear stress at $\mathcal{P}$ is zero, because there is no external force on the free-body diagram tangential to $\mathcal{P}$. The sum of the forces on the free-body diagram equals zero

Fig. 2.7 (**a**) Subjecting a bar to axial loads. (**b**) Obtaining a free-body diagram by passing a plane through the bar perpendicular to its axis

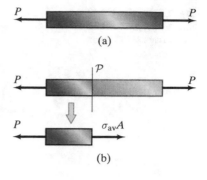

(a)

(b)

Fig. 2.8 Two views of a pin support

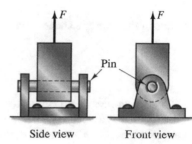

Side view   Front view

$$\sigma_{av}A - P = 0,$$

so we can solve for the average normal stress:

$$\sigma_{av} = \frac{P}{A}.$$

We see that the average normal stress is proportional to the external axial load and inversely proportional to the bar's cross-sectional area. Notice that the value of the average normal stress does not depend on the location of the plane $\mathcal{P}$ along the bar's axis. We can use this same approach to determine the average normal and shear stresses on a plane that is not perpendicular to the bar's axis (see Example 2.1).

2.1.2.2 Average Shear Stress in a Pin

The pin support in Fig. 2.8 holds a bar subjected to an axial load F. The support consists of a bracket supporting the cylindrical pin that passes through a hole in the bar. In Fig. 2.9 we pass vertical planes through the pin, one on each side of the bar, isolating the bar and part of the pin. In order for the resulting free-body diagram to be in equilibrium, we can see that shear stresses must act on the surfaces where the planes passed through the pin. We show the forces resulting from the average shear stresses in the pin at the two cutting planes, where A is the pin's cross-sectional area. From the equilibrium equation

Fig. 2.9 Passing two planes
through the pin to determine
the average shear stress

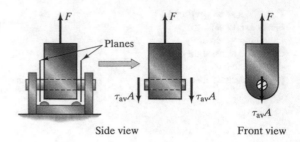

$$F - 2\tau_{\mathrm{av}}A = 0,$$

we obtain the average shear stress:

$$\tau_{\mathrm{av}} = \frac{F}{2A}.$$

These examples demonstrate how we can use free-body diagrams and equilibrium to determine the average stresses on a given plane within an object. But knowing the average stress on a plane is of limited usefulness in design, because it is the maximum, not the average, values of the normal and shear stresses that determine whether material failure will occur. Depending on the nature of the actual distributions of stress, the maximum stresses may be substantially greater than their average values. Moreover, it is not sufficient to determine the maximum stresses for a particular plane. The stresses acting on every plane must be considered. We examine the problem of determining maximum normal and shear stresses in Chap. 7.

There is one circumstance in which knowing the average stress on a given plane is useful in design. If experiments or analyses are first used to establish the relationship between the largest safe load on a given structural element and the value of the average normal or shear stress on a given plane, design can be carried out on that basis. For example, bolts made of a particular material that are to be used for a specific application can be tested to determine the largest safe value of average shear stress they will support.

Study Questions

1. If you know the distribution of the traction **t** on a plane $\mathcal{P}$, how can you determine the force exerted on an element of area dA of $\mathcal{P}$?
2. What are the definitions of the normal stress σ and shear stress τ?
3. If the normal stress σ at a particular point of a plane $\mathcal{P}$ is negative, what does that tell you?
4. If a straight prismatic bar is subjected to given axial loads at the ends, how can you determine the average normal stress on a plane perpendicular to the bar's axis?

Example 2.1 Determining Average Stresses
The truss in Fig. 2.10 supports a 10-kN force. Its members are solid cylindrical bars of 40-mm radius. Determine the average normal and shear stresses on the plane $\mathcal{P}$.

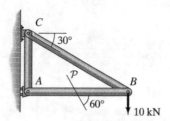

Fig. 2.10

Strategy
We must first determine the axial force in member AB, which we can do by drawing the free-body diagram of joint B. We can then use the free-body diagram of the part of member AB on either side of the plane $\mathcal{P}$ to determine the average normal and shear stresses.

Solution
We draw the free-body diagram of joint B of the truss in Fig. (a). From the equilibrium equations

$$\Sigma F_x = -P_{AB} - P_{BC}\cos 30° = 0,$$
$$\Sigma F_y = P_{BC}\sin 30° - 10\text{ kN} = 0,$$

we obtain $P_{AB} = -17.3$ kN, $P_{BC} = 20$ kN. Member AB is subject to a compressive axial load of 17.3 kN.

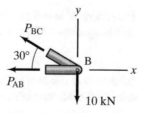

(a) Joint B

In Fig. (b) we isolate the part of member AB to the left of the plane $\mathcal{P}$ and complete the free-body diagram by showing the forces exerted by the average normal and shear stresses. The area A' is the intersection of the plane $\mathcal{P}$ with the bar, which is the area on which the average stresses σ_{av} and τ_{av} act

(Fig. (c)). The relationship between A' and the cross-sectional area A of the cylindrical bar is

$$A' \cos 30° = A = \pi(0.04 \text{ m})^2,$$

so $A' = \pi(0.04 \text{ m})^2 / \cos 30° = 0.00580 \text{ m}^2$.

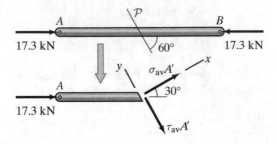

(b) Isolating the part of member AB on one side of the plane $\mathcal{P}$

(c) The cross-sectional area A of the bar and the area A' intersected by the plane $\mathcal{P}$

By aligning a coordinate system normal and tangential to the plane $\mathcal{P}$ as shown in Fig. (b) and summing forces in the x and y directions, we obtain equilibrium equations for the forces exerted by the stresses:

$$\Sigma F_x = \sigma_{av}A' + (17.3 \text{ kN}) \cos 30° = 0,$$
$$\Sigma F_y = -\tau_{av}A' - (17.3 \text{ kN}) \sin 30° = 0.$$

We find that $\sigma_{av}A' = -15$ kN and $\tau_{av}A' = -8.66$ kN. Solving for the average normal and shear stresses yields $\sigma_{av} = -2.58$ MPa and $\tau_{av} = -1.49$ MPa.

Discussion
The negative value of σ_{av} tells us that the average normal stress is compressive. Although we have established a convention for the positive direction of the normal stress, we have not yet done so for the shear stress. In drawing the free-body diagram in Fig. (b), we chose the direction of the stress τ_{av} arbitrarily, and the negative value we obtained indicates that it acts in the opposite direction.

Example 2.2 Average Shear Stress in a Pin

A truss and detailed views of the joint at B are shown in Fig. 2.11. The joint has a pin of 20-mm radius. What is the average shear stress in the pin?

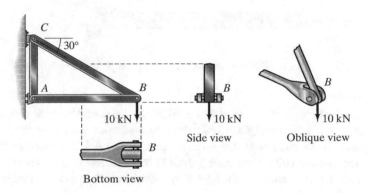

Fig. 2.11

Strategy

By passing planes through the pin on both sides of member BC, we can isolate either member AB or BC and obtain a free-body diagram from which we can determine the average shear stress in the pin. Because the 10-kN external load acts on member BC, we will obtain a simpler free-body diagram by isolating member AB.

Solution

In Example 2.1 we determined that member AB of this truss is subjected to a compressive axial load of 17.3 kN. In the bottom view shown in Fig. (a), we isolate member AB from member BC by passing two planes through the pin. From the resulting free-body diagram, we can see that the 17.3-kN compressive axial load in member AB must be supported by shear stresses acting on the surfaces where the pin was cut by the two planes. (The term A is the pin's cross-sectional area.) From the equilibrium equation

$$17.3 \text{ kN} - 2\tau_{av}A = 0,$$

we obtain $\tau_{av}A = 8.66$ kN. The pin's cross-sectional area is $A = \pi(0.02 \text{ m})^2 = 0.00126 \text{ m}^2$, so the average shear stress in the pin is $\tau_{av} = 6.89$ MPa.

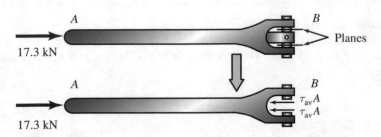

(a) Isolating member *AB* from member *BC* by passing two planes through the pin

Discussion

In the terminology of structural engineering, the pin connecting member *AB* to member *BC* in this example is said to be subjected to *double shear*. We needed to pass two planes through the pin to isolate member *AB* from member *BC*. In Fig. (b), the pin connecting member *AB* to member *BC* is in *single shear*. In this case we only need to cut the pin by a single plane to isolate member *AB* (Fig. (c)). From the equilibrium equation

$$17.3 \text{ kN} - \tau_{av}A = 0,$$

we obtain $\tau_{av} = 13.8$ MPa, twice the average shear stress to which the pin was subjected in the case of double shear.

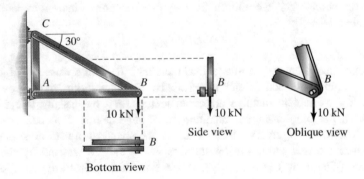

(b) Here the pin connecting members *AB* and *BC* is subjected to single shear

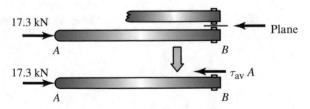

(c) Isolating member *AB* from member *BC* by passing a single plane through the pin

Example 2.3 Average Shear Stress in a Coining Process
Figure 2.12(a) is a simplified diagram of a machine that performs what is called a coining process. As its name implies, it can be used to manufacture coins and other disk-shaped objects. The cylindrical punch and the cylindrical hole under the plate have the same diameter D. The plate is of thickness t. By exerting a sufficiently large force F on the punch, a disk-shaped "blank" is cut out of the plate. Figure 2.12(b) shows a free-body diagram obtained by cutting the plate with a cylindrical surface of diameter D and isolating the punch and the blank that is to be cut out of the plate. If an average shear stress $\tau_{av} = 250$ MPa is necessary to cut the blank from the plate, $t = 1.5$ mm, and $D = 20$ mm, what force F is required to perform the operation?

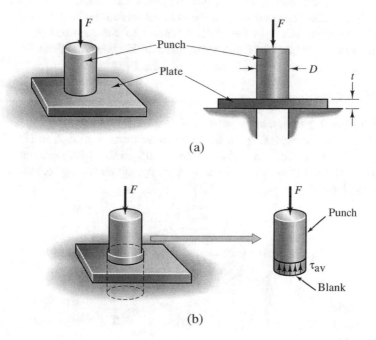

Fig. 2.12

Strategy
By writing the equilibrium equation for the free-body diagram in Fig. 2.12 (b) and assuming that the average shear stress equals 250 MPa, we can solve for the force F.

Solution
The surface area of the disk-shaped blank in Fig. 2.12(b) on which the average shear stress acts is equal to the product of the thickness t of the plate and the circumference πD. Summing the vertical forces on the free-body diagram

$$\tau_{av}(t)(\pi D) - F = 0,$$

we find that the required force is

$$F = \pi \tau_{av} t D$$
$$= \pi(250\text{E}6 \text{ N/m}^2)(0.0015 \text{ m})(0.02 \text{ m})$$
$$= 23{,}600 \text{ N } (5300 \text{ lb}).$$

Discussion
Notice how our determination of the shear stress induced by the punch
depended on a proper choice of free-body diagram. By isolating the punch
and the circular part of the plate directly below the punch, we obtained a free-
body diagram subjected only to the external force F and the force due to the
average shear stress in the plate. Learning to choose free-body diagrams
requires experience and exposure to different kinds of situations.

Example 2.4 Analyzing a Normal Stress Distribution
The normal stress acting on the rectangular cross section A in Fig. 2.13 is
given by the equation $\sigma = 200y^2$ lb/in^2. Determine (**a**) the maximum normal
stress, (**b**) the total normal force exerted by the stress, and (**c**) the average
normal stress.

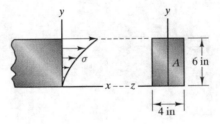

Fig. 2.13

Strategy
(**a**) The value of σ increases monotonically with increasing y, so its maximum
value occurs at $y = 6$ in. (**b**) The normal stress σ depends only upon y, so we
can integrate to determine the total normal force by using an element of area
dA in the form of a horizontal strip (Fig. (a)). (**c**) The average stress equals the
total normal force divided by A.

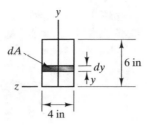

(a) Integration element dA

Solution

(a) The maximum normal stress is

$$\sigma_{max} = (200)(6 \text{ in})^2 = 7200 \text{ lb/in}^2.$$

(b) Using the strip element dA in Fig. (a), the total normal force is

$$\int_A \sigma \, dA = \int_0^6 (200y^2)(4dy)$$

$$= 800 \left[\frac{y^3}{3} \right]_0^6$$

$$= 57{,}600 \text{ lb.}$$

(c) The average stress is

$$\sigma_{av} = \frac{1}{A} \int_A \sigma \, dA$$

$$= \frac{1}{(4 \text{ in})(6 \text{ in})} (57{,}600 \text{ lb})$$

$$= 2400 \text{ lb/in}^2.$$

Discussion

Notice that the maximum normal stress is three times the average normal stress. This illustrates why knowledge of the average stress is usually insufficient for design. The actual distribution of the stress must be known to determine its maximum value.

2.1.3 Components of Stress

Suppose we are interested in the stresses acting on planes through a particular point p of an object. Let us introduce a coordinate system and pass a plane through p that is perpendicular to the x axis, isolating part of the object as shown in Fig. 2.14(a).

Fig. 2.14 (**a**) Passing a
plane perpendicular to the
x axis. (**b**) Normal stress and
components of the shear
stress at *p*

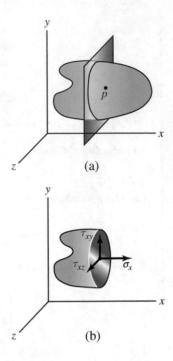

The normal stress acting on this plane at p is denoted by σ_x (Fig. 2.14(b)). The shear stress may act in any direction parallel to the y-z plane, so it may have components in both the y and z directions. These components are denoted by τ_{xy} (the shear stress on the plane perpendicular to the x axis that acts in the y direction) and τ_{xz} (the shear stress on the plane perpendicular to the x axis that acts in the z direction). Next, we pass a plane through p that is perpendicular to the y axis (Fig. 2.15(a)). The normal stress on this plane is denoted by σ_y and the shear stresses by τ_{yx} and τ_{yz} (Fig. 2.15 (b)). Finally, we pass a plane through p perpendicular to the z axis (Fig. 2.16(a)). The normal stress is denoted by σ_z and the shear stresses by τ_{zx} and τ_{zy} (Fig. 2.16(b)).

Let us assume that we know these normal and shear stresses at p acting on planes perpendicular to the x, y, and z axes. If we pass a different plane through p, the normal and shear stresses acting on it will generally be different from those acting on the planes perpendicular to the coordinate axes, so what have we achieved? We will show in Chap. 7 that if the normal and shear stresses acting on planes perpendicular to the x, y, and z axes are known for a given point p, the normal and shear stresses on *any* plane through p can be determined. For this reason, *the components of stress acting on the planes perpendicular to the three coordinate axes are called the state of stress* at p.

We will often find it useful to show the components of stress acting on the faces of a rectangular element as shown in Fig. 2.17. This diagram also provides a convenient graphical representation of the state of stress at a point.

The shear stresses $\tau_{xy} = \tau_{yx}$, $\tau_{yz} = \tau_{zy}$ and $\tau_{xz} = \tau_{zx}$. To prove this, let the element in Fig. 2.18 have dimension b. The sum of the moments about the z axis is

Fig. 2.15 (a) Passing a
plane perpendicular to the
y axis. (b) Normal stress and
components of the shear
stress at p

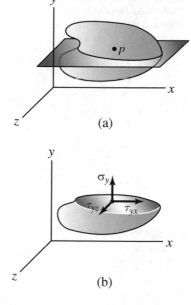

Fig. 2.16 (a) Passing a
plane perpendicular to the
z axis. (b) Normal stress and
components of the shear
stress at p

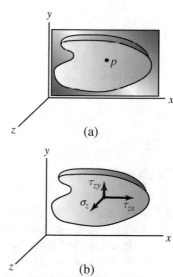

$$\left(\sigma_x b^2\right)(b/2) - \left(\sigma_x b^2\right)(b/2) + \left(\sigma_y b^2\right)(b/2) - \left(\sigma_y b^2\right)(b/2)$$
$$+ \left(\tau_{xy} b^2\right)(b) - \left(\tau_{yx} b^2\right)(b) + \left(\tau_{zy} b^2\right)(b) - \left(\tau_{zy} b^2\right)(b)$$
$$+ \left(\tau_{zx} b^2\right)(b) - \left(\tau_{zx} b^2\right)(b) = 0.$$

Therefore $\tau_{xy} = \tau_{yx}$. Summing moments about the x and y axes shows that $\tau_{yz} = \tau_{zy}$ and $\tau_{xz} = \tau_{zx}$. These results, which can be shown to hold even when the material is not in equilibrium, imply that *the state of stress at a point is completely described by the six components* σ_x, σ_y, σ_z, τ_{xy}, τ_{yz}, *and* τ_{xz}.

Fig. 2.17 Components of
stress on the faces of a
cubical element

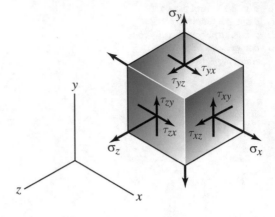

Fig. 2.18 Summing
moments about the z axis
confirms that $\tau_{xy} = \tau_{yx}$

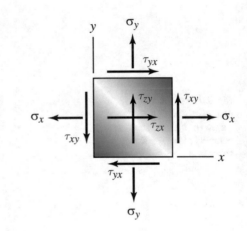

Problems

2.1 The bar BC in Fig. (a) has a 50 - mm × 50 - mm square cross section. It has a
  fixed support at B and is subjected to a 6-kN axial force at C. In Fig. (b), a free-
  body diagram of part of the bar is obtained by passing a plane $\mathcal{P}$ perpendicular
  to the bar's axis. The term σ_{av} is the average normal stress acting on the free-
  body diagram where it was cut by the plane, and A is the bar's cross-sectional
  area. Determine σ_{av} in pascals.

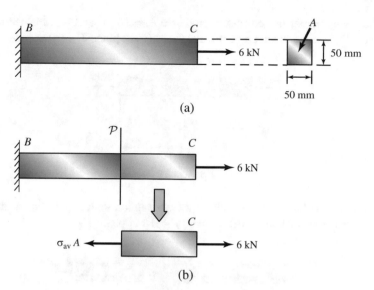

(a)

(b)

Problem 2.1

2.2 The bar BC has a 50 - mm $\times$ 50 - mm square cross section. A free-body diagram of part of the bar is obtained by passing the plane $\mathcal{P}$ shown. The terms σ_{av} and τ_{av} are the average normal stress and average shear stress acting on the free-body diagram where it was cut by the plane. The area A' is the intersection of the plane $\mathcal{P}$ with the bar, the area on which the average stresses σ_{av} and τ_{av} act. (**a**) What is the area A' in m^2? (**b**) Determine σ_{av} and τ_{av} in pascals.

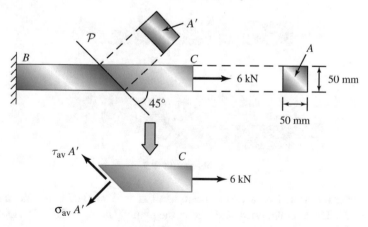

Problem 2.2

2.3 The prismatic bar has cross-sectional area $A = 30$ in^2 and is subjected to axial loads. Determine the average normal stress (**a**) at plane $\mathcal{P}_1$; (**b**) at plane $\mathcal{P}_2$.

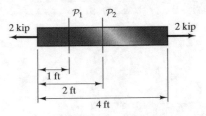

Problem 2.3

2.4 The prismatic bar has a solid circular cross section with 2-in radius. Determine the average normal stress (**a**) at plane $\mathcal{P}_1$; (**b**) at plane $\mathcal{P}_2$.

Problem 2.4

2.5 The cross-sectional area of the prismatic bar (the area obtained by passing a plane perpendicular to the bar's axis) is $A = 0.02$ m^2. The bar is subjected to axial loads $P = 6$ kN. Determine the average normal stress σ_{av} and the average shear stress τ_{av} on the plane $\mathcal{P}$ if $\theta = 30°$.

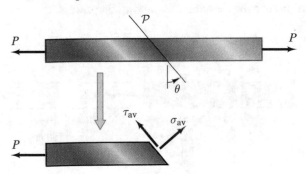

Problems 2.5–2.7

2.6 The cross-sectional area of the prismatic bar is $A = 0.024$ m^2 and the angle $\theta = 40°$. The average normal stress on the plane $\mathcal{P}$ is $\sigma_{av} = 200$ kPa. Determine the axial force P and the average shear stress τ_{av}.

2.7 The cross-sectional area of the prismatic bar is $A = 0.02$ m². The bar is subjected to axial loads $P = 4$ kN. Determine the average normal stress σ_{av} and the average shear stress τ_{av} on the plane $\mathcal{P}$ as functions of θ.

2.8 The rectangular 12 in × 6 in plate is 1-in thick and is subjected to forces $P = 8$ kip. Determine the average normal stress and the magnitude of the average shear stress at the plane $\mathcal{P}$.

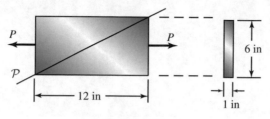

Problem 2.8

2.9 The rectangular plate has a hole with 10-mm radius drilled through it. The plate is subjected to loads $P = 4$ kN. Determine the average normal stress in the material (a) at plane $\mathcal{P}_1$ and (b) at plane $\mathcal{P}_2$.

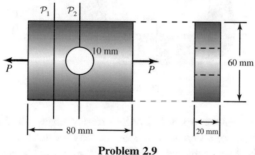

Problem 2.9

2.10 The chain is subjected to axial forces $P = 2000$ lb. The chain's links have circular cross sections of ¼-in diameter. What is the average normal stress in the material at the plane $\mathcal{P}$?

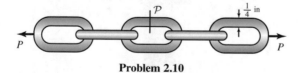

Problem 2.10

2.11 The C-clamp exerts 200-N forces on the cylindrical object. The cross-sectional area of the clamp at the plane $\mathcal{P}$ is $A = 175$ mm². What are the average normal and shear stresses at $\mathcal{P}$?

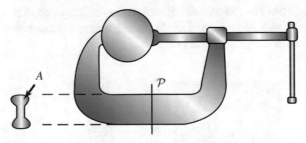

Problem 2.11

2.12 The prismatic bar has a solid circular cross section with 50-mm radius. It is suspended from one end and is loaded only by its own weight. The mass density of the homogeneous material is 2800 kg/m^3. Determine the average normal stress at the plane $\mathcal{P}$, where x is the distance from the bottom of the bar in meters.

Strategy: Draw a free-body diagram of the part of the bar below the plane $\mathcal{P}$.

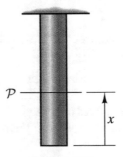

Problem 2.12

2.13 The bar is made of material 1 in thick. Its width varies linearly from 2 in at its left end to 4 in at its right end. If the axial load $P = 200$ lb, what is the average normal stress (**a**) at plane $\mathcal{P}_1$? (**b**) at plane $\mathcal{P}_2$?

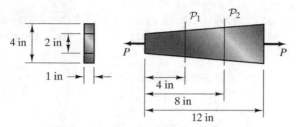

Problem 2.13

2.14 The cantilever beam has cross-sectional area $A = 12$ in^2. What are the average normal stress and the magnitude of the average shear stress at the plane $\mathcal{P}$?

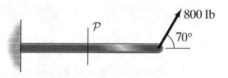

Problem 2.14

2.15 The beam has cross-sectional area $A = 0.02$ m^2. Determine the reactions at the supports, and then determine the average normal stress and the magnitude of the average shear stress at the plane $\mathcal{P}$ by drawing the free-body diagram of the part of the beam to the left of $\mathcal{P}$. Compare your answers to those of Problem 2.16.

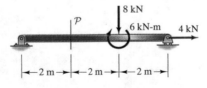

Problems 2.15–2.16

2.16 The beam has cross-sectional area $A = 0.02$ m². Determine the reactions at the supports, and then determine the average normal stress and the magnitude of the average shear stress at the plane $\mathcal{P}$ by drawing the free-body diagram of the part of the beam to the right of $\mathcal{P}$.

2.17 The beams have cross-sectional area $A = 60$ in². What are the average normal stress and the magnitude of the average shear stress at the plane $\mathcal{P}$ in cases (a) and (b)?

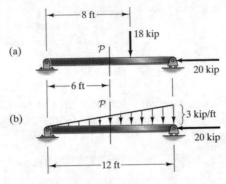

Problem 2.17

2.18 Figure (a) is a diagram of the bones and biceps muscle of a person's arm supporting a mass. Figure (b) is a biomechanical model of the arm in which the biceps muscle AB is represented by a bar with pin supports. The suspended mass is $m = 2$ kg and the weight of the forearm is 9 N. If the cross-sectional area of the tendon connecting the biceps to the forearm at A is 28 mm², what is the average normal stress in the tendon?

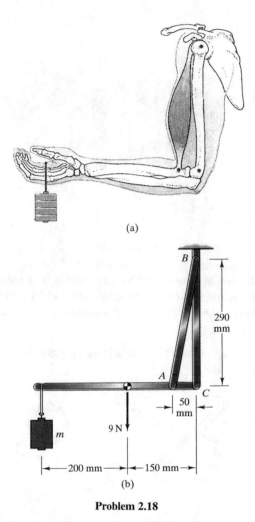

(a)

290
mm

B

A

50
mm

C

9 N

m

←——200 mm——→←—150 mm—→

(b)

Problem 2.18

2.19 The force **F** exerted on the bar is $20\mathbf{i} - 20\mathbf{j} - 10\mathbf{k}$ (lb). The plane $\mathcal{P}$ is parallel to the y–z plane and is 5 in from the origin O. The bar's cross-sectional area at $\mathcal{P}$ is 0.65 in^2. What is the average normal stress in the bar at $\mathcal{P}$?

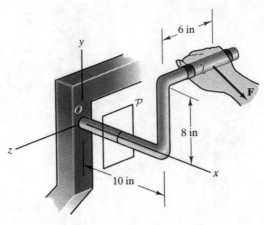

Problem 2.19

2.20 The truss is loaded by a vertical 10-kN force at B. The members have hollow circular cross sections with 20-mm outer radii and 15-mm inner radii. Determine the average normal stresses in the members (a) at plane $\mathcal{P}_1$ and (b) at plane $\mathcal{P}_2$.

Strategy: Begin by drawing the free-body diagram of joint B obtained by cutting the truss at $\mathcal{P}_1$ and $\mathcal{P}_2$.

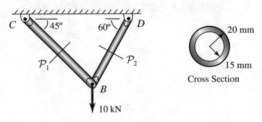

Problem 2.20

2.21 The plane $\mathcal{P}$ is parallel to the y-z plane of the coordinate system. The cross-sectional area of the tennis racquet at $\mathcal{P}$ is 400 mm^2. Including the force exerted on the racquet by the ball and inertial effects, the total force on the racquet above the plane $\mathcal{P}$ is $\mathbf{F} = 35\mathbf{i} - 16\mathbf{j} - 85\mathbf{k}$ (N). What is the magnitude of the average shear stress on the racquet at $\mathcal{P}$?

Problem 2.21

2.22 Figures (a) and (b) show two methods of supporting the end of a bar that is subjected to an axial load P. Method (a) subjects the supporting pin to *single shear*, and method (b) subjects the supporting pin to *double shear*. Draw free-body diagrams, and determine the average shear stress to which the pin is subjected in the two cases in terms of P and the cross-sectional area A of the pin.

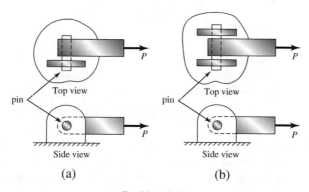

Problem 2.22

2.23 The bar in Fig. (a) is subjected to a downward force $F = 8$ kN. Figure (b) shows two views of the pin support at A. The support has a pin 40 mm in diameter. What is the magnitude of the average shear stress in the pin?

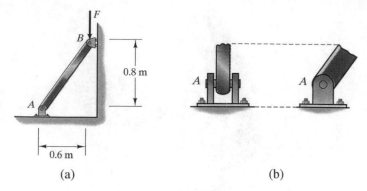

(a)   (b)

Problem 2.23

2.24 The fixture shown connects a 50-mm diameter bridge cable to a flange that is attached to the bridge. A 60-mm diameter circular pin connects the fixture to the flange. If the average normal stress in the cable is $\sigma_{av} = 120$ MPa, what average shear stress τ_{av} must the pin support?

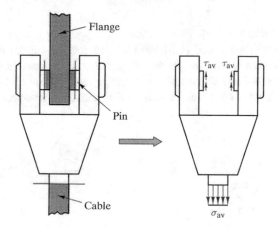

Problem 2.24

2.25 The person exerts 20-N forces on the pliers. The pin at C that connects the two members of the pliers is 5 mm in diameter. A bottom view of the pin connection is shown. What is the average shear stress in the pin?

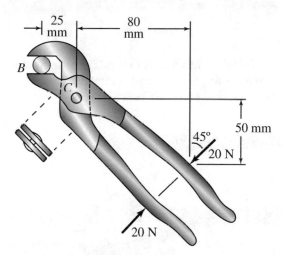

Problem 2.25

2.26 The truss is made of prismatic bars with cross-sectional area $A = 0.25$ ft^2. Determine the average normal stress in member BE acting on a plane perpendicular to the axis of the member.

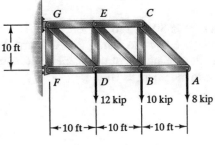

Problems 2.26–2.27

2.27 The truss is made of prismatic bars with cross-sectional area $A = 0.25 \text{ ft}^2$. Determine the average normal stress in member BD acting on a plane perpendicular to the axis of the member.

2.28 Three views of joint A of the truss are shown. The joint is supported by a cylindrical pin of 2-in diameter. What is the magnitude of the average shear stress in the pin?

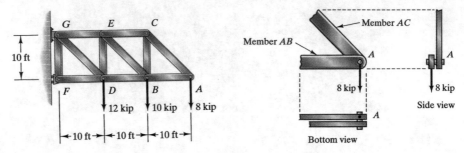

Problem 2.28

2.29 The top view of pin A of the pliers is shown. The cross-sectional area of the pin is 12.0 mm^2. What is the magnitude of the average shear stress in the pin when 150-N forces are applied to the pliers as shown?

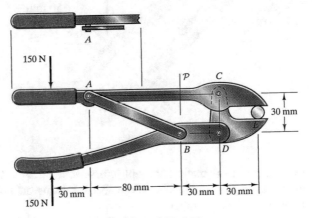

Problems 2.29–2.30

2.30 The vertical plane $\mathcal{P}$ is 30 mm to the left of C. The cross-sectional area of member AC of the pliers at the plane $\mathcal{P}$ is 60 mm^2. Determine the average normal stress and the magnitude of the average shear stress at $\mathcal{P}$ when 150-N forces are applied to the pliers as shown.

2.31 The suspended crate weighs 2000 lb. and the angle $\alpha = 40°$. The top view of the pin support A of the crane's boom is shown. The cross-sectional area of the pin is 8.0 in^2. What is the average shear stress in the pin?

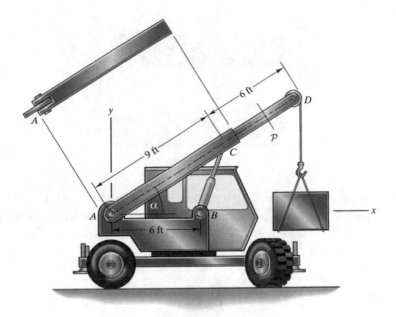

Problems 2.31–2.32

2.32 The suspended crate weighs 2000 lb. and the angle $\alpha = 40°$. The plane $\mathcal{P}$ is 3 ft. from end D of the crane's boom and is perpendicular to the boom. The cross-sectional area of the boom at $\mathcal{P}$ is 10.0 in^2. Determine the average normal stress and the magnitude of the average shear stress in the boom at $\mathcal{P}$.

2.33 Three rectangular boards are glued together and subjected to axial loads as shown. What is the average shear stress on each glued surface?

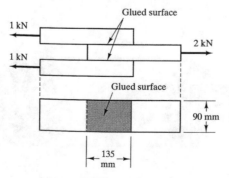

Problem 2.33

2.34 Two boards with 4 in × 4 in square cross sections are mitered and glued together as shown. If the axial forces $P = 600$ lb, what average shear stress must the glue support?

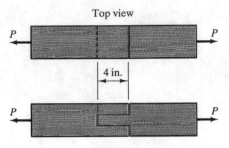

Problem 2.34

2.35 A $\frac{1}{8}$-in-diameter punch is used to cut blanks out of a $\frac{1}{16}$-in-thick plate of aluminum. If an average shear stress of 20,000 psi must be induced in the plate to create a blank, what force F must be applied?

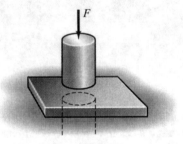

Problem 2.35

2.36 Two pipes are connected by bolted flanges. The bolts are 20 mm in diameter. One pipe has a fixed support, and the other is subjected to a torque $T = 6$ kN · m about its axis. Estimate the resulting average shear stress in each bolt.

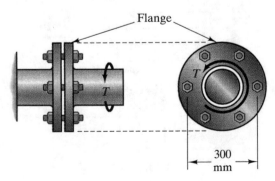

Problem 2.36

2.37 The cylindrical shaft of the bolt in Fig. (a) extends through a hole in the fixed vertical plate that is slightly larger than the shaft. An axial force P acts on the right end of the shaft. In a test of the strength of the bolt, P is progressively increased. When it reaches 46 kN, the head of the bolt shears off of the shaft as shown in Fig. (b). What was the average shear stress between the head of the bolt and the shaft just before the head sheared off?

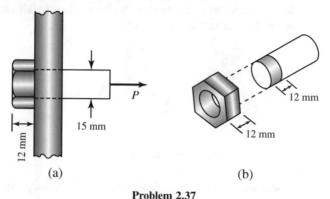

(a)   (b)

Problem 2.37

2.38 "Shears" such as the familiar scissors have two blades that subject a material to shear stress. For the shearing process shown, draw a suitable free-body diagram, and determine the average shear stress the blades exert on the sheet of material of thickness t and width b. (*Strategy:* Obtain a free-body diagram by passing a vertical plane through the material between the two blades.)

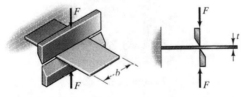

Problem 2.38

2.39 A 2-in diameter cylindrical steel bar is attached to a 3-in thick fixed plate by a cylindrical rubber grommet. If the axial load $P = 60$ lb, what is the average shear stress on the cylindrical surface of contact between the bar and the grommet?

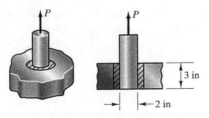

Problems 2.39–2.40

2.40   A 2-in diameter cylindrical steel bar is attached to a 3-in thick fixed plate by a cylindrical rubber grommet. The outer diameter of the grommet is 3.5 in. The axial load $P = 60$ lb. What is the average shear stress on the cylindrical surface of contact between the grommet and the fixed plate?

2.41   A 2-in diameter cylindrical steel bar is attached to a 3-in thick fixed plate by a cylindrical rubber grommet. The bar is subjected to a torque $T = 100$ in - lb about its axis. What is the average shear stress on the cylindrical surface of contact between the bar and the grommet?

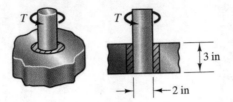

Problem 2.41

2.42   A traction distribution **t** acts on a plane surface A. The value of **t** at a given point on A is $\mathbf{t} = 50\mathbf{i} + 45\mathbf{j} - 20\mathbf{k}$ (kPa). The unit vector **i** is perpendicular to A and points away from the material. What is the normal stress σ at the given point?

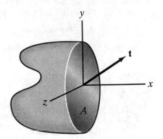

Problems 2.42–2.43

2.43   A traction distribution **t** acts on a plane surface A. The value of **t** at a given point on A is $\mathbf{t} = 50\mathbf{i} + 45\mathbf{j} - 20\mathbf{k}$ (kPa). The unit vector **i** is perpendicular to A and points away from the material. What is the magnitude of the shear stress τ at the given point?

2.44∗ A traction distribution **t** acts on a plane surface A. The value of **t** at a given point on A is $\mathbf{t} = 3000\mathbf{i} - 2000\mathbf{j} + 4000\mathbf{k}$ (psi). The unit vector $\mathbf{e} = \frac{6}{7}\mathbf{i} + \frac{3}{7}\mathbf{j} + \frac{2}{7}\mathbf{k}$ is perpendicular to A and points away from the material. What is the normal stress σ at the given point?

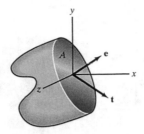

Problems 2.44–2.45

2.45∗ A traction distribution **t** acts on a plane surface A. The value of **t** at a given point on A is $\mathbf{t} = 3000\mathbf{i} - 2000\mathbf{j} + 4000\mathbf{k}$ (psi). The unit vector $\mathbf{e} = \frac{6}{7}\mathbf{i} + \frac{3}{7}\mathbf{j} + \frac{2}{7}\mathbf{k}$ is perpendicular to A and points away from the material. What is the magnitude of the shear stress τ at the given point?

2.46 An object in equilibrium is subjected to three forces. (The object's weight is negligible in comparison.) The forces $\mathbf{F}_1 = -40\mathbf{i} + 10\mathbf{j} - 10\mathbf{k}$ (kN) and $\mathbf{F}_2 = -30\mathbf{i} + 15\mathbf{k}$ (kN). The area of intersection of the plane $\mathcal{P}$ with the object is A. **(a)** If you draw the free-body diagram of the part of the object to the left of the plane $\mathcal{P}$, what total force is exerted by the traction distribution on A? **(b)** If you draw the free-body diagram of the part of the object to the right of the plane $\mathcal{P}$, what total force is exerted by the traction distribution on A?

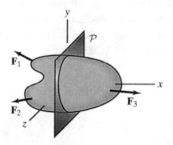

Problems 2.46–2.47

2.47 An object in equilibrium is subjected to three forces. (The object's weight is negligible in comparison.) The forces $\mathbf{F}_1 = -40\mathbf{i} + 10\mathbf{j} - 10\mathbf{k}$ (kN) and $\mathbf{F}_2 = -30\mathbf{i} + 15\mathbf{k}$ (kN). The plane $\mathcal{P}$ is perpendicular to the x axis, and the area of intersection of $\mathcal{P}$ with the object is $A = 0.04$ m^2. Draw the free-body diagram of the part of the object to the left of the plane $\mathcal{P}$, and determine **(a)** the average traction on A and **(b)** the average normal stress on A.

2.48∗ The rectangular cross section A is subjected to the traction distribution

$$\mathbf{t} = \left(4y - 5y^2\right)\mathbf{i} + 6y^2\mathbf{j} + (3 + 2y)\mathbf{k} \ (\text{MPa}).$$

What is the average normal stress on A?

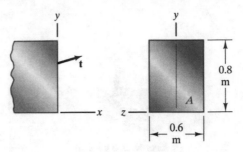

Problems 2.48–2.50

2.49* The rectangular cross section A is subjected to the traction distribution

$$\mathbf{t} = \left(4y - 5y^2\right)\mathbf{i} + 6y^2\mathbf{j} + (3 + 2y)\mathbf{k}\ (\text{MPa}).$$

What is the magnitude of the average shear stress on A?

2.50* The rectangular cross section A is subjected to the traction distribution

$$\mathbf{t} = (2 - 2y + z)\mathbf{i} + 3yz\mathbf{j}\ (\text{MPa}).$$

What is the average normal stress on A?

2.51* The rectangular cross section A is subjected to a normal stress given by the equation $\sigma = ay$, where a is a constant. At $y = 0.8$ m, $\sigma = 20$ MPa. Determine **(a)** the average normal stress on A and **(b)** the magnitude of the moment about the z axis exerted by the normal stress.

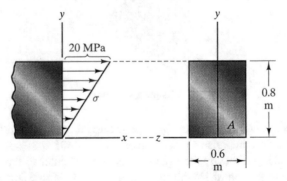

Problem 2.51

2.52 The beam has a rectangular cross section with dimensions $b = 4$ in, $h = 6$ in. It is subjected to a counterclockwise couple $M = 8000$ in - lb at its left end and at its right end is subjected to a normal stress distribution $\sigma = ay$, where a is a constant. The origin of the coordinate system is at the center of the beam's cross section, and the beam is in equilibrium. Determine the constant a.

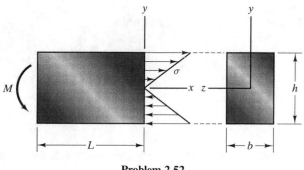

Problem 2.52

2.2   Strains

What happens to an object when it is subjected to external loads? In addition to possibly changing its state of motion, we are all familiar from everyday experiments—stretching rubber bands, bending popsicle sticks—that objects can *deform*, or change shape, under the action of loads. Our ultimate objective is to be able to determine the deformations of simple structural elements resulting from given loads. For now, we merely want to introduce quantities that are used to describe an object's change in shape.

If we put ourselves in the place of the pioneers of this subject, the task before us seems very difficult. How can we even describe the shape of a given object analytically, much less a change in its shape? To make the problem tractable, we don't approach it in this global way, asking "What is the object's new shape?" Instead, we begin by considering what happens to the material near a single point of the object.

2.2.1   Normal Strain

Consider a sample of a solid material such as the volume of iron with which we began discussing stresses. We can imagine taking a pen and drawing a line of infinitesimal length dL somewhere within the material (Fig. 2.19(a)). Suppose that we then subject the material to some set of loads, causing it to undergo a deformation. What happens to the line? We don't know, because we don't know what the loads are and cannot yet predict the effects those loads have on the material, but the line of original length dL may contract or stretch to some new length dL' and may also change direction (Fig. 2.19(b)). The *normal strain* ε is a measure of the change

Fig. 2.19 (a) Infinitesimal line within a volume of material. (b) Volume and line after a deformation

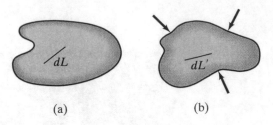

(a)   (b)

in length of the line dL. It is defined to be the change in length divided by the original length:

$$\varepsilon = \frac{dL' - dL}{dL}.$$ (2.2)

For a given point of the material and a given direction (the direction of the line dL), ε measures how much the material contracts or stretches in the subsequent deformation. Notice that ε is negative if the material contracts and positive if it stretches. Because ε is a change in length divided by a length, it is dimensionless.

It is important to realize that the value of the normal strain corresponds to a given point and a given direction in the material prior to the deformation. To define it, we had to choose a point in the material to place the line dL and choose a direction for dL. *In general, the value of the normal strain may vary from point to point in a material and may be different in different directions.* For example, a material may stretch in one direction but contract in a different direction, the way a stretched rubber band becomes thinner (contracts) in the direction perpendicular to its axis. Also, notice that the normal strain measures the contraction or stretch of the material relative to some initial state, which we call the *reference state*.

Suppose that an object undergoes a deformation, and we know the value of ε corresponding to a given point of the material and a given direction prior to the deformation. What does that tell us? Consider a line in the material that has length dL and is in that direction prior to the deformation. We can solve Eq. (2.2) for the length of the line following the deformation:

$$dL' = (1 + \varepsilon)\, dL.$$ (2.3)

We can determine the length of a finite line following a deformation if we know the value of the normal strain ε in the direction tangent to the line at each point of the line. Let L be the length of a finite line in a volume of material in a reference state (Fig. 2.20(a)). We obtain the length L' of the line in a deformed state (Fig. 2.20(b)) by integrating Eq. (2.3)

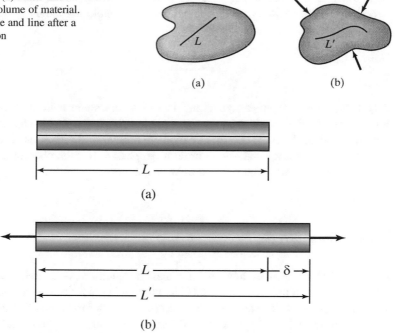

Fig. 2.20 (a) Finite line within a volume of material. (b) Volume and line after a deformation

(a)   (b)

(a)

(b)

Fig. 2.21 Applying axial loads to a bar: (a) The reference state. (b) The deformed state after application of the axial loads

$$L' = \int_L (1 + \varepsilon)\, dL = L + \int_L \varepsilon\, dL, \qquad (2.4)$$

where the subscript L on the integral signs means that the integrations are carried out over the entire length of the line. We denote the change in length of a finite line by δ:

$$\delta = L' - L = \int_L \varepsilon\, dL. \qquad (2.5)$$

The change in length δ is positive if the line L increases in length and negative if L decreases in length. If the value of ε is uniform (constant) along the line, the change in length is

$$\delta = L' - L = \varepsilon L \qquad \text{(if } \varepsilon \text{ is constant along } L\text{).} \qquad (2.6)$$

Notice that Eqs. (2.4)–(2.6) apply even if the line is not straight in the reference state as long as ε is the normal strain in the direction tangent to the line at each point.

As an example, consider a bar of material of length L, and imagine drawing a straight line along the bar's axis (Fig. 2.21(a)). If the bar is stretched by axial loads, the line increases in length (Fig. 2.21(b)). *If we assume that the normal strain ε in the direction parallel to the bar's axis is uniform throughout the length of the bar*, we can solve Eq. (2.6) for the normal strain:

$$\varepsilon = \frac{L' - L}{L} = \frac{\delta}{L}.$$

This example is easy to visualize and interpret, and some authors introduce the normal strain by defining it to be the change in length of a bar divided by its length in a reference state. We have chosen to define the normal strain in terms of the change in length of an infinitesimal line in order to emphasize from the beginning that the normal strain can vary from point to point in a material.

2.2.2   Shear Strain

The normal strain tells us how much a material contracts or stretches in a given direction at a given point of a material. We will find in the following chapters that an additional type of strain is necessary for describing the deformation of a material. Let us return to our volume of material and imagine drawing two perpendicular infinitesimal lines dL_1 and dL_2 somewhere within a reference state (Fig. 2.22(a)). The angle between the two lines is 90° or $\pi/2$ radians. After the material undergoes a deformation, these two lines may no longer be perpendicular. Let us denote the angle in radians between the two lines after the deformation as $\pi/2 - \gamma$ (Fig. 2.22(b)). The angle γ will be positive, zero, or negative if the angle between the two lines decreases, remains the same, or increases, respectively.

The angle γ is called the *shear strain* referred to the directions dL_1 and dL_2. Thus the shear strain is a measure of the change in the angle between two particular lines that were perpendicular in the reference state. In general, its value may vary from point to point in a material and may be different for different orientations of the perpendicular lines dL_1 and dL_2. Because γ is an angle in radians, the shear strain is dimensionless.

If we consider an infinitesimal rectangle in the reference state with sides dL_1 and dL_2 (Fig. 2.23(a)), the shear strain referred to the directions dL_1 and dL_2 tells us the

Fig. 2.22 (a) Perpendicular infinitesimal lines at a point in the reference state. **(b)** The lines in a deformed state

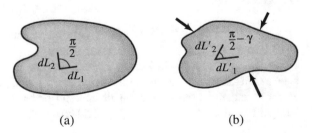

(a)   (b)

Fig. 2.23 (a) Rectangle in the reference state. (b) Shear strain causes the rectangle to become a parallelogram

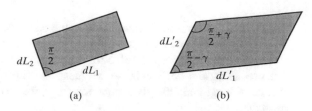

(a)   (b)

angles between the sides of the resulting parallelogram in the deformed state (Fig. 2.23(b)).

Notice that in defining the normal and shear strains, we have made no assumption about the state of motion of the material. In the examples and problems presented in this introductory treatment of mechanics of materials, we assume that the material in its deformed state is in equilibrium. But our definitions of strain are not limited to that situation. The deformed state can instead represent the state at an instant in time of a material undergoing an arbitrary motion.

Study Questions

1. What is the definition of the normal strain ε? What are the dimensions of the normal strain?
2. Consider a line with finite length L within a reference state of a material. What do you need to know to determine the line's length L' in a deformed state? How do you determine L'?
3. What is the definition of the shear strain γ?

Example 2.5 Normal Strain in a Bar
A bar has length $L = 2$ m when it is unloaded. (a) In Fig. 2.24(a), the bar is subjected to axial loads that increase its length to 2.04 m. If the normal strain ε in the direction of the bar's axis is assumed to be uniform throughout the bar's length, what is ε? (b) In Fig. 2.24(b), the bar is suspended from one end, causing its length to increase to 2.01 m. The resulting normal strain in the direction of the bar's axis is given as a function of x by $\varepsilon = ax/L$, where a is a constant. What is the value of a?

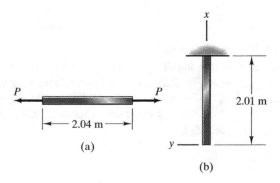

Fig. 2.24

Strategy
In part (a) the normal strain is uniform, so the bar's change in length δ is given in terms of its reference length L and the strain ε by Eq. (2.6). We know the change in length, so we can solve for ε. In part (b) the strain is a function of distance along the bar's axis, so we must use Eq. (2.5). Using the known change in length and the given equation for ε, we can solve Eq. (2.5) for the constant a.

Solution
(a) From Eq. (2.6), with the lengths in meters, we obtain

$$\delta = L' - L = \varepsilon L:$$

$$2.04 - 2 = \varepsilon(2)$$

$$0.04 = \varepsilon(2).$$

The normal strain in the direction of the bar's axis is $\varepsilon = 0.02$.
(b) From Eq. (2.5), the bar's change in length is

$$\delta = L' - L = \int_L \varepsilon \, dL:$$

$$2.01 - 2 = \int_0^2 \frac{ax}{2} dx.$$

Integrating and solving for the constant a, we obtain $a = 0.01$.

Discussion
In part (b) of this example, the form of the equation for the distribution of the normal strain along the axis of a suspended bar was specified. In Chap. 3 we relate the normal strain to the normal stress in a linear elastic bar subjected to axial load. You will then be able to show that the normal strain is indeed a linear function of x in a suspended prismatic bar.

Example 2.6 Determining a Shear Strain
Two infinitesimal lines dL_1 and dL_2 within a material are parallel to the x and y axes in a reference state (Fig. 2.25). After a deformation of the material, they point in the directions of the unit vectors $\mathbf{e}_1 = 0.976\,\mathbf{i} + 0.098\mathbf{j} + 0.195\mathbf{k}$ and $\mathbf{e}_2 = 0.228\,\mathbf{i} + 0.911\mathbf{j} - 0.342\mathbf{k}$, respectively. What is the shear strain referred to the directions dL_1 and dL_2?

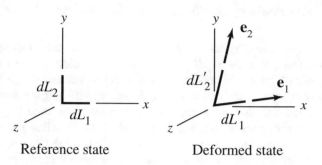

Fig. 2.25

Strategy

To determine the shear strain, we must calculate the angle between the infinitesimal lines in the deformed state. The components of the unit vectors $\mathbf{e}_1$ and $\mathbf{e}_2$ are given, so we can use the dot product to obtain the angle between them.

Solution

Let β be the angle between the unit vectors $\mathbf{e}_1$ and $\mathbf{e}_2$. From the definition of the dot product

$$\mathbf{e}_1 \cdot \mathbf{e}_2 = |\mathbf{e}_1||\mathbf{e}_2| \cos \beta$$
$$= \cos \beta.$$

Evaluating the dot product in terms of the components of the vectors

$$\mathbf{e}_1 \cdot \mathbf{e}_2 = (0.976)(0.228) + (0.098)(0.911) + (0.195)(-0.342)$$
$$= 0.245.$$

We see that $\beta = \arccos(0.245) = 1.32$ rad. The shear strain γ is related to the angle β in radians by

$$\frac{\pi}{2} - \gamma = \beta.$$

Solving, the shear strain referred to the directions dL_1 and dL_2 is $\gamma = 0.248$.

Discussion

As this example demonstrates, the shear strain is a measure of the change in the angle between line elements that are perpendicular in the reference state. To calculate the shear strain, we needed to determine the angle between the lines dL_1 and dL_2 in the deformed state.

2.2.3 Components of Strain and the Stress-Strain Relations

Consider a material line element of length dL in a reference state of a material whose length in a deformed state is dL' (Fig. 2.26). The value of the normal strain $\varepsilon = (dL' - dL)/dL$ generally depends on the direction chosen for the line element dL. The shear strain γ is defined to be the decrease in the angle between two line elements dL_1 and dL_2 that are perpendicular in a reference state (Fig. 2.27). The value of γ generally depends on the orientation chosen for the line elements dL_1 and dL_2. The *state of strain* at a point p of a material is defined by making particular choices of the directions of the line elements used to evaluate the normal and shear strains. Let us introduce a coordinate system (Fig. 2.28(a)) and define six *components of strain*:

ε_x The normal strain determined with the element dL parallel to the x axis
     (Fig. 2.28(b))
ε_y The normal strain determined with the element dL parallel to the y axis
ε_z The normal strain determined with the element dL parallel to the z axis
γ_{xy} The shear strain determined with the elements dL_1 and dL_2 in the positive x and
     y directions (Fig. 2.28(c))
γ_{yz} The shear strain determined with the elements dL_1 and dL_2 in the positive y and
     z directions
γ_{xz} The shear strain determined with the elements dL_1 and dL_2 in the positive x and
     z directions

If these six components are known at a point, the normal strain in *any* direction and the shear strain for *any* orientation of the two perpendicular elements can be determined. (As we have stated it, this holds only when the components of strain are *small*, meaning that they are sufficiently small that products of the components are negligible in comparison to the components themselves.) Therefore *the state of*

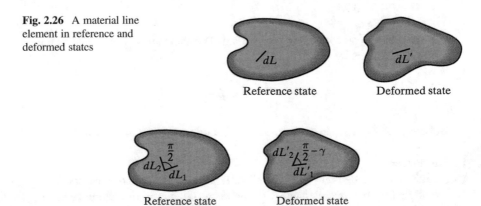

Fig. 2.26 A material line element in reference and deformed states

Reference state   Deformed state

Reference state   Deformed state

Fig. 2.27 Line elements that are perpendicular in the reference state. The decrease in the angle between them in the deformed state is the shear strain

Fig. 2.28 (a) Introducing a coordinate system. (b) Line element for determining the value of ε_x at p. (b) Line elements for determining the value of γ_{xy} at p

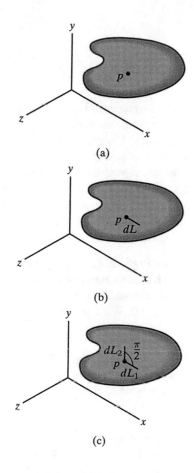

(a)

(b)

(c)

strain at a point is completely described by the six components ε_x, ε_y, ε_z, γ_{xy}, γ_{yz}, and γ_{xz}.

For the model of material behavior called *isotropic linear elasticity*, the components of the state of strain can be expressed as linear functions of the components of the state of stress in the forms

$$\varepsilon_x = \frac{1}{E}\sigma_x - \frac{\nu}{E}(\sigma_y + \sigma_z), \tag{2.7}$$

$$\varepsilon_y = \frac{1}{E}\sigma_y - \frac{\nu}{E}(\sigma_x + \sigma_z), \tag{2.8}$$

$$\varepsilon_z = \frac{1}{E}\sigma_z - \frac{\nu}{E}(\sigma_x + \sigma_y), \tag{2.9}$$

$$\gamma_{xy} = \frac{1}{G}\tau_{xy}, \tag{2.10}$$

$$\gamma_{yz} = \frac{1}{G}\tau_{yz}, \tag{2.11}$$

$$\gamma_{xz} = \frac{1}{G}\tau_{xz}. \tag{2.12}$$

The constant E is called the *modulus of elasticity*, ν is called *Poisson's ratio*, and G is called the *shear modulus*. The dimensions of E and G are (force)/(area). Poisson's ratio ν is dimensionless. We discuss the origin and forms of these *stress-strain relations* in Chap. 8. The most common structural materials, including steel and aluminum, exhibit isotropic linear elastic behavior when subjected to small strains. Values of E, ν, and G are given for a variety of materials in Appendix B. Notice that if the six components of strain are known at a point, Eqs. (2.7)–(2.12) can be solved for the six components of stress (see Example 2.7).

Example 2.7 Applying the Stress-Strain Relations
The components of strain at a point of a structural member made of 2014-T6 aluminum alloy are $\varepsilon_x = 0.002$, $\varepsilon_y = 0.001$, $\varepsilon_z = 0$, $\gamma_{xy} = 0.003$, $\gamma_{yz} = 0$, $\gamma_{xz} = 0$. What are the six components of stress at the point in psi (lb/in^2)?

Strategy
We can solve Eqs. (2.7)–(2.12) for the six components of stress if we know the modulus of elasticity E, the Poisson's ratio ν, and the shear modulus G. The values of E, ν, and G for 2014-T6 aluminum are given in Appendix B.

Solution
From Appendix B, we see that for 2014-T6 aluminum alloy, $E = 10.6E6$ psi, $\nu = 0.33$, and $G = 4E6$ psi. (Because we want to determine the values of the stress components in psi, we use the values of E and G in psi.) We substitute these values and the state of strain into Eqs. (2.7)–(2.12):

$$0.002 = \frac{1}{10.6E6 \text{ psi}}\left[\sigma_x - 0.33\left(\sigma_y + \sigma_z\right)\right],$$

$$0.001 = \frac{1}{10.6E6 \text{ psi}}\left[\sigma_y - 0.33(\sigma_x + \sigma_z)\right],$$

$$0 = \frac{1}{10.6E6 \text{ psi}}\left[\sigma_z - 0.33\left(\sigma_x + \sigma_y\right)\right],$$

$$0.003 = \frac{1}{4E6 \text{ psi}}\tau_{xy},$$

$$0 = \frac{1}{4E6 \text{ psi}}\tau_{yz},$$

$$0 = \frac{1}{4E6 \text{ psi}}\tau_{xz}.$$

Solving these equations for the components of stress, we obtain $\sigma_x = 39$, 100 psi, $\sigma_y = 31$, 200 psi, $\sigma_z = 23$, 200 psi, $\tau_{xy} = 12$, 000 psi, $\tau_{yz} = 0$, and $\tau_{xz} = 0$.

Discussion
If a material exhibits isotropic linear elastic behavior (discussed in Chap. 8) and the values of E, ν, and G are known, the state of stress can be determined when the state of strain is known and vice versa. But how can the state of strain or the state of stress at a point of an object that is subjected to loads be determined? In the following chapters, we discuss the states of stress and strain in structural members subjected to loads, beginning in Chap. 3 with bars subjected to axial loads.

Problems

2.53 A line of length dL at a particular point of a material in a reference state has length $dL' = 1.2dL$ in a deformed state. What is the normal strain corresponding to that particular point and in the direction of the line dL?

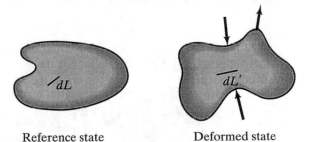

Reference state   Deformed state

Problems 2.53–2.54

2.54 The normal strain corresponding to a point of a material and the direction of a line of length dL in the reference state is $\varepsilon = 0.15$. What is the length dL' of the line in the deformed state?

2.55 The length of the straight line within the reference state of the object is $L = 50$ mm. In the deformed state, the length of the line is $L' = 54$ mm. If the normal strain ε in the direction tangent to the line is uniform throughout the line's length, what is ε?

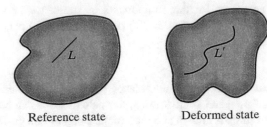

Reference state   Deformed state

Problems 2.55–2.57

2.56 The length of the line within the material in the reference state is $L = 6$ in. The material then undergoes a deformation such that the value of the normal strain ε in the direction tangent to the line is $\varepsilon = 0.04$ at each point of the line. What is the length L' of the line in the deformed state?

2.57 The length of the line L in the reference state is 0.5 m. Let a coordinate s measure distance along the line, from $s = 0$ at the left end to $s = L$ at the right end. The material then undergoes a deformation such that the normal strain in the direction tangent to the line is $\varepsilon = 0.002 + 0.004s$. What is the length L' of the line in the deformed state?

2.58 The length of the straight line within the reference state of the object is $L = 0.2$ m. The material then undergoes a deformation such that the value of the normal strain in the direction tangent to the line is $\varepsilon = 0.03 + 2x^2$, where x is in meters. What is the length L' of the line in the deformed state?

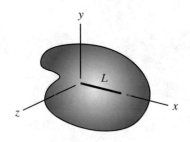

Reference state

Problems 2.58–2.59

2.59 The length of the straight line within the reference state of the object is $L = 0.2$ m. The material then undergoes a deformation such that the value of the normal strain in the direction tangent to the line is $\varepsilon = a[1 + (x/L)^3]$, where x is in meters and a is a constant. The length of the line in the deformed state is $L' = 0.21$ m. Determine the constant a.

2.60 A 4-ft prismatic bar is subjected to axial loads that increase its length to 4.025 ft. If you assume that the normal strain ε in the direction parallel to the bar's axis is uniform (constant), what is ε?

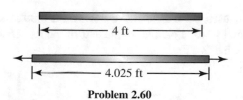

Problem 2.60

2.61 Suppose that you subject a 2-m prismatic bar to compressive axial loads that cause a uniform normal strain $\varepsilon = -0.003$. What is the length of the deformed bar?

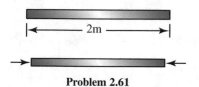

Problem 2.61

2.62 The prismatic bar is subjected to axial loads that cause uniform normal strains $\varepsilon_{AB} = 0.002$ in part AB and $\varepsilon_{BC} = -0.006$ in part BC. What is the resulting change in length of the 1400-mm bar?

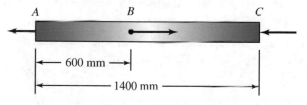

Problems 2.62–2.63

2.63 The prismatic bar is subjected to axial loads that cause a uniform normal strain $\varepsilon_{AB} = 0.003$ in part AB and a uniform normal strain ε_{BC} in part BC. As a result, the length of the 1400-mm bar decreases by 10 mm. What is ε_{BC}?

2.64 When it is unloaded, the nonprismatic bar is 16-in long. The loads cause normal strain given by the equation $\varepsilon = 0.04/(x + 16)$, where x is the distance from the left end of the bar in inches. What is the change in length of the bar?

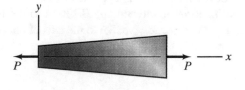

Problem 2.64

2.65 The force F causes point B to move downward 0.002 m. If you assume the resulting normal strain ε parallel to the axis of the bar AB is uniform along the bar's length, what is ε?

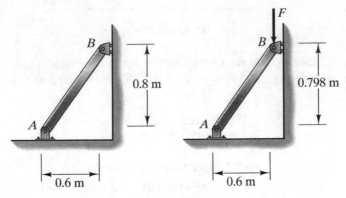

Problem 2.65

2.66 When the truss is subjected to the vertical force F, joint A moves a distance $v = 0.1$ m vertically and a distance $u = 0.3$ m horizontally. If the normal strain ε_{AB} in the direction parallel to member AB is uniform throughout the length of the member, what is ε_{AB}?

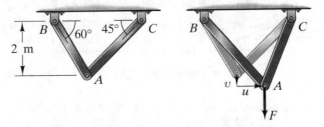

Problems 2.66–2.67

2.67 When the truss is subjected to the vertical load F, joint A moves a distance $v = 0.1$ m vertically and a distance $u = 0.3$ m horizontally. If the normal strain ε_{AC} in the direction parallel to member AC is uniform throughout the length of the member, what is ε_{AC}?

2.68 Suppose that a downward force is applied at point A of the truss, causing point A to move 0.360 in downward and 0.220 in to the left. If the resulting normal strain ε_{AB} in the direction parallel to the axis of bar AB is uniform, what is ε_{AB}?

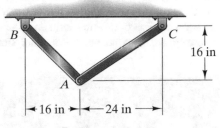

Problems 2.68–2.69

2.69 A downward force is applied at point A of the truss, causing point A to move 0.360 in downward and 0.280 in to the left. If the resulting normal strain ε_{AC} in the direction parallel to the axis of bar AC is uniform, what is ε_{AC}?

2.70 A steel tube (Fig. (a)) has an outer radius $r = 80$ mm. The tube is then pressurized, increasing its outer radius to $r' = 80.6$ mm (Fig. (b)). What is the resulting normal strain of the bar's outer circumference in the direction tangent to the circumference?

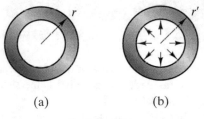

(a)   (b)

Problem 2.70

2.71 The angle between two infinitesimal lines dL_1 and dL_2 that are perpendicular in a reference state is $120°$ in a deformed state. What is the shear strain at this point corresponding to the directions dL_1 and dL_2?

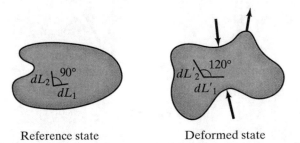

Reference state   Deformed state

Problem 2.71

2.72 When the airplane's wing is unloaded (the reference state), the perpendicular lines L_1 and L_2 on the upper surface of the left wing are each 600 mm long. In the loaded state shown, L_1 is 600.2-mm long and L_2 is 595-mm long. If you assume that they are uniform, what are the normal strains in the L_1 and L_2 directions?

Problems 2.72–2.73 (*Photograph from iStock by Thordur_A*)

2.73 The angle between the perpendicular lines L_1 and L_2 at the point where they intersect is 90.2° in the loaded state. What is the shear strain referred to the directions L_1 and L_2 at that point?

2.74 Two infinitesimal lines dL_1 and dL_2 are shown in a reference state and in a deformed state. (The lines dL_1, dL_2, dL'_1, and dL'_2 are contained in the x-y plane.) What is the shear strain at this point corresponding to the directions dL_1 and dL_2?

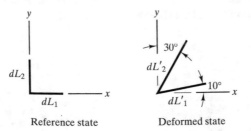

Problem 2.74

2.75* In the upper part of Fig. (a), a cylindrical bar with 12-m length and 1-m radius has a fixed support at the left end. Axial and circumferential lines are drawn on the bar's surface. The lower part of Fig. (a) shows the bar's curved surface "unwrapped," illustrating the geometry of the axial and circumferential lines. The infinitesimal line dL_1 lies on an axial line, and the infinitesimal line dL_2 lies on a circumferential line. A torque T is then applied about the axis of the bar at its right end, causing the end to rotate (Fig. (b)). The bar's length and radius do not change when T is applied. The resulting geometry of the axial and circumferential lines is shown in the lower part of Fig. (b); in this unwrapped view, they remain straight but are no longer orthogonal. The resulting shear strain referred to the directions dL_1 and dL_2 is 0.13. Use this value to determine the angle through which the end of the bar rotates.

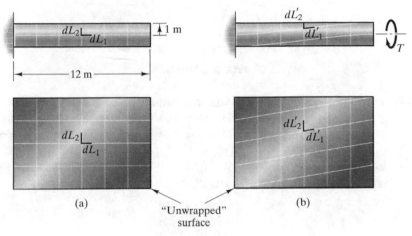

(a)   "Unwrapped"   (b)
                              surface

Problems 2.75–2.76

2.76∗ In the upper part of Fig. (a), a cylindrical bar with 12-m length and 1-m radius has a fixed support at the left end. Axial and circumferential lines are drawn on the bar's surface. The lower part of Fig. (a) shows the bar's curved surface "unwrapped," illustrating the geometry of the axial and circumferential lines. The infinitesimal line dL_1 lies on an axial line, and the infinitesimal line dL_2 lies on a circumferential line. A torque T is then applied about the axis of the bar at its right end, causing the end to rotate (Fig. (b)). The bar's length and radius do not change when T is applied. The resulting geometry of the axial and circumferential lines is shown in the lower part of Fig. (b); in this unwrapped view, they remain straight but are no longer orthogonal. If the applied torque causes the end of the bar to rotate through an angle of 60°, what is the shear strain referred to the directions dL_1 and dL_2?

2.77* Two infinitesimal lines dL_1 and dL_2 within a material are parallel to the x and y axes in a reference state (Fig. (a)). After a deformation of the material, dL_1 points in the direction of the unit vector $e_1 = 0.667i + 0.667j + 0.333k$, and dL_2 points in the direction of the unit vector $e_2 = -0.408i + 0.816j - 0.408k$ (Fig. (b)). What is the shear strain referred to the directions dL_1 and dL_2?

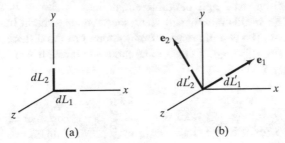

(a)   (b)

Problems 2.77–2.78

2.78* Two infinitesimal lines dL_1 and dL_2 within a material are parallel to the x and y axes in a reference state (Fig. (a)). After a motion and deformation of the material, dL_1' points in the direction of the unit vector $e_1 = 0.717i + 0.535j - 0.447k$, and dL_2' points in the direction of the unit vector $e_2 = -0.514i + 0.686j + 0.514k$ (Fig. (b)). What is the shear strain referred to the directions dL_1 and dL_2?

Problem 2.79

2.79* An infinitesimal rectangle at a point in a reference state of a material is shown. In a deformed state, the normal strains in the dL_1 and dL_2 directions are $\varepsilon_1 = 0.04$ and $\varepsilon_2 = -0.02$, and the shear strain referred to the dL_1 and dL_2 directions is $\gamma = 0.02$. What is the normal strain in the dL direction?

2.80 The state of stress at a point of a structural member made of 7075-T6 aluminum alloy is $\sigma_x = 530$ MPa, $\sigma_y = 480$ MPa, $\sigma_z = 480$ MPa, $\tau_{xy} = -100$ MPa, $\tau_{yz} = 0$, $\tau_{xz} = 0$. What is the state of strain at the point?

Strategy: Use Eqs. (2.7)–(2.12) to calculate the components of the state of strain. Determine E, ν, and G from Appendix B.

2.81 The state of strain at a point of a structural member made of pure titanium is $\varepsilon_x = 0.0038$, $\varepsilon_y = -0.0018$, $\varepsilon_z = 0.0016$, $\gamma_{xy} = 0.0034$, $\gamma_{yz} = 0$, $\gamma_{xz} = 0$. What is the state of stress at the point in ksi (kip/in²)?

Chapter Summary

We have introduced fundamental measures of the stresses and strains in a material. We defined average stresses on an area and showed how to evaluate them in particular cases. In Chap. 3 we apply these definitions to structural applications involving axially loaded bars.

Stresses

The *traction* **t** (Fig. (a)) is defined such that the force exerted on each infinitesimal element dA of the area A is **t** dA. The normal component σ is the *normal stress*, and the tangential component τ is the *shear stress*. The normal stress σ is defined to be positive if it points outward, or away from the material, and is then said to be *tensile*. If σ is negative, the normal stress points toward the material and is said to be *compressive*. The dimensions of **t**, σ, and τ are (force)/(area).

The average value of the traction **t** on the area A in Fig. (a) is

$$\mathbf{t}_{\mathrm{av}} = \frac{1}{A} \int_A \mathbf{t}\, dA. \tag{2.1}$$

The components of the average traction normal and tangential to A are the average normal stress σ_{av} and average shear stress τ_{av}. The total normal and tangential forces exerted on A are $\sigma_{\mathrm{av}}A$ and $\tau_{\mathrm{av}}A$, respectively.

The state of stress at a point is completely described by the six components σ_x, σ_y, σ_z, τ_{xy}, τ_{yz}, *and* τ_{xz}. The directions of these stresses, and the orientations of the planes on which they act, are shown in Fig. (b). The shear stresses $\tau_{xy} = \tau_{yx}$, $\tau_{yz} = \tau_{zy}$, and $\tau_{xz} = \tau_{zx}$.

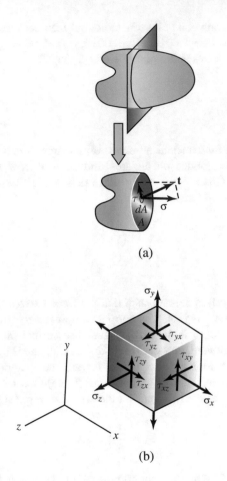

(a)

(b)

Strains and the Stress-Strain Relations

Consider an infinitesimal line dL in a volume of material in a reference state (Fig. (c)). Let the length of the line in a deformed state be dL' (Fig. (d)). The *normal strain* ε is defined by

$$\varepsilon = \frac{dL' - dL}{dL}. \tag{2.2}$$

Let L be the length of a finite line within the material in a reference state. The length L' in a deformed state is

$$L' = L + \int_L \varepsilon \, dL, \tag{2.4}$$

where ε is the normal strain in the direction tangent to the line at each point. The change in length of the line is denoted by δ:

$$\delta = L' - L = \int_L \varepsilon \, dL. \tag{2.5}$$

If ε is uniform along the line, the change in length is

$$\delta = L' - L = \varepsilon L. \tag{2.6}$$

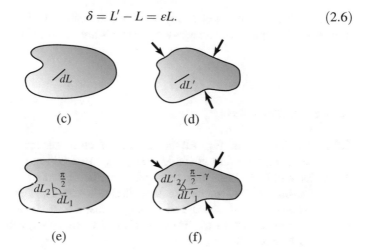

(c) (d)

(e) (f)

Consider two perpendicular infinitesimal lines dL_1 and dL_2 in a volume of material in a reference state (Fig. (e)). Let the angle in radians between the two lines in a deformed state be denoted by $\pi/2 - \gamma$ (Fig. (f)). The angle γ is the *shear strain* referred to the directions dL_1 and dL_2.

The *state of strain* at a point is completely described by the six components $\varepsilon_x, \varepsilon_y, \varepsilon_z$, γ_{xy}, γ_{yz}, and γ_{xz}. The components $\varepsilon_x, \varepsilon_y$, and ε_z are the normal strains in the x, y, and z directions. The component $\gamma_{xy} = \gamma_{yx}$ is the shear strain referred to the positive directions of the x and y axes, and the components $\gamma_{yz} = \gamma_{zy}$ and $\gamma_{xz} = \gamma_{zx}$ are defined similarly.

When the most common structural materials are subjected to small strains, the components of the state of strain are given in terms of the components of the state of stress by the *stress-strain relations*

$$\varepsilon_x = \frac{1}{E}\sigma_x - \frac{\nu}{E}(\sigma_y + \sigma_z), \tag{2.7}$$

$$\varepsilon_y = \frac{1}{E}\sigma_y - \frac{\nu}{E}(\sigma_x + \sigma_z), \tag{2.8}$$

$$\varepsilon_z = \frac{1}{E}\sigma_z - \frac{\nu}{E}(\sigma_x + \sigma_y), \qquad (2.9)$$

$$\gamma_{xy} = \frac{1}{G}\tau_{xy}, \qquad (2.10)$$

$$\gamma_{yz} = \frac{1}{G}\tau_{yz}, \qquad (2.11)$$

$$\gamma_{xz} = \frac{1}{G}\tau_{xz}, \qquad (2.12)$$

where E is the modulus of elasticity, ν is Poisson's ratio, and G is the shear modulus. Values of E, ν, and G for a variety of materials are given in Appendix B.

Review Problems

2.82 The bar BC in Fig. (a) has a circular cross section with 4-in radius. It has a fixed support at B and is subjected to a 20-kip (20,000 lb) downward force at C. In Fig. (b), a free-body diagram of part of the bar is obtained by passing a plane $\mathcal{P}$ perpendicular to the bar's axis. The terms σ_{av} and τ_{av} are the average normal stress and average shear stress acting on the free-body diagram where it was cut by the plane. Determine σ_{av} and τ_{av} in psi (pounds per square inch).

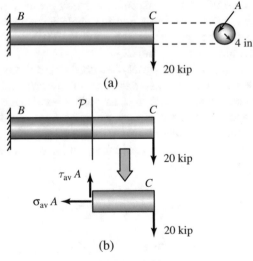

(a)

(b)

Problem 2.82

2.83 The bar BC has a circular cross section with 4-in radius. It has a fixed support at B and is subjected to a 20-kip (20,000 lb) downward force at C. A free-body diagram of part of the bar is obtained by passing the plane $\mathcal{P}$ shown. The terms σ_{av} and τ_{av} are the average normal stress and average shear stress acting on the free-body diagram where it was cut by the plane. The area A' is the intersection of the plane $\mathcal{P}$ with the bar, which is the area on which the average stresses σ_{av} and τ_{av} act. **(a)** What is the area A' in in^2? **(b)** Determine σ_{av} and τ_{av} in psi (pounds per square inch).

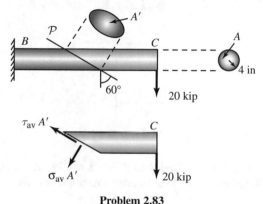

Problem 2.83

2.84 The bar is made of material 1-in thick. Its width varies linearly from 2 in at its left end to 4 in at its right end. The bar is subjected to axial loads P. The average normal stress at plane $\mathcal{P}_1$ is $\sigma_{av} = 400$ psi (lb/in^2). Determine the average normal stress at plane $\mathcal{P}_2$.

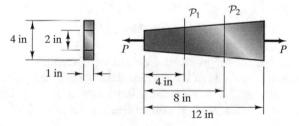

Problem 2.84

2.85* The 12-in bar is suspended as shown and is loaded only by its own weight. It is made of material 1-in thick. Its width varies linearly from 2 in at the bottom to 4 in at the top. The weight density of the homogeneous material is 0.42 lb/in^3. Determine the average normal stress at the plane $\mathcal{P}$, where x is the distance from the bottom of the bar in inches.

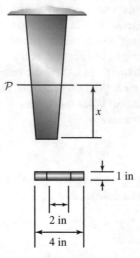

Problem 2.85

2.86 The beam has cross-sectional area $A = 0.1$ m^2. What are the average normal stress and the magnitude of the average shear stress at the plane $\mathcal{P}$ if $x = 2$ m?

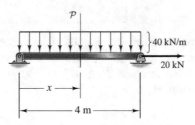

Problems 2.86–2.87

2.87 The beam has cross-sectional area $A = 0.1$ m^2. Determine the average normal stress and the magnitude of the average shear stress at the plane $\mathcal{P}$ if $x = 3$ m.

2.88 The prismatic bar BC has cross-sectional area $A = 0.01$ m^2. If the force $F = 6$ kN, what is the average normal stress at the plane $\mathcal{P}$?

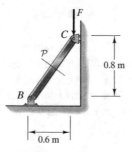

Problems 2.88–2.89

2.89 The prismatic bar BC has cross-sectional area $A = 0.01$ m^2. The bar will safely support an average compressive normal stress of 1.2 MPa on the plane $\mathcal{P}$. Based on this criterion, what is the largest downward force F that can safely be applied?

2.90 Member ABC of the frame has a pin at C that is supported by a smooth horizontal slot in member DC. The cross-sectional area of member ABC at the plane $\mathcal{P}_1$ is $A = 0.03$ m^2. Determine the average normal stress and the magnitude of the average shear stress at $\mathcal{P}_1$.

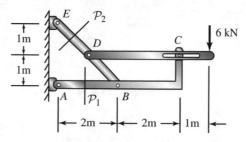

Problems 2.90–2.91

2.91 Member ABC of the frame has a pin at C that is supported by a smooth horizontal slot in member DC. The cross-sectional area of member BDE of the frame at the plane $\mathcal{P}_2$ is $A = 0.03$ m^2. Determine the average normal stress and the magnitude of the average shear stress at $\mathcal{P}_2$.

2.92 The members of the pliers are attached to each other by pin connections at B, C, D, and E. The bottom view of the pin connection at C is shown. The cylindrical pin is 3 mm in diameter. The reactions the members exert on each other at C were determined in Example 1.4. Determine the average shear stress in the pin resulting from the 40-N forces exerted on the pliers.

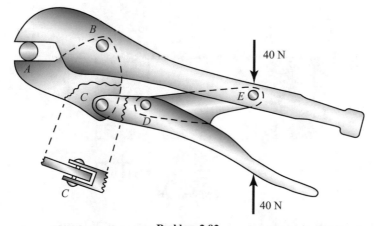

Problem 2.92

2.93 The jaws of the bolt cutter are connected by two links AB. The cross-sectional area of each link is 750 mm^2. What average normal stress is induced in each link by the 90-N forces exerted on the handles?

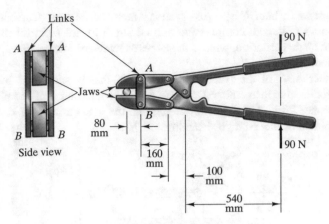

Problems 2.93–2.94

2.94 The pins connecting the links *AB* to the jaws of the bolt cutter are 20 mm in diameter. What average shear stress is induced in the pins by the 90-N forces exerted on the handles?

2.95∗ The rectangular cross section *A* is subjected to a shear stress in the *y* axis direction given by the equation

$$\tau = \frac{6V}{bh^3}\left(\frac{h^2}{4} - y^2\right),$$

where *V* is a constant. Determine **(a)** the total force exerted on *A*, **(b)** the average shear stress on *A*, and **(c)** the ratio of the maximum shear stress on *A* to the average shear stress.

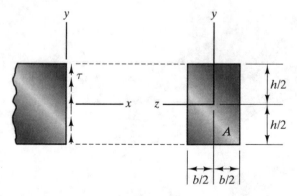

Problem 2.95

2.96 The length of the line within the reference state of the material is *L* = 0.2 m. What is the length *L′* of the line in the deformed state of the material if the normal strain in the direction tangent to the line at each point of the line is ε = 0.004?

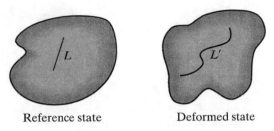

Reference state   Deformed state

Problem 2.96

2.97 The length of the straight line within the reference state of the material is $L = 0.2$ m. The material then undergoes a deformation such that the value of the normal strain in the direction tangent to the line is $\varepsilon = 0.004(1 + 6x)$, where x is in meters. What is the length L' of the line in the deformed state?

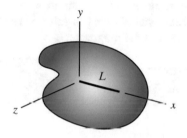

Reference state

Problem 2.97

2.98 The prismatic bar is subjected to axial loads that cause uniform normal strains $\varepsilon = 0.002$ in part AB and $\varepsilon = -0.004$ in part BC. What is the resulting change in length of the 28-in bar?

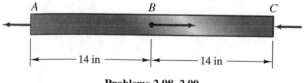

Problems 2.98–2.99

2.99 The prismatic bar is subjected to loads that cause a uniform normal strain $\varepsilon_L = 0.006$ in part AB and a uniform normal strain ε_R in part BC. As a result, the length of the 28-in bar increases by 0.032 in. What is ε_R?

2.100 A spherical pressure vessel is used in a fuel cell carried aboard a satellite. Before it is pressurized, the outer diameter of the vessel is $D = 1$ m. When the vessel is pressurized, its outer diameter is $D' = 1.02$ m. What is the resulting normal strain of the vessel material at its outer surface in the direction tangent to the surface?

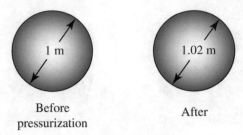

Before
pressurization

After

Problem 2.100

2.101 The infinitesimal rectangle at a point in a reference state of a material becomes the parallelogram shown in a deformed state. Determine (**a**) the normal strain in the dL_1 direction, (**b**) the normal strain in the dL_2 direction, and (**c**) the shear strain corresponding to the dL_1 and dL_2 directions.

2.102∗ The infinitesimal rectangle at a point in a reference state of a material becomes the parallelogram shown in a deformed state. Determine the normal strain in the direction of the diagonal dL.

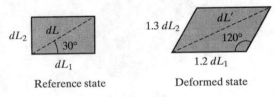

Reference state   Deformed state

Problems 2.101–2.102

Chapter 3
Axially Loaded Bars

Truss structures—widely used to support bridges, buildings, vehicles, and other mechanical devices—are frameworks of bars that support external loads by subjecting their members to axial loads. Applying axial loads to a bar and measuring the resulting change in its length are common methods of testing the response of materials to stress. Not only is this test relatively straightforward to carry out, it subjects the material to a simple and easily determined distribution of stress. Because of this simplicity and the great importance of axially loaded bars from the standpoint of applications, we analyze them in detail in this chapter.

3.1 Stresses in Prismatic Bars

In Chap. 2 we determined *average* normal and shear stresses acting on planes within objects. However, we emphasized that the actual distributions of stress, not their average values, must usually be known for design. We now discuss these distributions for axially loaded bars.

3.1.1 The State of Stress

Recall that a prismatic bar is one whose cross section is uniform throughout its length. Suppose we could somehow load a prismatic bar by applying uniform normal stresses σ to its ends (Fig. 3.1). For this loading, it can be shown that the stress distribution on *every* plane perpendicular to the axis of the bar consists of the same uniform normal stress and no shear stress (Fig. 3.2). If we isolate a rectangular element of the bar that is oriented as shown in Fig. 3.3, the faces of the element that are perpendicular to the bar's axis are subjected to the normal stress σ, and the other faces are free of stress. In terms of the coordinate system shown, the components of

© Springer Nature Switzerland AG 2020
A. Bedford, K. M. Liechti, *Mechanics of Materials*,
https://doi.org/10.1007/978-3-030-22082-2_3

Fig. 3.1 Loading a bar by
normal tractions at its ends

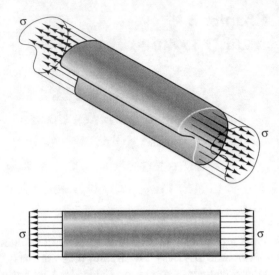

Fig. 3.2 The same uniform
normal stress acts on every
plane perpendicular to the
bar's axis

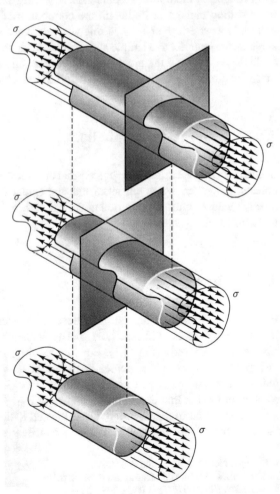

Fig. 3.3 Isolating an
element within the bar

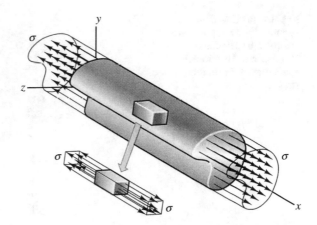

stress are $\sigma_x = \sigma$, $\sigma_y = 0$, $\sigma_z = 0$, $\tau_{xy} = 0$, $\tau_{yz} = 0$, and $\tau_{xz} = 0$. Proving these
results, which is beyond our scope in this discussion involves showing that no
other distribution of stress can satisfy both the boundary conditions (the normal
stresses applied to the ends and the stress-free lateral surfaces of the bar) and
equilibrium.

The state of stress we have described applies to a prismatic bar that is subjected to
axial loads in a very idealized way. What does this tell us about the states of stress in
bars that are subjected to axial loads in other ways? To answer this question, we first
need to demonstrate that applying a uniform normal stress to the end of a bar is
equivalent to applying a single axial force at the centroid of its cross section.
(Remember that we define two systems of forces and moments to be equivalent if
the sums of the forces in the two systems are equal and the sums of the moments
about any point due to the two systems are equal.) The total force exerted by a
uniform normal stress σ on the bar's cross-sectional area A is σA. Let us place a
coordinate system with its origin at the centroid of the cross section (Fig. 3.4(a)), so
that the y and z coordinates of the centroid are zero:

$$\bar{y} = \frac{\int_A y \, dA}{A} = 0, \qquad \bar{z} = \frac{\int_A z \, dA}{A} = 0.$$

Then the moment about the z axis due to the uniform normal stress distribution
is zero

$$\int_A - y\sigma \, dA = -\sigma \int_A y \, dA = 0,$$

and the moment about the y axis is zero

$$\int_A z\sigma \, dA = \sigma \int_A z \, dA = 0.$$

Fig. 3.4 (**a**) Coordinate
system with its origin at the
centroid. (**b**) The force
P acting at the centroid is
equivalent to the uniform
stress σ

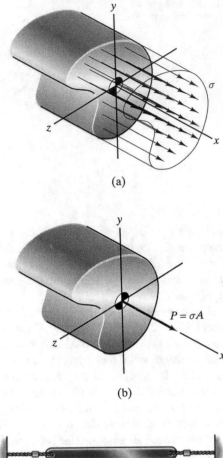

(a)

(b)

Fig. 3.5 Applying axial
loads to a bar

If we place a force $P = \sigma A$ at the centroid of the bar's cross section (Fig. 3.4(b)), the
moments due to *P* about the *y* and *z* axes are zero, so *the force P is equivalent to the
uniform normal stress distribution.*

Now suppose that we apply loads to the ends of a prismatic bar that are equivalent
to axial forces *P* applied at the centroid of the cross section, but do it in some
practical way (Fig. 3.5). On planes at increasing distances from the ends of the bar,
the stress distribution approaches a uniform normal stress

$$\sigma = \frac{P}{A}. \tag{3.1}$$

No matter how we apply the axial loads, the stress distribution at axial distances
greater than a few times the bar's width is approximately the same one obtained by
subjecting the ends of the bar to uniform normal stresses (Fig. 3.6).

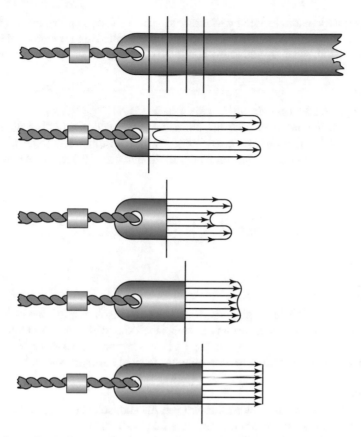

Fig. 3.6 Stress distributions at increasing distances from the end of the bar

This result was stated by Barré de Saint-Venant (1797–1886), who conducted experiments with rubber bars to demonstrate it for particular cases. In a more general form, it is known as *Saint-Venant's principle* and has been proven analytically in recent years. Based on this result, we assume in examples and problems that the stress distribution on a plane perpendicular to the axis of an axially loaded bar consists of uniform normal stress and no shear stress. But remember that this assumption is invalid near the ends of the bar. A separate analysis, again beyond the scope of our discussion, is necessary to determine the distribution of stress near the ends.

Study Questions

1. If a prismatic bar could be loaded by applying uniform normal stresses σ at its ends, what is the resulting stress distribution on planes perpendicular to the bar's axis?

2. If a prismatic bar is loaded by arbitrary systems of forces and couples at the ends that are each equivalent to a tensile axial load P applied at the centroid of the bar's cross section, what do you know about the resulting stress distribution on planes perpendicular to the bar's axis?

Example 3.1 Stresses in an Axially Loaded Bar
Part AB of the bar in Fig. 3.7 has a circular cross section with 2-in diameter, and part BC has a circular cross section with 4-in diameter. Determine the normal stress on a plane perpendicular to the bar's axis (**a**) in part AB and (**b**) in part BC.

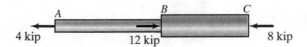

Fig. 3.7

Strategy
By passing a plane through part AB of the bar and isolating the part of the bar on one side of the plane, we can determine the axial load in part AB required to satisfy equilibrium and then use Eq. (3.1) to determine the normal stress. We can use the same method to determine the normal stress in part BC.

Solution
(**a**) In Fig. (a) we pass a plane through part AB and isolate the part of the bar to the left of the plane. From the resulting free-body diagram, the axial load in part AB is $P_{AB} = 4$ kip. The cross-sectional area of part AB of the bar is

$$A_{AB} = \frac{\pi(2 \text{ in})^2}{4} = 3.14 \text{ in}^2.$$

The normal stress in part AB is therefore

$$\sigma_{AB} = \frac{P_{AB}}{A_{AB}}$$
$$= \frac{4000 \text{ lb}}{3.14 \text{ in}^2}$$
$$= 1270 \text{ psi}.$$

(**b**) In Fig. (b) we pass a plane through part BC and isolate the part of the bar to the left of the plane. From the equilibrium equation $P_{BC} + 12 \text{ kip} - 4 \text{ kip} = 0$, the axial load in part BC of the bar is $P_{BC} = -8$ kip. The cross-sectional area of part BC of the bar is

$$A_{BC} = \frac{\pi(4 \text{ in})^2}{4} = 12.6 \text{ in}^2.$$

The normal stress in part BC is

$$
\begin{aligned}
\sigma_{BC} &= \frac{P_{BC}}{A_{BC}} \\
&= \frac{-8000 \text{ lb}}{12.6 \text{ in}^2} \\
&= -637 \text{ psi.}
\end{aligned}
$$

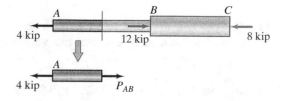

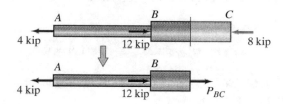

(a) Obtaining a free-body diagram by passing a plane through part AB. (b) Obtaining a free-body diagram by passing a plane through part BC

Discussion
In Figs. (a) and (b), notice that we defined P_{AB} and P_{BC} so that they would be positive if the axial force is tensile and negative if it is compressive. As a consequence, the values of the normal stress we obtained with Eq. (3.1) were positive if the axial force is tensile and negative if it is compressive. From the results, we see that the normal stress in part AB of the bar is tensile and the normal stress in part BC is compressive.

Example 3.2 Stresses in Truss Members
The members of the truss in Fig. 3.8 each have a cross-sectional area of 400 mm^2. The mass of the suspended object is $m = 3400$ kg. What are the normal stresses in the members?

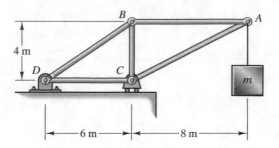

Fig. 3.8

Strategy
We can use the method of joints to determine the axial forces in the members and divide by the cross-sectional area to determine the normal stresses.

Solution
The weight of the suspended mass is $mg = (3400 \text{ kg})(9.81 \text{ m/s}^2) = 33.4 \text{ kN}$. In Fig. (a) we draw the free-body diagram of joint A of the truss. The angle $\theta = \arctan(4/8) = 26.6°$. From the equilibrium equations

$$\Sigma F_x = -P_{AB} - P_{AC} \cos 26.6° = 0,$$
$$\Sigma F_y = -P_{AC} \sin 26.6° - 33.4 \text{ kN} = 0,$$

we obtain $P_{AB} = 66.7 \text{ kN}$, $P_{AC} = -74.6 \text{ kN}$. Continuing in this way, we obtain the axial forces in the members:

Member :	AB	AC	BC	BD	CD
Axial force (kN) :	66.7	−74.6	−44.5	80.2	−66.7

Dividing these values by the cross-sectional area 400E - 6 m², the normal stresses in the members are

Member :	AB	AC	BC	BD	CD
Normal stress (MPa) :	167	−186	−111	200	−167

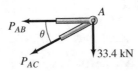

(a) Joint A

Discussion
How could you determine whether the truss in this example would safely support the suspended object? The normal stresses in the members must not exceed

allowable values that are chosen based on the properties of the material the
members are made of. We discuss allowable stresses in Sect. 3.6. You will also
need to insure that members subjected to compressive loads do not fail by
buckling, which is discussed in Chap. 10.

3.1.2 Stresses on Oblique Planes

In structural design it is not sufficient to consider the stress distribution on only one
particular plane or subset of planes. We can demonstrate this for an axially loaded
bar. Let us determine the normal and shear stresses acting on a plane $\mathcal{P}$ oriented as
shown in Fig. 3.9. The angle θ is measured from the bar's axis to a line *normal*
(perpendicular) to the plane $\mathcal{P}$. To determine the stresses on $\mathcal{P}$, we define a free-body
diagram as shown in Fig. 3.10. We first isolate a rectangular element from the bar
and then isolate the part of the element to the left of plane $\mathcal{P}$. Our objective is to
obtain a geometrically simple free-body diagram subjected to the normal and shear
stresses on $\mathcal{P}$. We denote the normal stress by σ_θ and the shear stress by τ_θ, indicating

Fig. 3.9 Passing a plane $\mathcal{P}$
through a bar at an arbitrary
angle relative to its axis

Fig. 3.10 Obtaining a free-
body diagram for
determining the stresses on
plane $\mathcal{P}$

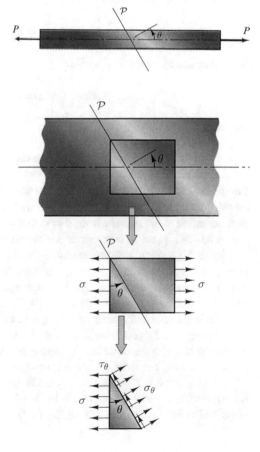

Fig. 3.11 (a) Areas of the
faces. (b) Forces on the free-
body diagram

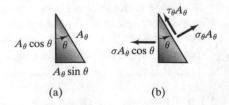

(a)   (b)

that they act upon the plane defined by the angle θ. We have arbitrarily chosen the
direction of τ_θ.

We will determine σ_θ and τ_θ by writing equilibrium equations for the triangular
free-body diagram we cut from the element. Letting A_θ be the area of the slanted
face, we can express the areas of the horizontal and vertical faces in terms of A_θ and θ
(Fig. 3.11(a)). Then the forces on the free-body diagram are the products of the areas
and the uniform stresses (Fig. 3.11(b)). Summing forces in the σ_θ direction

$$\sigma_\theta A_\theta - (\sigma A_\theta \cos \theta) \cos \theta = 0,$$

we determine the normal stress on $\mathcal{P}$:

$$\sigma_\theta = \sigma \cos^2 \theta. \tag{3.2}$$

Summing forces in the τ_θ direction

$$\tau_\theta A_\theta + (\sigma A_\theta \cos \theta) \sin \theta = 0,$$

we determine the shear stress on $\mathcal{P}$:

$$\tau_\theta = -\sigma \sin \theta \cos \theta. \tag{3.3}$$

Equations (3.2) and (3.3) give the normal and shear stresses acting on a plane
oriented at an arbitrary angle relative to the axis of a prismatic bar. In Fig. 3.12 we
plot the ratios σ_θ/σ and τ_θ/σ from $\theta = 0$ to $\theta = 180°$. We also show the elements and
the values of σ_θ and τ_θ corresponding to particular values of θ.

Figure 3.12 permits us to examine the normal and shear stresses acting on all
planes through an axially loaded prismatic bar. Notice that there is no plane for
which the normal stress is larger in magnitude than the normal stress on the plane
perpendicular to the bar's axis ($\theta = 0$). When we use the equation $\sigma = P/A$ to
determine the normal stress on the plane perpendicular to a bar's axis, we are
determining the largest tensile or compressive normal stress acting on any plane
within the bar. On the other hand, there is no shear stress on the plane perpendicular
to the bar's axis, but there are shear stresses on oblique planes. On planes oriented at
45° relative to the bar's axis, the magnitude of the shear stress is 1/2 the magnitude of
the maximum normal stress: $|\tau_\theta| = |\sigma/2|$. We can confirm that this is the maximum
magnitude of the shear stress from Eq. (3.3). The derivative of τ_θ with respect to θ is

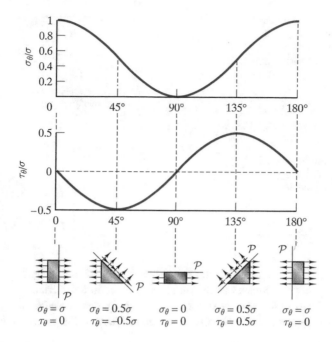

Fig. 3.12 Normal and shear stresses as functions of θ

Fig. 3.13 A column loaded
by a weight W

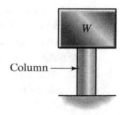

$$\frac{d\tau_\theta}{d\theta} = \sigma\left(\sin^2\theta - \cos^2\theta\right).$$

Equating this expression to zero to determine the angles at which τ_θ has a maximum or minimum, we obtain $\theta = 45°$ and $\theta = 135°$. At $\theta = 45°$, $\tau_\theta = -\sigma/2$, and at $\theta = 135°$, $\tau_\theta = \sigma/2$.

We have shown that the maximum tensile or compressive normal stress on any plane through an axially loaded prismatic bar is $\sigma = P/A$, and the magnitude of the maximum shear stress is $|\sigma/2|$. As an example, suppose that a column of concrete with cross-sectional area A supports a weight W (Fig. 3.13). Drawing a free-body diagram by passing a plane through the column perpendicular to its axis (Fig. 3.14), we see that the compressive stress is $\sigma = -W/A$. This is the maximum compressive stress on any plane through the column. The magnitude of the maximum shear stress, which occurs on planes oriented at 45° relative to the column's axis, is $|\sigma/2| = W/2A$. If W is increased sufficiently, a column of brittle material such as concrete tends to fail along a

Fig. 3.14 Obtaining a free-
body diagram by passing a
plane through the column
perpendicular to its axis

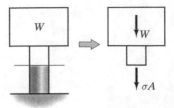

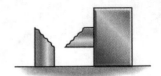

Fig. 3.15 A brittle material
would tend to fail along a
plane at 45° relative to
the axis

plane on which the magnitude of the shear stress is a maximum (Fig. 3.15). In design,
we must consider the stresses acting on all planes within a material.

*When we refer to the normal stress in an axially loaded bar without specifying the
plane, we will mean the normal stress on a plane perpendicular to the bar's axis.*

Study Questions

1. If a bar is subjected to axial loads, how can you determine the normal and shear
   stresses acting on a plane that is not perpendicular to the bar's axis?
2. If a bar with cross-sectional area A is subjected to tensile axial loads P, what is the
   maximum tensile stress to which the material is subjected? What is the orientation
   of the planes on which the maximum tensile stress occurs?
3. If a bar with cross-sectional area A is subjected to tensile axial loads P, what is the
   magnitude of the maximum shear stress to which the material is subjected? What
   is the orientation of the planes on which the magnitude of the shear stress is a
   maximum?

Example 3.3 Stresses on an Oblique Plane
To test a glue, two plates are glued together as shown in Fig. 3.16. The bar
formed by the joined plates is then subjected to tensile axial loads of 200 N.
What normal and shear stresses act on the plane where the plates are glued
together? (In other words, what stresses must the glue support?)

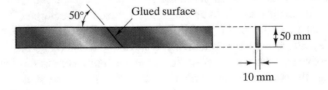

Fig. 3.16

Strategy

We know the axial load and the bar's cross-sectional area, so we can determine the normal stress $\sigma = P/A$ on a plane perpendicular to the bar's axis. Then we can use Eqs. (3.2) and (3.3) to determine the normal and shear stresses on the glued plane.

Solution

The normal stress on a plane perpendicular to the bar's axis is

$$\sigma = \frac{P}{A} = \frac{200 \text{ N}}{(0.01 \text{ m})(0.05 \text{ m})} = 400,000 \text{ Pa}.$$

Our objective is to determine the normal stress σ_θ and shear stress τ_θ on the plane where the plates are glued (Fig. (a)).

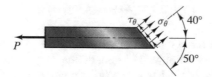

(a) Stresses on the glued plane

The angle between the normal to the glued plane and the bar's axis is 40°. From Eq. (3.2)

$$\sigma_\theta = \sigma \cos^2\theta = (400,000 \text{ Pa}) \cos^2 40° = 235,000 \text{ Pa},$$

and from Eq. (3.3)

$$\tau_\theta = -\sigma \sin\theta \cos\theta = -(400,000 \text{ Pa}) \sin 40° \cos 40° = -197,000 \text{ Pa}.$$

The glued surface is subjected to a tensile normal stress of 235 kPa and a shear stress of magnitude 197 kPa. The negative value of the shear stress indicates that the shear stress acts in the direction opposite to the defined direction of τ_θ.

Discussion

You could have predicted that the stresses σ_θ and τ_θ would be of opposite sign. Notice that the sum of the forces in the vertical direction on the free-body diagram in Fig. (a) is not zero if σ_θ and τ_θ are either both positive or both negative. You can often avoid errors in solutions to problems in mechanics by confirming that your results are consistent with the free-body diagrams you have used.

Problems

3.1 A prismatic bar with cross-sectional area $A = 0.1$ m^2 is loaded at the ends in two ways: (**a**) by 100-Pa uniform normal stresses and (**b**) by 10-N axial forces acting at the centroid of the beam's cross section. What is the normal stress at the plane $\mathcal{P}$ in the two cases?

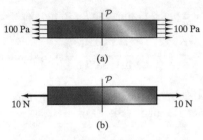

Problem 3.1

3.2 Part AB of the bar has a circular cross section with 15-mm radius, and part BC has a circular cross section with 45-mm radius. Determine the normal stress on a plane perpendicular to the bar's axis (**a**) in part AB of the bar and (**b**) in part BC of the bar.

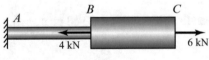

Problem 3.2

3.3 The prismatic bar is subjected to a 12-kN axial load. Determine the normal stress on a plane perpendicular to the bar's axis if the bar has the cross section (a) and if the bar has the cross section (b).

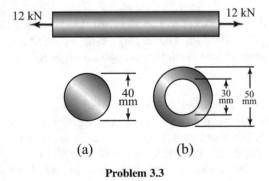

Problem 3.3

3.4 The mass of the ceiling fan is 12 kg. The cross section of the pipe that suspends the fan is shown. What is the normal stress on a plane perpendicular to the pipe? (Assume that the weight of the pipe is negligible in comparison to the weight of the fan.)

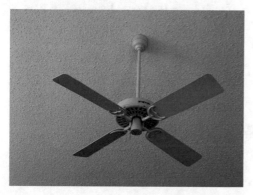
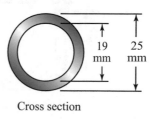

Cross section

Problem 3.4

3.5 The connecting rod *AB* is subjected to a tensile axial load of 4 kN. The rod's cross section is shown. If you don't want the rod to be subjected to a normal stress greater than 800 kPa, what is the maximum acceptable value of the cross section's inner radius *r*?

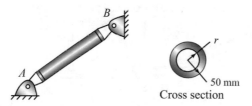

Cross section

Problem 3.5

3.6 The cross-sectional area of bar *AB* is 2 in². If the force $F = 6$ kip, what is the normal stress on a plane perpendicular to the axis of bar *AB*?

Strategy: You can determine the axial force in bar *AB* by drawing a free-body diagram of the horizontal bar and summing moments about point *C*.

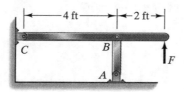

Problems 3.6–3.7

3.7 Bar *AB* consists of a material that will safely support a tensile normal stress of 60 ksi. If you want to design the frame to support forces *F* as large as 50 kip, what is the minimum required cross-sectional area of bar *AB*?

3.8 The cross section of bars *BC* and *DG* is shown. If a downward force $F = 520$ kN is applied at *H*, what are the normal stresses in bars *BC* and *DG*?

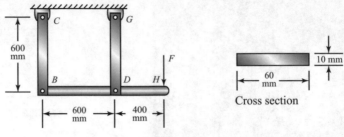

Problem 3.8

3.9 The truss is subjected to a downward load $F = 25$ kN at *B*. The cross section of the members *BC* and *BD* is shown. What are the normal stresses in the members on planes perpendicular to their axes?

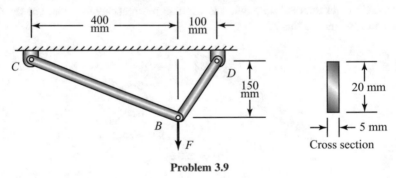

Problem 3.9

3.10 The mass of the suspended box is 1800 kg. The mass of the crane's arm (not including the hydraulic actuator *BC*) is 200 kg, and its center of mass is 2 m to the right of *A*. The cross-sectional area of the upper part of the hydraulic actuator is 0.006 m². What is the normal stress on a plane perpendicular to the axis of the upper part of the actuator?

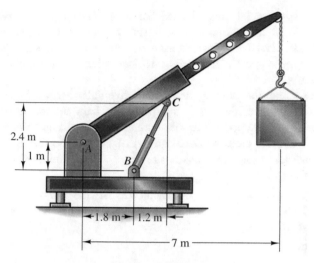

Problems 3.10–3.11

3.11 The mass of the suspended box is 1800 kg. The mass of the crane's arm (not including the hydraulic actuator *BC*) is 200 kg, and its center of mass is 2 m to the right of *A*. The cross-sectional area of the lower part of the hydraulic actuator *BC* is 0.032 m². What is the normal stress on a plane perpendicular to the axis of the lower part of the actuator?

3.12 The cross-sectional area of each bar is 0.4 in². The angle $\beta = 50°$. If the force $F = 3$ kip, what is the normal stress on a plane perpendicular to the axis of one of the bars?

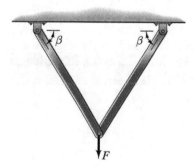

Problems 3.12–3.14

3.13 The cross-sectional area of each bar is A. The angle β is $60°$. The bars are made of a material that will safely support a tensile normal stress of 12 ksi. Based on this criterion, if you want to design the system so that it will support a force $F = 15$ kip, what is the minimum necessary value of A?

3.14* Suppose that the load F and the horizontal distance between the two supports are specified, and the prismatic bars are made of a material that will safely support a tensile normal stress σ_0. You want to choose the angle β and the cross-sectional area A of the bars so that the total volume of material used to make the bars is a minimum. Show that $\beta = 45°$ and $A = F/(\sqrt{2}\sigma_0)$.

3.15 The cross-sectional area of each bar of the truss is 400 mm². If $F = 30$ kN, what is the normal stress on a plane perpendicular to the axis of member AC?

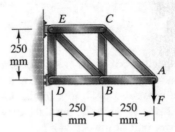

Problems 3.15–3.17

3.16 The cross-sectional area of each bar of the truss is 400 mm². If $F = 30$ kN, what is the normal stress on a plane perpendicular to the axis of member BE?

3.17* The cross-sectional area of each bar of the truss is 400 mm². The bars are made of a material that will safely support a normal stress (tension or compression) of 340 MPa. Based on this criterion, what is the largest safe value of the force F?

3.18 The structure shown supports a downward force F at G. The cross section of the connecting link BC is shown. If the tensile normal stress in the link BC is 1.2 MPa, what is the force F?

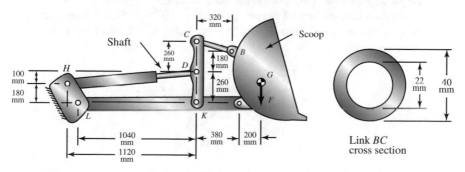

Problem 3.18

3.19 The cross section of the shaft of the hydraulic piston *DH* in Problem 3.18 is shown. A downward force $F = 1600$ N acts at *G*. What is the normal stress in the shaft *DH*?

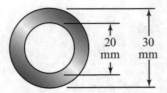

Shaft cross section

Problem 3.19

3.20 The cross-sectional area of the prismatic bar is $A = 4$ in^2 and the axial force $P = 60$ kip. Determine the normal and shear stresses on the plane $\mathcal{P}$. Draw a diagram isolating the part of the bar to the left of plane $\mathcal{P}$ and show the stresses.

Problems 3.20–3.21

3.21 The cross-sectional area of the prismatic bar is $A = 4$ in^2. If the normal stress on the plane $\mathcal{P}$ is 1500 psi, what is the axial force P?

3.22 The cross-sectional area of the prismatic bar is 0.02 m^2. If the normal and shear stresses on the plane $\mathcal{P}$ are $\sigma_\theta = 1.25$ MPa and $\tau_\theta = -1.50$ MPa, what are the angle θ and the axial force P? (Assume that $0 \le \theta \le 90°$.)

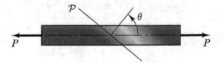

Problem 3.22

3.23 The architectural column is subjected to an 800-kN compressive axial load. The cross section of the column is shown. Determine the normal stresses acting on planes (a), (b), and (c).

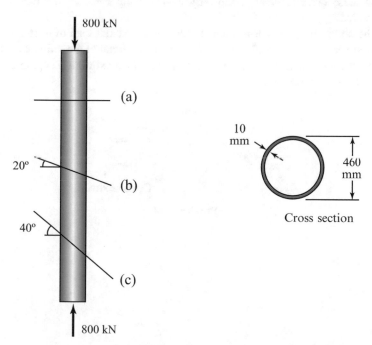

Problems 3.23–3.24

3.24 The architectural column is subjected to an 800-kN compressive axial load. The cross section of the column is shown. Determine the magnitudes of the shear stresses acting on planes (a), (b), and (c).

3.25 The cross-sectional area of the bars is $A = 0.5$ in^2 and the force $F = 3000$ lb. Determine the normal stresses and the magnitudes of the shear stresses on the planes (a) and (b).

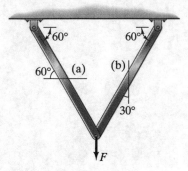

Problems 3.25–3.26

3.26 The cross-sectional area of the bars is $A = 0.5$ in^2. They are made of a material that will safely support a normal stress of 8 ksi and a shear stress of 3 ksi. Based on these criteria, what is the largest downward force F that can safely be applied?

3.27 The connecting rod AB is subjected to a tensile axial load of 4 kN. The rod's cross section is shown. If you don't want the material of the rod to be subjected to a shear stress greater than 400 kPa, what is the maximum acceptable value of the cross section's inner radius r?

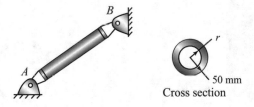

Cross section

Problem 3.27

3.28 The cross section of bars *BC* and *DG* is shown. Suppose that bars *BC* and *DG* consist of a material that will safely support a maximum *shear* stress of 240 MPa. Based on this criterion, what is the maximum safe value of the downward force *F*?

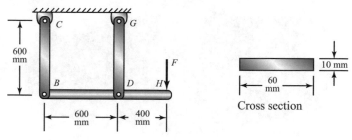

Problem 3.28

3.29 The cross section of members *BC* and *BD* of the truss is shown. They are made of an aluminum alloy that will support a maximum *shear* stress of 240 MPa. Based on this criterion, what is the maximum downward force *F* that can be applied at *B*?

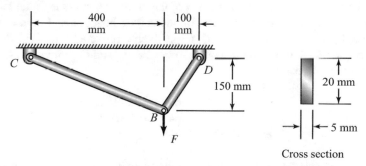

Problem 3.29

3.30∗ The space truss supports a vertical 800-lb load. The cross-sectional area of the bars is 0.2 in². Determine the normal stress in member *AB*.

Strategy: Draw a free-body diagram of joint *A*.

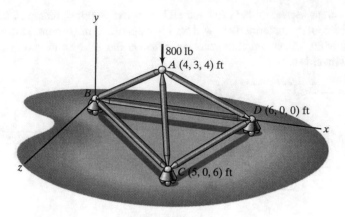

800 lb
A (4, 3, 4) ft

B

D (6, 0, 0) ft

x

y

z

C (5, 0, 6) ft

Problems 3.30–3.31

3.31* The space truss supports a vertical 800-lb load and has roller supports at B, C, and D. The cross-sectional area of the bars is 0.2 in². Determine the normal stress in member BC.

3.32* The free-body diagram of the part of the construction crane to the left of the plane is shown. The coordinates (in meters) of the joints A, B, and C are (1.5, 1.5, 0), (0, 0, 1), and (0, 0, −1), respectively. The axial forces P_1, P_2 and P_3 are parallel to the x axis. The axial forces P_4, P_5, and P_6 point in the directions of the unit vectors

$$\mathbf{e}_4 = 0.640\,\mathbf{i} - 0.640\,\mathbf{j} - 0.426\,\mathbf{k},$$
$$\mathbf{e}_5 = 0.640\,\mathbf{i} - 0.640\,\mathbf{j} + 0.426\,\mathbf{k},$$
$$\mathbf{e}_6 = 0.832\,\mathbf{i} - 0.555\,\mathbf{k}.$$

The total force exerted on the free-body diagram by the weight of the crane and the load it supports is $-F\,\mathbf{j} = -44\,\mathbf{j}$ kN acting at the point $(-20, 0, 0)$ m. The cross-sectional area of members 1, 2, and 3 is 5000 mm², and the cross-sectional area of members 4, 5, and 6 is 1600 mm². What is the normal stress in member 3?

Strategy: Calculate the moment about the line that passes through joints A and B.

Problem 3.32

3.2   Strains and the Elastic Constants

Pulling on a rubber band causes it to stretch. If you pull on a bar of steel in the same way, it will also stretch. You would not see the change in its length—in fact, the deformations of most structural elements under their operating loads are too small to be seen—but we show in this section how you can calculate the change in length of a bar subjected to axial loads. We also show how to determine the change in the bar's dimensions perpendicular to its axis.

3.2.1   Axial and Lateral Strains

If a prismatic bar of length L is loaded at its ends by uniform normal stresses σ as shown in Fig. 3.17, its length will obviously change by some amount δ. Its dimensions in the lateral direction will also change, as can be seen when we observe a stretched rubber band become thinner. But how can we determine the bar's change in length and its new lateral dimensions? We indicated in Sect. 3.2 that the components of stress at each point of the bar in terms of the coordinates system shown in Fig. 3.17 are $\sigma_x = \sigma$, $\sigma_y = 0$, $\sigma_z = 0$, $\tau_{xy} = 0$, $\tau_{yz} = 0$, and $\tau_{xz} = 0$. If we assume that the bar consists of material for which the stress-strain relations (2.7)–(2.12) apply, we can use them to determine the components of strain in terms of the coordinate system shown. Substituting the components of stress into Eqs. (2.7)–(2.9) yields

$$\varepsilon_x = \frac{1}{E}\sigma, \qquad \varepsilon_y = -\frac{\nu}{E}\sigma, \qquad \varepsilon_z = -\frac{\nu}{E}\sigma,$$

where E is the modulus of elasticity and ν is the Poisson's ratio of the material. The component ε_x is the normal strain parallel to the bar's axis. We call it the *axial strain* of the bar and denote it by ε:

$$\varepsilon = \frac{1}{E}\sigma. \tag{3.4}$$

Fig. 3.17 Loading a bar by normal stresses at its ends

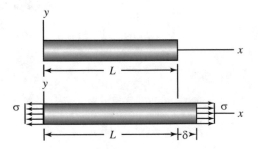

The change in the bar's length is

$$\delta = \varepsilon L = \frac{\sigma L}{E}. \tag{3.5}$$

The components ε_y and ε_z are the normal strains in the directions of the y and z axes, which are perpendicular to the bar's axis. Notice that they are equal. The components of stress are the same for any orientation of the y and z axes relative to the bar, which means that *the value of the normal strain is the same in every direction perpendicular to the bar's axis.* We call the normal strain perpendicular to the bar's axis the *lateral strain* and denote it by ε_{lat}:

$$\varepsilon_{\text{lat}} = -\frac{\nu}{E}\sigma.$$

If the bar is subjected to tensile stresses at the ends (σ is positive), the lateral strain ε_{lat} is negative, meaning that the bar contracts laterally. Notice that *the negative of the ratio of the lateral strain to the axial strain is Poisson's ratio*:

$$\nu = -\frac{\varepsilon_{\text{lat}}}{\varepsilon}. \tag{3.6}$$

If the bar is subjected to *any* loads at the ends that are equivalent to axial loads P acting at the centroid of the cross section, we have seen that Saint-Venant's principle states that the state of stress everywhere except near the ends of the bar is approximately the same as if the bar were loaded by uniform normal stresses $\sigma = P/A$, where A is the bar's cross-sectional area. *If the length of the bar is large in comparison to its lateral dimensions*, the axial strain over most of the bar's length can be approximated by Eq. (3.4), and the bar's change in length can be approximated by Eq. (3.5).

If the material exhibits linear elastic behavior and the modulus of elasticity E and Poisson's ratio ν are known, we have seen that the change in length and the lateral strain of a prismatic bar subjected to known axial loads can be determined. But how can we know whether a given material will exhibit linear elastic behavior, and how can its elastic constants E and ν be determined?

3.2.2 Material Behavior

Leonardo da Vinci (1452–1519) and Galileo (1564–1642) investigated the strengths of wires and bars by subjecting them to axial tension. In a modern *tension test*, still the most common test of the mechanical behavior of materials, a bar of material is mounted in a machine that subjects it to a tensile axial load (Fig. 3.18). For a given value of tensile load P, the normal stress $\sigma = P/A$ can be determined. (The normal

Fig. 3.18 (**a**) Machine for subjecting a bar of material to axial loads. (**b**) A specimen mounted in the machine. The extensometer measures the specimen's change in length

Fig. 3.19 Stress-strain diagram for a tension test of low-carbon steel

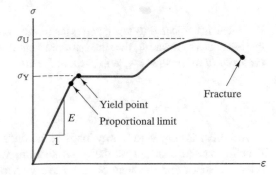

stress is defined in terms of the cross-sectional area of the bar in its reference state. The term *true stress* denotes the stress calculated using the cross-sectional area of the loaded bar. In typical engineering applications, the difference between these definitions of stress is negligible because the change in the bar's cross-sectional area is small.) The bar's axial strain ε is determined by measuring the change in the distance between two axial positions on the bar. The distance between these axial positions before the axial load is applied is called the *gauge length*. When the test proceeds until the bar fractures (breaks), the *elongation*, the final change in the gauge length expressed as a percentage of the gauge length, can also be measured. Because the axial strain becomes nonuniform as fracture approaches, the gauge length used in determining the elongation must be specified.

3.2.2.1 Ductile Materials

We first describe some phenomena that occur in a tension test of a low-carbon steel. Progressively increasing the axial strain and recording the corresponding stress, the graph of the normal stress as a function of the axial strain, or *stress-strain diagram*, appears qualitatively as shown in Fig. 3.19. For small values of the axial strain, the relationship is linear. For that portion of the graph, the relationship between the

Fig. 3.20 Necking in a
tension test specimen

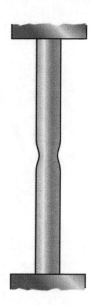

normal stress and axial strain can be written as $\sigma = E\varepsilon$, where E is the modulus of elasticity of the material. At a point called the *proportional limit*, the graph deviates from a straight line. (The stress is no longer a linear function of the strain.) At a point called the *yield point*, the strain begins increasing with no change in stress. The corresponding stress is called the *yield stress* σ_Y. Eventually, the stress again begins increasing with increasing strain, a phenomenon called *strain hardening*, and reaches a maximum value, the *ultimate stress* σ_U.

As the strain continues to increase beyond the occurrence of the ultimate stress, the stress decreases until the bar fractures or breaks. The decreasing stress is associated with the formation of a region of decreased cross-sectional area in the bar, which is called *necking* (Fig. 3.20). The apparently decreasing stress is an artifact caused by defining σ in terms of the cross-sectional area of the undeformed bar. The true stress in the necked portion of the bar continues to increase with strain until fracture occurs.

If the strain to which the steel is subjected remains below the proportional limit, so that $\sigma = E\varepsilon$, we say that its stress-strain relationship is *linearly elastic*, and if the stress is removed, the bar returns to its original length. If the bar is strained beyond the yield point and the stress is then decreased, the resulting path in the σ - ε plane does not return to the origin along the original curve. Instead, it returns to zero stress along a straight line with slope E, and a residual strain remains (Fig. 3.21(a)). If the bar is then reloaded, it follows the new straight path until it reaches the stress-strain curve obtained with monotonically increasing strain (Fig. 3.21(b)). Strain beyond the value of strain corresponding to the yield point, which remains as residual strain if the bar is unloaded, is called *plastic strain*. Thus the yield stress σ_Y is the stress at which plastic strain begins. Materials with stress-strain relationships of this type are

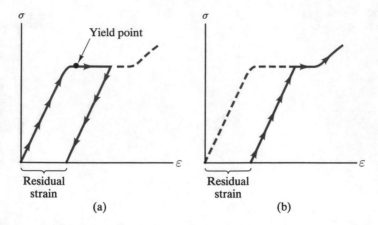

Fig. 3.21 (**a**) Straining the bar beyond the yield point and then unloading it. (**b**) Reloading the bar

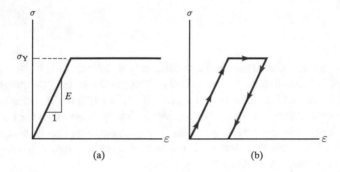

Fig. 3.22 (**a**) Model of an elastic-perfectly plastic material. (**b**) Loading-unloading behavior

sometimes modeled by representing their stress-strain relationship as shown in Fig. 3.22, which is referred to as an *elastic-perfectly plastic material*. ("Perfectly" plastic means it is assumed that there is no strain hardening.)

Materials that undergo significant plastic strain before fracture are said to be *ductile*. Our discussion of tensile behavior has focused on low-carbon steel in part due to its importance in structural applications, but also because describing the stress-strain behavior of steel permitted us to introduce concepts and terminology—linearly elastic, yield stress, ultimate stress, plastic strain, fracture—that apply to many other ductile materials. Some ductile materials, such as aluminum and copper, do not exhibit an easily identifiable yield stress but undergo a smooth transition from linearly elastic to plastic behavior with strain hardening (Fig. 3.23). However, the yield stress is a very useful parameter in design, so it has become traditional to artificially designate a yield point and yield stress for such materials. This is done by drawing a line parallel to the linear part of the stress-strain diagram which intersects the strain axis at some arbitrarily chosen value, often $\varepsilon = 0.002$. The point at which this line intersects the stress-strain diagram is defined to be the yield

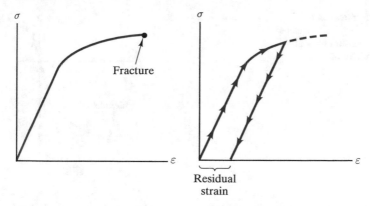

Fig. 3.23 (a) Stress-strain relationship for a ductile material that does not exhibit an identifiable yield stress. (b) Loading-unloading behavior

Fig. 3.24 Designating a yield point and yield stress by the offset method

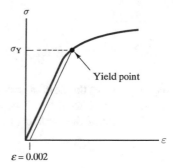

point, and the corresponding stress is defined to be the yield stress (Fig. 3.24). The arbitrary strain chosen is called the *offset*, and this technique is called the offset method of determining the yield point.

3.2.2.2   Brittle Materials

Materials such as cast iron, high-carbon steel, and masonry that exhibit relatively little plastic strain prior to fracture are said to be *brittle*. In a tension test, the stress increases monotonically, so that the ultimate stress occurs at fracture. A notable feature of brittle materials is that their ultimate stress in compression is considerably greater in magnitude than their ultimate stress in tension. This is illustrated in Fig. 3.25, in which we show the stress as a function of strain for a typical brittle material in both tension and compression. Unlike ductile materials, brittle materials subjected to a compression test undergo fracture. In different materials the sample may crush or may fracture along a plane of maximum shear stress, as shown in Fig. 3.15.

Fig. 3.25 Stress as a
function of strain for tension
and compression tests of a
typical brittle material

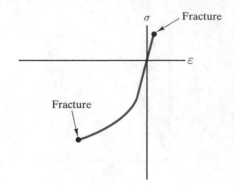

Thus for many structural materials, the modulus of elasticity and also Poisson's
ratio can be determined by subjecting a bar to axial loads and measuring the resulting
axial and lateral strains (see Example 3.4).

3.2.3 Applications

By expressing the normal stress in Eq. (3.5) in terms of the axial load and the bar's
cross-sectional area, $\sigma = P/A$, we obtain an equation for the change in length of the
bar in terms of the applied axial load:

$$\delta = \frac{PL}{EA}. \qquad (3.7)$$

For a given axial load, the change in length is inversely proportional to the bar's
cross-sectional area. If the axial load is tensile (P is positive), the bar increases in
length, and if the axial load is compressive (P is negative), the bar decreases in
length. Because the modulus of elasticity E depends on the material, the change in
length resulting from a given axial load depends on what the bar is made of. For
example, notice in Appendix B that the modulus of elasticity for copper is less than
the modulus of elasticity for titanium. That means that for a given axial load, the
change in length of a copper bar would be greater than the change in length of a
titanium bar of the same dimensions.

In many engineering applications, the changes in length of bars subjected to axial
loads are small in comparison to their lengths. For example, the modulus of elasticity
of pure aluminum, a relatively weak structural material, is $E = 70$ GPa, and the
largest normal stress a bar of pure aluminum can support without fracturing (the
ultimate stress) is approximately 70 MPa. If a bar of pure aluminum that is 1 m in
length is subjected to this normal stress, the change in length given by Eq. (3.5) is
$\delta = (70E6 \text{ N/m}^2)(1 \text{ m})/(70E9 \text{ N/m}^2) = 0.001$ m, or 1 mm. This often simplifies the
analysis of structures. For example, when we draw free-body diagrams of a truss to
determine the axial forces in its members, we can often ignore the changes in the

geometry of the truss due to the changes in length of its members. We demonstrate this in Example 3.6 and also show how determining the deformation of a truss is simplified if the changes in the lengths of the members are small.

Study Questions

1. A prismatic bar with length L is subjected to axial loads, and as a result, its length increases by an amount δ. What is the axial strain (the normal strain ε parallel to the bar's axis)?
2. If the modulus of elasticity of the bar in question 1 is E, what is the normal stress σ on a plane perpendicular to the bar's axis?
3. If a prismatic bar with length L and cross-sectional area A is subjected to tensile axial loads P at its ends, what additional information do you need to determine the resulting change in the bar's length?
4. A prismatic bar has length L and a circular cross section with diameter D. Suppose that you subject the bar to axial loads and measure the resulting changes in its length and diameter. How can you determine the Poisson's ratio ν of the material?

Example 3.4 Determining Elastic Constants
The prismatic bar in Fig. 3.26 has length $L = 200$ mm and a circular cross section with diameter $D = 10$ mm. In a tensile test, the bar is subjected to tensile loads $P = 16$ kN. The length and diameter of the loaded bar are measured and determined to be $L' = 200.60$ mm and $D' = 9.99$ mm. What are the modulus of elasticity and Poisson's ratio of the material?

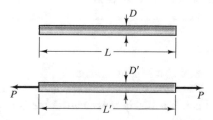

Fig. 3.26

Strategy
We can calculate the normal stress $\sigma = P/A$. Because we know L, L', D, and D', we can also calculate the axial and lateral strains. Then we can use Eqs. (3.6) and (3.7) to determine E and v.

Solution
The normal stress is

$$\sigma = \frac{P}{A} = \frac{16,000 \text{ N}}{\pi(0.01 \text{ m})^2/4} = 204 \text{ MPa}.$$

The axial strain is

$$\varepsilon = \frac{L' - L}{L} = \frac{200.60 \text{ mm} - 200 \text{ mm}}{200 \text{ mm}} = 0.003,$$

and the lateral strain is

$$\varepsilon_{\text{lat}} = \frac{D' - D}{D} = \frac{9.99 \text{ mm} - 10 \text{ mm}}{10 \text{ mm}} = -0.001.$$

The modulus of elasticity is the ratio of the normal stress to the axial strain

$$E = \frac{\sigma}{\varepsilon} = \frac{204\text{E}6 \text{ Pa}}{0.003} = 67.9 \text{ GPa},$$

and Poisson's ratio is

$$\nu = -\frac{\varepsilon_{\text{lat}}}{\varepsilon} = -\frac{-0.001}{0.003} = 0.333.$$

Discussion
In Chap. 8 we show that the elastic properties of the class of materials called isotropic linear elastic materials are completely characterized by the modulus of elasticity and Poisson's ratio. In this example we demonstrate that a simple test, subjecting a bar to axial load and measuring its axial and lateral strains, is sufficient to determine both of these constants.

Example 3.5 Determining a Displacement
The bar in Fig. 3.27 has cross-sectional area $A = 0.4 \text{ in}^2$ and modulus of elasticity $E = 12\text{E}6$ psi. If a 10-kip downward force is applied at B, how far down does point B move?

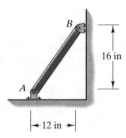

Fig. 3.27

Strategy
By drawing the free-body diagram of joint B, we can determine the axial force in the bar. We can then use Eq. (3.7) to determine the bar's change in length.

Knowing the change in length, we must use geometry to determine how far down point B moves. To simplify the analysis, we will take advantage of the fact that the change in length of the bar is small in comparison to its length.

Solution
We draw the free-body diagram of joint B showing the 10-kip force in Fig. (a). The angle $\beta = \arctan(12/16) = 36.9^\circ$.

(a) Joint B

From the equilibrium equation in the vertical direction

$$\Sigma F_y = -10 \text{ kip} - P \cos \beta = 0,$$

we obtain $P = -12.5$ kip. (The bar is in compression.) The bar's length is $L = 20$ in, so from Eq. (3.7), its change in length is

$$\delta = \frac{PL}{EA} = \frac{(-12,500 \text{ lb})(20 \text{ in})}{(12\text{E}6 \text{ lb/in}^2)(0.4 \text{ in}^2)} = -0.0521 \text{ in.}$$

Let v be the distance point B moves downward when the force is applied (Fig. (b)). The bar's deformed length is $(20 + \delta)$ in. Applying the Pythagorean theorem

$$(12)^2 + (16 - v)^2 = (20 + \delta)^2,$$

we obtain an equation relating v to δ:

$$-32v + v^2 = 40\delta + \delta^2. \tag{1}$$

Because the change in length of the bar is small in comparison to its length, we neglect the second-order terms v^2 and δ^2, obtaining the linear equation

$$-4v = 5\delta. \tag{2}$$

The vertical distance point B moves is $v = (-5/4)(-0.0521 \text{ in}) = 0.0651 \text{ in.}$

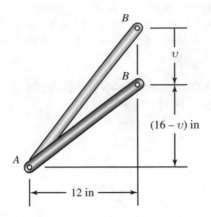

(b) Analyzing the bar's geometry

We can also derive Eq. (2) using an approximate geometric approach that is more intuitive and direct. From the right triangle in Fig. (c), we obtain the relation

$$|\delta| = v\cos\beta.$$

Because $\cos\beta = 16/20 = 4/5$ and δ is negative, we obtain Eq. (2).

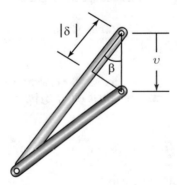

(c) Triangle for obtaining an approximate relation between v and δ

Discussion
There doesn't appear to be any motivation to neglect the second-order terms in the quadratic equation (1). Why didn't we simply solve it for v and obtain a more accurate answer? The reason is that we had already introduced an approximation based on the assumption of a small change in length—we neglected the bar's change in length when we determined P—so retaining the second-order terms in Eq. (1) would not be justified.

Example 3.6 Displacement of a Truss Joint
Bars AB and AC in Fig. 3.28 each have a cross-sectional area of 60 mm^2 and modulus of elasticity $E = 200$ GPa. The dimension $h = 200$ mm. If a downward force $F = 40$ kN is applied at A, what are the resulting horizontal and vertical displacements of point A?

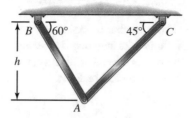

Fig. 3.28

Strategy
By drawing the free-body diagram of joint A, we can determine the axial forces in the two bars. Knowing the axial forces, we can use Eq. (3.7) to determine the change in length of each bar. We must then use geometry to determine the horizontal and vertical displacements of point A. We simplify the analysis by taking advantage of the fact that the changes in length of the bars are small in comparison to their lengths (see Example 3.5).

Solution
In Fig. (a) we draw the free-body diagram of joint A showing the force F. The equilibrium equations are

$$\Sigma F_x = -P_{AB} \cos 60^\circ + P_{AC} \cos 45^\circ = 0,$$
$$\Sigma F_y = P_{AB} \sin 60^\circ + P_{AC} \sin 45^\circ - F = 0.$$

Solving these equations for the two axial loads yields

$$P_{AB} = \frac{F \cos 45^\circ}{D}, \qquad P_{AC} = \frac{F \cos 60^\circ}{D}, \qquad (1)$$

where $D = \sin 45^\circ \cos 60^\circ + \cos 45^\circ \sin 60^\circ$.

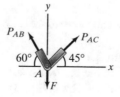

(a) Joint A

The lengths of the bars are $L_{AB} = h/\sin 60°$ and $L_{AC} = h/\sin 45°$. Using Eqs. (3.7) and (1), the changes in length of the bars are

$$\delta_{AB} = \frac{P_{AB}L_{AB}}{EA} = \frac{FL_{AB}\cos 45°}{EAD},$$

$$\delta_{AC} = \frac{P_{AC}L_{AC}}{EA} = \frac{FL_{AC}\cos 60°}{EAD}. \tag{2}$$

To determine the displacement of point A, we first consider bar AB, denoting the horizontal and vertical displacements of point A by u and v (Fig. (b)). The direction of u is chosen arbitrarily; we don't know beforehand whether point A will move to the left or right.

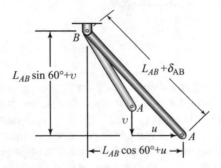

(b) Analyzing the geometry of bar AB

The deformed length of the bar is $L_{AB} + \delta_{AB}$. Applying the Pythagorean theorem, we obtain an equation relating δ_{AB}, u, and v:

$$(L_{AB}\sin 60° + v)^2 + (L_{AB}\cos 60° + u)^2 = (L_{AB} + \delta_{AB})^2.$$

Squaring these expressions and using the identity $\sin^2 60° + \cos^2 60° = 1$, we obtain

$$2vL_{AB}\sin 60° + v^2 + 2uL_{AB}\cos 60° + u^2 = 2\delta_{AB}L_{AB} + \delta_{AB}^2.$$

Assuming that u, v, and δ_{AB} are small in comparison to L_{AB}, we neglect the second-order terms v^2, u^2, and δ_{AB}^2, obtaining a linear equation relating δ_{AB}, u, and v:

$$v\sin 60° + u\cos 60° = \delta_{AB}. \tag{3}$$

We can also derive this result geometrically, as shown in Fig. (c). Approximating the change in length of the bar by the sum of the components of u and v parallel to the unloaded bar, we obtain Eq. (3).

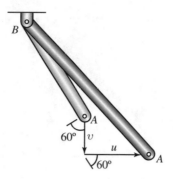

(c) Obtaining the linear equation relating u, v, and δ_{AB}

We next consider bar AC (Fig. (d)). Applying the Pythagorean theorem, we obtain an equation relating δ_{AC}, u, and v:

$$(L_{AC} \sin 45° + v)^2 + (L_{AC} \cos 45° - u)^2 = (L_{AC} + \delta_{AC})^2.$$

Squaring these expressions, we obtain

$$2vL_{AC} \sin 45° + v^2 - 2u\, L_{AC} \cos 45° + u^2 = 2\delta_{AC}L_{AC} + \delta_{AC}^2.$$

Neglecting the second-order terms v^2, u^2, and δ_{AC}^2 gives a linear equation relating δ_{AC}, u, and v:

$$v \sin 45° - u \cos 45° = \delta_{AC}. \qquad (4)$$

The geometrical derivation of this result is shown in Fig. (e). The change in length of the bar is approximated by the sum of the components of u and v parallel to the unloaded bar.

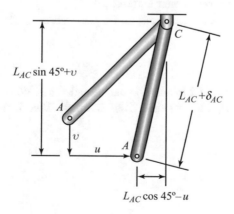

(d) Analyzing the geometry of bar AC

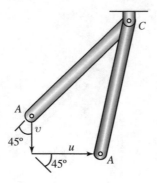

(e) Obtaining the linear equation relating u, v, and δ_{AC}

Solving Eqs. (3) and (4) for u and v and substituting Eq. (2), we obtain

$$u = \frac{F\left(L_{AB} \sin 45° \cos 45° - L_{AC} \sin 60° \cos 60°\right)}{EAD^2},$$

$$v = \frac{F\left(L_{AB} \cos^2 45° + L_{AC} \cos^2 60°\right)}{EAD^2}.$$

Substituting the numerical values, the displacements are

$$u = -0.0250 \text{ mm}, \qquad v = 0.665 \text{ mm}.$$

Point A moves 0.0250 mm to the left and 0.665 mm downward.

Discussion
Notice that in drawing the free-body diagram of joint A in Fig. (a), we disregarded the changes in the 45° and 60° angles due to the changes in length of the bars. This approximation requires that the changes in length of the bars be small in comparison to their lengths.

Problems

3.33 Two marks are made 2 in apart on the unloaded bar. When the bar is subjected to axial forces P, the marks are 2.004 in apart. What is the axial strain of the loaded bar?

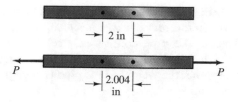

Problems 3.33–3.35

3.34 Two marks are made 2 in apart on the unloaded bar. When the bar is subjected to axial forces P, the marks are 2.004 in apart. The total length of the unloaded bar is 10 in. What is the total length of the loaded bar? What assumption are you making in determining it?

3.35 Two marks are made 2 in apart on the unloaded bar. When the bar is subjected to the axial forces P, the marks are 2.004 in apart. The bar's cross-sectional area is $A = 1.5 \text{ in}^2$, and the modulus of elasticity of its material is $E = 28\text{E}6$ psi. Determine P.

3.36 The bolt extends through an annular sleeve. The cross sections of the bolt and sleeve are shown. The modulus of elasticity of the sleeve is $E = 70$ GPa. There is no normal stress in the sleeve when the nut is in the position shown. If the nut is rotated so that it moves 1 mm to the left relative to the bolt and *the deformation of the bolt is neglected*, what is the resulting normal stress in the sleeve?

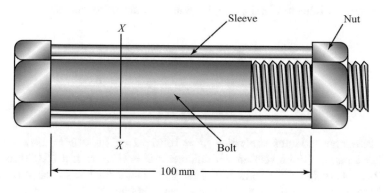

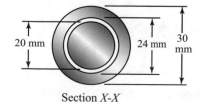

Section X-X

Problem 3.36

3.37 The mass of the ceiling fan is 12 kg. The cross section of the pipe that suspends the fan is shown. Before the fan was installed, the pipe suspending it was 500 mm in length. The pipe consists of steel with modulus of elasticity $E = 190$ GPa. What is the increase in the pipe's length when the fan is installed? (Assume that the weight of the pipe is negligible in comparison to the weight of the fan.)

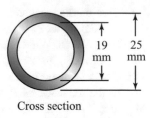

Cross section

Problem 3.37

3.38 A prismatic bar with length $L = 6$ m and a circular cross section with diameter $D = 0.01$ m is subjected to 20-kN compressive forces at its ends. The length and diameter of the deformed bar are $L' = 5.98$ m and $D' = 0.0100117$ m. What are the modulus of elasticity and Poisson's ratio of the material?

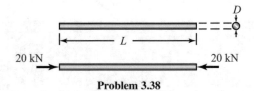

Problem 3.38

3.39 The bar has modulus of elasticity $E = 16E6$ psi and Poisson's ratio $\nu = 0.33$. It has a circular cross section with diameter $D = 0.75$ in. If a 16,000-lb tensile force is applied at the right end of the bar, what is the bar's change in length? What is the bar's diameter after this load is applied?

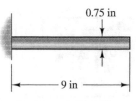

Problems 3.39–3.40

3.40 The bar has modulus of elasticity $E = 16E6$ psi and Poisson's ratio $v = 0.33$. It has a circular cross section with diameter $D = 0.75$ in. What compressive force would have to be exerted on the right end of the bar to increase its diameter to 0.752 in?

3.41 The connecting rod AB has a solid circular cross section with 2-in diameter. Its modulus of elasticity is 30E6 psi, and its Poisson's ratio is 0.3. The rod's supports subject it to an axial force that causes its diameter to decrease to 1.985 in. What is the resulting normal stress on a plane perpendicular to the rod's axis?

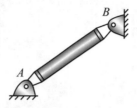

Problem 3.41

3.42 When unloaded, bars AB and AC are each 36-in long and have a cross-sectional area of 2 in². Their modulus of elasticity is $E = 1.6E6$ psi. If the weight $W = 10,000$ lb, what are the changes in length of the bars AB and AC?

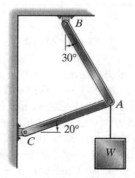

Problem 3.42

3.43 The cross section of bars BC and DG is shown. Their modulus of elasticity is $E = 72$ GPa. If a downward force $F = 520$ kN is applied at H, what is the resulting downward deflection of point H? Assume that the deformation of bar BDH is negligible.

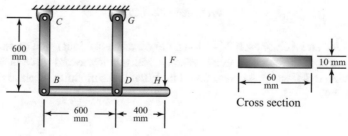

Problem 3.43

3.44 Bars BC and DG have the same cross-sectional area A but have different moduli of elasticity E_{BC} and E_{DG}. Assume that the deformation of bar BHD is negligible. If the bar BHD remains horizontal when a downward force F is applied at H, what is the distance x?

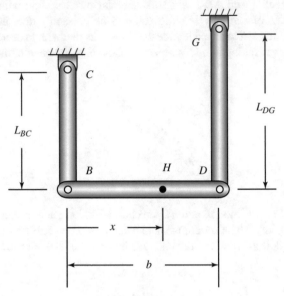

Problems 3.44–3.45

3.45 Bars BC and DG have cross-sectional area $A = 2000$ mm^2 and moduli of elasticity $E_{BC} = 200$ GPa and $E_{DG} = 70$ GPa. The dimensions $L_{BC} = 600$ mm, $L_{DG} = 800$ mm, and $b = 800$ mm. Assume that the deformation of bar BHD is negligible. If a 200-kN downward force is applied at H and $x = 600$ mm, what is the resulting downward deflection of point H?

3.46 Bars AB and AC are each 20 in long, have a cross-sectional area of 4 in^2, and have modulus of elasticity $E = 12$E6 psi. If a 12,000-lb downward force is applied at A, what is the resulting downward displacement of point A?

Problems 3.46–3.47

3.47 Bars AB and AC are each 20-in long, have a cross-sectional area of 1.5 in^2, and are made of the same material. When a 1500-lb downward force is applied at point A, it deflects downward 0.003 in. What is the modulus of elasticity of the material?

3.48 Bar *AB* has cross-sectional area $A = 100 \text{ mm}^2$ and modulus of elasticity $E = 102$ GPa. The distance $H = 400$ mm. If a 200-kN downward force is applied to bar *CD* at *D*, through what angle in degrees does bar *CD* rotate? (You can neglect the deformation of bar *CD*.)

Strategy: Because the change in length of bar *AB* is small, you can assume that the downward displacement v of point *B* is vertical and that the angle (in radians) through which bar *CD* rotates is v/H.

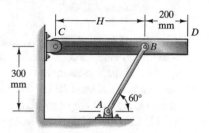

Problems 3.48–3.49

3.49 Bar *AB* has cross-sectional area $A = 150 \text{ mm}^2$ and modulus of elasticity $E = 90$ GPa. The distance $H = 400$ mm. If you don't want to subject bar *AB* to a normal stress (in tension or compression) greater than 5 GPa, what is the largest angle through which bar *CD* can be rotated relative to the position shown? (Neglect the deformation of bar *CD*.)

3.50 Bar *BE* of the frame has cross-sectional area $A = 0.5 \text{ in}^2$ and modulus of elasticity $E = 14\text{E}6$ psi. If a 1000-lb downward force is applied at *C*, what is the resulting change in the length of bar *BE*?

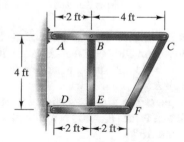

Problems 3.50–3.51

3.51 Bar *CF* of the frame has cross-sectional area $A = 0.5 \text{ in}^2$ and modulus of elasticity $E = 14\text{E}6$ psi. After a downward force is applied at *C*, the length of bar *CF* is measured and determined to have decreased by 0.125 in. What force was applied at *C*?

3.52 Both bars have a cross-sectional area of 0.002 m^2 and modulus of elasticity $E = 70$ GPa. If an 80-kN downward force is applied at *A*, what are the resulting changes in length of the bars?

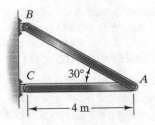

Problems 3.52–3.53

3.53 Both bars have a cross-sectional area of 0.002 m² and modulus of elasticity $E = 70$ GPa. If a 200-kN downward force is applied at joint A, what are the resulting horizontal and vertical displacements of A?

3.54 The cross section of the members BC and BD of the truss is shown. They are made of 7075-T6 aluminum alloy (see Appendix B). If the truss is subjected to a downward force $F = 25$ kN at B, what are the changes in length of members BC and BD?

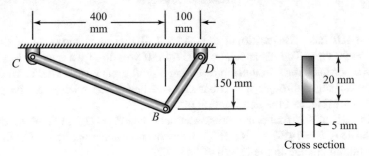

Problems 3.54–3.55

3.55 The cross section of the members BC and BD of the truss is shown. They are made of 7075-T6 aluminum alloy (see Appendix B). If the truss is subjected to a downward force $F = 25$ kN at B, what are the horizontal and vertical displacements of point B?

3.56 The link AB of the pliers has a cross-sectional area of 40 mm² and elastic modulus $E = 210$ GPa. If forces $F = 200$ N are applied to the pliers, what is the change in length of link AB?

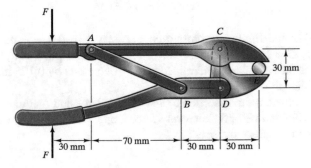

Problems 3.56–3.57

3.57 Suppose that you want to design the pliers so that forces F as large as 450 N can be applied. The link AB is to be made of a material that will support a compressive normal stress of 200 MPa. Based on this criterion, what minimum cross-sectional area must link AB have?

3.58∗ The bar AB in Fig. (a) is rigid. If it rotates through a *small* angle θ (in radians), assuming the change in length of the bar BC to be equal to $b\theta$ appears to be a valid approximation (Fig. (b)). Prove that this is true. Assume that bar AB rotates through an arbitrary angle θ (Fig. (c)) and solve for L' as a function of θ. Then show that if θ is sufficiently small that second- and higher-order terms can be neglected, $L' - L = b\theta$.

Strategy: Express L' as a Taylor series in terms of θ and neglect terms of second and higher order.

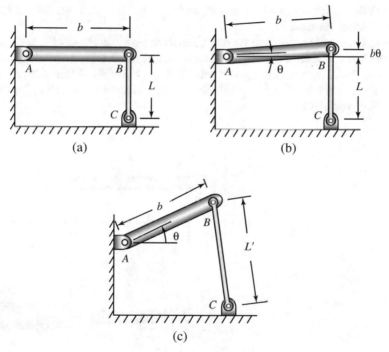

(a)

(b)

(c)

Problem 3.58

3.3 Statically Indeterminate Problems

Objects in equilibrium for which the number of unknown reactions exceeds the number of independent equilibrium equations are said to be *statically indeterminate*. They are common in engineering, because safe and conservative design frequently

requires the use of redundant supports—that is, more supports than the minimum
number necessary for equilibrium. We now have the means to solve statically
indeterminate problems involving axially loaded bars, by supplementing the equilibrium equations with the relationships between the axial loads in bars and their
changes in length.

For example, the bar ABC in Fig. 3.29(a) has parts AB and BC with different
lengths and cross-sectional areas. The ends A and C are fixed. Both parts consist of
the same material with modulus of elasticity E. Suppose that the bar is subjected to
a given axial force at B (Fig. 3.29(b)). We draw the free-body diagram of the entire
bar in Fig. 3.30. The equilibrium equation is

$$F - P_A + P_C = 0. \tag{3.8}$$

We can't determine the two reactions P_A and P_C from this equation. This problem is
statically indeterminate.

Let us disregard the fact that we don't know the reactions P_A and P_C, and proceed
to determine the change in length of each part of the bar as if we did know the
reactions. The axial force throughout part AB of the bar is P_A. (If this does not seem
obvious, confirm it by drawing a suitable free-body diagram.) Therefore the change
in length of part AB is

Fig. 3.29 (a) Bar fixed at
both ends. (b) Applying an
axial load at B

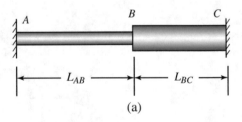

(a)

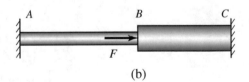

(b)

Fig. 3.30 Free-body
diagram of the bar

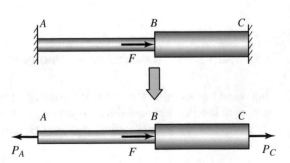

$$\delta_{AB} = \frac{P_A L_{AB}}{EA_{AB}}, \tag{3.9}$$

where A_{AB} is the cross-sectional area of part AB. The axial force throughout part BC is P_C, so its change in length is

$$\delta_{BC} = \frac{P_C L_{BC}}{EA_{BC}}. \tag{3.10}$$

But we know that the change in length of the entire bar ABC is zero, because it is fixed at both ends. The changes in length of the two parts must therefore satisfy the equation

$$\delta_{AB} + \delta_{BC} = 0. \tag{3.11}$$

This is called a *compatibility condition*. The changes in length of the two parts of the bar must be compatible with the constraint that the total length of the bar cannot change. Notice that Eq. (3.11) states that the change in length of one part of the bar must be positive and the change in length of the other part must be negative. We can see intuitively that the force F will cause the length of part AB to increase and the length of part BC to decrease.

We substitute Eqs. (3.9) and (3.10) into Eq. (3.11), obtaining

$$\frac{P_A L_{AB}}{EA_{AB}} + \frac{P_C L_{BC}}{EA_{BC}} = 0. \tag{3.12}$$

This equation together with the equilibrium equation (3.8) gives us two equations in terms of the unknown reactions P_A and P_C. Solving them, we obtain

$$P_A = \frac{F}{1 + \dfrac{L_{AB} A_{BC}}{L_{BC} A_{AB}}}, \qquad P_C = -\frac{F}{1 + \dfrac{L_{BC} A_{AB}}{L_{AB} A_{BC}}}.$$

Now that we know the reactions, we can determine the change in length of each part of the bar from Eqs. (3.9) and (3.10). We can also determine the normal stress in each part of the bar: $\sigma_{AB} = P_A/A_{AB}$ and $\sigma_{BC} = P_C/A_{BC}$.

Notice that our solution was based on three elements: (1) equilibrium, (2) relations between the axial forces in the parts of the bar and their changes in length or deformations, and (3) compatibility. Other statically indeterminate problems involving axially loaded bars, as well as more elaborate problems in structural analysis, can be solved using these three elements.

Study Questions

1. What is a statically indeterminate problem?
2. For the axially loaded bar in Fig. 3.29, what is the compatibility condition?
3. What are the three essential elements used in solving statically indeterminate problems?

Example 3.7 Statically Indeterminate Bar

Bar *ABC* in Fig. 3.31 has modulus of elasticity $E = 82$ GPa. Part *AB* has length $L_{AB} = 250$ mm and a circular cross section 50 mm in diameter. Part *BC* has length $L_{BC} = 200$ mm and a circular cross section 100 mm in diameter. The bar is fixed at *A*, and there is a gap $b = 0.5$ mm between the end of the bar at *C* and the rigid wall. If a 700-kN axial force pointing to the right is applied to the bar at *B*, what are the normal stresses in parts *AB* and *BC*?

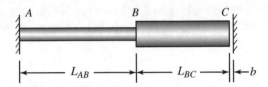

Fig. 3.31

Strategy

Imagine applying a slowly increasing force toward the right at *B*. Part *AB* of the bar will begin stretching. Until the bar comes into contact with the wall at *C*, no force is exerted on the bar at *C*. We will first assume that the bar is not in contact with the wall at *C* and calculate the change in length of part *AB* due to the 700-kN force. If the resulting change in length of part *AB* is less than the gap *b*, the bar does not come into contact with the right wall, part *AB* is subjected to a 700-kN tensile load, and the axial load in part *BC* is zero. But if the resulting change in length of part *AB* is greater than *b*, we know that the bar does contact the right wall and the problem is statically indeterminate. In that case, we must apply force-deformation relations, compatibility, and equilibrium to determine the axial forces in parts *AB* and *BC*. The compatibility condition is that the change in the length of the entire bar *ABC* must equal *b*.

Solution

In Fig. (a) we draw the free-body diagram of the bar under the assumption that it doesn't contact the wall at *C*. The axial force in part *AB* is $P_A = 700,000$ N, and there is no axial force in part *BC*. The change in length of part *AB* is

$$\delta_{AB} = \frac{P_A L_{AB}}{EA_{AB}} = \frac{(700,000 \text{ N})(0.25 \text{ m})}{(82 \times 10^9 \text{ N/m}^2)\left[\frac{\pi(0.05 \text{ m})^2}{4}\right]} = 0.00109 \text{ m}.$$

Because $\delta_{AB} = 1.09$ mm $> b$, we know that the bar will come into contact with the wall at *C* and the problem is statically indeterminate.

(a) Free-body diagram of the bar assuming it doesn't contact the right wall

In Fig. (b) we draw the free-body diagram of the bar including the reaction exerted by the wall at C.

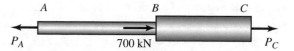

(b) Free-body diagram showing the reactions at both walls

Equilibrium   The equilibrium equation is

$$-P_A + P_C + 700{,}000 \text{ N} = 0.$$

Force-deformation relations   The changes in length of the two parts of the bar are

$$\delta_{AB} = \frac{P_A L_{AB}}{EA_{AB}}, \qquad \delta_{BC} = \frac{P_C L_{BC}}{EA_{BC}}.$$

Compatibility   The change in length of the entire bar equals b:

$$\delta_{AB} + \delta_{BC} = b.$$

There are four equations in terms of the four unknowns P_A, P_C, δ_{AB}, and δ_{BC}. We substitute the forces-deformation relations into the compatibility equation, obtaining

$$\frac{P_A L_{AB}}{EA_{AB}} + \frac{P_C L_{BC}}{EA_{BC}} = b : \frac{P_A(0.25 \text{ m})}{(82 \times 10^9 \text{ N/m}^2)\left[\pi(0.05 \text{ m})^2/4\right]} + \frac{P_C(0.2 \text{ m})}{(82 \times 10^9 \text{ N/m}^2)\left[\pi(0.1 \text{ m})^2/4\right]}$$

$$= 0.0005 \text{ m}.$$

We can solve this equation together with the equilibrium equation to determine the axial forces. The results are $P_A = 385{,}000$ N and $P_C = -315{,}000$ N. The resulting normal stresses in the two parts of the bar are

$$\sigma_{AB} = \frac{P_A}{A_{AB}} = \frac{385{,}000 \text{ N}}{\pi(0.05 \text{ m})^2/4} = 196 \text{ MPa},$$

$$\sigma_{BC} = \frac{P_C}{A_{BC}} = \frac{-315{,}000 \text{ N}}{\pi(0.1 \text{ m})^2/4} = -40.1 \text{ MPa}.$$

Discussion

Once we determined that part AB of the bar was going to contact the wall, we knew that part BC would be in compression due to the force exerted on it by the wall. Why did we show a tensile reaction P_C on the free-body diagram in Fig. (b)? The reason is that doing so made it easier to cope with the signs of terms in the solution. If P_C denotes the *tensile* axial load in part BC, the *increase* in the length of part BC is $\delta_{BC} = P_C L_{BC}/EA_{BC}$, and the compatibility condition is that the sum of the increases in length of parts AB and BC must equal b: $\delta_{AB} + \delta_{BC} = b$. Notice that our solution for P_C is negative, which means that δ_{BC} is also negative. Part BC is in compression.

Example 3.8 Statically Indeterminate Structure
In Fig. 3.32 two aluminum bars (Modulus of elasticity $E_{Al} = 10E6$ psi) are attached to a rigid support at the left and a rigid cross-bar at the right. An iron bar ($E_{Fe} = 28.5E6$ psi) is attached to the rigid support at the left, and there is a gap $b = 0.02$ in between the right end of the iron bar and the cross-bar. The cross-sectional area of each bar is $A = 0.5$ in^2 and $L = 10$ in. If the iron bar is stretched until it contacts the cross-bar and is welded to it, what are the normal stresses in the bars afterward?

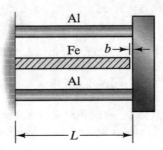

Fig. 3.32

Strategy
By drawing a free-body diagram of the cross-bar, we can obtain one equilibrium equation, but the problem is statically indeterminate because the axial forces in the aluminum and iron bars may be different. (Notice, however, that the axial forces in the two aluminum bars are equal because of the symmetry of the system.) The compatibility condition is that the change in length of the iron bar must equal the change in length of the aluminum bars plus the gap b.

Solution
Equilibrium   In Fig. (a) we assume the iron bar has been welded to the cross-bar and obtain a free-body diagram of the cross-bar by passing a plane through the right ends of the three bars. The equilibrium equation for the cross-bar is

$$2P_{Al} + P_{Fe} = 0.$$

Force-deformation relations   The change in length of one of the aluminum bars due to its axial force is

$$\delta_{Al} = \frac{P_{Al}L}{E_{Al}A},$$

and the change in length of the iron bar due to its axial force is

$$\delta_{Fe} = \frac{P_{Fe}(L - b)}{E_{Fe}A}.$$

Compatibility   The compatibility condition is

$$\delta_{Fe} = \delta_{Al} + b.$$

We substitute the force-deformation relations into this equation, obtaining

$$\frac{P_{Fe}(L - b)}{E_{Fe}A} = \frac{P_{Al}L}{E_{Al}A} + b.$$

We can solve this equation together with the equilibrium equation for the axial forces P_{Al} and P_{Fe}. The results are $P_{Al} = -5880$ lb, $P_{Fe} = 11,760$ lb. The normal stresses in the bars are

$$\sigma_{Al} = \frac{P_{Al}}{A} = \frac{-5880 \text{ lb}}{0.5 \text{ in}^2} = -11,760 \text{ psi},$$

$$\sigma_{Fe} = \frac{P_{Fe}}{A} = \frac{11,760 \text{ lb}}{0.5 \text{ in}^2} = 23,520 \text{ psi}.$$

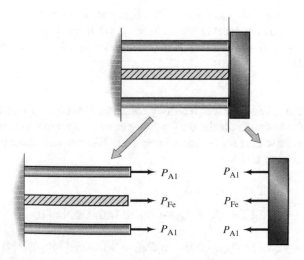

(a) Isolating the cross-bar

Discussion

When the stretched iron bar is welded to the cross-bar and then released, it contracts, compressing the aluminum bars. The forces exerted on the cross-bar by the compressed aluminum bars prevent the iron bar from returning to its original length, so an equilibrium state is reached in which the aluminum bars are in compression and the iron bar is in tension.

Example 3.9 Statically Indeterminate Frame
Bars *AB* and *CD* of the frame in Fig. 3.33 each have a cross-sectional area of 600 mm^2 and modulus of elasticity $E = 200$ GPa. The dimensions $b = h = 400$ mm. If a 140-kN upward force is applied at H, what are the resulting normal stresses in bars *AB* and *CD*? The deformation of bar *GH* can be neglected.

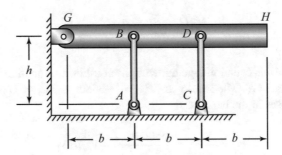

Fig. 3.33

Strategy
By drawing free-body diagrams of the bars, we will determine whether the problem is statically indeterminate. If it is, we must use force-deformation relations, compatibility, and equilibrium to determine the axial forces in bars *AB* and *CD*.

Solution
Equilibrium   We draw the free-body diagrams of the bars in Fig. (a). Notice that no information is obtained by writing equilibrium equations for bars *AB* and *CD*, because they are two-force members. We can write three equilibrium equations for bar *GH*

$$\Sigma F_x = G_x = 0, \tag{1}$$

$$\Sigma F_y = G_y - P_{AB} - P_{CD} + 140{,}000 \text{ N} = 0, \tag{2}$$

$$\Sigma M_{\text{point } G} = -(0.4 \text{ m})P_{AB} - 2(0.4 \text{ m})P_{CD} + 3(0.4 \text{ m})(140{,}000 \text{ N}) = 0, \tag{3}$$

but we cannot solve Eqs. (2) and (3) for the three unknowns G_y, P_{AB}, and P_{CD}. The problem is statically indeterminate.

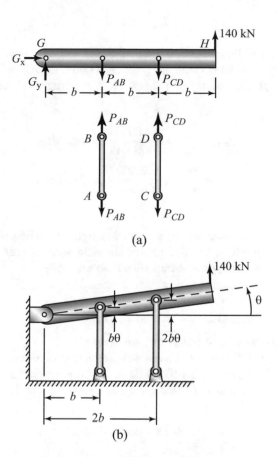

(a) Free-body diagrams of the bars. (b) Determining the changes in length of bars *AB* and *CD*

Force-deformation relations   The changes in length of bars *AB* and *CD* are

$$\delta_{AB} = \frac{P_{AB}L_{AB}}{EA_{AB}} = \frac{P_{AB}(0.4 \text{ m})}{(200E9 \text{ N}/\text{m}^2)(600E\text{-}6 \text{ m}^2)}, \tag{4}$$

$$\delta_{CD} = \frac{P_{CD}L_{CD}}{EA_{CD}} = \frac{P_{CD}(0.4 \text{ m})}{(200E9 \text{ N}/\text{m}^2)(600E\text{-}6 \text{ m}^2)}. \tag{5}$$

Compatibility   The changes in length of bars *AB* and *CD* are not independent because both bars are connected to bar *GH*. When the 140-kN force is
applied, bar *GH* will rotate through some counterclockwise angle θ. If the
angle θ is small, the changes in length of bars *AB* and *CD* can be approximated
by $\delta_{AB} = b\theta$ and $\delta_{CD} = 2b\theta$ (Fig. (b)). From these expressions we see that

$$\delta_{CD} = 2\delta_{AB}. \tag{6}$$

This is the compatibility condition we need. We can now solve Eqs. (2)–(6), obtaining $P_{AB} = 84{,}000$ N, $P_{CD} = 168{,}000$ N, $\delta_{AB} = 0.00028$ m, $\delta_{CD} = 0.00056$ m, and $G_y = 112{,}000$ N. The normal stresses in bars AB and CD are

$$\sigma_{AB} = \frac{P_{AB}}{A_{AB}} = \frac{84{,}000 \text{ N}}{600\text{E-6 m}^2} = 140 \text{ MPa},$$

$$\sigma_{CD} = \frac{P_{CD}}{A_{CD}} = \frac{168{,}000 \text{ N}}{600\text{E-6 m}^2} = 280 \text{ MPa}.$$

Discussion
The expressions we used for the changes in length of the bars in terms of the angle θ through which bar GH rotates are valid approximations when θ is small, but we have not proven this. (See Problem 3.58.)

Example 3.10 Statically Indeterminate Truss
The bars in Fig. 3.34 each have a cross-sectional area of 60 mm^2 and modulus of elasticity $E = 200$ GPa. If a 40-kN downward force is applied at A, what are the resulting horizontal and vertical displacements of point A?

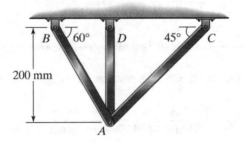

Fig. 3.34

Strategy
In Example 3.6 we solved this same problem except that the vertical bar AD was absent. (You should review that solution.) In this case, the problem is statically indeterminate, because we cannot determine the axial forces in the bars from equilibrium alone. We can obtain two equilibrium equations from the free-body diagram of joint A, but there are three unknown axial forces. However, we can approach the problem in exactly the same way that we approached Example 3.6. We can express the change in length of each bar in

terms of its unknown axial force. Compatibility conditions arise from the fact that the bars are pinned together. The horizontal and vertical displacements of point A must be the same for each bar. The equilibrium equations, force-deformation equations, and compatibility conditions will provide a complete system of equations for the axial forces, the changes in length of the bars, and the horizontal and vertical displacements of point A.

Solution

Equilibrium   In Fig. (a), we draw the free-body diagram of joint A showing the 40-kN force.

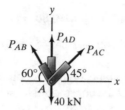

(a) Joint A

The equilibrium equations are

$$\Sigma F_x = -P_{AB}\cos 60^\circ + P_{AC}\cos 45^\circ = 0,$$
$$\Sigma F_y = P_{AB}\sin 60^\circ + P_{AD} + P_{AC}\sin 45^\circ - 40{,}000\ \text{N} = 0.$$

Force-deformation relations   The lengths of the bars are $L_{AB} = 0.2/\sin 60^\circ$ m, $L_{AC} = 0.2/\sin 45^\circ$ m, and $L_{AD} = 0.2$ m. We can express the changes in length of the bars in terms of their lengths and the unknown axial forces:

$$\delta_{AB} = \frac{P_{AB}L_{AB}}{EA}, \qquad \delta_{AC} = \frac{P_{AC}L_{AC}}{EA}, \qquad \delta_{AD} = \frac{P_{AD}L_{AD}}{EA}. \qquad (1)$$

Compatibility   We now relate the changes in length of the bars to the horizontal and vertical displacements of point A. Consider bar AD, denoting the horizontal and vertical displacements of point A by u and v (Fig. (b)). Applying the Pythagorean theorem, we obtain the equation

$$(L_{AD} + v)^2 + u^2 = (L_{AD} + \delta_{AD})^2,$$

which we can write as

$$2vL_{AD} + v^2 + u^2 = 2\delta_{AD}L_{AD} + \delta_{AD}^2.$$

Neglecting the second-order terms v^2, u^2, and δ_{AD}^2 yields the equation

$$v = \delta_{AD}. \tag{2}$$

Just as in Example 3.6, we find that the change in length of the bar is approximated by the sum of the components of u and v in the direction parallel to the unloaded bar. The relations we obtained in Example 3.6 for the changes in length of bars AB and AC are

$$v\sin 60^\circ + u\cos 60^\circ = \delta_{AB}, \tag{3}$$

$$v\sin 45^\circ - u\cos 45^\circ = \delta_{AC}. \tag{4}$$

Equations (2)–(3) are the compatibility conditions for this problem. The changes in length of the bars are constrained because they are pinned together at A. These constraints are enforced by these equations, which require the end of each bar to undergo the same horizontal and vertical displacement.

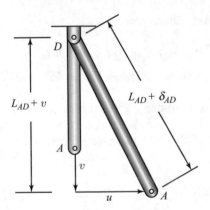

(b) Analyzing the geometry of bar AD

The equilibrium equations together with Eqs. (1)–(4) provide eight equations in the eight unknowns P_{AB}, P_{AC}, P_{AD}, δ_{AB}, δ_{AC}, δ_{AD}, u, and v. Solving them, we obtain

$$u = -0.0125 \text{ mm}, \qquad v = 0.333 \text{ mm}.$$

Discussion
Compare this example to Example 3.6, which was identical except that the bar AD was absent. In Example 3.6 we were able to determine the axial loads in the bars by statics alone. The truss was statically determinate. We say that the

truss in this example has a "redundant support" to help support the vertical load applied at A. The added bar introduces a new unknown into the analysis, the axial force P_{AD}, *but there is also a new compatibility condition*. This is characteristic of statically indeterminate problems in general. Redundant supports introduce new unknowns but also corresponding new compatibility conditions.

Problems

3.59 The bar has cross-sectional area A and modulus of elasticity E. The left end of the bar is fixed. There is initially a gap b between the right end of the bar and the rigid wall (Fig. 1). The bar is stretched until it comes into contact with the rigid wall and is welded to it (Fig. 2). Notice that this problem is statically indeterminate because the axial force in the bar after it is welded to the wall cannot be determined from statics alone. (**a**) What is the compatibility condition in this problem? (**b**) What is the axial force in the bar after it is welded to the wall?

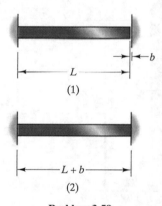

(1)

(2)

Problem 3.59

3.60 The bar has cross-sectional area A and modulus of elasticity E. If an axial force F directed toward the right is applied at C, what is the normal stress in the part of the bar to the left of C? (*Strategy:* Draw the free-body diagram of the entire bar and write the equilibrium equation. Then apply the compatibility condition that the increase in length of the part of the bar to the left of C must equal the decrease in length of the part to the right of C.)

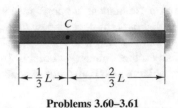

Problems 3.60–3.61

3.61 The bar has cross-sectional area A and modulus of elasticity E. If an axial force F directed toward the right is applied at C, what is the resulting displacement of point C?

3.62 A cylindrical fuel rod is surrounded by an annular sleeve of aluminum. There is an annular space between them. The cross sections of the rod and sleeve are shown. The moduli of elasticity of the fuel rod and aluminum sleeve are $E_{Ro} = 380$ GPa and $E_{Al} = 70$ GPa, respectively. If an 80-kN downward force is applied to the cap, what are the normal stresses in the fuel rod and the sleeve?

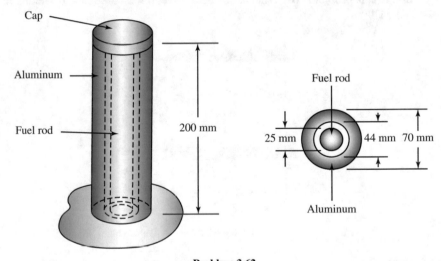

Problem 3.62

3.63 The steel bolt extends through an annular aluminum alloy sleeve. The cross sections of the bolt and sleeve are shown. The moduli of elasticity of the steel bolt and aluminum sleeve are $E_{St} = 200$ GPa and $E_{Al} = 70$ GPa, respectively. There is no normal stress in the sleeve when the nut is in the position shown. If the nut is rotated so that it moves 1 mm to the left relative to the bolt, what is the resulting normal stress in the sleeve? Compare your answer to the answer to Problem 3.36.

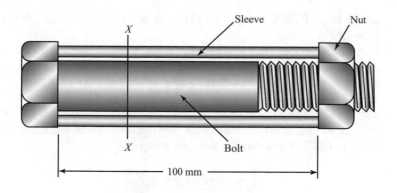

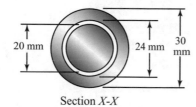

Section X-X

Problem 3.63

3.64 The bar has a circular cross section and modulus of elasticity $E = 70$ GPa. Parts AB and CD are 40 mm in diameter and part BC is 80 mm in diameter. If $F_1 = 60$ kN and $F_2 = 30$ kN, what is the normal stress in part BC?

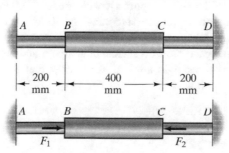

Problems 3.64–3.66

3.65 The bar has a circular cross section and modulus of elasticity $E = 70$ GPa. Parts AB and CD are 40 mm in diameter and part BC is 80 mm in diameter. If $F_1 = 100$ kN, what force F_2 will cause the normal stress in part CD to be $\sigma_{CD} = -20$ MPa?

3.66* The bar has a circular cross section and modulus of elasticity $E = 70$ GPa. Parts AB and CD are 40 mm in diameter and part BC is 80 mm in diameter. Suppose that you don't want to subject the bar to a normal stress of magnitude greater than 40 MPa. If $F_2 = 20$ kN, what is the largest acceptable positive value of F_1?

3.67 Two aluminum bars ($E_{Al} = 10E6$ psi) are attached to a rigid support at the left and a cross-bar at the right. An iron bar ($E_{Fe} = 28.5E6$ psi) is attached to the rigid support at the left, and there is a gap b between the right end of the iron bar and the cross-bar. The cross-sectional area of each bar is $A = 0.5$ in^2 and $L = 10$ in. The iron bar is stretched until it contacts the cross-bar and welded to it. Afterward, the axial strain of the iron bar is measured and determined to be $\varepsilon_{Fe} = 0.002$. What was the size of the gap b?

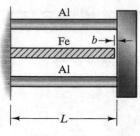

Problems 3.67–3.68

3.68 Two aluminum bars ($E_{Al} = 10E6$ psi) are attached to a rigid support at the left and a cross-bar at the right. An iron bar ($E_{Fe} = 28.5E6$ psi) is attached to the right support at the left, and there is a gap b between the right end of the iron bar and the cross-bar. The cross-sectional area of each bar is $A = 0.5$ in^2 and $L = 10$ in. The iron bar is stretched until it contacts the cross-bar and welded to it. The iron will safely support a tensile stress of 100 ksi, and the aluminum will safely support a compressive stress of 40 ksi. What is the largest safe value of the gap b?

3.69 Bars AB and AC each have cross-sectional area A and modulus of elasticity E. If a downward force F is applied at A, show that the resulting downward displacement of point A is

$$\frac{Fh}{EA}\left(\frac{1}{1 + \cos^3\theta}\right).$$

Problems 3.69–3.70

3.70 Bars *AB* and *AC* each have cross-sectional area *A* and modulus of elasticity *E*. If a downward force *F* is applied at *A*, what are the resulting normal stresses in the bars?

3.71 Each bar has a 500-mm^2 cross-sectional area and modulus of elasticity *E* = 72 GPa. If a 160-kN downward force is applied at *A*, what is the resulting displacement of point *A*?

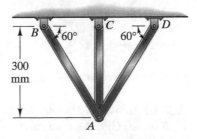

Problems 3.71–3.72

3.72 Each bar has a 500-mm^2 cross-sectional area and modulus of elasticity *E* = 72 GPa. If you don't want to subject the bars to a tensile stress greater than 270 MPa, what is the largest downward force that be applied at *A*?

3.73 Each bar has a 500-mm^2 cross-sectional area and modulus of elasticity *E* = 72 GPa. If there is a gap *h* = 2 mm between the hole in the vertical bar *AC* and the pin *A* connecting bars *AB* and *AD*, what are the normal stresses in the three bars after bar *AC* is connected to the pin at *A*?

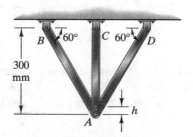

Problems 3.73–3.74

3.74 Each bar has a 500-mm^2 cross-sectional area and modulus of elasticity *E* = 72 GPa. There is a gap *h* between the hole in the vertical bar *AC* and the pin *A* connecting bars *AB* and *AD* before bar *AC* is connected to the pin. The bars are made of material that will safely support a normal stress (tension or compression) of 400 MPa. Based on this criterion, what is the largest safe value of *h*?

3.75 The bars each have cross-sectional area *A* = 180 mm^2 and modulus of elasticity *E* = 72 GPa. If a 50-kN downward force is applied at *B*, what is the resulting downward deflection of point *B*?

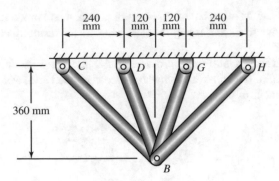

Problem 3.75

3.76 The bars *BC*, *DG*, and *HI* each have cross-sectional area $A = 180 \text{ mm}^2$ and modulus of elasticity $E = 220$ GPa. Assume that the deformation of bar *BDH* is negligible. If an 80-kN downward force is applied at *H*, what are the resulting normal stresses in the bars?

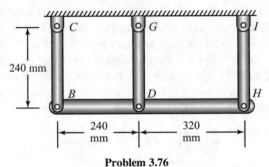

Problem 3.76

3.77 Bars *AB*, *CD*, and *EF* each have a cross-sectional area of 25 mm^2 and modulus of elasticity $E = 200$ GPa. If a 5-kN upward force is applied at *H*, what are the normal stresses in the three bars? (You can neglect the deformation of bar *GH*.)

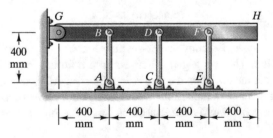

Problems 3.77–3.78

3.78 Bars *AB*, *CD*, and *EF* each have a cross-sectional area of 25 mm² and modulus of elasticity *E* = 200 GPa. If you don't want to subject the bars to a tensile normal stress greater than 340 MPa, what is the largest upward force that can be applied at *H*?

3.79 Bars *AB* and *AC* have cross-sectional area *A* = 100 mm², modulus of elasticity *E* = 102 GPa, and are pinned at *A*. If a 200-kN downward force is applied to bar *DE* at *E*, through what angle in degrees does bar *DE* rotate? (Neglect the deformation of bar *DE*.)

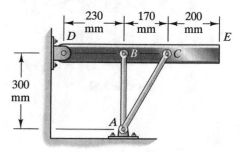

Problems 3.79–3.80

3.80 Bars *AB* and *AC* have cross-sectional area *A* = 100 mm², modulus of elasticity *E* = 70 GPa, and are pinned at *A*. If you don't want to subject the two bars to a normal stress larger than 270 MPa, what is the largest upward force you can apply at *E*? (Neglect the deformation of bar *DE*.)

3.81 Each bar has a cross-sectional area of 3 in² and modulus of elasticity *E* = 12E6 lb/in². If a 40-kip horizontal force directed toward the right is applied at *A*, what are the normal stresses in the bars?

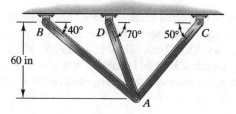

Problem 3.81

3.82∗ Each bar is 400-mm long and has cross-sectional area *A* = 100 mm² and modulus of elasticity *E* = 102 GPa. Each bar is 2 mm too short to reach point *G*. (This distance is exaggerated in the figure.) If the bars are pinned together, what are the horizontal and vertical distances from point *G* to the equilibrium position of the ends of the bars?

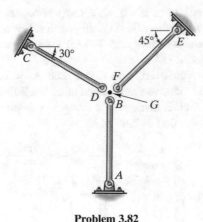

Problem 3.82

3.4 Nonprismatic Bars and Distributed Loads

We now extend our analysis of prismatic bars to additional important applications. We consider bars whose cross sections vary with distance along their axes and also bars subjected to axial loads that are distributed along their axes.

3.4.1 Bars with Gradually Varying Cross Sections

Let us consider a bar whose cross-sectional area varies with distance along its axis, as shown in Fig. 3.35(a). The cross-sectional area is a function of x, which we indicate by writing it as $A(x)$. If we subject the bar to axial loads and the change in $A(x)$ with x is gradual ($dA(x)/dx$ is small), the stress distribution on a plane perpendicular to the bar's axis can be approximated by a uniform normal stress (Fig. 3.35(b)):

$$\sigma = \frac{P}{A(x)}. \tag{3.13}$$

In other words, for a given perpendicular plane, the stress distribution is approximately the same as for the case of a prismatic bar. But because $A(x)$ varies with distance along the bar's axis, the normal stress does also.

Thus the cross-sectional area and normal stress vary along the length of the bar. How can we determine the change in the bar's length due to the axial loads? We begin by considering an infinitesimal element of the bar whose length is dx in

Fig. 3.35 (**a**) Bar with a
varying cross-sectional area.
(**b**) Approximate stress
distribution

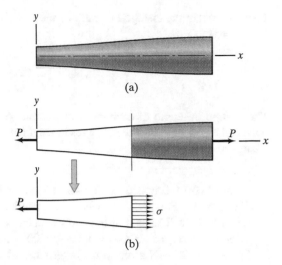

Fig. 3.36 (**a**) Element of
the unloaded bar. (**b**) Length
of the element in the loaded
state

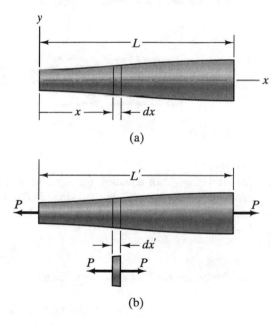

the unloaded state (Fig. 3.36(a)). We can use Eq. (3.7) to determine the element's
change in length (Fig. 3.36(b)):

$$dx' - dx = \frac{P\,dx}{EA(x)}.$$

Then we obtain the change in length of the entire bar by integrating this expression from $x = 0$ to $x = L$:

$$\delta = \int_0^L \frac{P\,dx}{EA(x)}. \tag{3.14}$$

The change in length of each element depends on its cross-sectional area, and we must add up, or integrate, the changes in length of the elements to determine the total change in length.

Example 3.11 Bar with a Varying Cross Section
The bar in Fig. 3.37 consists of material with modulus of elasticity $E = 120$ GPa. The area of the bar's circular cross section in square meters is given by the equation $A(x) = 0.03 + 0.008x^2$. If the bar is subjected to axial forces $P = 20$ MN at the ends, determine (a) the normal stress in the bar at $x = 1$ m; (b) the bar's change in length.

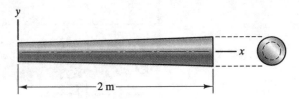

Fig. 3.37

Strategy
(a) From the given equation for $A(x)$, we can determine the cross-sectional area at $x = 1$ m. Then the normal stress is $P/A(x)$. (b) Because the bar's cross-sectional area is a function of x, we must determine the bar's change in length from Eq. (3.14).

Solution
(a) The cross-sectional area at $x = 1$ m is

$$A(1) = 0.03 + 0.008(1)^2 = 0.038 \text{ m}^2,$$

so the normal stress at $x = 1$ m is

$$\sigma = \frac{P}{A(1)} = \frac{20\text{E6 N}}{0.038 \text{ m}^2} = 526 \text{ MPa}.$$

(b) From Eq. (3.14), the change in length of the bar is

$$\delta = \int_0^L \frac{P\,dx}{EA(x)} = \int_0^2 \frac{(20\text{E}6\ \text{N})\,dx}{(120\text{E}9\ \text{Pa})(0.03 + 0.008x^2\ \text{m}^2)} = 0.00862\ \text{m}.$$

Discussion

We evaluated the integral in part (b) by using the tabulated expression in Appendix A. Depending on the complexity of the function $A(x)$, it may be necessary to use numerical integration to determine the change in length of a bar with a varying cross section.

3.4.2 Distributed Axial Loads

In some situations bars are subjected to axial forces that are continuously distributed along some part of the bar's axis or to axial forces that can be modeled as continuous distributions. For example, a pile driven into the ground is subjected to a resisting friction force that is distributed along the pile's length (Fig. 3.38).

To describe a distributed axial force, we introduce a function q defined such that the force on each element dx of the bar is $q\,dx$ (Fig. 3.39). Because the product of q and dx is a force, the dimensions of q are force/length.

For example, the bar in Fig. 3.40(a) is fixed at the left end and subjected to a distributed axial force throughout its length. In Fig. 3.40(b) we pass a plane through the bar at an arbitrary position x and draw the free-body diagram of the part of the bar to the right of the plane. From the equilibrium equation

$$-P + \int_x^L q_0 \left(\frac{x}{L}\right)^2 dx = 0,$$

we determine the axial load in the bar at the position x:

$$P = \frac{q_0}{3}\left(L - \frac{x^3}{L^2}\right).$$

The distributed axial force causes the internal axial force in the bar to vary with axial position. To determine the bar's change in length, we consider an element of the bar of length dx (Fig. 3.40(c)). From Eq. (3.7), its change in length is

$$\delta_{\text{element}} = \frac{P\,dx}{EA} = \frac{q_0}{3EA}\left(L - \frac{x^3}{L^2}\right) dx.$$

Integrating this result from $x = 0$ to $x = L$, we obtain the change in length of the entire bar:

Fig. 3.38 A driven pile is subjected to a distributed axial force (*Photograph from iStock by SergeyVButorin*)

Fig. 3.39 Describing a distributed axial force by a function. The force on an element of length *dx* is *q dx*

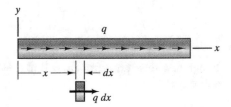

Fig. 3.40 (a) Bar subjected
to a distributed load. (b)
Determining the internal
axial force at x. (c) Element
of length dx

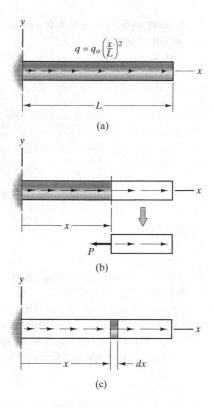

$$\delta = \int_0^L \frac{q_0}{3EA}\left(L - \frac{x^3}{L^2}\right)dx = \frac{q_0 L^2}{4EA}.$$

Notice the parallel between the definition of the distributed axial force q and the definitions of other kinds of distributed loads: we have represented distributed lateral loads on beams by a function w defined such that the lateral force on an element dx of the beam's axis is $w\,dx$, and we represented the stress distribution on a surface by introducing a function $\mathbf{t}$ defined such that the force on an element dA of the surface is $\mathbf{t}\,dA$.

Example 3.12 Bar with a Distributed Axial Load
The pile in Fig. 3.41 has a circular cross section with 12-in diameter, and its modulus of elasticity is $E = 2\mathrm{E}6$ psi. A downward force with initial value $F = 480$ kip is applied to the top of the pile, driving it into the ground at a constant rate. The resisting force on the bottom of the pile is negligible in comparison to the axial frictional force on its lateral surface. Assume that the frictional force is uniformly distributed. As the driving process begins,

determine (a) the maximum compressive normal stress and (b) the change in length of the pile.

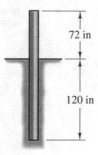

72 in

120 in

Fig. 3.41

Strategy
The part of the pile below the ground is subjected to a uniform distributed force q (Fig. (a)). From the equilibrium equation for the entire pile, we can determine the value of q. Then, by passing a plane through the pile at an arbitrary position x, we can determine the internal axial force in the pile (Fig. (b)). Once we know the axial force, we can determine the maximum compressive stress and the change in the pile's length.

Solution
(a) From the equilibrium equation for the entire pile (Fig. (a))

$$\int_0^{120} q \, dx - F = 0,$$

we determine that $q = F/120 = 4000$ lb/in. Then from the free-body diagram in Fig. (b), we obtain an equation for the internal axial force at an arbitrary position x:

$$\int_0^x q \, dx + P = 0.$$

Solving this equation, the internal axial load from $x = 0$ to $x = 120$ in is $P = -qx = -4000x$ lb. We show the distribution of the internal axial load in the pile in Fig. (c).

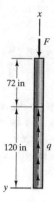

(a) Free-body diagram of the entire pile

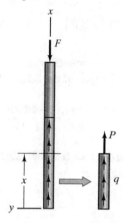

(b) Passing a plane at an arbitrary position x

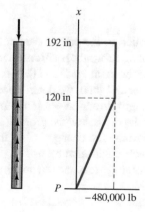

(c) Axial load P as a function of x

The maximum compressive load is $P = -480,000$ lb, so the maximum compressive stress in the pile is

$$\sigma = \frac{P}{A} = \frac{-480,000 \text{ lb}}{\pi(12 \text{ in})^2/4} = -4240 \text{ psi}.$$

(b) The internal axial force is given by the equation $P = -4000x$ lb from $x = 0$ to $x = 120$ in (the part of the pile below the ground) and has the constant value $P = -480,000$ lb from $x = 120$ in to $x = 192$ in (the part above the ground). To determine the pile's change in length, we need to analyze the parts above and below the ground separately.

Above the ground   The change in length is

$$\delta_{\text{above}} = \frac{PL}{EA} = \frac{(-480,000 \text{ lb})(72 \text{ in})}{(2\text{E}6 \text{ lb/in}^2)\left[\pi(12 \text{ in})^2/4\right]} = -0.153 \text{ in}.$$

Below the ground   Below the ground the internal axial load varies with x. The change in length of an element of the beam of length dx is

$$\delta_{\text{element}} = \frac{P \, dx}{EA} = \frac{-4000x \, dx}{EA}.$$

Integrating this expression from $x = 0$ to $x = 120$ in, the change in length of the part of the pile below the ground is

$$\delta_{\text{below}} = \int_0^{120} \frac{-4000x \, dx}{EA} = \frac{(-4000)(120 \text{ in})^2}{(2)(2\text{E}6 \text{ lb/in}^2)\left[\pi(12 \text{ in})^2/4\right]} = -0.127 \text{ in}.$$

The change in length of the pile is $\delta_{\text{above}} + \delta_{\text{below}} = -0.280$ in.

Discussion

Notice the method we applied to the part of the bar below the ground in this example. We cut the bar at an arbitrary position x and used the resulting free-body diagram to determine the axial load (and therefore the normal stress) as a function of x. Knowing the axial load, we determined the change in length of an element of the bar of length dx and integrated it to determine the change in length of the entire part of the bar below the ground. In Example 3.11 we used the same approach to determine the change in length of a bar with a varying cross section. This approach could also be used to determine the change in length of a bar that has a varying cross section and is subjected to a distributed axial load.

Example 3.13 Length of a Suspended Bar
The prismatic bar in Fig. 3.42 is suspended from the ceiling and loaded only by its own weight. If the unloaded bar has length L, cross-sectional area A, weight density (weight per unit volume) γ, and modulus of elasticity E, what is the bar's length when it is suspended?

Fig. 3.42

Strategy
We can treat the weight as an axial force distributed along the bar's axis. The force exerted on an element of the bar of length dx by its weight is equal to the product of the weight density and the volume of the element (Fig. (a)).

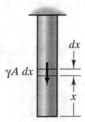

(a) Axial force on an element of length dx

Solution
To determine the internal axial force, we obtain a free-body diagram by passing a plane through the bar at an arbitrary distance x from the bottom (Fig. (b)).

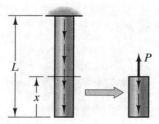

(b) Determining the internal axial force

From the equilibrium equation

$$P - \int_0^x \gamma A \, dx = 0,$$

we determine that $P = \gamma A x$. The internal axial force increases linearly from zero at the bottom of the bar to $\gamma A L = W$, the bar's weight, at the top. The change in length of the element dx in Fig. (a) is

$$\delta_{\text{element}} = \frac{P \, dx}{EA} = \frac{\gamma A x \, dx}{EA}.$$

We integrate this expression from $x = 0$ to $x = L$ to determine the bar's change in length:

$$\delta = \int_0^L \frac{\gamma A x \, dx}{EA} = \frac{\gamma A L^2}{2EA}.$$

In terms of the bar's weight $W = \gamma A L$, the change in length is

$$\delta = \frac{WL}{2EA}.$$

The length of the suspended bar is $L + \delta = L + WL/2EA$.

Discussion
The bar in this example is subjected to a distributed axial load due to its own weight. What is the function q in this case? The force exerted by the bar's weight on an element of length dx is $\gamma A \, dx$ in the negative x direction. We defined the function q such that the force acting on an element of length dx in the positive x direction is $q \, dx$. Therefore $q = -\gamma A$.

Problems

3.83 The bar's cross-sectional area is given as a function of the axial coordinate x in inches by $A = (1 + 0.1x)$ in². The modulus of elasticity of the material is $E = 12E6$ psi. If the bar is subjected to tensile axial loads $P = 20$ kip at its ends, what is the normal stress at $x = 6$ in?

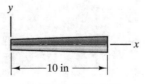

Problems 3.83–3.85

3.84 The bar's cross-sectional area is given as a function of the axial coordinate x in inches by $A = (1 + 0.1x)$ in^2. The modulus of elasticity of the material is $E = 12E6$ psi. If the bar is subjected to tensile axial loads $P = 20$ kip at its ends, what is its change in length?

3.85 The bar's cross-sectional area is given as a function of the axial coordinate x in inches by $A = (1 + ax)$ in^2, where a is a constant. The modulus of elasticity of the material is $E = 8E6$ psi. When the bar is subjected to tensile axial loads $P = 14$ kip at its ends, its change in length is $\delta = 0.01$ in. What is the value of the constant a?

Strategy: Estimate the value of a by drawing a graph of δ as a function of a, or use appropriate equation-solving software.

3.86 From $x = 0$ to $x = 100$ mm, the bar's height is 20 mm. From $x = 100$ mm to $x = 200$ mm, its height varies linearly from 20 to 40 mm. From $x = 200$ mm to $x = 300$ mm, its height is 40 mm. The flat bar's thickness is 20 mm. The modulus of elasticity of the material is $E = 70$ GPa. If the bar is subjected to tensile axial forces $P = 50$ kN at its ends, what is its change in length?

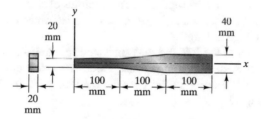

Problem 3.86

3.87 From $x = 0$ to $x = 10$ in, the bar's cross-sectional area is $A = 1$ in^2. From $x = 10$ in to $x = 20$ in, $A = (0.1x)$in^2. The modulus of elasticity of the material is $E = 12E6$ psi. There is a gap $b = 0.02$ in between the right end of the bar and the rigid wall. If the bar is stretched so that it contacts the rigid wall and is welded to it, what is the axial force in the bar afterward?

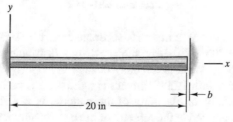

Problems 3.87–3.88

3.88 From $x = 0$ to $x = 10$ in, the bar's cross-sectional area is $A = 1$ in². From $x = 10$ in to $x = 20$ in, $A = (0.1x)$ in². The modulus of elasticity of the material is $E = 12\text{E}6$ psi. There is a gap $b = 0.02$ in. between the right end of the bar and the rigid wall. If a 40-kip axial force toward the right is applied to the bar at $x = 10$ in, what is the resulting normal stress in the left half of the bar?

3.89 The diameter of the bar's circular cross section varies linearly from 10 mm at its left end to 20 mm at its right end. The modulus of elasticity of the material is $E = 45$ GPa. If the bar is subjected to tensile axial forces $P = 6$ kN at its ends, what is the normal stress at $x = 80$ mm?

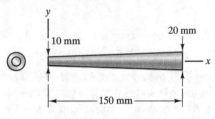

Problems 3.89–3.90

3.90 The diameter of the bar's circular cross section varies linearly from 10 mm at its left end to 20 mm at its right end. The modulus of elasticity of the material is $E = 45$ GPa. If the bar is subjected to tensile axial forces $P = 6$ kN at its ends, what is its change in length?

3.91 The bar is fixed at the left end and subjected to a uniformly distributed axial force. It has cross-sectional area A and modulus of elasticity E. (**a**) Determine the internal axial force P in the bar as a function of x. (**b**) What is the bar's change in length?

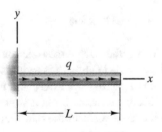

Problems 3.91–3.92

3.92 The bar has length $L = 2$ m, cross-sectional area $A = 0.03$ m², and modulus of elasticity $E = 200$ GPa. It is subjected to a distributed axial force $q = 12(1 + 0.4x)$ MN/m. What is the bar's change in length?

3.93 A cylindrical bar with 1-in diameter fits tightly in a circular hole in a 5-in thick plate. The modulus of elasticity of the material is $E = 14\text{E}6$ psi. A 1000-lb tensile force is applied at the left end of the bar, causing it to begin slipping out of the hole. At the instant slipping begins, determine (**a**) the magnitude of the uniformly distributed axial force exerted on the bar by the plate and **(b)** the total change in the bar's length.

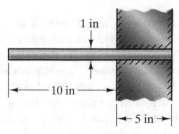

1 in

10 in

5 in

Problem 3.93

3.94 The bar has a circular cross section with 0.002-m diameter, and its modulus of elasticity is $E = 86.6$ GPa. It is subjected to a uniformly distributed axial force $q = 75$ kN/m and an axial force $F = 10$ kN. What is its change in length?

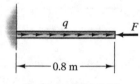

q F

0.8 m

Problems 3.94–3.96

3.95 The bar has a circular cross section with 0.002-m diameter, and its modulus of elasticity is $E = 86.6$ GPa. It is subjected to a uniformly distributed axial force $q = 75$ kN/m. What axial force F would cause the bar's change in length to be zero?

3.96 The bar has a circular cross section with 0.002-m diameter, and its modulus of elasticity is $E = 86.6$ GPa. It is subjected to a distributed axial force $q = 75$ $(1 + 0.2x)$ kN/m and an axial force $F = 10$ kN. What is its change in length?

3.97 The bar has a cross-sectional area of 0.0025 m² and modulus of elasticity $E = 200$ GPa. The bar is fixed at both ends and is subjected to a distributed axial force $q = 80x^2$ kN/m. What are the maximum tensile and compressive stresses in the bar?

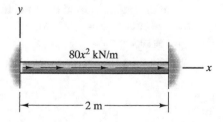

y

$80x^2$ kN/m

x

2 m

Problems 3.97–3.98

3.98 The bar has a cross-sectional area of 0.0025 m² and modulus of elasticity $E = 200$ GPa. The bar is fixed at both ends and is subjected to a distributed axial force $q = 80x^2$ kN/m. What point of the bar undergoes the largest horizontal displacement, and what is the displacement?

3.99 The bar has cross-sectional area A and modulus of elasticity E. It is fixed at A, and there is a gap b between the right end and the rigid wall B. If the bar is subjected to a uniformly distributed axial force q directed toward the right, what force is exerted on the bar by the wall B?

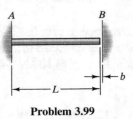

Problem 3.99

3.100 Bar A has a cross-sectional area of 2 in², and its modulus of elasticity is 10E6 psi. Bar B has a cross-sectional area of 4 in², and its modulus of elasticity is 16E6 psi. There is a gap $b = 0.01$ in between the bars. A uniformly distributed load of 20 kip/in directed to the right is applied to bar A, and a uniformly distributed load of 20 kip/in directed to the left is applied to bar B. What is the normal stress in bar A at the wall?

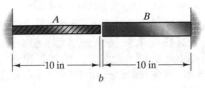

Problem 3.100

3.101 The mass of the lamp is 2 kg. The cross section of the pole supporting it is shown. The pole is 2.2 m in length, its modulus of elasticity is $E = 200$ GPa, and the density of the material is $\rho = 7800$ kg/m³. What is the maximum compressive normal stress in the pole?

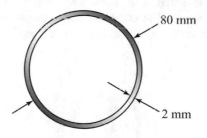

Cross section

Problems 3.101–3.102

3.102∗ The mass of the lamp is 2 kg. The cross section of the pole supporting it is shown. The pole is 2.2 m in length when it is unloaded, its modulus of elasticity is $E = 200$ GPa, and the density of the material is $\rho = 7800$ kg/m^3. What is the decrease in the pole's length when it is supporting the lamp?

3.103∗ A conical indentor is used to test samples of soil. A force pushes the indentor into the soil at a constant rate. The soil exerts a distributed axial force on the indentor that is proportional to the circumference of the indentor, which means that the distributed force is given by a linear equation $q = cx$. **(a)** Determine the value of c in terms of F and d. **(b)** Assuming the angle α to be known, determine the average normal stress in the indentor as a function of x for $0 < x < d$. (Assume that the indentor's weight is negligible.)

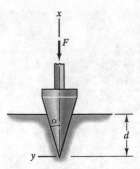

Problems 3.103–3.104

3.104* A conical indentor is used to test samples of soil. A force pushes the indentor into the soil at a constant rate. The soil exerts a distributed axial force on the indentor that is proportional to the circumference of the indentor, which means that the distributed force is given by a linear equation $q = cx$. (You must determine the constant c.) What is the change in length of the part of the indentor that is below the surface of the soil? Assume that the indentor's weight is negligible.

3.105* The column is suspended and loaded only by its own weight. The mass density of the material is $\rho = 2300$ kg/m^3. (The weight density is $g\rho = 9.81\rho$ N/m^3.) The column's width varies linearly from 1 m at the bottom to 2 m at the top. What is the normal stress at $x = 3$ m?

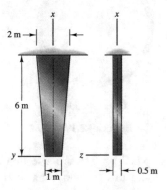

Problems 3.105–3.106

3.106* The column is suspended and loaded only by its own weight. The mass density of the material is $\rho = 2300$ kg/m^3, and its modulus of elasticity is $E = 25$ GPa. The column's width varies linearly from 1 m at the bottom to 2 m at the top. The length of the unloaded column is 6 m. What is its change in length when it is suspended?

3.107* Suppose that you want to design a column to support a uniform compressive stress σ_0. If the column's cross-sectional area was uniform, the compressive stress at increasing distances x from the top would increase due to the column's weight. You can optimize your use of material by varying the column's cross-sectional area in such a way that the compressive stress at *every* position x equals σ_0. Prove that this is accomplished if the column's cross-sectional area is

$$A = A_0 e^{(\gamma/\sigma_0)x},$$

where A_0 is the cross-sectional area at $x = 0$ and γ is the weight density of the material.

Problem 3.107

3.5 Thermal Strains

All structures that are not provided with a controlled environment are subject to changes in temperature. Materials tend to expand and contract when their temperatures rise and fall, which can have important effects on structures and mechanisms. This is illustrated by such dramatic effects as heated glass dishes breaking when suddenly cooled and concrete sidewalks developing cracks due to environmental temperature changes. Changes in temperature can also alter the mechanical properties of materials. Structural engineers must be aware of thermal effects and account for them in design. In this section we introduce the concept of thermal strain and demonstrate its effects in axially loaded bars.

Consider a volume of solid material that is at a uniform temperature T and is unconstrained—not subject to any forces or couples. Imagine drawing an infinitesimal line of length dL within the material (Fig. 3.43(a)). If the temperature of the material changes by an amount ΔT, the material will expand or contract and the length of the line will change to a value dL' (Fig. 3.43(b)). The *thermal strain* of the material at the location and in the direction of the line dL is

Fig. 3.43 (a) Infinitesimal line within a volume of material. (b) Volume and line after a temperature change

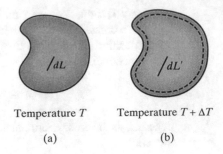

Temperature T Temperature $T + \Delta T$

(a)   (b)

$$\varepsilon_T = \frac{dL' - dL}{dL}. \tag{3.15}$$

If ΔT is positive, most materials will expand, and ε_T will be positive, and if ΔT is negative, the material will contract, and ε_T will be negative. For many materials used in engineering, the relationship between the thermal strain and the change in temperature can be approximated by a linear equation over some range of temperature:

$$\varepsilon_T = \alpha\,\Delta T. \tag{3.16}$$

This equation gives the thermal strain relative to the unconstrained material at temperature T. The constant α is a material property called the *coefficient of thermal expansion*. Because strain is dimensionless, the dimensions of α are (temperature)$^{-1}$. Typical values are given in Appendix B. Some materials exhibit different thermal strains in different directions (the value of α depends on direction), but we consider only materials in which the thermal strain in an unconstrained volume of material is the same in every direction.

The bar in Fig. 3.44(a) is fixed at both ends. It is at a uniform temperature T, and the walls do not exert axial forces on it. What happens if we raise the temperature of the bar by an amount ΔT (Fig. 3.44(b))? Although the bar would increase in length if it was unconstrained, the walls prevent it by exerting compressive forces on the ends that subject the bar to compressive stress, a *thermal stress*. What is that stress? We draw the free-body diagram of the bar in Fig. 3.44(c), where P is the unknown axial load exerted by the walls. (Because the bar is in compression, we know that P will be negative.) Because the ends of the bar are fixed, the total change in length of the bar resulting from the combined effects of the axial load and the change in temperature is zero. The change in length due to the change in temperature is $\varepsilon_T L = \alpha\,\Delta TL$, so the total change in length of the bar is

$$\frac{PL}{EA} + \alpha \Delta TL = 0.$$

From this equation we see that the axial load exerted by the walls is $P = -\alpha\,\Delta TEA$, implying that the thermal stress induced by the walls is

Fig. 3.44 (a) A bar with uniform temperature T and no axial load. (b) Changing the temperature subjects the bar to axial load. (c) Free-body diagram of the bar

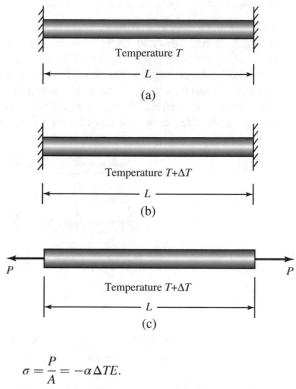

Temperature T

L

(a)

Temperature $T+\Delta T$

L

(b)

P P

Temperature $T+\Delta T$

L

(c)

$$\sigma = \frac{P}{A} = -\alpha \Delta T E.$$

Notice that P and σ are negative, confirming that the bar is in compression.

This example explains one of the reasons concrete sidewalks sometimes develop cracks. Concrete is relatively weak in tension, and if the sidewalk is not designed so that it can contract freely when its temperature decreases, it is subjected to tensile stresses and cracks form.

Study Questions

1. What is the definition of the coefficient of thermal expansion?
2. Consider an infinitesimal line dL within an unconstrained volume of material with temperature T. If the temperature of the material is increased an amount ΔT, how can you determine the resulting length of the infinitesimal line?
3. Consider an unconstrained bar of length L and temperature T. If the temperature of the bar is increased an amount ΔT, how can you determine the resulting change in length of the bar?
4. After the temperature of the unconstrained bar in question 3 is increased, what is the normal stress on a plane perpendicular to the bar's axis? (Draw a free-body diagram of the part of the bar on one side of the plane.)

Example 3.14 Thermal Stresses in a Structure
In Fig. 3.45 two aluminum bars ($E_{Al} = 10E6$ psi, $\alpha_{Al} = 13.3E\text{-}6\,°F^{-1}$) and an iron bar ($E_{Fe} = 28.5E6$ psi, $\alpha_{Fe} = 6.5E\text{-}6\,°F^{-1}$) are attached to a rigid support at the left and a cross-bar at the right. The cross-sectional area of each bar is $A = 0.5$ in², and $L = 10$ in. The normal stresses in the bars are initially zero. If the temperature is increased by $\Delta T = 100\,°F$, what are the resulting normal stresses in the three bars? The deformation of the cross-bar can be neglected.

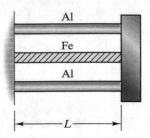

Fig. 3.45

Strategy
Because the two materials have different coefficients of thermal expansion, the aluminum and iron bars tend to lengthen by different amounts. As a result, they will develop internal axial forces when the temperature increases. By drawing a free-body diagram of the cross-bar, we can obtain one equilibrium equation, but the problem is statically indeterminate because the axial forces in the aluminum and iron bars may be different. (Notice, however, that the axial forces in the two aluminum bars are equal because of the symmetry of the system.) We can express the changes in length of the bars as the sum of their changes in length due to their internal axial forces and their changes in length due to thermal strain. We can then apply the compatibility condition that the changes in length of the bars must be equal because they are attached to the rigid cross-bar.

Solution
Equilibrium   In Fig. (a) we obtain a free-body diagram of the cross-bar by passing a plane through the right ends of the three bars. The equilibrium equation for the cross-bar is

$$2P_{Al} + P_{Fe} = 0.$$

Deformation relations   The change in length of one of the aluminum bars due to its axial force and the change in temperature is

$$\delta_{Al} = \frac{P_{Al}L}{E_{Al}A} + \alpha_{Al}\,\Delta T\,L,$$

and the change in length of the iron bar is

$$\delta_{Fe} = \frac{P_{Fe}L}{E_{Fe}A} + \alpha_{Fe}\,\Delta T\,L.$$

Compatibility The compatibility condition is that the changes in length of the bars are equal:

$$\delta_{Al} = \delta_{Fe}.$$

Substituting the deformation relations into this equation gives

$$\frac{P_{Al}L}{E_{Al}A} + \alpha_{Al}\,\Delta T\,L = \frac{P_{Fe}L}{E_{Fe}A} + \alpha_{Fe}\,\Delta T\,L.$$

This equation can be solved together with the equilibrium equation for the axial forces P_{Al} and P_{Fe}. Solving them, we obtain $P_{Al} = -2000$ lb, $P_{Fe} - 4000$ lb. The normal stresses in the bars are

$$\sigma_{Al} = \frac{P_{Al}}{A} = \frac{-2000\ \text{lb}}{0.5\ \text{in}^2} = -4000\ \text{psi},$$

$$\sigma_{Fe} = \frac{P_{Fe}}{A} = \frac{4000\ \text{lb}}{0.5\ \text{in}^2} = 8000\ \text{psi}.$$

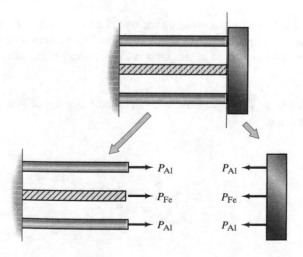

(a) Isolating the cross-bar

Discussion
Beyond being an exercise to illustrate thermal stresses, this example demonstrates an important problem in structural design. The elements of structures and machines constructed of various materials are subject to different thermal strains due to their different coefficients of thermal expansion. As a result, temperature changes can cause both stresses and deformations that must often be considered in design.

Example 3.15 Stresses in an Axially Loaded Bar
The bars in Fig. 3.46 each have a cross-sectional area 60 mm^2, modulus of elasticity $E = 200$ GPa, and coefficient of thermal expansion $\alpha = 12\text{E-6}\,^{\circ}\text{C}^{-1}$. If a 40-kN downward force is applied at A and the temperature of the bars is raised $30\,^{\circ}\text{C}$, what are the resulting horizontal and vertical displacements of point A?

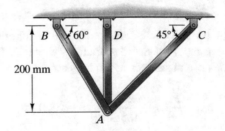

Fig. 3.46

Strategy
We can solve this problem by using the equilibrium equations for the joint A, the equations for the changes in length of the individual bars in terms of their axial loads and the change in temperature, and compatibility conditions.

Solution
Equilibrium   In Fig. (a) we draw the free-body diagram of joint A showing the 40-kN force.

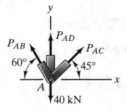

(a) Joint A

The equilibrium equations are

$$\Sigma F_x = -P_{AB} \cos 60° + P_{AC} \cos 45° = 0,$$
$$\Sigma F_y = P_{AB} \sin 60° + P_{AD} + P_{AC} \sin 45° - 40,000 = 0. \tag{1}$$

Deformation relations   The lengths of the bars are $L_{AB} = 0.2/\sin 60°$ m, $L_{AC} = 0.2/\sin 45°$ m and $L_{AD} = 0.2$ m. The change in length of each bar will be the sum of its change in length due to its axial force and the change in length due to thermal strain:

$$\delta_{AB} = \frac{P_{AB}L_{AB}}{EA} + \alpha \, \Delta T L_{AB},$$
$$\delta_{AC} = \frac{P_{AC}L_{AC}}{EA} + \alpha \, \Delta T L_{AC}, \tag{2}$$
$$\delta_{AD} = \frac{P_{AD}L_{AD}}{EA} + \alpha \, \Delta T L_{AD}.$$

Compatibility   We denote the horizontal and vertical displacements of joint A by u and v (Fig. (b)). From Example 3.10, the approximate relationships between the changes in length of the bars and the displacements u and v are

$$v \sin 60° + u \cos 60° = \delta_{AB},$$
$$v \sin 45° - u \cos 45° = \delta_{AC}, \tag{3}$$
$$v = \delta_{AD}.$$

These equations are the compatibility conditions. Equations (1)–(3) provide eight equations in the unknowns P_{AB}, P_{AC}, P_{AD}, δ_{AB}, δ_{AC}, δ_{AD}, u, and v. Solving them, we obtain

$$u = -0.0425 \text{ mm}, \qquad v = 0.426 \text{ mm}.$$

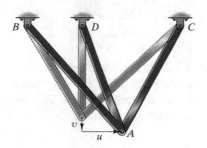

(b) Displacements of joint A

Discussion
Compare this example to Example 3.10, which was identical except for the change in temperature. The solutions are identical except that this example includes terms in the equations for the changes in length of the bars that account for the changes in length due to ΔT.

Problems

3.108 A line L within an unconstrained volume of material is 200-mm long. The coefficient of thermal expansion of the material is $\alpha = 22\text{E-}6 \, °\text{C}^{-1}$. If the temperature of the material is increased by 30 °C, what is the length of the line?

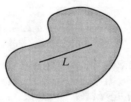

Problem 3.108

3.109 Consider a 1 in $\times$ 1 in $\times$ 1 in cube within an unconstrained volume of material. The coefficient of thermal expansion of the material is $\alpha = 14\text{E-}6 \, °\text{F}^{-1}$. If the temperature of the material is decreased by 40 °F, what is the volume of the cube?

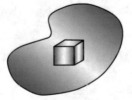

Problem 3.109

3.110 The prismatic bar is 200-mm long and has a circular cross section with 30-mm diameter. After the temperature of the unconstrained bar is increased, its length is measured and determined to be 200.160 mm. What is the bar's diameter after the increase in temperature?

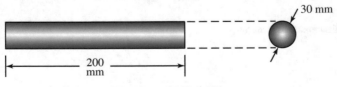

30 mm

200
mm

Problems 3.110–3.112

3.111 The prismatic bar is 200-mm long and has a circular cross section with 30-mm diameter. After the temperature of the unconstrained bar is increased 20°C, its length is measured and determined to be 200.160 mm. What is the coefficient of thermal expansion of the material?

3.112 The prismatic bar is 200 mm long and has a circular cross section with 30-mm diameter. Its modulus of elasticity is $E = 72$ GPa. After the temperature of the unconstrained bar is increased 20°C, its length is measured and determined to be 200.160 mm. What is the normal stress on a plane perpendicular to the bar's axis after the increase in temperature?

Strategy: Obtain a free-body diagram by passing a plane through the bar.

3.113 The prismatic bar is made of material with modulus of elasticity $E = 28E6$ psi and coefficient of thermal expansion $\alpha = 8E\text{-}6\,°F^{-1}$. The temperature of the unconstrained bar is increased by 50 °F above its initial temperature T. **(a)** What is the change in the bar's length? **(b)** What is the change in the bar's diameter? **(c)** What is the normal stress on a plane perpendicular to the bar's axis after the increase in temperature?

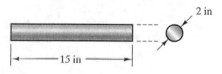

2 in

15 in

Problems 3.113 3.114

3.114 The prismatic bar is made of material with modulus of elasticity $E = 28E6$ psi and coefficient of thermal expansion $\alpha = 8E\text{-}6°F^{-1}$. The temperature of the unconstrained bar is increased by 50°F above its initial temperature T, and the bar is also subjected to 30,000-lb tensile axial forces at the ends. What is the resulting change in the bar's length? Determine the change in length assuming that **(a)** the temperature is first increased and then the axial forces are applied and **(b)** the axial forces are first applied and then the temperature is increased.

3.115 The prismatic bar is made of material with modulus of elasticity $E = 28E6$ psi and coefficient of thermal expansion $\alpha = 8E\text{-}6\,°F^{-1}$. It is constrained between rigid walls. If the temperature is increased by 50 °F above the bar's initial temperature T, what is the normal stress on a plane perpendicular to the bar's axis?

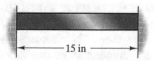

15 in

Problems 3.115–3.117

3.116 The prismatic bar is made of material with modulus of elasticity $E = 28E6$ psi and coefficient of thermal expansion $\alpha = 8E\text{-}6°F^{-1}$. It is compressed between rigid walls. If you don't want to subject the walls to a compressive normal stress greater than 30,000 psi, what is the largest acceptable temperature increase to which the bar can be subjected?

3.117 The prismatic bar has a cross-sectional area $A = 3$ in^2 and is made of material with modulus of elasticity $E = 28E6$ psi and coefficient of thermal expansion $\alpha = 8E\text{-}6\,°F^{-1}$. It is constrained between rigid walls. The temperature is increased by 50 °F above the bar's initial temperature T, and a 20,000-lb axial force to the right is applied midway between the two walls. What is the normal stress on a plane perpendicular to the bar's axis to the right of the point where the force is applied?

3.118 The prismatic bar is made of material with modulus of elasticity $E = 28E6$ psi and coefficient of thermal expansion $\alpha = 8E\text{-}6\,°F^{-1}$. It is fixed to a rigid wall at the left. There is a gap $b = 0.002$ in between the bar's right end and the rigid wall. If the temperature is increased by 50 °F above the bar's initial temperature T, what is the normal stress on a plane perpendicular to the bar's axis?

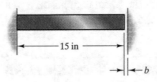

$\longleftarrow$ 15 in $\longrightarrow$

$\rightarrow|\!|\!\leftarrow b$

Problem 3.118

3.119 Bar A has a cross-sectional area of 0.04 m^2, modulus of elasticity $E = 70$ GPa, and coefficient of thermal expansion $\alpha = 14E\text{-}6\ °C^{-1}$. Bar B has a cross-sectional area of 0.01 m^2, modulus of elasticity $E = 120$ GPa, and coefficient of thermal expansion $\alpha = 16E\text{-}6\ °C^{-1}$. There is a gap $b = 0.4$ mm between the ends of the bars. What minimum increase in the temperature of the bars above their initial temperature T is necessary to cause them to come into contact?

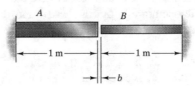

A B

$\longleftarrow$ 1 m $\longrightarrow$ $\longleftarrow$ 1 m $\longrightarrow$

$\rightarrow|\!|\!\leftarrow b$

Problems 3.119–3.120

3.120 Bar A has a cross-sectional area of 0.04 m^2, modulus of elasticity $E = 70$ GPa, and coefficient of thermal expansion $\alpha = 14E\text{-}6°C^{-1}$. Bar B has a cross-sectional area of 0.01 m^2, modulus of elasticity $E = 120$ GPa, and coefficient of thermal expansion $\alpha = 16E\text{-}6°C^{-1}$. There is a gap $b = 0.4$ mm between the ends of the bars. If the temperature of the bars is increased by 40°C above their initial temperature T, what are the normal stresses in the bars?

3.121 A cylindrical fuel rod is surrounded by an annular sleeve of aluminum. There is an annular space between them. The cross sections of the rod and sleeve are shown. The moduli of elasticity of the fuel rod and aluminum sleeve are $E_{Ro} = 380$ GPa and $E_{Al} = 70$ GPa, and their coefficients of thermal expansion are $\alpha_{Ro} = 8.5\text{E-}6°\text{C}^{-1}$ and $\alpha_{Al} = 23\text{E-}6°\text{C}^{-1}$. If an 80-kN downward force is applied to the cap and the temperatures of the rod and sleeve are increased 10°C, what are the normal stresses in the fuel rod and the sleeve?

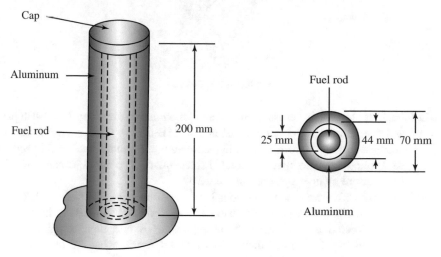

Problem 3.121

3.122 The steel bolt extends through an annular aluminum alloy sleeve. The cross sections of the bolt and sleeve are shown. The moduli of elasticity of the steel bolt and aluminum sleeve are $E_{St} = 200$ GPa and $E_{Al} = 70$ GPa, respectively. The coefficients of thermal expansion of the steel bolt and aluminum sleeve are $\alpha_{St} = 16\text{E-}6°\text{C}^{-1}$ and $\alpha_{Al} = 23\text{E-}6°\text{C}^{-1}$, respectively. There is no normal stress in the sleeve when the nut is in the position shown. If the nut is rotated so that it moves 1 mm to the left relative to the bolt and the temperature of the bolt and sleeve is increased 30°C, what is the resulting normal stress in the sleeve?

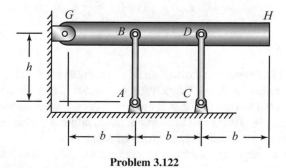

Problem 3.122

3.123 Each bar has a 2-in^2 cross-sectional area, modulus of elasticity $E = 14\text{E}6$ psi, and coefficient of thermal expansion $\alpha = 11\text{E-}6\ °\text{F}^{-1}$. If their temperature is increased by 40 °F from their initial temperature T, what is the resulting displacement of point A?

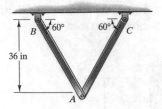

Problems 3.123–3.124

3.124 Each bar has a 2-in^2 cross-sectional area, modulus of elasticity $E = 14E6$ psi, and coefficient of thermal expansion $\alpha = 11E\text{-}6°F^{-1}$. If their temperature is decreased by 30°F from their initial temperature T, what force would need to be applied at A so that the total displacement of point A caused by the temperature change and the force is zero?

3.125 Both bars have cross-sectional area 3 in^2, modulus of elasticity $E = 12E6$ psi, and coefficient of thermal expansion $\alpha = 6.6E\text{-}6 °F^{-1}$. If a 20-kip horizontal force directed toward the right is applied at A and their temperature is increased by 30 °F from their initial temperature T, what are the resulting changes in length of the bars?

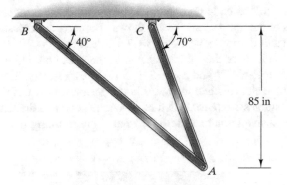

Problems 3.125–3.126

3.126 Both bars have cross-sectional area 3 in^2, modulus of elasticity $E = 12E6$ psi, and coefficient of thermal expansion $\alpha = 6.6E\text{-}6°F^{-1}$. A 20-kip horizontal force directed toward the right is applied at A and the temperature of the bars is increased by 30°F from their initial temperature T. What are the resulting horizontal and vertical displacements of point A?

3.127 Both bars have cross-sectional area 0.002 m^2, modulus of elasticity $E = 70$ GPa, and coefficient of thermal expansion $\alpha = 23E\text{-}6 °C^{-1}$. If the temperature of the bars is increased by 30 °C from their initial temperature T, what are the resulting horizontal and vertical displacements of joint A?

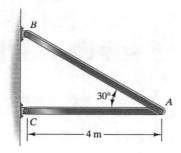

Problems 3.127–3.128

3.128 Both bars have cross-sectional area 0.002 m², modulus of elasticity $E = 70$ GPa, and coefficient of thermal expansion $\alpha = 23\text{E-}6°\text{C}^{-1}$. An 80-kN downward force is applied at A, and the temperature of the bars is increased by 30°C from their initial temperature T. What are the resulting horizontal and vertical displacements of point A?

3.129 Each bar has a 500-mm² cross-sectional area, modulus of elasticity $E = 72$ GPa, and coefficient of thermal expansion $\alpha = 25\text{E-}6°\text{C}^{-1}$. The normal stresses in the bars are initially zero. If their temperature is increased by 20 °C, what are the normal stresses in the bars?

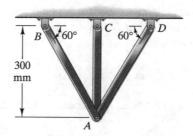

Problems 3.129–3.130

3.130 Each bar has a 500-mm² cross-sectional area, modulus of elasticity $E = 72$ GPa, and coefficient of thermal expansion $\alpha = 25\text{E-}6°\text{C}^{-1}$. The normal stresses in the bars are initially zero. If their temperature is increased by 20°C and a 15-kN downward force is applied at A, what are the normal stresses in the bars?

3.131 Bars AB and AC have cross-sectional area $A = 100$ mm², modulus of elasticity $E = 102$ GPa, and coefficient of thermal expansion $\alpha = 18\text{E-}6 °\text{C}^{-1}$. The two bars are pinned at A. If a 10-kN downward force is applied to bar DE at E and the temperature is lowered by 40 °C, through what angle in degrees does bar *DE* rotate? (You can neglect the deformation of bar *DE*.)

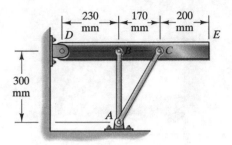

Problems 3.131–3.133

3.132 Bars *AB* and *AC* have cross-sectional area $A = 100 \text{ mm}^2$, modulus of elasticity $E = 102$ GPa, and coefficient of thermal expansion $\alpha = 18\text{E-}6°\text{C}^{-1}$. A 10-kN downward force is applied to bar *DE* at *E*, and the temperature is raised by 30°C. What are the resulting normal stresses in bars *AB* and *AC*?

3.133 Bars *AB* and *AC* have cross-sectional area $A = 100 \text{ mm}^2$, modulus of elasticity $E = 102$ GPa, and coefficient of thermal expansion $\alpha = 18\text{E-}6°\text{C}^{-1}$. A 10-kN downward force is applied to bar *DE* at *E*. What increase in temperature would be necessary for bar *DE* to be horizontal?

3.134 The bars each have cross-sectional area $A = 180 \text{ mm}^2$, modulus of elasticity $E = 72$ GPa, and coefficient of thermal expansion $\alpha = 23\text{E-}6°\text{C}^{-1}$. If a 50-kN downward force is applied at *B* and the temperature of the bars is increased 30°C, what is the resulting downward deflection of point *B*?

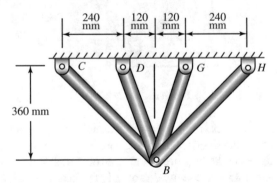

Problem 3.134

3.6   Design Issues

We have established some of the essential background for designing bars to support axial loads and structures made up of such bars. If there is a dictum for the structural designer equivalent to the physician's "First, do no harm," it is "First, prevent failure." Although *failure* must be defined in different ways for different applications, at a minimum the normal and shear stresses in a structural member must not be allowed to exceed the values the material will support.

3.6.1   Allowable Stress

In a prismatic bar subjected to a specified axial load P, we have shown in Sect. 3.1 that there is no plane on which the normal stress is larger in magnitude than the stress $\sigma = P/A$ on a plane perpendicular to the bar's axis. The magnitude of the maximum shear stress, which occurs on planes oriented at 45° relative to the bar's axis, is $|\sigma/2| = |P/2A|$. A criterion frequently used in design is to ensure that σ does not exceed the yield stress σ_Y of the material or a specified fraction of the yield stress called the *allowable stress* σ_{allow}. The ratio of the yield stress to the allowable stress is called the *factor of safety*, which we denote by FS:

$$\text{FS} = \frac{\sigma_Y}{\sigma_{allow}}. \tag{3.17}$$

In choosing the factor of safety, the design engineer must consider how accurately the loads to which the bar will be subjected in normal use can be estimated. Whenever possible, a conservatively large factor of safety would be desirable, but compromises are usually necessary. For example, cost may require compromise in the properties of the materials used. Excessive structural weight can discourage the use of high factors of safety, especially in vehicle applications. Contingencies beyond normal use must also be considered, such as potential earthquake loads in the structure of a building or loads exerted on the members of a car's suspension during emergency maneuvers. If the item being designed will be mass produced, the factor of safety must be chosen to account for anticipated variations in dimensions and material properties. The engineer must balance these considerations within the essential constraint of arriving at a reliable and safe design.

3.6.2   Other Design Considerations

A bar subjected to compression can fail by buckling (geometric instability) at a much smaller axial load that is necessary to cause yielding of the material. We analyze buckling in Chap. 10. Furthermore, even when the stresses to which a structural member is subjected are small compared to the yield stress, failure can occur if they are applied repeatedly. We discuss failure due to cyclic loads in Chap. 12. Also, our analysis of the stress distribution in an axially loaded bar does not apply near the ends where the loads are applied. Those regions normally require a detailed stress analysis that is beyond our scope.

Our examples and problems are limited primarily to designing axially loaded members and structures to meet the objective of supporting given loads. But in addition to the overriding concern of preventing failure, the structural designer is usually confronted with a broad array of decisions relevant to a particular application, and the finest designs are achieved by successfully meeting a spectrum of requirements. Material cost and availability, cost and feasibility of processing and manufacture, resistance to corrosion in the expected environment, compatibility with other materials, and the effect of aging on the material's properties may be important considerations. Decisions on a given design can also be influenced to a greater or lesser extent by concern for safety, ease of maintenance, and aesthetics.

Example 3.16 Design of Truss Members
The truss in Fig. 3.47 is to be constructed of members with yield stress $\sigma_Y = 700$ MPa and equal cross-sectional areas A. If the structure must support a mass m as large as 3400 kg with factor of safety FS $= 3$, what should the cross-sectional area of the members be?

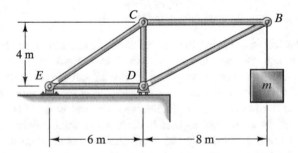

Fig. 3.47

Strategy
The method of joints can be used to determine the axial forces in the members when $m = 3400$ kg. We must then choose the cross-sectional area so that the member subjected to the largest axial load has a factor of safety of 3.

Solution
With $m = 3400$ kg, the axial forces in the members are:

Member :	BC	BD	CD	CE	DE
Axial force (N) :	66,700	−74,600	−44,500	80,200	−66,700

The largest axial force, 80,200 N, occurs in member *CE*. With a factor of safety of 3, the allowable stress is

$$\sigma_{allow} = \frac{\sigma_Y}{FS} = \frac{700 \text{ MPa}}{3} = 233 \text{ MPa}.$$

Equating the normal stress in member *CE* to the allowable stress

$$\frac{80,200 \text{ N}}{A} = 233\text{E}6 \text{ N/m}^2,$$

we obtain $A = 0.000344 \text{ m}^2$.

Discussion

Truss members that are in compression can fail by buckling, and we have not accounted for that possibility in this example. Buckling is discussed in Chap. 10.

Problems

3.135 You are designing a bar with a solid circular cross section that is to support a 4-kN tensile axial load. You have decided to use 6061-T6 aluminum alloy (see Appendix B), and you want the factor of safety to be FS = 2. Based on this criterion, what should the bar's diameter be?

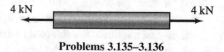

4 kN ← → 4 kN

Problems 3.135–3.136

3.136 You are designing a bar with a solid circular cross section with 5-mm diameter that is to support a 4-kN tensile axial load, and you want the factor of safety to be at least FS = 2. Choose an aluminum alloy from Appendix B that satisfies this requirement.

3.137 You are designing a bar with a solid circular cross section that is to support a 6000-lb tensile axial load. You have decided to use ASTM-A572 structural steel (see Appendix B), and you want the factor of safety to be FS = 1.5. Based on this criterion, what should the bar's diameter be?

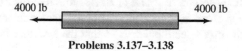

4000 lb ← → 4000 lb

Problems 3.137–3.138

3.138   You are designing a bar with a solid circular cross section with $\frac{1}{2}$-in. diameter that is to support a 4000-lb tensile axial load, and you want the factor of safety to be at least FS $= 3$. Choose a structural steel from Appendix B that satisfies this requirement.

3.139   The horizontal beam of length $L = 2$ m supports a load $F = 30$ kN. The beam is supported by a pin support and the brace BC. The dimension $h = 0.54$ m. Suppose that you want to make the brace out of existing stock that has cross-sectional area $A = 0.0016$ m^2 and yield stress $\sigma_Y = 400$ MPa. If you want the brace to have a factor of safety FS $= 1.5$, what should the angle θ be?

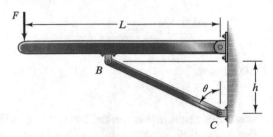

Problems 3.139–3.142

3.140   The horizontal beam of length $L = 4$ ft supports a load $F = 20$ kip. The beam is supported by a pin support and the brace BC. The dimension $h = 1$ ft and the angle $\theta = 60°$. Suppose that you want to make the brace out of existing stock that has yield stress $\sigma_Y = 50$ ksi. If you want to design the brace BC to have a factor of safety FS $= 2$, what should its cross-sectional area be?

3.141* The horizontal beam is of length L and supports a load F. The beam is supported by a pin support and the brace BC. Suppose that the brace is to consist of a specified material for which you have chosen an allowable stress σ_{allow}, and you want to design the brace so that its weight is a minimum. You can do this by assuming that the brace is subjected to the allowable stress and choosing the angle θ so that the volume of the brace is a minimum. What is the necessary angle θ?

3.142* In Problem 3.141, draw a graph showing the dependence of the volume of the brace on the angle θ for $5° \leq \theta \leq 85°$. Notice that the graph is relatively flat near the optimum angle, meaning that the designer can choose θ within a range of angles near the optimum value and still obtain a nearly optimum design.

3.143   The truss is a preliminary design for a structure to attach one end of a stretcher to a rescue helicopter. Based on dynamic simulations, the design engineer estimates that the downward forces the stretcher will exert will be no greater than 360 lb. at A and at B. Assume that the members of the truss have the same cross-sectional area. Choose a material from Appendix B, and determine the cross-sectional area so that the structure has a factor of safety FS $= 2.5$.

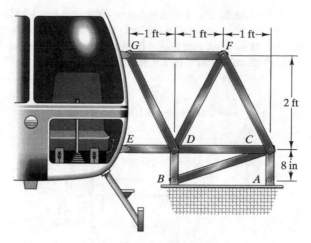

Problems 3.143–3.144

3.144 Upon learning of an upgrade in the engine of the helicopter in Problem 3.143, the engineer designing the truss does new simulations and concludes that the downward forces the stretcher will exert at A and at B may be as large as 400 lb. He also decides the truss will be made of existing stock with cross-sectional area $A = 0.1$ in^2. Choose an aluminum alloy from Appendix B so that the structure will have a factor of safety of at least FS = 5.

3.145 Two candidate truss designs to support the load F are shown. Members of a given cross-sectional area A and yield stress σ_Y are to be used. Compare the factors of safety and weights of the two designs and discuss reasons that might lead you to choose one design over the other. (The weights can be compared by calculating the total lengths of their members.)

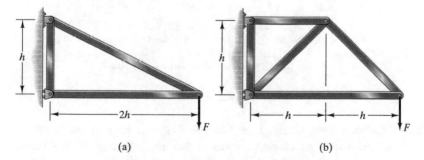

(a)   (b)

Problems 3.145–3.146

3.146∗ Two trusses that support a load F that is a horizontal distance $2\,h$ from the roller support are shown. They consist of members of a given cross-sectional area A and yield stress σ_Y. Design (b) has a higher factor of safety than design (a). Try to design a truss with the same pin and roller supports that supports the load F and has a higher factor of safety than design (b).

3.147   Design a truss attached at A and B that will support the 2-kN loads at C and D. Assume that the members of the truss have the same cross-sectional area. Choose a material from Appendix B and determine the cross-sectional area so that your structure has a factor of safety FS $= 2$.

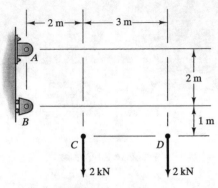

Problem 3.147

3.148   Design a truss attached at A and B that will support the loads at C and D. Assume that the members of the truss have the same cross-sectional area. Choose a material from Appendix B and determine the cross-sectional area so that your structure has a factor of safety FS $= 3$.

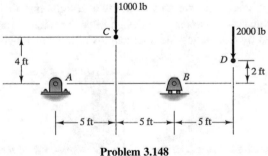

Problem 3.148

3.149   Design a truss attached at A and B that will support the 2-kN load at C and clear the obstacle. Assume that the members of the truss have the same cross-sectional area. Choose a material from Appendix B and determine the cross-sectional area so that your structure has a factor of safety FS $= 2$.

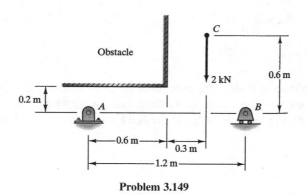

Problem 3.149

3.150* The space truss is going to be constructed of 7075-T6 aluminum alloy to support the 800-lb vertical load. If you want to design member *AD* to have a factor of safety FS = 4, what should its cross-sectional area be?

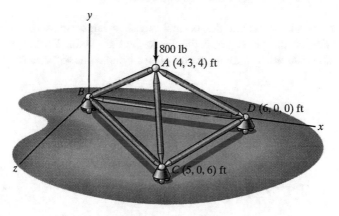

Problems 3.150–3.151

3.151* The space truss has roller supports at *B*, *C*, and *D*. It is going to be constructed of 7075-T6 aluminum alloy to support the 800-lb vertical load, and each member is to have the same cross-sectional area. If you want the structure to have a factor of safety FS = 4, what should the cross-sectional area of the bars be?

3.152* The construction crane in Problem 3.32 is to be made of ASTM-A572 structural steel. If you want to design member 3 to have a factor of safety FS = 2, what should its cross-sectional area be?

3.153* The construction crane in Problem 3.32 is to be made of ASTM-A572 structural steel. Members 1, 2, and 3 are to have the same cross-sectional area, and members 4, 5, and 6 are to have the same cross-sectional area. If you want to choose the cross-sectional areas of these members to obtain a factor of safety FS = 2 for the load shown, what should they be?

Chapter Summary

Stresses in Prismatic Bars

If axial forces P are applied at the centroid of the cross section of a prismatic bar, the stress distribution on any plane perpendicular to the bar's axis that is not near the ends of the bar (Fig. (a)) can be approximated by a uniform normal stress

$$\sigma = \frac{P}{A}. \tag{3.1}$$

The normal and shear stresses on a plane $\mathcal{P}$ oriented as shown in Fig. (b) are

$$\sigma_\theta = \sigma \cos^2\theta, \tag{3.2}$$

$$\tau_\theta = -\sigma \sin\theta \cos\theta. \tag{3.3}$$

There is no plane on which the normal stress is larger in magnitude than the stress $\sigma = P/A$ on the plane perpendicular to the bar's axis. The magnitude of the maximum shear stress, which occurs on planes oriented at 45° relative to the bar's axis, is $|\sigma/2|$.

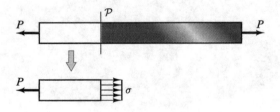

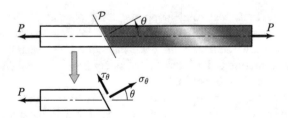

Strains and the Elastic Constants

The axial strain ε of a prismatic bar subjected to axial load is given in terms of the normal stress by

$$\varepsilon = \frac{1}{E}\sigma, \tag{3.4}$$

where E is the modulus of elasticity. The negative of the ratio of the lateral strain to the axial strain is Poisson's ratio:

$$\nu = -\frac{\varepsilon_{lat}}{\varepsilon}. \tag{3.6}$$

The change in length of a bar of length L and cross-sectional area A subjected to axial load P is (Fig. (c))

$$\delta = \frac{PL}{EA}. \tag{3.7}$$

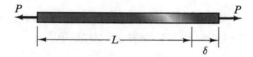

Statically Indeterminate Problems

Solutions to problems involving axially loaded bars in which the number of unknown reactions exceeds the number of independent equilibrium equations involve three elements: (1) relations between the axial forces in bars and their changes in length; (2) compatibility relations imposed on the changes in length by the geometry of the problem; and (3) equilibrium.

Bars with Gradually Varying Cross Sections

If the cross-sectional area of a bar is a gradually varying function of axial position $A(x)$ (Fig. (d)), the stress distribution on a plane perpendicular to the bar's axis can be approximated by a uniform normal stress:

$$\sigma = \frac{P}{A(x)}. \tag{3.13}$$

The bar's change in length is

$$\delta = \int_0^L \frac{P\,dx}{EA(x)}.$$

(3.14)

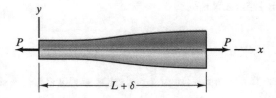

Distributed Axial Loads

A distributed axial force on a bar can be described by a function q defined such that the force on each element dx of the bar is $q\,dx$ (Fig. (e)).

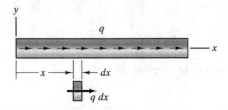

Thermal Strains

Consider an infinitesimal line dL within an unconstrained material at a uniform temperature T. If the temperature of the material changes by an amount ΔT, the material will expand or contract and the length of the line will change to a value dL' (Fig. (f)). The *thermal strain* of the material at the location and in the direction of the line dL is

$$\varepsilon_T = \frac{dL' - dL}{dL}.$$

(3.15)

The thermal strain is related to the change in temperature of the material by

$$\varepsilon_T = \alpha \Delta T,$$

(3.16)

where α is the *coefficient of thermal expansion*. Typical values of α are given in Appendix B.

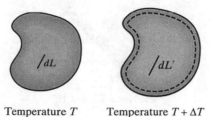

Temperature T Temperature $T + \Delta T$

Design Issues

A criterion frequently used in the design of axially loaded bars is to ensure that the normal stress σ does not exceed the yield stress σ_Y of the material or a specified fraction of the yield stress called the *allowable stress* σ_{allow}. The ratio of the yield stress to the allowable stress is called the *factor of safety*:

$$\text{FS} = \frac{\sigma_Y}{\sigma_{\text{allow}}}. \tag{3.17}$$

Review Problems

3.154 The cross-sectional area of bar AB is 0.015 m^2. If the force $F = 20$ kN, what is the normal stress on a plane perpendicular to the axis of bar AB?

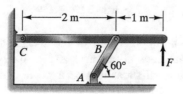

Problems 3.154–3.155

3.155 The cross-sectional area of bar AB is 0.015 m^2. It consists of a material that will safely support a tensile normal stress of 20 MPa. Based on this criterion, what is the largest safe value of the force F?

3.156 The cross-sectional area of each bar is 60 mm^2. If $F = 40$ kN, what are the normal stresses on planes perpendicular to the axes of the bars?

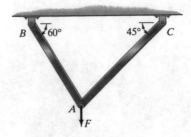

Problems 3.156–3.157

3.157 The cross-sectional area of each bar is 60 mm². If you don't want the material to be subjected to a normal stress greater than 600 MPa, what is the largest acceptable value of the downward force F?

3.158 The system shown supports half of the weight of the 680-kg excavator. The cross-sectional area of member AB is 0.0016 m². If the system is stationary, what normal stress acts on a plane perpendicular to the axis of member AB?

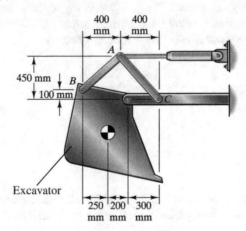

Problems 3.158–3.159

3.159 The system shown supports half of the weight of the 680-kg excavator. The cross-sectional area of member AC is 0.0018 m². If the system is stationary, what normal stress acts on a plane perpendicular to the axis of member AC?

3.160 The cross-sectional area of bar BE is 3 in². Determine the normal stress and the magnitude of the shear stress on the indicated plane through BE.

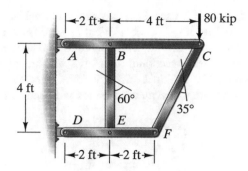

Problems 3.160–3.161

3.161 The cross-sectional area of bar CF is 4 in². Determine the normal stress and the magnitude of the shear stress on the indicated plane through CF.

3.162 The bar has modulus of elasticity $E = 30E6$ psi, Poisson's ratio $\nu = 0.32$, and a circular cross section with diameter $D = 0.75$ in. There is a gap $b = 0.02$ in between the right end of the bar and the rigid wall. If the bar is stretched so that it contacts the rigid wall and is welded to it, what is the bar's diameter afterward?

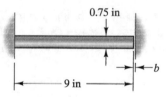

Problems 3.162–3.163

3.163 The bar has modulus of elasticity $E = 30E6$ psi and Poisson's ratio $\nu = 0.32$. It has a circular cross section with diameter $D = 0.75$ in. There is a gap $b = 0.02$ in between the right end of the bar and the rigid wall. The bar is stretched so that it contacts the rigid wall and is welded to it. What is the resulting normal stress on a plane perpendicular to the bar's axis?

3.164 If an upward force is applied at H that causes bar GH to rotate 0.02° in the counterclockwise direction, what are the axial strains in bars AB, CD, and EF? (Neglect the deformation of bar GH.)

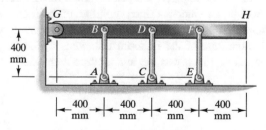

Problems 3.164–3.165

3.165 Bars *AB*, *CD*, and *EF* each have a cross-sectional area of 25 mm² and modulus of elasticity $E = 200$ GPa. What upward force applied at *H* would cause bar *GH* to rotate 0.02° in the counterclockwise direction?

3.166 Both bars have a cross-sectional area of 2 in² and modulus of elasticity $E = 12E6\text{lb/in}^2$. If a 40-kip horizontal force directed toward the right is applied at *A*, what are the resulting changes in length of the bars?

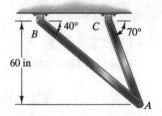

Problems 3.166–3.167

3.167 Both bars have a cross-sectional area of 2 in² and modulus of elasticity $E = 12E6$ lb/in². If a 40-kip horizontal force directed toward the right is applied at joint *A*, what are the resulting horizontal and vertical displacements of *A*?

3.168 The bar is fixed at *A* and *B*. It has length $L = 4$ m, cross-sectional area 0.08 m², and modulus of elasticity $E = 100$ GPa. It is subjected to a uniformly distributed axial force $q = 200$ MN/m. What are the reactions at *A* and *B*?

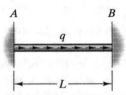

Problems 3.168–3.169

3.169 The bar is fixed at *A* and *B*. It has length $L = 4$ m, cross-sectional area 0.08 m², and modulus of elasticity $E = 100$ GPa. It is subjected to a uniformly distributed axial force $q = 200$ MN/m. What point of the bar undergoes the largest horizontal displacement, and what is the displacement?

3.170 A line *L* within an unconstrained volume of material is 200-mm long. The coefficient of thermal expansion of the material is $\alpha = 24E\text{-}6\,°C^{-1}$. After the temperature of the material is increased, the length of the line is 200.192 mm. How much was the temperature increased?

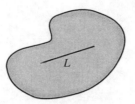

Problem 3.170

3.171 Each bar has a cross-sectional area of 120 mm² and modulus of elasticity $E = 190$ GPa. If a 10-kN horizontal force directed toward the right is applied at A, what are the normal stresses in the bars?

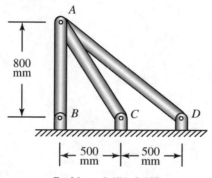

Problems 3.171–3.172

3.172 Each bar has a cross-sectional area of 120 mm², modulus of elasticity $E = 190$ GPa, and coefficient of thermal expansion $\alpha = 14\text{E-}6°\text{C}^{-1}$. The normal stresses in the bars are initially zero. If their temperature is increased by 35°C above their initial temperature T and a 10-kN horizontal force directed toward the right is applied at A, what are the normal stresses in the bars?

Chapter 4
Torsion

A car's drive shaft is subjected to torque by its engine, transmitting power to the wheels. The power produced by turbine engines and electrical generators is also transmitted by shafts subjected to torques about their longitudinal axes. Bars are subjected to axial torques in many engineering applications, and we analyze the resulting stresses and deformations in this chapter. To do so, we must first introduce the concept of a state of pure shear stress.

4.1 Pure Shear Stress

We have seen in Chap. 3 that subjecting a bar to axial forces results in a state of uniform normal stress and no shear stress on planes perpendicular to the bar's axis. We begin this chapter by describing how, at least in principle, a uniform distribution of shear stress and no normal stress can be achieved on particular planes within a material.

4.1.1 Components of Stress and Strain

If we were to subject opposite faces of a cube of material to uniform shear stresses τ as shown in Fig. 4.1(a), the cube would clearly not be in equilibrium because the stresses exert a couple on it. However, if we also apply uniform shear stresses to the top and bottom faces as shown in Fig. 4.1(b), the cube is in equilibrium. If a cube could be loaded in this way, it can be shown that the stress on any plane parallel to the loaded faces consists of the same uniform shear stress and no normal stress, as shown in Fig. 4.2. For this reason, we say that the cube is in a state of *pure shear stress*. If we isolate a rectangular element of the cube that is oriented as shown in Fig. 4.3, the element is subjected to the same state of pure shear stress. In terms of the

© Springer Nature Switzerland AG 2020
A. Bedford, K. M. Liechti, *Mechanics of Materials*,
https://doi.org/10.1007/978-3-030-22082-2_4

Fig. 4.1 (a) A cube is not in equilibrium if only two faces are subjected to shear stress. (b) A cube subjected to shear stresses in this way is in equilibrium

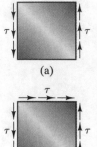

(a)

(b)

Fig. 4.2 Planes parallel to the faces of the cube are subjected to the same uniform shear stress

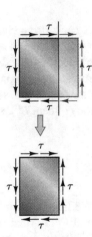

coordinate system shown in Fig. 4.3, the components of stress are $\sigma_x = 0$, $\sigma_y = 0$, $\sigma_z = 0$, $\tau_{xy} = \tau$, $\tau_{yz} = 0$, and $\tau_{xz} = 0$. Assuming that the stress-strain relations (2.7)–(2.12) apply, we can use them to determine the corresponding components of strain, obtaining

$$\varepsilon_x = 0, \ \varepsilon_y = 0, \ \varepsilon_z = 0,$$
$$\gamma_{xy} = \frac{1}{G}\tau, \ \gamma_{yz} = 0, \ \gamma_{xz} = 0,$$

where G is the shear modulus. The component γ_{xy} is the shear strain induced by the shear stress τ (Fig. 4.4), which we denote by γ:

$$\gamma = \frac{1}{G}\tau. \tag{4.1}$$

With this equation, we can determine the shear strain resulting from a given state of pure shear stress if the shear modulus G is known. In Chap. 8 we show that the

Fig. 4.3 An element with faces parallel to the faces of the cube is subjected to the same state of pure shear stress

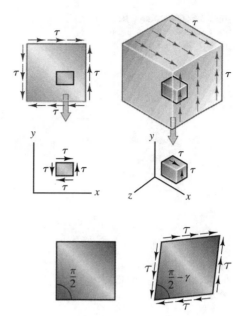

Fig. 4.4 Shear strain resulting from the shear stresses on the element

shear modulus is not independent of the modulus of elasticity E and Poisson's ratio v, but is related to them by the equation

$$G = \frac{E}{2(1+v)}. \tag{4.2}$$

4.1.2 Stresses on Oblique Planes

We have described the stresses acting on the faces of a cube subjected to pure shear stress, but in design, we cannot consider only the stresses on particular planes. In our discussion of bars subjected to axial forces in Chap. 3, we found that the values of the normal and shear stresses were different for planes having different orientations. Let us consider a cube of material subjected to pure shear stress and obtain a free-body diagram as shown in Fig. 4.5. The normal and shear stresses on the plane $\mathcal{P}$ are denoted by σ_θ and τ_θ. Letting A_θ be the area of the slanted face of the free-body diagram, the areas of the horizontal and vertical faces in terms of A_θ and θ are shown in Fig. 4.6(a). The forces on the free-body diagram are the products of the areas and the uniform stresses (Fig. 4.6(b)). Summing forces in the σ_θ direction

Fig. 4.5 Obtaining a free-body diagram by passing a plane $\mathcal{P}$ through the cube

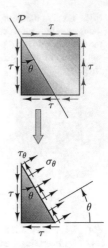

Fig. 4.6 (a) Areas of the faces. (b) Forces on the free-body diagram

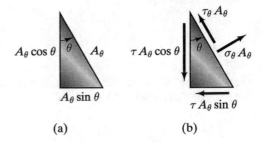

(a)   (b)

$$\sigma_\theta A_\theta - (\tau A_\theta \cos\theta)\sin\theta - (\tau A_\theta \sin\theta)\cos\theta = 0,$$

we determine the normal stress on $\mathcal{P}$:

$$\sigma_\theta = 2\tau \sin\theta \cos\theta. \tag{4.3}$$

Then by summing forces in the τ_θ direction

$$\tau_\theta A_\theta - (\tau A_\theta \cos\theta)\cos\theta + (\tau A_\theta \sin\theta)\sin\theta = 0,$$

we determine the shear stress on $\mathcal{P}$:

$$\tau_\theta = \tau\left(\cos^2\theta - \sin^2\theta\right). \tag{4.4}$$

In Fig. 4.7, we plot the ratios σ_θ/τ and τ_θ/τ obtained from Eqs. (4.3) and (4.4). We also show the free-body diagrams and values of σ_θ and τ_θ corresponding to particular values of θ. There is no value of θ for which the shear stress is larger in magnitude than τ. Notice that although there are no normal stresses on the faces of the element in Fig. 4.3, normal stresses do occur on oblique planes. On the plane oriented at

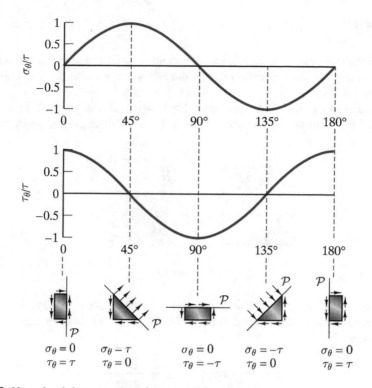

Fig. 4.7 Normal and shear stresses as functions of θ.

$\theta = 45°$, $\sigma_\theta = \tau$, and at $\theta = 135°$, $\sigma_\theta = -\tau$. These are the maximum tensile and compressive stresses for any value of θ. Although we have considered only a subset of planes, it can be shown that these are the maximum shear, tensile, and compressive stresses.

In summary, *if a material is subjected to a state of pure shear stress of magnitude* τ, *the maximum shear stress, maximum tensile stress, and maximum compressive stress each equal* τ.

Study Questions

1. What is the definition of a state of pure shear stress?
2. What is the definition of the shear modulus G of a material? If you know the modulus of elasticity E and the Poisson's ratio ν of a material, how can you determine the shear modulus?
3. If a material is subjected to a state of pure shear stress τ, what is the magnitude of the maximum shear stress to which the material is subjected? What are the values of the maximum tensile and compressive stresses to which the material is subjected?

Problems

4.1 The modulus of elasticity of pure aluminum is 70 GPa, and its Poisson's ratio is 0.33. What is its shear modulus? Compare your result to the value given in Appendix B.

4.2 A cube of material with shear modulus $G = 26$ GPa is subjected to a state of pure shear stress $\tau = 450$ MPa. What is the resulting angle β in degrees?

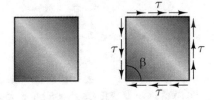

Problem 4.2

4.3 The cube of material is subjected to a pure shear stress $\tau = 9$ MPa. The angle β is measured and determined to be 89.98°. What is the shear modulus G of the material?

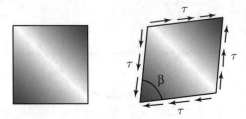

Problems 4.3–4.5

4.4 If the cube consists of material with shear modulus $G = 4.6E6$ psi and the shear stress $\tau = 8000$ psi, what is the angle β in degrees?

4.5 The cube consists of material with shear modulus $G = 26.3$ GPa. If you don't want to subject the material to a shear stress greater than $\tau = 270$ MPa, what is the largest shear strain to which the cube can be subjected?

4.6 The prismatic bar is 800 mm in length and has a circular cross section with 100-mm diameter. In a test to determine its elastic properties, the bar is subjected to 4.8-MN axial loads. After the loads are applied, the bar's length is 802.44 mm, and its diameter is 99.902 mm. What is the shear modulus of the material?

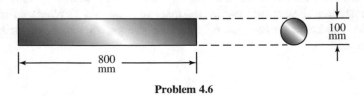

Problem 4.6

4.7 The cube in Fig. (a) consists of material with shear modulus $G = 40$ GPa. In Fig. (b), the cube is subjected to a state of pure shear stress $\tau = 720$ MPa. Determine the length of the diagonal AD before and after the shear stress is applied. (The shear strain shown is exaggerated. The lengths of the sides AB, AC, BD, and CD do not change when the shear stress is applied.)

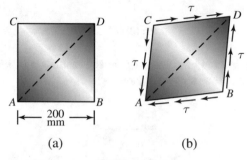

(a)   (b)

Problem 4.7

4.8 The cube of material is subjected to a pure shear stress $\tau = 12$ MPa. What are the normal stress and the magnitude of the shear stress on the plane $\mathcal{P}$?

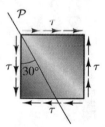

Problems 4.8–4.11

4.9 The cube of material is subjected to a pure shear stress $\tau = 12$ MPa. What are the magnitudes of the maximum tensile, compressive, and shear stresses to which the material is subjected?

4.10 The cube of material is subjected to a pure shear stress τ. If the normal stress on the plane $\mathcal{P}$ is 14 MPa, what is τ?

4.11 The cube of material is subjected to a pure shear stress τ. The shear modulus of the material is $G = 28$ GPa. If the normal stress on the plane $\mathcal{P}$ is 80 MPa, what is the shear strain of the cube?

4.12 The cube of material is subjected to a pure shear stress $\tau = 20$ ksi. **(a)** What are the normal stress and the magnitude of the shear stress on the plane $\mathcal{P}$? **(b)** What are the magnitudes of the maximum tensile, compressive, and shear stresses to which the material is subjected?

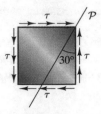

Problem 4.12

4.13 The cube is subjected to a state of pure shear stress $\tau = 3$ MPa and the angle $\theta = 35°$. What are the stresses σ_θ and τ_θ?

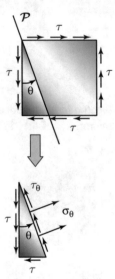

Problems 4.13–4.14

4.14∗ If the stresses $\sigma_\theta = 2.6$ MPa and $\tau_\theta = 3.0$ MPa, what are the stress τ and the angle θ? Assume that $0 \leq \theta \leq 90°$.

4.2 Prismatic Circular Bars

Our analysis of prismatic bars subjected to axial forces in Chap. 3 applied to bars of arbitrary cross section. In contrast, simple analytical solutions for the deformation and stresses in a bar subjected to axial torsion exist only for bars with circular cross sections. We analyze such bars in this section.

Fig. 4.8 Subjecting a bar to
an axial torque

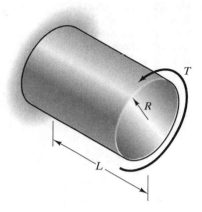

4.2.1   Stresses and Strains

Our objective is to describe the deformation and stresses that result when a cylindrical bar fixed at one end is subjected to an axial torque T at the free end (Fig. 4.8). To describe the bar's deformation, we consider a cylindrical shell of the material within the bar that has inner radius r and infinitesimal thickness dr. In Fig. 4.9(a) we draw a radial line from the center of the end of the bar to the cylindrical shell and extend it along the length of the shell parallel to the bar's axis. When a torque T is applied (Fig. 4.9(b)), the radial line rotates through an angle ϕ, the bar's *angle of twist*, and the line along the length of the shell rotates through an angle γ. If ϕ is expressed in radians, the circumferential distance $c = r\phi$. If c is small in comparison to the bar's length L, we can approximate the angle γ in radians by $\gamma = c/L$. Therefore

$$\gamma = \frac{r\phi}{L}. \tag{4.5}$$

Now let us consider an infinitesimal rectangular element of the shell with sides parallel and perpendicular to the bar's axis before the torque is applied (Fig. 4.10(a)). When the torque is applied (Fig. 4.10(b)), the element is subjected to a shear strain γ, implying that it is subjected to a shear stress

$$\tau = G\gamma = \frac{Gr\phi}{L}. \tag{4.6}$$

We have given a suggestive argument (not a proof) that the applied torque subjects infinitesimal rectangular elements of the cylindrical shell oriented as shown in Fig. 4.11 to shear stress given by Eq. (4.6). If we pass a plane perpendicular to the shell's axis (Fig. 4.12), every element around the circumference is acted upon by the same state of shear stress, so the exposed surfaces are subjected to a uniform circumferential stress τ. By using Fig. 4.12 and Eq. (4.6), we can describe the stress

Fig. 4.9 (a) Cylindrical
shell within the bar. (b)
Rotations of the radial and
longitudinal lines when
torque is applied

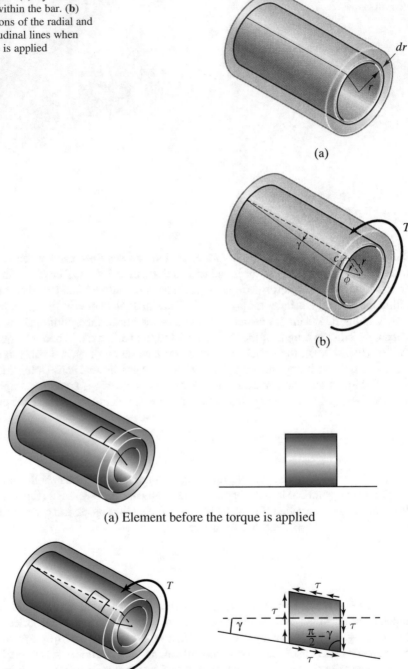

(a)

(b)

(a) Element before the torque is applied

(b) Element after the torque is applied

Fig. 4.10 An element of the shell before and after the torque is applied

Fig. 4.11 Stress
distribution on an element of
the shell

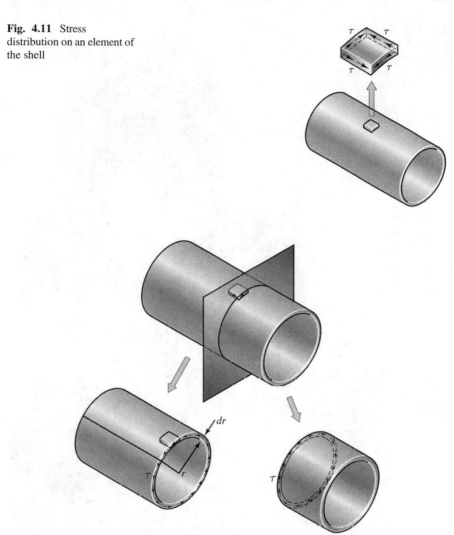

Fig. 4.12 Passing a plane perpendicular to the shell's axis

distribution on a plane perpendicular to the axis of the cylindrical bar. It is subjected
to a circumferential shear stress distribution τ whose magnitude is proportional to the
distance from the center of the bar (Fig. 4.13).

We have now described the stress distribution in the bar, but Eq. (4.6) gives the
shear stress in terms of the bar's angle of twist. Our next objective is to determine the
shear stress in terms of the applied torque T. From the free-body diagram in
Fig. 4.13, equilibrium requires that the moment about the bar's axis due to the
shear stress distribution must equal T. Let dA be an infinitesimal element of the bar's
cross-sectional area at a distance r from the center. The moment about the bar's axis
due to the shear stress acting on dA is $r\,\tau\,dA$ (Fig. 4.14). Integrating to determine the
total moment due to the shear stress, we obtain the equilibrium equation

Fig. 4.13 Stress distribution on a plane perpendicular to the axis of the bar

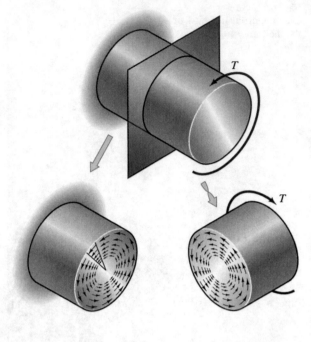

Fig. 4.14 Calculating the moment about the bar's axis due to the shear stress distribution

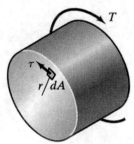

$$\int_A r\tau\, dA = T.$$

Substituting Eq. (4.6) into this equation, we can solve for the bar's angle of twist ϕ. The result is

$$\phi = \frac{TL}{GJ},\tag{4.7}$$

where

$$J = \int_A r^2\, dA\tag{4.8}$$

is the polar moment of inertia of the bar's cross-sectional area about its axis. [Notice the similarity between Eq. (4.7) and the equation $\delta = PL/EA$ for the change in length of a bar subjected to axial forces.] We can now substitute Eq. (4.7) back into Eq. (4.6), obtaining an expression for the shear stress in terms of the applied torque T and the distance r from the bar's axis:

$$\tau = \frac{Tr}{J}. \tag{4.9}$$

We have attempted to make Eqs. (4.7) and (4.9) plausible, but we have not proven that they determine the angle of twist and distribution of stress in a solid or hollow circular bar. Although the proof is beyond our scope, it can be shown that if such a bar could be loaded at the ends by shear stress distributions satisfying Eq. (4.9), our expressions for the angle of twist and the state of pure shear stress within the bar are the exact and unique solutions. Furthermore, no matter how the torques are applied, the stress distribution at axial distances from the ends greater than a few times the bar's width is given approximately by Eq. (4.9). This is another example of Saint-Venant's principle (see Sect. 3.1). As a consequence, Eq. (4.7) approximates the angle of twist of a slender bar no matter how the torques are applied at the ends. These results require that the material have the property of isotropy, which is discussed in Chap. 8.

A simple experiment clearly demonstrates the effect of the state of stress in a bar subjected to torsion. Apply torque to a piece of blackboard chalk as shown in Fig. 4.15(a). The maximum tensile stress resulting from the state of pure shear stress occurs on planes at 45° relative to the chalk's axis (see Figs. 4.7 and 4.15(b)). Brittle materials such as chalk are weak in tension, and as a result, the chalk fails along a clearly defined 45° spiral line (Fig. 4.15(c)).

4.2.2 Polar Moment of Inertia

To evaluate J, we can use an annular element of area with radius r and thickness dr (Fig. 4.16). The area of the element is the product of its circumference and thickness, $dA = 2\pi r\,dr$, so the polar moment of inertia is

$$J = \int_A r^2\,dA = \int_0^R r^2(2\pi r\,dr).$$

Integrating, we obtain

$$J = \frac{\pi}{2}R^4. \qquad \text{(solid circular cross section)} \tag{4.10}$$

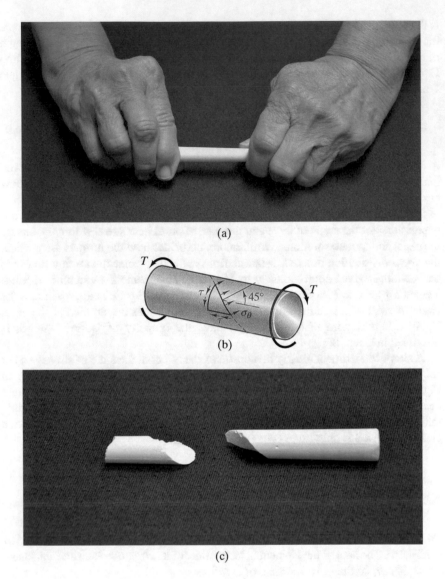

Fig. 4.15 (a) Applying axial torques to a piece of chalk. (b) The maximum tensile stress occurs at 45° relative to the longitudinal axis. (c) Failure occurs along the plane subjected to the maximum tensile stress

Equations (4.7) and (4.9) also apply to a bar with a hollow circular cross section. In that case, the polar moment of inertia is given by

$$J = \frac{\pi}{2} \left(R_o^4 - R_i^4 \right), \qquad \text{(hollow circular cross section)} \qquad (4.11)$$

where R_i and R_o are the bar's inner and outer radii.

Fig. 4.16 Element of area for calculating J

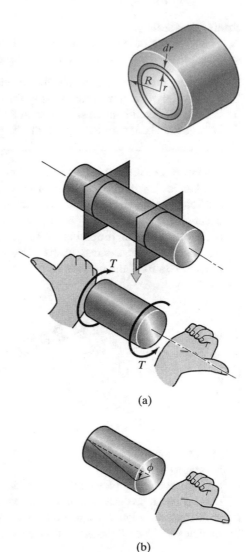

Fig. 4.17 (**a**) Positive directions of the torque. (**b**) Segment with a positive angle of twistϕ.

(a)

(b)

4.2.3 Positive Directions of the Torque and Angle of Twist

Equations (4.7) and (4.9), with the polar moment of inertia J evaluated using Eq. (4.10) or (4.11), determine the angle of twist and distribution of shear stress for a circular bar subjected to a torque T. For some applications, it will be convenient to have sign conventions for the torque and angle of twist.

In Fig. 4.17(a) we isolate a segment of a bar and indicate the right-hand rule for the positive directions of the torque acting on the segment: pointing the thumb of the right hand outward from the cross section under consideration, the fingers point in the direction of positive torque. The angle of twist of a segment is defined to be positive if it is in the direction resulting from positive torque (Fig. 4.17(b)).

Study Questions

A cylindrical bar of length L, radius R, and shear modulus G is fixed at one end and subjected to an axial torque T at the other end.

1. How can you determine the resulting angle of twist of the end of the bar?
2. What is the maximum shear stress in the bar, and where does it occur?
3. What is the maximum tensile stress in the bar, and where does it occur?
4. What is the shear stress at a point on the bar's axis?

Example 4.1 Bar Subjected to Torsion

The bar in Fig. 4.18 has a circular cross section with 1-in diameter. From a previous tension test of the bar, it is known that the modulus of elasticity of the material is $E = 14E6$ psi. When a torque $T = 1800$ in - lb is applied to the end of the bar as shown, the end of the bar rotates $2°$. (**a**) What is the magnitude of the maximum shear stress in the bar due to the torque T? (**b**) What is the Poisson's ratio of the material?

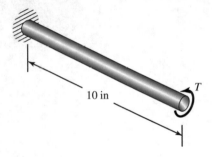

10 in

Fig. 4.18

Strategy

(**a**) After using Eq. (4.10) to calculate the polar moment of inertia of the bar's cross section, we can determine the maximum shear stress from Eq. (4.9).
(**b**) Because we know the angle of twist, we can solve Eq. (4.7) for the shear modulus G. Then Poisson's ratio can be determined from Eq. (4.2).

Solution

(**a**) The radius of the bar's cross section is $R = 0.5$ in. The polar moment of inertia is

$$J = \frac{\pi}{2}R^4 = \frac{\pi}{2}(0.5 \text{ in})^4 = 0.0982 \text{ in}^4.$$

The shear stress in the bar is given by Eq. (4.9). The maximum shear stress occurs at $r = R = 0.5$ in. The magnitude of the maximum shear stress is

$$|\tau| = \frac{TR}{J}$$

$$= \frac{(1800 \text{ in-lb})(0.5 \text{ in})}{0.0982 \text{ in}^4}$$

$$= 9170 \text{ psi.}$$

(b) The angle of twist of the bar in radians is $\phi = (\pi/180°)2° = 0.0349$ rad. From Eq. (4.7)

$$\phi = \frac{TL}{GJ}:$$

$$0.0349 \text{ rad} = \frac{(1800 \text{ in-lb})(10 \text{ in})}{G(0.0982 \text{ in}^4)},$$

we obtain $G = 5.25E6$ psi. Then from Eq. (4.2)

$$G = \frac{E}{2(1 + \nu)}:$$

$$5.25E6 \text{ psi} = \frac{14E6 \text{ psi}}{2(1 + \nu)},$$

we obtain $\nu = 0.333$.

Discussion
In Example 3.4 we showed how the modulus of elasticity E and Poisson's ratio ν of a material can be determined experimentally by subjecting a bar to axial loads and measuring the axial and lateral strains. As this example demonstrates, an alternative procedure is to determine E from a tension test and the shear modulus G from a torsion test and then use Eq. (4.2) to calculate Poisson's ratio.

Example 4.2 Stepped Bar Subjected to Torsion
Part AB of the bar in Fig. 4.19 has a circular cross section with radius $R_{AB} = 20$ mm, and part BC has a circular cross section with radius $R_{BC} = 10$ mm. The bar consists of material with shear modulus $G = 28$ GPa. (a) Determine the magnitudes of the maximum shear stresses in parts AB and BC of the bar. (b) What is the angle of twist of the end C of the bar relative to the wall?

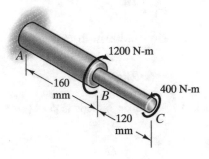

Fig. 4.19

Strategy

By passing a plane through part *AB* of the bar and isolating the part of the bar on one side of the plane, we can determine the internal torque in part *AB*. We can then use Eq. (4.9) to determine the maximum shear stress in part *AB* and use Eq. (4.7) to determine the angle of twist of part *AB*. We will determine the maximum shear stress in part *BC* and the angle of twist of part *BC* in the same way.

Solution

(**a**) In Fig. (a) we draw the free-body diagram of the entire bar, where T_A is the torque exerted on the bar by the wall. From the equilibrium equation 400 N-m − 1200 N-m − T_A = 0, we find that T_A = − 800 N-m. In Fig. (b) we pass a plane through part *AB* of the free-body diagram of the entire bar and isolate the part to the left of the plane. From the equilibrium equation T_{AB} + 800 N-m = 0, the torque in part *AB* of the bar is T_{AB} = − 800 N-m. The polar moment of inertia of the cross section of part *AB* is

$$J_{AB} = \frac{\pi}{2} R_{AB}^4 = \frac{\pi}{2}(0.02 \text{ m})^4 = 2.51\text{E-7 m}^4.$$

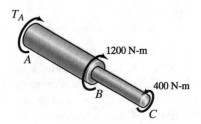

(a) Isolating the bar from the wall

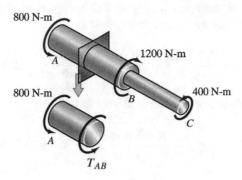

(b) Obtaining a free-body diagram by passing a plane through part *AB*

Because the shear stress is proportional to the radial distance from the axis of the bar, the maximum shear stress in part *AB* occurs at $r = R_{AB} = 0.02$ m. The magnitude of the maximum shear stress is

$$|\tau_{AB}| = \frac{|\,T_{AB}\,|\,R_{AB}}{J_{AB}} = \frac{(800 \text{ N-m})\,(0.02 \text{ m})}{2.51\text{E-}7 \text{ m}^4} = 63.7 \text{ MPa}.$$

In Fig. (c) we pass a plane through part *BC* of the free-body diagram of the entire bar and isolate the part to the left of the plane. From the equilibrium equation $T_{BC} - 1200$ N-m + 800 N-m = 0, the torque in part *BC* of the bar is $T_{BC} = 400$ N-m. The polar moment of inertia of the cross section of part *BC* is

$$J_{BC} = \frac{\pi}{2} R_{BC}^4 = \frac{\pi}{2}(0.01 \text{ m})^4 = 1.57\text{E-}8 \text{ m}^4.$$

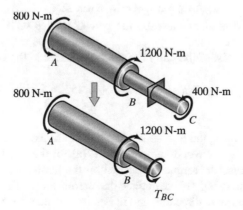

(c) Obtaining a free-body diagram by passing a plane through part *BC*

The maximum shear stress in part BC occurs at $r = R_{BC} = 0.01$ m. The magnitude of the maximum shear stress is

$$|\tau_{BC}| = \frac{|T_{BC}| R_{BC}}{J_{BC}} = \frac{(400 \text{ N-m})(0.01 \text{ m})}{1.57\text{E-8 m}^4} = 255 \text{ MPa}.$$

(b) The angle of twist of part AB of the bar is

$$\phi_{AB} = \frac{T_{AB} L_{AB}}{GJ_{AB}} = \frac{(-800 \text{ N-m})(0.16 \text{ m})}{(28\text{E9 N/m}^2)(2.51\text{E-7 m}^4)} = -0.0182 \text{ rad} = -1.04°.$$

This is the angle of twist of the bar at B relative to the wall. The fingers of the right hand in Fig. (d) indicate the positive direction of the angle of twist, so the negative value of ϕ_{AB} means that at B the bar rotates $1.04°$ in the opposite direction. The angle of twist of part BC of the bar is

$$\phi_{BC} = \frac{T_{BC} L_{BC}}{GJ_{BC}} = \frac{(400 \text{ N-m})(0.12 \text{ m})}{(28\text{E9 N/m}^2)(1.57\text{E-8 m}^4)} = 0.1091 \text{ rad} = 6.25°.$$

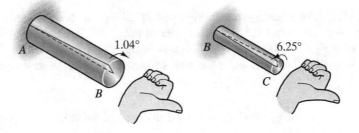

(d) Angle of twist of part AB. (e) Angle of twist of part BC

Relative to point B of the bar, the end C rotates $6.25°$ in the positive direction (Fig. (e)).

The angle of twist of the end D of the bar relative to the wall is therefore

$$\phi_{AB} + \phi_{BC} = -1.04° + 6.25° = 5.21°.$$

Discussion
Notice that in drawing the free-body diagrams in this example, we consistently adhered to the sign conventions in Fig. 4.17(a). In the free-body diagram in Fig. (a), we defined the torque T_A exerted on the bar by the wall to be in the positive direction. In Fig. (b) we defined the internal torque T_{AB} in part AB of the bar to be in the positive direction, and in Fig. (c), we defined the internal torque T_{BC} in part BC of the bar to be in the positive direction. Why was this worth doing? For one thing, it meant that when we solved for the angles of twist of parts AB and BC, the signs of the results corresponded to the sign convention in Fig. 4.17(b), which helped us in determining the angle of twist of the entire bar.

Problems

4.15 If a bar has a solid circular cross section with 15-mm diameter, what is the polar moment of inertia of its cross section in m^4?

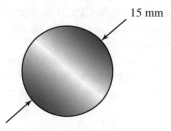

15 mm

Problem 4.15

4.16 If a bar has a hollow circular cross section with 2-in outer radius and 1-in inner radius, what is the polar moment of inertia of its cross section?

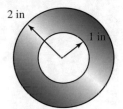

2 in

1 in

Problem 4.16

4.17 The bar has a circular cross section with 15-mm diameter, and the shear modulus of the material is $G = 26$ GPa. If the torque $T = 10$ N-m, determine (**a**) the magnitude of the maximum shear stress in the bar; (**b**) the angle of twist of the end of the bar in degrees.

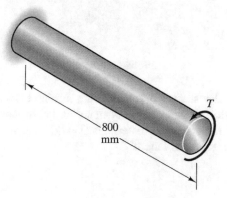

T

800 mm

Problems 4.17–4.19

4.18 The bar has a circular cross section with 15-mm diameter, and the shear modulus of the material is $G = 26$ GPa. If the torque T causes the end of the bar to rotate $4°$, what is the magnitude of the maximum shear stress in the bar?

4.19 The bar has a circular cross section with 15-mm diameter, and the shear modulus of the material is $G = 26$ GPa. If the bar is to be used in an application that requires that it be subjected to an angle of twist no greater than $1°$, what is the maximum allowable value of the torque T?

4.20 The solid circular shaft that connects the turbine blades of the hydroelectric power unit to the generator has a 0.4-m radius and supports a torque $T = 2$ MN-m. What is the magnitude of the maximum shear stress in the shaft?

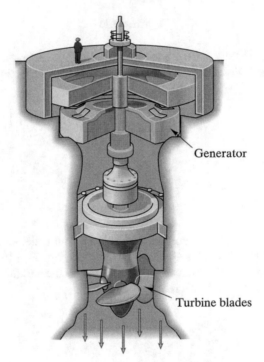

Generator

Turbine blades

Problems 4.20–4.22

4.21 The solid circular shaft that connects the turbine blades of the hydroelectric power unit to the generator has a 0.4-m radius and supports a torque $T = 2$ MN-m. The shear modulus of the material is $G = 80$ GPa. What is the angle of twist of the shaft per meter of length?

4.22 The shaft that connects the turbine blades of the hydroelectric power unit to the generator has a hollow circular cross section with 0.5-m outer radius and 0.3-m inner radius. It supports a torque $T = 2$ MN-m. What is the magnitude of the maximum shear stress in the shaft?

4.23 The cross section of a submarine's propeller shaft is shown. If the engine exerts a 600-kN-m torque on the shaft, what is the magnitude of the maximum shear stress in the shaft?

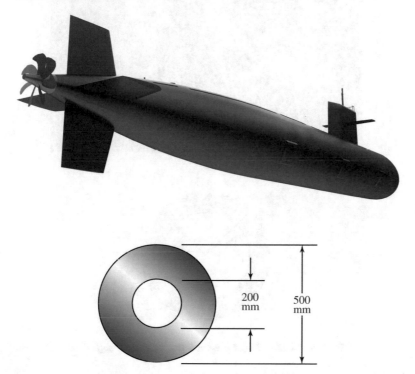

Problems 4.23–4.24 (*Photograph from iStock by Nerthuz*)

4.24 A submarine's engine exerts a 600-kN-m torque on the propeller shaft. If the shaft had a solid circular cross section whose cross-sectional area was equal to the cross-sectional area of the hollow circular cross section shown (i.e., the weight per unit length of the solid shaft would be the same as that of the hollow shaft), what would be the magnitude of the maximum shear stress in the shaft? Compare your answer to that of Problem 4.23.

4.25 The cross section of a truck's drive shaft is shown. The outer radius is $R_o = 50$ mm. If the shaft is subjected to a 200 N-m torque, what is the magnitude of the maximum resulting shear stress?

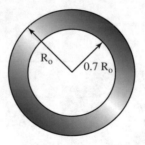

Problems 4.25–4.26 (*Photograph from iStock by coffeekai*)

4.26 The cross section of a truck's drive shaft is shown. The greatest torque expected to act on the shaft is 1600 N-m. If you don't want the shaft to be subjected to a shear stress greater than 4 MPa, what is the minimum necessary value of the outer radius R_o?

4.27 One type of high-strength steel drill pipe used in drilling oil wells has a 5-in outside diameter and a 4.28-in inside diameter. If you don't want the steel to be subjected to a shear stress greater than 65 ksi (65, 000 lb/in^2), what is the largest torque to which the pipe can be subjected?

Problems 4.27–4.28 (*Photograph from iStock by HHakim*)

4.28 A steel drill pipe has a 5-in outside diameter and a 4.28-in inside diameter. The shear modulus of the steel is $G = 12E6$ psi. If it is used to drill an oil well 15,000 ft deep and the drilling operation subjects the bottom of the pipe to a torque $T = 7500$ in - lb, what is the resulting angle of twist (in degrees) of the 15,000-ft pipe?

4.29 The propeller of the wind generator is supported by a hollow circular shaft with 0.4-m outer radius and 0.3-m inner radius. If the propeller exerts an 840-kN-m torque on the shaft, what is the resulting maximum shear stress?

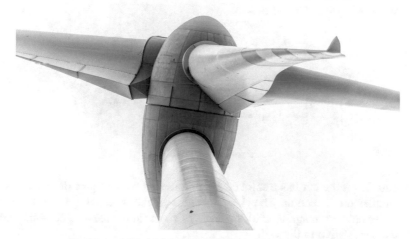

Problems 4.29–4.31 (*Photograph from iStock by Imikeee*)

4.30 The propeller of a wind turbine is supported by a hollow circular shaft with 0.4-m outer radius and 0.3-m inner radius. The shear modulus of the material is $G = 80$ GPa. If the propeller exerts an 840-kN-m torque on the shaft, what is the angle of twist of the shaft per meter of length?

4.31 A wind turbine's propeller exerts an 840-kN-m torque on its shaft. In designing a new shaft with a hollow circular cross section, the engineer wants to limit the maximum shear stress in the shaft to 25 MPa. Design constraints require that the outer radius of the shaft be 0.3 m. What maximum inner radius is necessary?

4.32 The bar has a circular cross section with 1-in radius, and the shear modulus of the material is $G = 5.8E6$ psi. If the torque $T = 1000$ in - lb, determine **(a)** the magnitude of the maximum shear stress in the bar and **(b)** the magnitude of the angle of twist of the right end of the bar relative to the wall in degrees.

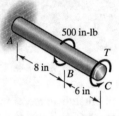

Problems 4.32–4.33

4.33 The bar has a circular cross section with 1-in radius, and the shear modulus of the material is $G = 5.8E6$ psi. What value of the torque T would cause the angle of twist of the right end of the bar relative to the wall to be zero?

4.34 Part AB of the bar has a solid circular cross section, and part BC has a hollow circular cross section. Determine the magnitudes of the maximum shear stresses in parts AB and BC.

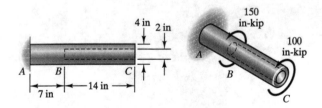

Problems 4.34–4.35

4.35 Part AB of the bar has a solid circular cross section, and part BC has a hollow circular cross section. The shear modulus of the material is $G = 3.8E6$ psi. Determine the magnitude of the angle of twist (in degrees) of the right end of the bar relative to the wall.

4.36 The cross section of the hollow bar is shown. The bar has a fixed support at the left end A and is subjected to torques at B, C, and D. Determine the magnitudes of the maximum shear stresses in parts AB, BC, and CD of the bar.

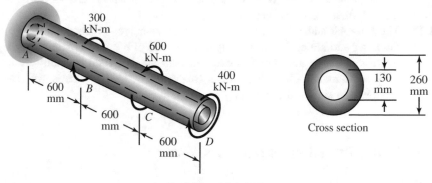

Cross section

Problems 4.36–4.37

4.37 The cross section of the hollow bar is shown. The bar has a fixed support at the left end A and is subjected to torques at B, C, and D. The bar is made of a material with shear modulus $G = 72$ GPa. What is the magnitude of the angle of twist (in degrees) of the end D of the bar relative to the fixed end at A?

4.38 The bars AB and CD each have a solid circular cross section with 30-mm diameter and are each 1 m in length. The radii of the gears are $r_B = 120$ mm and $r_C = 90$ mm. If the torque $T_A = 200$ N-m, what are the magnitudes of the maximum shear stresses in the bars?

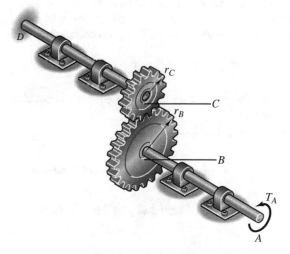

Problems 4.38–4.40

4.39∗ The bars AB and CD each have a solid circular cross section with 30-mm
diameter, are each 1 m in length, and consist of a material with shear
modulus $G = 28$ GPa. The radii of the gears are $r_B = 120$ mm and $r_C = 90$ mm.
If the torque $T_A = 200$ N-m, what is the angle of rotation at A? (Assume that
the deformations of the gears are negligible.)

4.40∗ The bars AB and CD each have a solid circular cross section with 30-mm
diameter and are each 1 m in length. The radii of the gears must satisfy the
relation $r_B + r_C = 210$ mm. If the torque $T_A = 200$ N-m and you don't want to
subject the material to a shear stress greater than 40 MPa, what is the largest
acceptable value of the radius r_C?

4.3   Statically Indeterminate Problems

The approach we used in Chap. 3 to solve statically indeterminate problems involving axially loaded bars—applying equilibrium, force-deformation relations, and
compatibility—applies to virtually all statically indeterminate problems. In this
section we demonstrate the solution of statically indeterminate problems involving
bars subjected to torsion. The force-deformation relations will now be relations
between torques and angles of twist, and the compatibility conditions will be
constraints imposed on the angles of twist of torsionally loaded bars.

To emphasize that the same procedure used for axially loaded bars can be applied
to bars with torsional loading, we present an example equivalent to the one discussed
in Sect. 3.3 and solve it in the same way. The bar in Fig. 4.20(a) consists of material
with shear modulus G. It has two segments AB and BC with different lengths and
diameters. The bar is subjected to a torque T_0 at B.

We draw the free-body diagram in Fig. 4.20(b), where T_A and T_C are the torques
exerted on the bar by the walls at A and C. The equilibrium equation is

$$-T_A + T_0 + T_C = 0. \quad \textbf{Equilibrium} \tag{4.12}$$

We cannot determine the reactions T_A and T_C from this equation. The problem is
statically indeterminate.

The angle of twist of part AB of the bar is

$$\phi_{AB} = \frac{T_A L_{AB}}{G J_{AB}}, \quad \textbf{Force} - \textbf{deformation} \tag{4.13}$$

where J_{AB} is the polar moment of inertia of part AB. The angle of twist of part BC of
the bar is

Fig. 4.20 (a) Torsionally
loaded bar fixed at both
ends. (b) Free-body diagram
of the bar

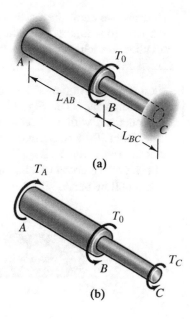

$$\phi_{BC} = \frac{T_C L_{BC}}{G J_{BC}}. \qquad \textbf{Force} - \textbf{deformation} \qquad (4.14)$$

Because the bar is fixed at both ends, the compatibility condition is that the angle
of twist of the entire bar is zero:

$$\phi_{AB} + \phi_{BC} = 0. \qquad \textbf{Compatibility} \qquad (4.15)$$

We substitute Eqs. (4.13) and (4.14) into this equation, obtaining

$$\frac{T_A L_{AB}}{G J_{AB}} + \frac{T_C L_{BC}}{G J_{BC}} = 0. \qquad (4.16)$$

We can solve this equation together with the equilibrium equation to determine
the reactions T_A and T_C. The results are

$$T_A = \frac{T_0}{1 + \dfrac{L_{AB} J_{BC}}{L_{BC} J_{AB}}}, \quad T_B = -\frac{T_0}{1 + \dfrac{L_{BC} J_{AB}}{L_{AB} J_{BC}}}.$$

Now that we know the reactions, we can determine the angle of twist of each part of the bar from Eqs. (4.13) and (4.14). We can also determine the shear stress distribution in each part of the bar: $\tau_{AB} = T_A r / J_{AB}$ and $\tau_{BC} = T_C r / J_{BC}$.

Example 4.3 Torsion of a Composite Bar

The composite bar in Fig. 4.21 consists of a cylindrical core A with radius R_A contained within a hollow cylindrical bar B of a different material with outer radius R_B. The circular end plate is bonded to the end of the bar. Suppose that an axial torque T_0 is applied to the end plate. Determine the resulting angle of twist of the composite bar and the distributions of shear stress in the core A and hollow bar B.

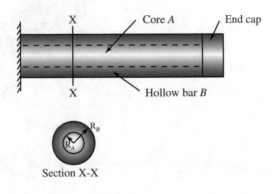

Fig. 4.21

Strategy

We assume that the geometry of deformation of the core A is identical to what it would be if the hollow bar B was not there and make the same assumption for the hollow bar B. That means that Eqs. (4.7) and (4.9) apply to each bar individually. But the torques applied to the individual bars by the end plate are not known, which makes the problem statically indeterminate. The compatibility condition is that the angles of twist of the core A and the hollow bar B are equal.

Solution

Equilibrium   In Fig. (a) we obtain a free-body diagram by isolating the end plate from the composite bar. The terms T_A and T_B are the torques exerted on the core A and the hollow bar B by the end plate. The equilibrium equation for the end plate is

$$T_0 - T_A - T_B = 0.$$

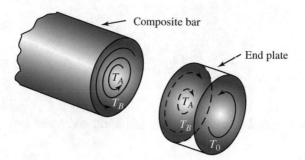

(a) Free-body diagram of the end plate

Force-deformation relations   The angles of twist of the core A and hollow bar B are

$$\phi_A = \frac{T_A L}{G_A J_A}, \qquad \phi_B = \frac{T_B L}{G_B J_B},$$

where the polar moments of inertia are

$$J_A = \frac{\pi}{2} R_A^4, \qquad J_B = \frac{\pi}{2} \left(R_B^4 - R_A^4 \right).$$

Compatibility   The angles of twist of the core A and the hollow bar B are equal:

$$\phi_A = \phi_B.$$

Substituting the force-deformation relations into the compatibility equation yields

$$\frac{T_A L}{G_A J_A} = \frac{T_B L}{G_B J_B}.$$

Solving this equation together with the equilibrium equation, we determine T_A and T_B:

$$T_A = \frac{T_0}{1 + \frac{G_B J_B}{G_A J_A}}, \qquad T_B = \frac{T_0}{1 + \frac{G_A J_A}{G_B J_B}}.$$

Now we can use either of the force-deformation relations to determine the angle of twist:

$$\phi_A = \phi_B = \frac{T_0 L}{G_A J_A + G_B J_B}.$$

The distributions of shear stress in the core A and hollow bar B are

$$\tau_A = \frac{T_A r}{J_A} = \frac{T_0 r}{J_A + \dfrac{G_B}{G_A} J_B},$$

$$\tau_B = \frac{T_B r}{J_B} = \frac{T_0 r}{J_B + \dfrac{G_A}{G_B} J_A},$$

where r is the radial distance from the axis of the composite bar.

Discussion
Hollow circular bars are often used to support and transmit torque, because their polar moments of inertia are larger than a solid circular bar of the same cross sectional area. When a thin-walled hollow bar is used, the bar is sometimes filled with a core of lighter material. For example, a hollow steel bar might be filled with a polymer core. The role of the lighter material is to stabilize the walls of the hollow bar and give it additional bending stiffness, but most of the torque is supported by the hollow bar. Observe in our solutions for the torques T_A and T_B supported by the core and the hollow bar that when $G_A J_A$ is small in comparison to $G_B J_B$, the torque T_B supported by the hollow bar is nearly equal to T_0.

Problems

4.41 The bar has a circular cross section with 1-in diameter. If the torque $T_0 = 1000$ in - lb, determine the magnitudes of the maximum shear stresses in parts AB and BC.

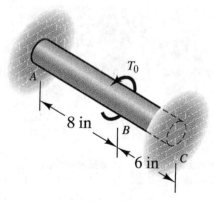

Problems 4.41–4.43

4.42 The bar has a circular cross section with 1-in diameter. It consists of a material that will safely support a maximum shear stress of 40 ksi. Based on this criterion, what is the maximum safe magnitude of the torque T_0?

4.43 The bar has a circular cross section with 1-in diameter. It is to be subjected to a torque $T_0 = 10,000$ in - lb. If the bar consists of a material that will safely support a maximum shear stress of 40 ksi, what is the largest distance from the left end of the bar at which the torque can safely be applied?

4.44 The bar is fixed at both ends. It has a solid circular cross section. Part AB is 40 mm in diameter, and part BC is 20 mm in diameter. Determine the magnitudes of the torques exerted on the bar at A and C by the walls.

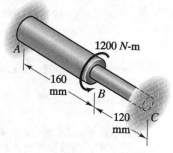

Problems 4.44–4.45

4.45 The bar is fixed at both ends. It consists of material with shear modulus $G = 28$ GPa and has a solid circular cross section. Part AB is 40 mm in diameter, and part BC is 20 mm in diameter. Determine the magnitude of the angle (in degrees) through which the bar rotates at B when the 1200-N-m torque is applied.

4.46 The composite bar consists of a cylindrical steel core with a shear modulus of 75 GPa contained within a hollow cylindrical bar of titanium. An axial torque $T_0 = 1000$ N-m is applied to the right end of the bar. What is the resulting angle of twist of the bar?

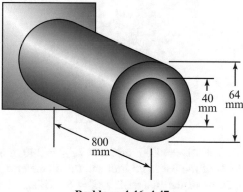

Problems 4.46–4.47

4.47   The composite bar consists of a cylindrical steel core with a shear modulus of 75 GPa contained within a hollow cylindrical bar of titanium. An axial torque $T_0 = 1000$ N-m is applied to the right end of the bar. What are the magnitudes of the maximum shear stresses in the steel core and the hollow titanium bar?

4.48   Each bar is 10 in long and has a solid circular cross section. Bar A has a diameter of 1 in and its shear modulus is 6E6 psi. Bar B has a diameter of 2 in and its shear modulus is 3.8E6 psi. The ends of the bars are separated by a small gap. The free end of bar A is rotated 2° about the bar's axis, and the bars are welded together. What are the magnitudes of the angles of twist (in degrees) of the two bars afterward?

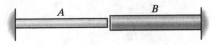

Problems 4.48–4.49

4.49   Each bar is 10 in long and has a solid circular cross section. Bar A has a diameter of 1 in and its shear modulus is 6E6 psi. Bar B has a diameter of 2 in and its shear modulus is 3.8E6 psi. The ends of the bars are separated by a small gap. The free end of bar A is rotated 2° about its axis, the free end of bar B is rotated 2° about its axis in the opposite direction, and the bars are welded together. What are the magnitudes of the maximum shear stresses in the two bars afterward?

4.50   The lengths $L_{AB} = L_{BC} = 200$ mm and $L_{CD} = 240$ mm. The diameter of parts AB and CD of the bar is 25 mm and the diameter of part BC is 50 mm. The shear modulus of the material is $G = 80$ GPa. What is the magnitude of the maximum shear stress in the bar?

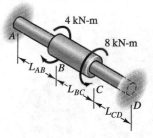

Problems 4.50–4.51

4.51   The lengths $L_{AB} = L_{BC} = 200$ mm and $L_{CD} = 240$ mm. The diameter of parts AB and CD of the bar is 25 mm and the diameter of part BC is 50 mm. The shear modulus of the material is $G = 80$ GPa. What is the magnitude (in degrees) of the angle through which the bar rotates at C when the torques are applied?

4.52* The collar is rigidly attached to bar A. The cylindrical bar A is 80 mm in diameter and its shear modulus is $G = 66$ GPa. There are gaps $b = 2$ mm between the arms of the collar and the ends of the identical bars B and C. Bars B and C are 30 mm in diameter and their modulus of elasticity is $E = 170$ GPa. If the bars B and C are stretched until they come into contact with the arms of the collar and are welded to them, what is the magnitude of the maximum shear stress in bar A afterward?

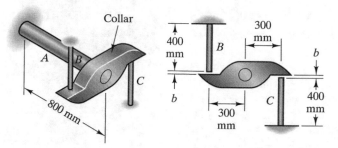

Problems 4.52–4.53

4.53* The collar is rigidly attached to bar A. The cylindrical bar A is 80 mm in diameter and its shear modulus is $G = 66$ GPa. There are gaps $b = 2$ mm between the arms of the collar and the ends of the identical bars B and C. Bars B and C are 30 mm in diameter and their modulus of elasticity is $E = 170$ GPa. The bars B and C are stretched until they come into contact with the arms of the collar and are welded to them. If bar A is made of a material that will safely support a tensile or compressive normal stress of 170 MPa, what is the largest safe value of b?

4.4 Nonprismatic Bars and Distributed Loads

In this section we discuss torsion of circular bars whose diameters vary with distance along their axes and also bars subjected to torsional loads that are distributed along their axes. The derivations closely follow our treatment in Sect. 3.4 of analogous applications involving axially loaded bars.

4.4.1 Bars with Gradually Varying Cross Sections

Figure 4.22(a) shows a bar with a hollow circular cross section whose inner and outer radii vary with distance along the bar's axis. The polar moment of inertia of the cross section depends upon x, which we indicate by writing it as $J(x)$. If we subject the bar to a torsional load (Fig. 4.22(b)) and the change in $J(x)$ with x is gradual, we can approximate the stress distribution on a plane at axial position x by using Eq. (4.9):

Fig. 4.22 (a) Bar with
a varying cross section.
(b) Subjecting the bar
to a torque T

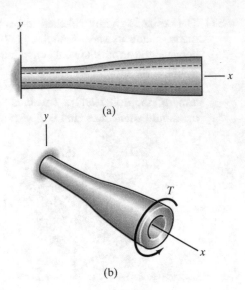

$$\tau = \frac{Tr}{J(x)}. \tag{4.17}$$

For a given perpendicular plane, the stress distribution is approximately the same as for the case of a prismatic bar. But because $J(x)$ varies with distance along the bar's axis, the shear stress distribution does also.

To determine the bar's angle of twist, we begin by considering an infinitesimal element of the bar of length dx (Fig. 4.23). We can use Eq. (4.7) to determine the element's angle of twist:

$$\phi_{\text{element}} = \frac{T\,dx}{GJ(x)}.$$

We obtain the angle of twist of the entire bar by integrating this expression from $x = 0$ to $x = L$:

$$\phi = \int_0^L \frac{T\,dx}{GJ(x)}. \tag{4.18}$$

The angle of twist of each element depends on its polar moment of inertia, and we must add up the angles of twist of the elements to determine the angle of twist of the bar.

Fig. 4.23 Determining
the angle of twist of an
infinitesimal element
of the bar

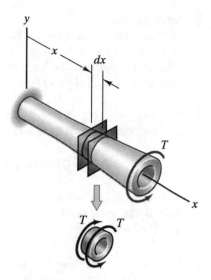

Example 4.4 Torsion of a Nonprismatic Bar
The bar in Fig. 4.24 has a solid circular cross section and consists of material
with shear modulus $G = 47$ GPa. The bar's polar moment of inertia in m^4 is
given by the equation $J(x) = 0.00016 + 0.0006x^2$. If the bar is subjected to a
torque $T = 200$ kN-m at its free end, determine **(a)** the magnitude of the
maximum shear stress in the bar at $x = 1$ m; **(b)** the magnitude of the angle of
twist of the entire bar in degrees.

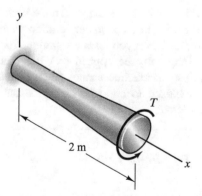

Fig. 4.24

Strategy

(a) From the given equation for $J(x)$, we can determine the polar moment of inertia and the bar's radius R at $x = 1$ m. Then the maximum shear stress is $TR/J(x)$. (b) Because the bar's polar moment of inertia is a function of x, we must determine the bar's angle of twist from Eq. (4.18).

Solution

(a) The polar moment of inertia at $x = 1$ m is

$$J(1) = 0.00016 + 0.0006(1)^2 \text{ m}^4 = 0.00076 \text{ m}^4.$$

Solving the equation $J = (\pi/2)R^4$ for the bar's radius at $x = 1$ m, we obtain $R = 0.148$ m. Therefore, the magnitude of the maximum shear stress at $x = 1$ m is

$$|\tau| = \frac{TR}{J(1)} = \frac{(200,000 \text{ N-m}) (0.148 \text{ m})}{0.00076 \text{ m}^4} = 39.0 \text{ MPa}.$$

(b) From Eq. (4.18), the magnitude of the bar's angle of twist is

$$|\phi| = \int_0^L \frac{T\,dx}{GJ(x)} = \int_0^2 \frac{(200,000)\,dx}{(47\text{E}9)(0.00016 + 0.0006x^2)} = 0.0181 \text{ rad},$$

which is $1.04°$.

Discussion

Review Example 3.11, in which we analyzed a bar with varying cross section subjected to an axial load. The steps in our solution were identical to the ones we used in this example. When you learn a technique for solving a problem in engineering, it can frequently be applied to other kinds of problems. The problems you will meet in practice cannot be predicted, but in engineering education, you are exposed to concepts and procedures that can often be adapted to new situations.

4.4.2 Distributed Torsional Loads

We can describe a distributed torsional load on a bar by introducing a function c defined such that the axial torque on each element of length dx is $c\,dx$ (Fig. 4.25). Because the product of c and dx is a moment, the dimensions of c are (moment/length). For example, the bar in Fig. 4.26(a) is fixed at the left end and subjected to a

Fig. 4.25 Describing a
distributed torque by a
function. The torque on an
element of length dx is $c\,dx$.

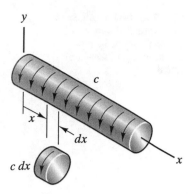

distributed axial torque throughout its length. In Fig. 4.26(b) we pass a plane through
the bar at an arbitrary position x and draw the free-body diagram of the part of the bar
to the right of the plane. From the equilibrium equation

$$-T + \int_x^L c_0 \left(\frac{x}{L}\right)^2 dx = 0,$$

we determine the internal torque in the bar at the position x:

$$T = \frac{c_0}{3} \left(L - \frac{x^3}{L^2}\right).$$

The distributed torque causes the internal torque in the bar to vary with axial
position. To determine the angle of twist of the right end of the bar relative to the
wall, we consider an element of the bar of length dx (Fig. 4.26(c)). From Eq. (4.7), its
angle of twist is

$$\phi_{\text{element}} = \frac{T\,dx}{GJ} = \frac{c_0}{3GJ} \left(L - \frac{x^3}{L^2}\right) dx.$$

Integrating this result from $x = 0$ to $x = L$, we obtain the angle of twist of the
right end:

$$\phi = \int_0^L \frac{c_0}{3GJ} \left(L - \frac{x^3}{L^2}\right) dx = \frac{c_0 L^2}{4GJ}.$$

Fig. 4.26 (**a**) Bar subjected
to a distributed torque. (**b**)
Determining the internal
torque at x. (**c**) Element of
length dx

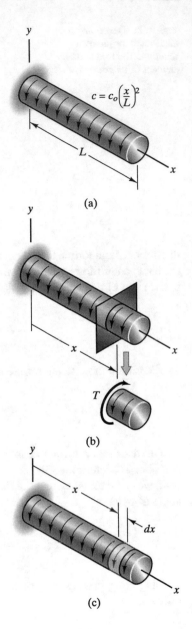

$$c = c_o \left(\frac{x}{L}\right)^2$$

(a)

(b)

(c)

Example 4.5 Bar Subjected to a Distributed Torsional Load
The cylindrical bar in Fig. 4.27 is fixed at both ends and is subjected to a uniform distributed torque c from $x = 0$ to $x = L_{AB}$. What torques are exerted on the bar by the walls?

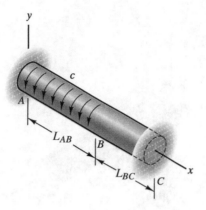

Fig. 4.27

Strategy
By using the equilibrium equation for the entire bar and the compatibility condition that the angle of twist of the right end of the bar relative to the left end is zero, we can solve for the two torques exerted on the bar by the walls.

Solution
In Fig. (a) we draw the free-body diagram of the entire bar, where T_A and T_C are the torques exerted by the walls at A and C. Because c is constant, the total moment exerted on the bar by the distributed torque is the product of c and L_{AB}. The equilibrium equation is

$$-T_A + T_C + cL_{AB} = 0.$$

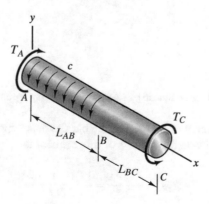

(a) Free-body diagram of the bar

This equation cannot be solved for T_A and T_C, so the problem is statically indeterminate.

In Fig. (b), we pass a plane through part AB of the bar at an arbitrary position x. From the equilibrium equation $-T_A + T_{AB} + cx = 0$, the torque in part AB of the bar is

$$T_{AB} = T_A - cx.$$

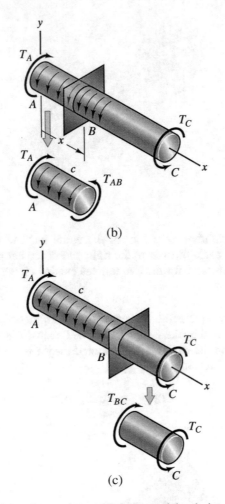

(b)

(c)

(b) Determining the internal torque in part AB. (c) Determining the internal torque in part BC

Consider an element of part AB of the bar that has length dx and is located at the position x. The angle of twist of this element is

$$d\phi_{AB} = \frac{T_{AB}\, dx}{GJ} = \frac{(T_A - cx)\, dx}{GJ}.$$

We integrate this expression to determine the angle of twist of part AB of the bar, obtaining

$$\phi_{AB} = \int_0^{L_{AB}} \frac{(T_A - cx)\,dx}{GJ} = \frac{T_A L_{AB}}{GJ} - \frac{cL_{AB}^2}{2GJ}.$$

The internal torque throughout part BC of the bar is T_C (Fig. (c)), so the angle of twist of part BC is

$$\phi_{BC} = \frac{T_C L_{BC}}{GJ}.$$

The compatibility condition is

$$\phi_{AB} + \phi_{BC} = \frac{T_A L_{AB}}{GJ} - \frac{cL_{AB}^2}{2GJ} + \frac{T_C L_{BC}}{GJ} = 0.$$

Solving this equation simultaneously with the equilibrium equation, we obtain the torques exerted on the bar by the walls:

$$T_A = \left(\frac{\frac{1}{2} + \frac{L_{BC}}{L_{AB}}}{1 + \frac{L_{BC}}{L_{AB}}} \right) cL_{AB}, \quad T_C = \left(\frac{-\frac{1}{2}}{1 + \frac{L_{BC}}{L_{AB}}} \right) cL_{AB}.$$

Discussion
One useful check of solutions to engineering problems is to see whether their limiting values are correct. In this example, when $L_{BC} = 0$ our solutions for the torques exerted on the bar by the walls give $T_A = cL_{AB}/2$ and $T_C = -cL_{AB}/2$. You can see from the free-body diagram in Fig. (a) that these values are correct. By evaluating the limits of our equations for T_A and T_C as $L_{AB} \to 0$, confirm that those limits are correct. This kind of check obviously does not guarantee that solutions are correct but can often indicate when errors have been made.

Problems

4.54 The bar has a solid circular cross section. Its polar moment of inertia in in^4 is given by $J = 0.1 + 0.15x$, where x is the axial position in inches. If the bar is subjected to an axial torque $T = 20$ in - kip, what is the magnitude of the maximum shear stress at $x = 6$ in?

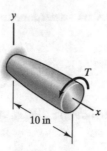

Problems 4.54–4.55

4.55 The bar has a solid circular cross section. Its polar moment of inertia in in^4 is given by $J = 0.1 + 0.15x$, where x is the axial position in inches. The shear modulus of the material is $G = 4.6\text{E}6$ psi. If the bar is subjected to an axial torque $T = 20$ in - kip, what is the bar's angle of twist in degrees?

4.56 The bar has a solid circular cross section. The radius of the cross section in inches is given by $R = 2 + 0.1x$, where x is the axial position in inches. If the bar is subjected to an axial torque $T = 120$ in - kip, what is the magnitude of the maximum shear stress at $x = 10$ in?

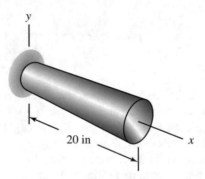

Problems 4.56–4.57

4.57 The bar has a solid circular cross section. The radius of the cross section in inches is given by $R = 2 + 0.1x$, where x is the axial position in inches. The shear modulus of the material is $G = 2\text{E}6$ psi. If the bar is subjected to an axial torque $T = 120$ in - kip, what is the bar's angle of twist in degrees?

4.58 The bar has a circular cross section whose radius in meters is given by the equation $R = 0.1e^{ax}$,

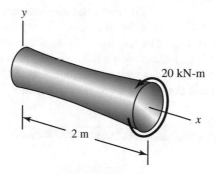

Problems 4.58–4.59

where x is in meters and a is a constant. The magnitude of the maximum shear stress in the bar at $x = 1$ m is 6 MPa. What is the value of the constant a?

4.59 The bar has a circular cross section whose radius in meters is given by the equation $R = 0.1e^{ax}$, where x is in meters and a is a constant. The shear modulus of the material is $G = 70$ GPa. The magnitude of the maximum shear stress in the bar at $x = 1$ m is 6 MPa. What is the bar's angle of twist?

4.60 The radius of the bar's circular cross section varies linearly from 10 mm at $x = 0$ to 5 mm at $x = 150$ mm. The shear modulus of the material is $G = 17$ GPa. What torque T would cause a maximum shear stress of 10 MPa at $x = 80$ mm?

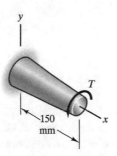

Problems 4.60–4.62

4.61 The radius of the bar's circular cross section varies linearly from 10 mm at $x = 0$ to 5 mm at $x = 150$ mm. The shear modulus of the material is $G = 17$ GPa. What torque T would cause the end of the bar to rotate 1°?

4.62* The radius of the bar's circular cross section varies linearly from 10 mm at $x = 0$ to 5 mm at $x = 150$ mm. The shear modulus of the material is $G = 17$ GPa. Suppose that the torque T at the end of the bar is 20 N-m and you want to apply a torque in the opposite direction at $x = 75$ mm so that the angle through which the end of the bar rotates is zero. What is the magnitude of the torque you must apply?

4.63 Bars A and B have solid circular cross sections and consist of material with shear modulus $G = 17$ GPa. Bar A is 150 mm long and its radius varies linearly from 10 mm at its left end to 5 mm at its right end. The prismatic bar B is 100 mm long and its radius is 5 mm. There is a small gap between the bars. The end of bar A is given an axial rotation of $1°$ and the bars are welded together. What is the torque in the bars afterward?

Problem 4.63

4.64 The aluminum alloy bar has a circular cross section with 20-mm diameter, length $L = 120$ mm, and a shear modulus of 28 GPa. If the distributed torque is uniform and causes the end of the bar to rotate $0.5°$, what is the magnitude of the maximum shear stress in the bar?

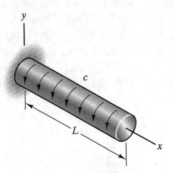

Problems 4.64–4.65

4.65 The aluminum alloy bar has a circular cross section with 20-mm diameter, length $L = 120$ mm, and a shear modulus of 28 GPa. If the distributed torque is given by the equation $c = c_0(x/L)^3$, where c_0 is a constant, and it causes the end of the bar to rotate $0.5°$, what is the magnitude of the maximum shear stress in the bar?

4.66∗ A cylindrical bar with 1-in diameter fits tightly in a circular hole in a 5-in thick plate. The shear modulus of the material is $G = 5.6\text{E}6$ psi. A 12,000-in-lb axial torque is applied at the left end of the bar. The distributed torque exerted on the bar by the plate is given by the equation

$$c = c_0\left[1 - \left(\frac{x}{5}\right)^{1/2}\right]\text{in-lb/in,}$$

where c_0 is a constant and x is the axial position in inches measured from the left side of the plate. Determine the constant c_0 and the magnitude of the maximum shear stress in the bar at $x = 2$ in.

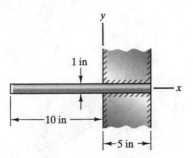

Problems 4.66–4.67

4.67* In Problem 4.66, what is the magnitude of the angle of twist of the left end of the bar relative to its right end?

4.68 The bar has a circular cross section with polar moment of inertia J and shear modulus G. The distributed torque $c = c_0(x/L)^2$, where c_0 is a constant. Show that the magnitudes of the torques exerted on the bar by the left and right walls are $c_0L/12$ and $c_0L/4$, respectively.

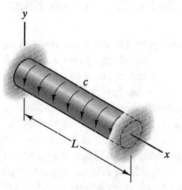

Problems 4.68–4.69

4.69 The bar has a circular cross section with polar moment of inertia J and shear modulus G. The distributed torque $c = c_0(x/L)^2$, where c_0 is a constant. At what axial position x is the magnitude of the bar's angle of twist relative to the walls the greatest, and what is its magnitude?

4.70 The bar has a circular cross section with polar moment of inertia J and shear modulus G. From $x = 0$ to $x = L/2$, it is acted upon by the distributed load $c = c_0(x/L)^2$, where c_0 is a constant. What are the magnitudes of the torques exerted on the bar by the left and right walls?

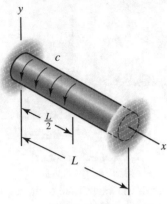

Problem 4.70

4.5   Elastic-Perfectly Plastic Circular Bars

In our discussion of bars subjected to torsional loads, we have assumed that the shear stress is a linear function of the shear strain: $\tau = G\gamma$. Many materials satisfy this relationship approximately if the shear strain does not become too large. At large values of the shear strain, the relationship between the shear stress and shear strain is for many materials qualitatively similar to the relationship between the normal stress and normal strain (see Sect. 3.6). In ductile materials, a yield stress occurs, and the material enters a plastic range in which strain increases with little or no increase in stress until the material fractures. The simplest model of this phenomenon is elastic-perfectly plastic behavior, in which it is assumed that plastic strain occurs with no increase in stress once the yield stress is reached (Fig. 4.28). Although more realistic models of plastic behavior must usually be used in actual design, understanding can be gained from analyses based on the elastic-perfectly plastic model. In this section we apply this model to torsion of a circular bar.

As in our analysis of the torsion of an elastic bar in Sect. 4.2, we begin with the assumption that each cross section of the bar undergoes a rigid rotation. This leads to

Fig. 4.28 Elastic-perfectly plastic model of shear stress-shear strain behavior

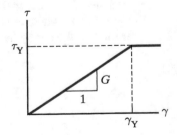

Eq. (4.5), which states that the shear strain is proportional to the radial distance r from the bar's axis:

$$\gamma = \frac{r\phi}{L}, \tag{4.19}$$

where ϕ is the angle of twist of the bar and L is its length. If the maximum shear strain in the bar (the shear strain at the bar's outer surface) does not exceed the value γ_Y corresponding to the yield stress, $\tau = G\gamma$ and the shear stress is also proportional to r (Fig. 4.29(a)). This is the elastic solution: the shear stress distribution is given by Eq. (4.9) and the angle of twist by Eq. (4.7).

If the shear strain does exceed the value corresponding to the yield stress at some radial position r_Y, the shear stress is proportional to r until it reaches the yield stress at $r = r_Y$, then remains constant (Fig. 4.29(b)). Equilibrium requires that the total moment about the axis of the bar due to the stress distribution be equal to T. We can use this condition to determine r_Y. The force due to the shear stress τ acting on an element dA at a radial distance r from the axis is $\tau\,dA$. The moment due to this force about the axis is $r(\tau\,dA)$, so

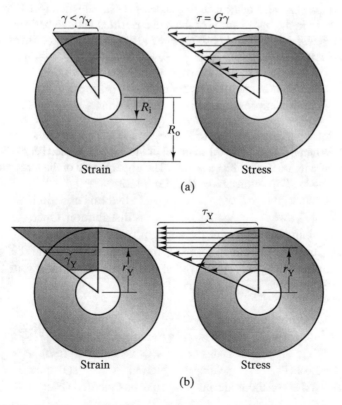

Fig. 4.29 (a) Strain and stress distributions when the maximum strain is less than the strain corresponding to the yield stress. (b) Strain and stress distributions when the maximum strain is greater than the strain corresponding to the yield stress

$$T = \int_A r\tau\, dA. \tag{4.20}$$

Let us apply this condition to the shear stress distribution shown in Fig. 4.29(b). For radial distances in the range $R_i \leq r \leq r_Y$, the shear stress $\tau = (\tau_Y/r_Y)r$. For radial distances greater than r_Y, the shear stress $\tau = \tau_Y$. If we use an annular element of area $dA = 2\pi r\, dr$ (Fig. 4.17), Eq. (4.20) becomes

$$T = \int_{R_i}^{r_Y} r\left(\frac{\tau_Y r}{r_Y}\right)(2\pi r\, dr) + \int_{r_Y}^{R_o} r\tau_Y(2\pi r\, dr). \tag{4.21}$$

Evaluating the integrals in this expression, we obtain

$$T = \pi\tau_Y R_o^3 \left[\frac{2}{3} - \frac{1}{6}\left(\frac{r_Y}{R_o}\right)^3 - \frac{1}{2}\left(\frac{R_i}{R_o}\right)^4 \frac{R_o}{r_Y}\right]. \tag{4.22}$$

For a given value of the external torque T, this nonlinear algebraic equation can be solved for the value of r_Y, which establishes the stress distribution. Once the value of r_Y is known, we can also determine the angle of twist. Equation (4.19), which gives the shear strain in the bar in terms of the radial position and the bar's angle of twist, applies if $r \leq r_Y$. At the radial position $r = r_Y$, the strain $\gamma = \gamma_Y$. Therefore we can solve Eq. (4.19) for ϕ:

$$\phi = \frac{\gamma_Y L}{r_Y} = \frac{\tau_Y L}{r_Y G}. \tag{4.23}$$

As the torque T increases beyond the value at which yielding first occurs, r_Y decreases. When $r_Y = R_i$, all of the material of the bar has yielded (Fig. 4.30), and the bar will twist with no further increase in T. The upper bound of the torque the bar will support is obtained by setting $r_Y = R_i$ in Eq. (4.22).

Fig. 4.30 Stress distribution when the entire cross section has yielded. This condition defines the upper bound of the torque an elastic-perfectly plastic bar can support

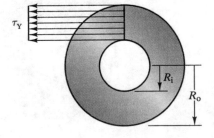

Example 4.6 Torsion of an Elastic-Perfectly Plastic Bar
A manganese bronze bar 2 m in length has the cross section shown in Fig. 4.31. The shear modulus $G = 39$ GPa and the yield stress $\tau_Y = 450$ MPa. The bar is fixed at one end and subjected to an axial torque T at the other end. Model the material as elastic-perfectly plastic. (a) If $T = 3$ kN-m, what is the bar's angle of twist? (b) What is the upper bound of the torque T the bar will support?

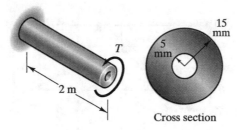

Cross section

Fig. 4.31

Strategy
(a) We will first determine whether we even need the elastic-perfectly plastic solution for the angle of twist. By substituting the yield stress $\tau_Y = 450$ MPa and $r = R_o = 0.015$ m into the elastic equation $\tau = Tr/J$, we will determine the maximum value of T for which the elastic solution applies. If the applied torque does not exceed that value, we can use the elastic solution to determine the bar's angle of twist. If it does exceed it, we must determine the value of r_Y from Eq. (4.22) and then determine the angle of twist from Eq. (4.23). (b) The upper bound of the torque the bar will support is given by Eq. (4.22) with $r_Y = R_i$.

Solution
(a) To determine the largest torque for which the elastic solution would apply, we substitute the yield stress and the bar's outer radius into the elastic equation for the stress distribution:

$$\tau_Y = \frac{TR_o}{J} :$$

$$450E6 \text{ N/m}^2 = \frac{T(0.015 \text{ m})}{(\pi/2)\left[(0.015 \text{ m})^4 - (0.005 \text{ m})^4\right]}.$$

Solving, we obtain $T = 2.36$ kN-m. The applied torque of 3 kN-m exceeds this value, so we must use the elastic-perfectly plastic solution. From Eq. (4.22), the relation between the applied torque T and r_Y is

$$T = \pi \left(450\text{E}6\,\text{N/m}^2\right)(0.015\,\text{m})^3 \left[\frac{2}{3} - \frac{1}{6}\left(\frac{r_Y}{0.015\,\text{m}}\right)^3 - \frac{1}{2}\left(\frac{0.005\,\text{m}}{0.015\,\text{m}}\right)^4 \frac{0.015\,\text{m}}{r_Y}\right].$$

(4.24)

Figure (a) shows the graph of T as a function of r_Y for $R_i \leq r_Y \leq R_o$. From the graph we estimate that $T = 3000$ N-m corresponds to a value of r_Y of 8.1 mm. By using software designed to solve nonlinear algebraic equations, we obtain $r_Y = 8.128$ mm.

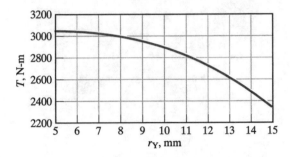

(a) Graph of T as a function of the radial position at which the yield stress is reached

From Eq. (4.23), the bar's angle of twist is

$$\phi = \frac{\tau_Y L}{r_Y G} = \frac{(450\text{E}6\,\text{N/m}^2)(2\,\text{m})}{(0.008128\,\text{m})(39\text{E}9\,\text{N/m}^2)} = 2.84 \ \text{rad},$$

which is 163°.

(b) To determine the upper bound of the torque T the bar will support, we set $r_Y = R_i = 0.005$ m in Eq. (4.24), obtaining $T = 3060$ N-m.

Discussion
As we discussed in Scct. 3.2, materials are permanently deformed when they are loaded beyond the yield stress. Structural designers usually seek to ensure that materials do not reach their yield stresses. So are solutions of the kind in this example only of academic interest? They have practical importance for two reasons. First, it is often useful to know the ultimate loads that structural elements will support before failing, even if in normal use their loads are significantly below those values. To understand the second reason, notice in Fig. (a) that the bar will support a substantially larger torque when it is partially yielded than the maximum torque it will support without causing the material to yield. For this reason, structural elements are sometimes intentionally designed so that their material is partially yielded when subjected to their normal loads. This may be done to satisfy weight constraints in particular applications but obviously requires great care. In such cases, designs should be validated by testing.

Problems

4.71 The steel bar has a circular cross section with radius $R = 20$ mm. If you model it as an elastic-perfectly plastic material with yield stress $\tau_Y = 400$ MPa, what is the upper bound of the torque T the bar will support?

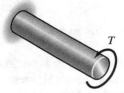

Problems 4.71–4.72

4.72 The steel bar has a circular cross section with radius $R = 20$ mm. It can be modeled as an elastic-perfectly plastic material with yield stress $\tau_Y = 400$ MPa. If the bar is subjected to an increasing torque T, for what value of T is $r_Y = 10$ mm? (i.e., the shear stress is equal to the yield stress from $r = 10$ mm to $r = 20$ mm.)

4.73 The cross section of the bar is shown. It consists of material with shear modulus $G = 39$ GPa and yield stress $\tau_Y = 450$ MPa. Model the material as elastic-perfectly plastic. What are the external torque T and the bar's angle of twist (in degrees) if $r_Y = 12$ mm?

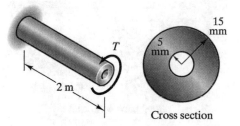

Cross section

Problems 4.73–4.74

4.74 The cross section of the bar is shown. It consists of material with shear modulus $G = 39$ GPa and yield stress $\tau_Y = 450$ MPa. Model the material as elastic-perfectly plastic. Determine the bar's angle of twist (in degrees) if the axial torque $T = 2.8$ kN-m.

4.75 The 48-in bar has a solid circular cross section with 4-in diameter. The shear modulus $G = 4E6$ psi and the yield stress $\tau_Y = 60,000$ psi. Model the material as elastic-perfectly plastic. What are the external torque T and the bar's angle of twist (in degrees) if $r_Y = 1$ in?

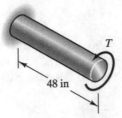

Problems 4.75–4.77

4.76 The 48-in bar has a solid circular cross section with 4-in diameter. The shear modulus $G = 4E6$ psi and the yield stress $\tau_Y = 60,000$ psi. Model the material as elastic-perfectly plastic. If the bar is subjected to a torque $T = 1E6$ in - lb, what are r_Y and the bar's angle of twist in degrees?

4.77 The 48-in bar has a solid circular cross section with 4-in diameter. The shear modulus $G = 4E6$ psi and the yield stress $\tau_Y = 60,000$ psi. Model the material as elastic-perfectly plastic. What is the upper bound of the torque T the bar will support?

4.78 The 4-m aluminum alloy bar has the cross section shown. The shear modulus $G = 28$ GPa and the yield stress $\tau_Y = 410$ MPa. Model the material as elastic-perfectly plastic. If the bar's angle of twist is $\phi = 130°$, what is the torque T?

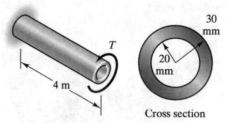

Cross section

Problems 4.78–4.80

4.79 The 4-m aluminum alloy bar has the cross section shown. The shear modulus $G = 28$ GPa and the yield stress $\tau_Y = 410$ MPa. Model the material as elastic-perfectly plastic. What is the upper bound of the torque T the bar will support?

4.80 The 4-m aluminum alloy bar has the cross section shown. The shear modulus $G = 28$ GPa and the yield stress $\tau_Y = 410$ MPa. Model the material as elastic-perfectly plastic. If the bar is subjected to a torque $T = 16$ kN-m, what is the resulting angle of twist in degrees?

4.81* The 4-m aluminum alloy bar has the cross section shown. The shear modulus $G = 28$ GPa and the yield stress $\tau_Y = 410$ MPa. Model the material as elastic-perfectly plastic. The bar is fixed at both ends and is subjected to an axial torque T as shown. If the bar's angle of twist where the torque is applied is $30°$, what is the torque T?

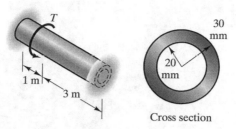

Problem 4.81

4.82 By evaluating the integrals in Eq. (4.21), derive the relation (4.22) between the torque T and the radial distance r_Y at which the shear stress becomes equal to the yield stress.

4.6 Thin-Walled Tubes

The results we have presented for the deformations and states of stress of bars subjected to torsional loads apply only to bars with circular cross sections. Extending these results to an arbitrary cross-sectional shape generally requires the use of advanced analytical techniques or a numerical approach such as the finite element method. In this section we describe a clever approximate analysis of a limited class of bars with noncircular cross sections that was presented by R. Bredt, a German engineer, in 1896.

4.6.1 Stress

Consider a prismatic tube whose wall thickness t is small compared to the lateral dimensions of the tube (Fig. 4.32). As we indicate in the figure, the wall thickness need not be uniform. In Fig. 4.33 we subject the tube to a torque T and obtain a free-

Fig. 4.32 Thin-walled prismatic tube

Fig. 4.33 Assumed shear
stress distribution on a plane
perpendicular to the
bar's axis

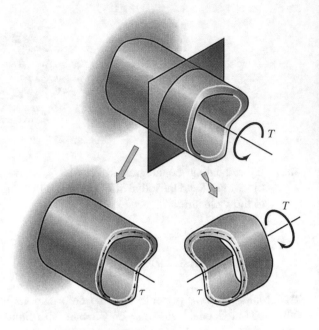

Fig. 4.34 Element of the
tube wall. The shear stress τ
may vary in the
circumferential direction

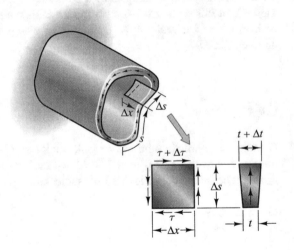

body diagram by passing a plane perpendicular to its axis. The stress distribution for
a circular bar suggests the assumption that the tube is subjected to a shear stress
parallel to the tube wall. Here we interpret τ as the average value of the shear stress
across the wall's thickness. We do not assume that τ is uniform around the tube's
circumference. In fact, we will show that its value must vary if the wall thickness
t varies around the circumference.

Our first step in determining τ is to consider the element in Fig. 4.34. The
coordinate s measures distance along the circumference of the wall relative to

some reference point. As s increases by the amount Δs, the wall thickness may change by an amount Δt, and the shear stress may change by an amount $\Delta \tau$. The sum of the forces on the element in the direction parallel to the bar's axis (the x direction) is

$$(\tau + \Delta \tau)(\Delta x)(t + \Delta t) - \tau \Delta x t = 0.$$

We divide this equation by $\Delta x \Delta s$ and write it as

$$\tau \frac{\Delta t}{\Delta s} + \frac{\Delta \tau}{\Delta s} t + \frac{\Delta \tau}{\Delta s} \frac{\Delta t}{\Delta s} \Delta s = 0.$$

The limit of this expression as $\Delta s \to 0$ is

$$\tau \frac{dt}{ds} + \frac{d\tau}{ds} t = \frac{d}{ds}(\tau t) = 0.$$

This result indicates that the product of the shear stress and the wall thickness, which we denote by

$$f = \tau t, \tag{4.25}$$

does not depend on s. That is, it is constant in the circumferential direction.

The quantity f is called the *shear flow*. This term originates from an interesting analogy with fluid flow. If we visualize the bar's cross section as a channel of steadily flowing incompressible fluid of uniform depth and width t, the product of the fluid's average velocity and the channel width, vt, is constant. Thus there is an exact analogy between the fluid velocity v and the shear stress τ (Fig. 4.35). The fluid flows more rapidly where the channel is narrow and more slowly where it is wide. The shear stress is greater where the wall is narrow and smaller where it is wide.

We now know that the shear flow $f = \tau t$ is constant around the tube's circumference, but we don't know its value. We must determine it by equating the couple exerted by the shear stress distribution to the external torque T (Fig. 4.33). Remarkably, we can determine the couple exerted by the shear stress distribution

Fig. 4.35 Analogy between the velocity of incompressible channel flow and the shear stress

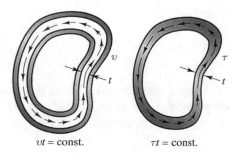

$vt = \text{const.}$ $\tau t = \text{const.}$

Fig. 4.36 (**a**) Determining
the moment due to the shear
stress on an element *ds*. (**b**)
The moment can be
expressed in terms of the
area *dA*. (**c**) *A* is the area
within the midline of the
tube wall

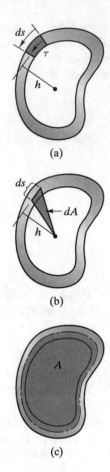

(a)

(b)

(c)

without specifying the shape of the tube's cross section. The force exerted by the
shear stress on an element of the cross section of length *ds* is $\tau t \, ds = f \, ds$ (Fig. 4.36
(a)). The moment due to this force about the tube's axis is $hf \, ds$, where *h* is the
perpendicular distance to the line of action of the force. (Because the moment
due to a couple is the same about any point, our choice of the location of the
axis is arbitrary.) The integral of this expression over the entire circumference
must equal *T*:

$$T = f \int_{s} h \, ds.$$

At this point, things don't look promising, because *h* depends on the shape of the
cross section. However, notice that the area of the triangle in Fig. 4.36(b) is
$dA = h \, ds/2$. Therefore, we can write the equilibrium equation as

$$T = 2f \int_A dA = 2f A,$$

where A is the cross-sectional area of the tube (Fig. 4.36(c)). From this result we obtain the value of the shear flow in terms of the external torque:

$$f = \tau t = \frac{T}{2A}. \tag{4.26}$$

Maximizing the tube's cross-sectional area minimizes the shear flow, which helps explain the popularity of circular tubes for supporting torsional loads.

We now know the stress distribution in the tube in terms of the external torque T. We can determine the shear flow from Eq. (4.26). Then for a given circumferential position on the tube wall with thickness t, we can evaluate the average shear stress τ from Eq. (4.25). (See Problem 4.101, in which the shear stress obtained in this way is compared to the exact shear stress for the case of a circular thin-walled tube.) Our next objective is to determine the tube's angle of twist in terms of T.

4.6.2   Angle of Twist

When the torque T is applied to the tube, twisting it through an angle ϕ (Fig. 4.37), work is done on the tube. The work done by a torque acting through an angle ϕ is

$$\text{work} = \int_0^\phi T \, d\phi.$$

Because the angle through which the elastic tube rotates is a linear function of the torque (Fig. 4.38), the work equals one-half the product of the torque and the angle of twist:

Fig. 4.37 The torque
T rotates the end of the tube
through an angle ϕ.

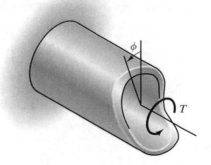

Fig. 4.38 The angle of
twist of the tube is a linear
function of the applied
torque. The triangular area
equals the work done

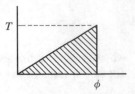

$$\text{work} = \frac{1}{2}T\phi. \tag{4.27}$$

The work done in stretching a linear spring is stored in the spring as an equal amount of potential energy. In the same way, the work done in twisting an elastic tube is stored within the material of the deformed tube as *strain energy*, which we discuss in Chap. 11. By equating the work done to the total strain energy, we can determine the tube's angle of twist.

Consider the infinitesimal element of the tube shown in Fig. 4.39(a). In Sect. 11.1 we show that the strain energy per unit volume of an element subjected to a pure shear stress τ is

$$u = \frac{1}{2}\tau\gamma,$$

where γ is the shear strain. The volume of the element in Fig. 4.39(a) is $t\,ds\,dx$ and the shear strain $\gamma = \tau/G$, so the strain energy of the element is

$$U_{\text{element}} = \frac{\tau^2 t\,ds\,dx}{2G}.$$

From Eq. (4.26), the shear stress $\tau = T/2At$. Using this expression, the strain energy of the element is

$$U_{\text{element}} = \frac{T^2\,ds\,dx}{8A^2Gt}.$$

By integrating this result with respect to x over the length of the tube, we obtain the strain energy of a strip element with circumferential dimension ds (Fig. 4.39(b)):

$$U_{\text{strip}} = \int_0^L \frac{T^2\,ds\,dx}{8A^2Gt} = \frac{T^2L\,ds}{8A^2Gt}.$$

To obtain the total strain energy of the tube, we must integrate this expression with respect to s over the tube's circumference:

Fig. 4.39 Integrating to
determine the strain energy
in the tube

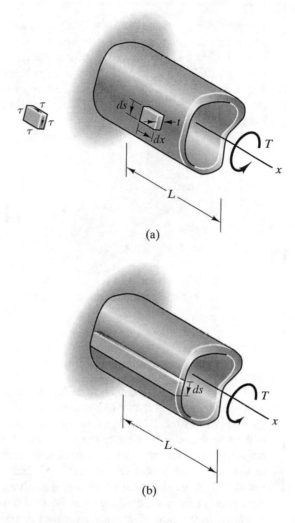

$$U = \frac{T^2 L}{8 A^2 G} \int_s \frac{ds}{t}.$$

We can now equate the strain energy stored within the material of the tube to the work done on the bar in twisting it, given by Eq. (4.27), and solve for the angle of twist:

$$\phi = \frac{TL}{4 A^2 G} \int_s \frac{ds}{t}. \tag{4.28}$$

To evaluate the integral in this expression, the dependence of the tube's wall thickness t on the circumferential coordinate s must be specified.

Example 4.7 Torsion of a Thin-Walled Tube

The prismatic tube in Fig. 4.40 consists of material with shear modulus $G = 28$ GPa and has the cross section shown. The straight part of the tube wall has a thickness of 4 mm, and the semicircular part has a thickness of 2 mm. The tube is subjected to a torque $T = 800$ N-m at the end. Determine (a) the magnitude of the maximum shear stress in the tube and (b) the angle of twist.

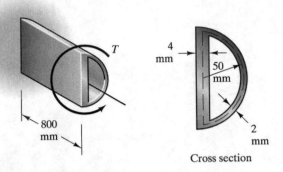

Cross section

Fig. 4.40

Strategy

(a) To determine the shear flow from Eq. (4.26), we must calculate the area A, which is the area within the dashed centerline of the tube wall. Once we know the shear flow, we can use Eq. (4.25) to determine the values of the shear stress in the straight and semicircular parts of the wall. The shear stress is inversely proportional to the wall thickness, so clearly the maximum shear stress occurs in the semicircular part of the wall. (b) The angle of twist is given by Eq. (4.28). Because the wall thickness is constant in the straight and semicircular parts of the wall, the integral in Eq. (4.28) can easily be evaluated if we express it as the sum of integrals over the two parts.

Solution

(a) The area A (Fig. (a)) is $\pi(0.05 \text{ m})^2/2 = 0.00393 \text{ m}^2$, so the shear flow is

$$f = \frac{T}{2A} = \frac{800 \text{ N-m}}{(2)(0.00393 \text{ m}^2)} = 102{,}000 \text{ N/m}.$$

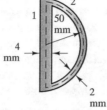

(a) Determining the area A. (b) Parts 1 and 2 of the tube wall

Let the straight vertical part of the tube's wall be called part 1, and let the semicircular part be called part 2 (Fig. (b)). The magnitude of the shear stress in part 1 is

$$\tau_1 = \frac{f}{t_1} = \frac{102{,}000 \text{ N-m}}{0.004 \text{ m}} = 25.5 \text{ MPa,}$$

and the magnitude of the shear stress in part 2 is

$$\tau_2 = \frac{f}{t_2} = \frac{102{,}000 \text{ N-m}}{0.002 \text{ m}} = 50.9 \text{ MPa.}$$

The straight part of the wall is twice as thick, so the magnitude of the shear stress in the straight part is one-half this value.

(b) The integral in Eq. (4.28) is

$$\int_s \frac{ds}{t} = \int_{s_1} \frac{ds}{t} + \int_{s_2} \frac{ds}{t}$$

$$= \frac{1}{t_1} \int_{s_1} ds + \frac{1}{t_2} \int_{s_2} ds$$

$$= \frac{s_1}{t_1} + \frac{s_2}{t_2}$$

$$= \frac{0.1 \text{ m}}{0.004 \text{ m}} + \frac{\pi(0.05 \text{ m})}{0.002 \text{ m}}$$

$$= 103.5.$$

The angle of twist of the bar is

$$\phi = \frac{TL}{4A^2G}\int_s \frac{ds}{t} = \frac{(800 \text{ N-m})(0.8 \text{ m})(103.5)}{(4)(0.00393 \text{ m}^2)^2(28\text{E}9 \text{ N/m}^2)} = 0.0384 \text{ rad},$$

which is 2.20°.

Discussion
When the cross section of a thin-walled tube subjected to torsion varies in thickness, it is much more common for the thickness to be constant over portions of the wall than for the thickness to vary continuously. When that is the case, the integral in Eq. (4.28) can be determined easily by expressing it as the sum of integrals over each portion of the wall for which the thickness is constant, as we demonstrate in this example.

Problems

4.83 The prismatic tube consists of material with shear modulus $G = 28$ GPa and has the cross section shown. The straight part of the tube wall has a thickness of 4 mm, and the semicircular part has a thickness of 2 mm. The tube is subjected to a torque $T = 800$ N-m at the end. What is the magnitude of the maximum shear stress in the tube?

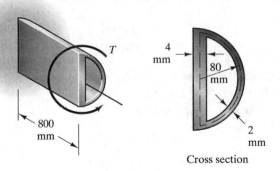

Cross section

Problems 4.83–4.84

4.84 The prismatic tube consists of material with shear modulus $G = 28$ GPa and has the cross section shown. The straight part of the tube wall has a thickness of

4 mm, and the semicircular part has a thickness of 2 mm. The tube is subjected to a torque $T = 800$ N-m at the end. What is the resulting angle of twist of the tube?

4.85 A thin-walled tube has wall thickness $t = \frac{1}{16}$ in and is subjected to a torque $T = 12$ in - kip. Determine the average shear stress in the tube if its cross-sectional shape is **(a)** circular with radius $R = 2$ in; **(b)** square with the same wall length as the circular cross section. (Notice that the weight of the tube is the same in the two cases.)

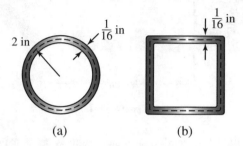

(a)   (b)

Problem 4.85

4.86 A thin-walled tube 4 m in length is subjected to an 8-kN-m torque. Determine the magnitude of the shear stress in the tube if it has the cross section (a) and if it has the cross section (b). The cross sections (a) and (b) have the same circumferential length.

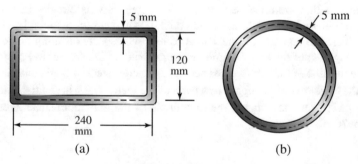

(a)   (b)

Problems 4.86–4.87

4.87 A thin-walled tube 4 m in length is subjected to an 8-kN-m torque. The shear modulus of the material is 80 GPa. Determine the angle of twist of the tube if it has the cross section (a) and if it has the cross section (b). The cross sections (a) and (b) have the same circumferential length.

4.88 A steel tube has the cross section shown. If it is subjected to an axial torque $T = 2$ kN-m, what is the magnitude of the average shear stress in the tube?

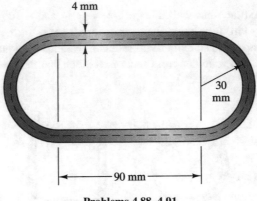

4 mm

30 mm

90 mm

Problems 4.88–4.91

4.89 Suppose that the cross section of a steel tube is circular instead of the shape shown but has the same wall thickness and circumferential length. If it is subjected to an axial torque $T = 2$ kN-m, what is the average shear stress in the tube? Compare your answer to the answer to Problem 4.88.

4.90 A 2-m steel tube has the cross section shown. The shear modulus of the steel is 80 GPa. If it is subjected to an axial torque $T = 2$ kN-m, what is the resulting angle of twist (in degrees) of the tube?

4.91 Suppose that the cross section of a 2-m steel tube is circular instead of the shape shown but has the same wall thickness and circumferential length. If it is subjected to an axial torque $T = 2$ kN-m, what is the resulting angle of twist (in degrees) of the tube? Compare your answer to the answer to Problem 4.90.

4.92 The midline of the tube's cross section is an equilateral triangle with 2-in sides. If the thickness of the upper two parts of the wall $t = \frac{1}{16}$ in and the axial torque $T = 1200$ in · lb, what is the magnitude of the maximum shear stress in the tube?

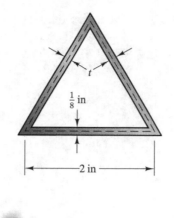

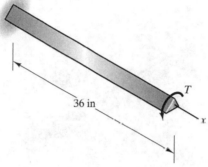

Problems 4.92–4.94

4.93 The midline of the tube's cross section is an equilateral triangle with 2-in sides. The thickness of the upper two parts of the wall is $t = \frac{1}{16}$ in. The shear modulus of the material is 5.8E6 psi. If the axial torque $T = 1200$ in - lb, what is the resulting angle of twist of the tube (in degrees)?

4.94 The midline of the tube's cross section is an equilateral triangle with 2-in sides. The shear modulus of the material is 5.8E6 psi. If the tube is subjected to an axial torque $T = 1000$ in - lb and a design requirement is that the average shear stress in each wall must not exceed 3600 psi, what is the minimum required value of the thickness t of the upper two parts of the wall? If t has the minimum value, what angle of twist (in degrees) is caused by the 1000-in-lb torque?

4.95 The midline of the 36-in tube's cross section is an equilateral triangle with 2-in sides. The thickness of the upper two parts of the wall is $t = \frac{1}{16}$ in. The tube is fixed at the left end and is subjected to a distributed torque $c = 200(x/36)^2$ in - lb/in. What is the average shear stress in the $\frac{1}{16}$-in thick walls at $x = 12$ in?

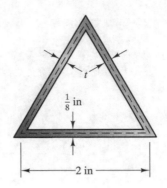

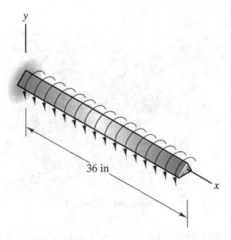

Problems 4.95–4.96

4.96 The midline of the 36-in tube's cross section is an equilateral triangle with 2-in sides. The thickness of the upper two parts of the wall is $t = \frac{1}{16}$ in. The tube is fixed at the left end and is subjected to a distributed torque $c = 200(x/36)^2$ in - lb/in. The shear modulus of the material is $G = 5.8\mathrm{E}6$ psi. What is the tube's angle of twist in degrees?

4.97 An engineering trainee is assigned to measure the cross-sectional area A of a prismatic conduit made of 2014-T6 aluminum. She measures the uniform wall thickness and determines it to be $t = 2$ mm, and measures the circumferential length of the wall and determines it to be 260 mm. She then takes a 1-m long section of the conduit, fixes one end, and applies a 1-kN-m axial torque to the other end. Measuring the resulting angle of twist, she finds it to be 5°. What is the conduit's cross-sectional area?

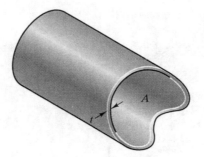

Problems 4.97–4.99

4.98 What average shear stress is exerted on the material as a result of the test described in Problem 4.97?

4.99 The cross-sectional area of the conduit is $A = 0.004$ m^2, and its wall thickness is $t = 3$ mm. It will safely support an average shear stress in the walls of 100 MPa. Based on this criterion, what is the largest axial torque T that can safely be applied to a section of the conduit?

4.100 A tube has the circular cross section shown with $R = 2$ in and $t = 0.1$ in. The tube is fixed at one end and is subjected to an axial torque $T = 4000$ in - lb at the other end. **(a)** Use Eqs. (4.25) and (4.26) to determine the magnitude of the average shear stress in the tube. **(b)** Use Eq. (4.9) to determine the magnitude of the shear stress in the tube at $r = R$.

Problems 4.100–4.101

4.101∗ A tube with the circular cross section shown is subjected to a torque T. By using Eqs. (4.25) and (4.26) to determine the average shear stress τ_{av} in the tube and using Eq. (4.9) to determine the exact shear stress τ_{ex} in the tube at $r = R$, show that

$$\frac{\tau_{av}}{\tau_{ex}} = 1 + \frac{1}{4}\left(\frac{t}{R}\right)^2.$$

4.102 The 800-mm tube with the cross section shown is fixed at both ends and is subjected to an axial torque $T = 800$ N-m at 300 mm from the left end. What is the magnitude of the maximum shear stress in the tube?

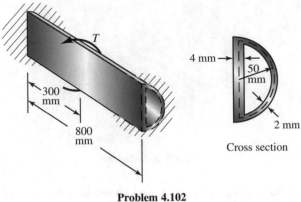

Cross section

Problem 4.102

4.7 Design Issues

The results we have presented in this chapter can be applied to the design of bars to support torsional loads.

4.7.1 Cross Sections

Except for our analysis of thin-walled cross sections in Sect. 4.6, we have restricted our study of torsion to bars with circular cross sections. Our primary reason for doing so is that consideration of other cross-sectional shapes requires advanced analytical methods, numerical solutions, or the use of empirical information. But this is not as serious a limitation for a discussion of design as it might appear. The reason is that in normal circumstances, a circular cross section is optimal for a bar supporting an axial torsional load. Figure 4.41 compares the ratios of the maximum shear stresses resulting from a given torque applied to bars with circular, elliptical, and square cross sections of equal area.

Furthermore, a hollow circular cross section is usually preferable to a solid one. In Fig. 4.42 a torque T is applied to bars with solid and hollow cross sections. The maximum shear stress in the solid bar is

Fig. 4.41 Ratio of the maximum shear stress to the value for a circular cross section of equal area

1.00   1.32   1.39

Fig. 4.42 Applying the same torque T to bars with solid and hollow circular cross sections

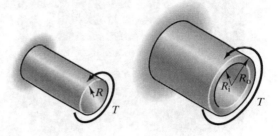

Fig. 4.43 Ratio of the maximum stress in a hollow circular bar to that in a solid circular bar

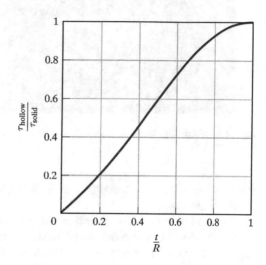

$$\tau_{\text{solid}} = \frac{TR}{(\pi/2)R^4}, \tag{4.29}$$

and the maximum shear stress in the hollow bar is

$$\tau_{\text{hollow}} = \frac{TR_{\text{o}}}{(\pi/2)\left(R_{\text{o}}^4 - R_{\text{i}}^4\right)}. \tag{4.30}$$

If the bars have equal cross-sectional areas, $\pi R^2 = \pi R_{\text{o}}^2 - \pi R_{\text{i}}^2$, and we denote the wall thickness of the hollow bar by $t = R_{\text{o}} - R_{\text{i}}$, we can write the ratio of the maximum shear stresses as

$$\frac{\tau_{\text{hollow}}}{\tau_{\text{solid}}} = \frac{\left[1 + \left(\frac{t}{R}\right)^2\right]\frac{t}{R}}{1 + \left(\frac{t}{R}\right)^4}.$$

From this relationship, plotted in Fig. 4.43, the case for using hollow bars to support torsional loads is clear. For a given torque, the maximum shear stress in a hollow bar is smaller than in a solid bar of the same cross-sectional area (which

Fig. 4.44 Applying torque to a bar with a thin wall (here an aluminum drink can) can cause it to fail by buckling

means a bar of the same weight). The angle of twist for a given torque is also smaller for the hollow bar.

The maximum stress in the hollow bar continues to decrease as the wall thickness decreases. But the wall must not be made too thin, as can readily be illustrated by applying torque to an aluminum soft drink can (Fig. 4.44). The wall of the can forms wrinkles and fails by buckling or geometric instability. The tendency of a hollow bar to buckle is sometimes prevented by putting some type of filler within it. Because of all these considerations, bars designed to support axial torsional loads usually have hollow circular cross sections.

4.7.2 Allowable Stress

We have seen that in a circular bar subjected to torsion there is no plane on which the shear stress is larger in magnitude than the values given by Eq. (4.29) or (4.30) and the maximum magnitude of the shear stress occurs at the outer surface of the bar. A criterion that can be used for the design of such bars is to insure that the maximum magnitude of the shear stress does not exceed a shear yield stress τ_Y, or some specified fraction of the shear yield stress, the *allowable shear stress* τ_{allow}. The factor of safety is defined by

$$\text{FS} = \frac{\tau_Y}{\tau_{allow}}. \tag{4.31}$$

Because the maximum shear stress in a tensile test is one-half of the maximum normal stress, we will assume here that the shear yield stress τ_Y is given in terms of the normal yield stress σ_Y measured in a tensile test by $\tau_Y = \frac{1}{2}\sigma_Y$. By doing so, we can express the factor of safety in terms of the allowable shear stress and the normal yield stress given in Appendix B:

$$FS = \frac{\sigma_Y}{2\tau_{\text{allow}}}. \tag{4.32}$$

We discuss failure criteria in more detail in Chap. 12. Some of the constraints that the design engineer must typically balance against the desire for a conservatively large factor of safety are discussed in Sect. 3.6.

Study Questions

1. Why is a hollow circular cross section advantageous for bars designed to support torsional loads?
2. What can happen if the walls of the hollow circular cross section of a bar subjected to torque are made too thin?
3. Why can the factor of safety of a bar subjected to torsion be defined in terms of the yield stress σ_Y of the material measured in a tensile test?

Example 4.8 Design of a Drive Shaft

The power (work per unit time) transmitted by a rotating shaft is $P = T\omega$, where T is the axial torque and ω is the shaft's angular velocity in radians per second. The maximum torque produced by the engine of the car in Fig. 4.45 occurs at 4750 rpm, when the engine is generating 286 horsepower (hp). Design a drive shaft for the car that is constructed of steel with yield stress $\sigma_Y = 80,000$ psi and has a factor of safety FS = 2. (Although the shaft is rotating, assume that the maximum shear stress can be adequately approximated by assuming that the material is in equilibrium.)

Fig. 4.45 (*Photograph from iStock by Sjo*)

Strategy

Knowing the power transmitted by the shaft and its angular velocity, we can determine the torque. We can use Eq. (4.32) to determine the allowable maximum shear stress and then use Eq. (4.29) or (4.30) to determine the dimensions of the cross section.

Solution

One horsepower is 550 ft - lb/s. Determining the torque from the expression

$$P = T\omega :$$

$$(286)(550) \text{ ft-lb/s} = T\left[4750\left(\frac{2\pi}{60}\right) \text{ rad/s}\right],$$

we obtain $T = 316$ ft - lb = 3790 in - lb. The allowable shear stress is

$$\tau_{allow} = \frac{\sigma_Y}{2FS} = \frac{80,000 \text{ psi}}{2(2)} = 20,000 \text{ psi.}$$

If we use a solid drive shaft, the radius of the shaft is determined from Eq. (4.29) with $\tau_{solid} = \tau_{allow}$:

$$\tau_{allow} = \frac{T}{\frac{\pi}{2}R^3} :$$

$$20,000 \text{ lb/in}^2 = \frac{3790 \text{ in-lb}}{\frac{\pi}{2}R^3}.$$

Solving, we obtain $R = 0.494$ in.

Suppose that we use a hollow drive shaft instead. For stability, we will require that the wall thickness of the hollow shaft be 30% of its outer radius. The inner and outer radii must satisfy Eq. (4.30) with $\tau_{hollow} = \tau_{allow}$:

$$\tau_{allow} = \frac{TR_o}{\frac{\pi}{2}(R_o^4 - R_i^4)} :$$

$$20,000 \text{ lb/in}^2 = \frac{(3790 \text{ in-lb})R_o}{\frac{\pi}{2}(R_o^4 - R_i^4)}.$$

Setting $R_i = 0.7R_o$ in this expression and solving, we obtain $R_o = 0.542$ in and $R_i = 0.379$ in.

Discussion

Using 15 slug/ft^3 as the density of steel, the solid drive shaft weighs 2.57 lb/ft, while the hollow drive shaft weighs 1.58 lb/ft. The hollow shaft also has superior dynamic (vibrational) properties.

Problems

4.103 Suppose you are designing a bar with a solid circular cross section that is to support an 800 N-m torsional load. The bar is to be made of 6061-T6 aluminum alloy (see Appendix B), and you want the factor of safety to be FS = 2. Based on these criteria, what should the bar's diameter be?

800 N–m

Problem 4.103

4.104 A hollow circular tube that is to support an airplane's aileron may be subjected to service torques as large as 520 N-m. It is to have a 20-mm outside radius and 14-mm inside radius. Choose an aluminum alloy from Appendix B so that the tube has a factor of safety of at least 4.

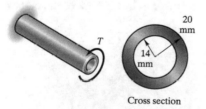

20 mm

14 mm

T

Cross section

Problem 4.104

4.105 Suppose you are designing a bar with a solid circular cross section that is to support a 6500-in-lb torsional load. The bar is to be made of ASTM-A572 structural steel (see Appendix B), and you want the factor of safety to be FS = 1.5. Based on these criteria, what should the bar's diameter be?

4.106 A bar with a solid circular cross section is to support a 1200 N-m torsional load. Choose a material from Appendix B and determine the bar's diameter so that the factor of safety is FS = 3.

4.107 A bar with a hollow circular cross section is to support a 1200 N-m torsional load. Choose a material from Appendix B and determine the bar's inner and outer radii so that the inner radius is one-half of the outer radius and the factor of safety is FS = 3.

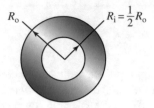

R_o $R_i = \frac{1}{2} R_o$

Problem 4.107

4.108 A bar with a solid circular cross section is to support an 8000-in-lb torsional load. Choose a material from Appendix B, and determine the bar's diameter so that the factor of safety is FS = 2.5.

8000 in–lb

Problem 4.108

4.109 A bar with a hollow circular cross section is to support an 8000-in-lb torsional load. Choose a material from Appendix B, and determine the bar's inner and outer radii so that the inner radius is 60% of the outer radius and the factor of safety is FS = 3.

R_i R_o

8000 in–lb

Problem 4.109

4.110 After being equipped with a turbocharger, the engine of the car in Example 4.8 produces its maximum torque at 5000 rpm when the engine is generating 320 horsepower. Choose a material from Appendix B and design a drive shaft for the car that has a solid circular cross section and a factor of safety FS = 3.

4.111 A Catalina 30 sailboat has an 11-hp Universal 5411 diesel engine. Assume that the propeller shaft transmits 11 hp at 1200 rpm. If the shaft is to have a solid circular cross section, be made of soft manganese bronze, and you want it to have a factor of safety FS = 3, what should its diameter be? (See Example 4.8.)

4.112 The shaft that connects the turbine blades of the hydroelectric power unit to the generator transmits 160 MW of power at 150 rpm. Choose a material from Appendix B and design a shaft with a solid circular cross section that has a factor of safety FS = 2. (See Example 4.8.)

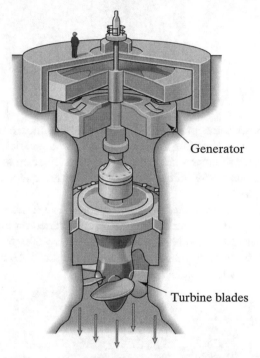

Generator

Turbine blades

Problem 4.112

4.113 At peak power, the propeller of the wind generator produces 3 MW of power at 34 rpm. Based on this criterion, choose a material from Appendix B and design a propeller shaft that has a solid circular cross section and a factor of safety FS = 2 (see Example 4.8). Notice that this analysis does not account for the load exerted on the shaft by the propeller's weight.

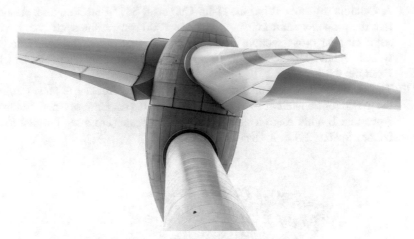

Problems 4.113–4.114

4.114 At peak power, the propeller of the wind generator produces 3 MW of power at 34 rpm. Based on this criterion, choose a material from Appendix B, and design a propeller shaft, with a hollow circular cross section in which the inner radius is 75% of the outer radius, that has a factor of safety FS = 2 (see Example 4.8).

4.115 The sum of the radii of the two gears is $r_B + r_C = 160$ mm. Suppose that you want to design the system so that the torque exerted on the fixed support at D by the bar CD is 200 N-m when a torque $T_A = 600$ N-m is applied at A. Determine the radii of the gears and design the shafts AB and CD so that the system has a factor of safety FS = 2.5.

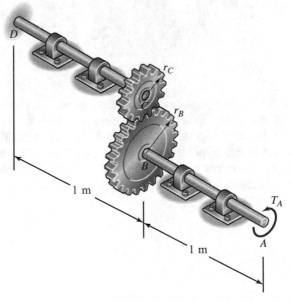

Problem 4.115

Chapter Summary

Pure Shear Stress

In Fig. (a), a cube of material is subjected to a state of pure shear stress. The shear strain is related to the shear stress by

$$\gamma = \frac{1}{G}\tau. \qquad (4.1)$$

The shear modulus G is given in terms of the modulus of elasticity E and Poisson's ratio ν by

$$G = \frac{E}{2(1+v)}. \qquad (4.2)$$

Stresses on Oblique Planes

The normal and shear stresses on a plane $\mathcal{P}$ oriented as shown in Fig. (b) are

$$\sigma_\theta = 2\tau \sin\theta \cos\theta, \qquad (4.3)$$

$$\tau_\theta = \tau\left(\cos^2\theta - \sin^2\theta\right). \qquad (4.4)$$

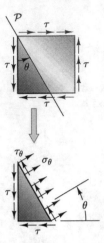

There is no value of θ for which the shear stress is larger in magnitude than τ. The maximum tensile and compressive stresses occur at $\theta = 45°$ where $\sigma_\theta = \tau$ and at $\theta = 135°$ where $\sigma_\theta = -\tau$.

Torsion of Prismatic Circular Bars

Consider a cylindrical bar of length L and radius R subjected to an axial torque T (Fig. (c)). The resulting angle of twist of the end of the bar is

$$\phi = \frac{TL}{GJ},\tag{4.7}$$

where

$$J = \frac{\pi}{2}R^4\tag{4.10}$$

is the polar moment of inertia of the bar's cross-sectional area about its axis. The shear stress on the plane shown at a distance r from the bar's axis is

$$\tau = \frac{Tr}{J}.\tag{4.9}$$

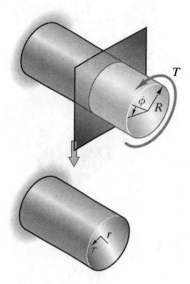

Equations (4.7) and (4.9) also apply to a bar with a hollow circular cross section. In that case the polar moment of inertia is given by

$$J = \frac{\pi}{2}\left(R_o^4 - R_i^4\right),$$

$$(4.11)$$

where R_i and R_o are the bar's inner and outer radii.

Statically Indeterminate Problems

Solutions to problems involving torsionally loaded bars in which the number of unknown reactions exceeds the number of independent equilibrium equations involve three elements: (1) relations between the torques in bars and their angles of twist, (2) compatibility relations imposed on the angles of twist by the geometry of the problem, and (3) equilibrium.

Bars with Gradually Varying Cross Sections

If the polar moment of inertia is a gradually varying function of axial position $J(x)$ (Fig. (d)), the stress distribution on a plane perpendicular to the bar's axis at axial position x can be approximated by

$$\tau = \frac{Tr}{J(x)}. \qquad (4.17)$$

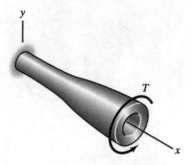

The angle of twist of the bar is

$$\phi = \int_0^L \frac{T\,dx}{GJ(x)}. \qquad (4.18)$$

Distributed Torsional Loads

A distributed torsional load on a bar can be described by a function c defined such that the axial torque on each element dx of the bar is $c\,dx$ (Fig. (e)).

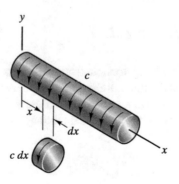

Torsion of an Elastic-Perfectly Plastic Circular Bar

The elastic-perfectly plastic model of shear stress-shear strain behavior is shown in Fig. (f). If the shear strain at the bar's outer surface does not exceed the value γ_Y corresponding to the yield stress, the shear stress distribution is given by Eq. (4.9) and the angle of twist by Eq. (4.7). If the shear strain does exceed the value corresponding to the yield stress at some radial position r_Y, the shear stress distribution is as shown in Fig. (g). The torque T and radial distance r_Y satisfy the equilibrium equation:

$$T = \pi \tau_Y R_o^3 \left[\frac{2}{3} - \frac{1}{6} \left(\frac{r_Y}{R_o} \right)^3 - \frac{1}{2} \left(\frac{R_i}{R_o} \right)^4 \frac{R_o}{r_Y} \right].$$ (4.22)

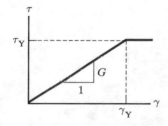

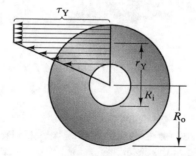

The upper bound of the torque the bar will support is obtained by setting $r_Y = R_i$ in Eq. (4.22). The bar's angle of twist is

$$\phi = \frac{\gamma_Y L}{r_Y} = \frac{\tau_Y L}{r_Y G}$$ (4.23)

Torsion of Thin-Walled Tubes

Consider a prismatic tube whose wall thickness t is small compared to the lateral dimensions of the tube (Fig. (h)). The *shear flow f* is given by

$$f = \tau t = \frac{T}{2A}, \tag{4.26}$$

where A is the cross-sectional area of the tube (Fig. (i)). The angle of twist of the tube is

$$\phi = \frac{TL}{4A^2 G} \int_s \frac{ds}{t}, \tag{4.28}$$

where s measures distance along the circumference of the wall relative to some reference point.

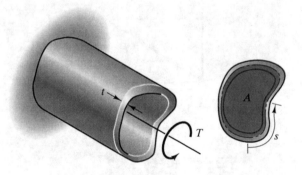

Design Issues

A criterion that can be used for the design of bars subjected to torsion is to insure that the maximum magnitude of the shear stress does not exceed a shear yield stress τ_Y, or some specified fraction of the shear yield stress, the *allowable shear stress* τ_{allow}. The factor of safety is defined by

$$\text{FS} = \frac{\tau_Y}{\tau_{\text{allow}}}. \tag{4.31}$$

If it is assumed that the shear yield stress is given in terms of the normal yield stress σ_Y measured in a tensile test by $\tau_Y = \frac{1}{2}\sigma_Y$, the factor of safety can be expressed in terms of the allowable shear stress and the normal yield stress given in Appendix B:

$$FS = \frac{\sigma_Y}{2\tau_{allow}}.$$ (4.32)

Review Problems

4.116 The prismatic bar is 12 in long and has a circular cross section with 1-in diameter. In a test to determine its elastic properties, the bar is subjected to 113-kip axial loads. After the loads are applied, the bar's length is 12.054 in and its diameter is 0.9986 in. What is the shear modulus of the material?

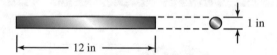

Problem 4.116

4.117 The cube of material is subjected to a pure shear stress τ. If the normal stress on the plane $\mathcal{P}$ is -20 ksi, what is τ?

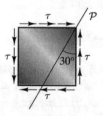

Problems 4.117–4.118

4.118 The cube of material is subjected to a pure shear stress τ. The shear modulus of the material is $G = 4E6$ psi. If the normal stress on the plane $\mathcal{P}$ is -12 ksi, what is the shear strain of the cube?

4.119 The cross section of a portion of a propeller's shaft is shown. If the engine exerts a 2200 N-m torque on the shaft, what is the magnitude of the maximum shear stress in the shaft?

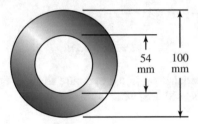

| | 54 mm | 100 mm |

Problems 4.119–4.120 (*Photograph from iStock by vm*)

4.120 The cross section of a portion of the propeller's shaft is shown. The shaft is made of material with shear modulus $G = 220$ GPa. The engine exerts a 2200 N-m torque on the shaft. If the length of the portion of shaft is 0.6 m, what is its angle of twist?

4.121 The lengths $L_{AB} = L_{BC} = 200$ mm and $L_{CD} = 240$ mm. The diameter of parts AB and CD of the bar is 25 mm and the diameter of part BC is 50 mm. The shear modulus of the material is $G = 80$ GPa. If the torque $T = 2.2$ kN-m, determine the magnitude of the angle of twist of the right end of the bar relative to the wall in degrees.

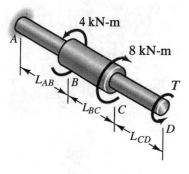

Problems 4.121–4.122

4.122 The lengths $L_{AB} = L_{BC} = 200$ mm and $L_{CD} = 240$ mm. The diameter of parts AB and CD of the bar is 25 mm and the diameter of part BC is 50 mm. The shear modulus of the material is $G = 80$ GPa. What value of the torque T would cause the angle of twist of the right end of the bar relative to the wall to be zero?

4.123 The radius $R = 200$ mm. The infinitesimal element is at the surface of the bar. What are the normal stress and the magnitude of the shear stress on the plane $\mathcal{P}$?

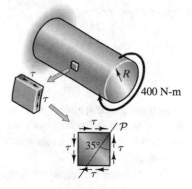

Problem 4.123

4.124 The radius $R = 200$ mm. The infinitesimal element is at the surface of the bar. What are the normal stress and the magnitude of the shear stress on the plane $\mathcal{P}$?

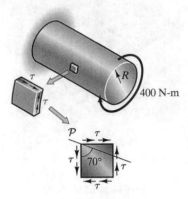

Problem 4.124

4.125 Part *AB* of the bar has a solid circular cross section, and part *BC* has a hollow circular cross section. The bar is fixed at both ends and a 150-in-kip axial torque is applied at *B*. Determine the magnitudes of the torques exerted on the bar by the walls at *A* and *C*.

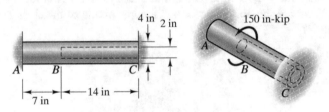

Problems 4.125–4.127

4.126 Part *AB* of the bar has a solid circular cross section and part *BC* has a hollow circular cross section. The bar is fixed at both ends and a 150-in-kip axial torque is applied at *B*. Determine the magnitudes of the maximum shear stresses in parts *AB* and *BC*.

4.127* Part *AB* of the bar has a solid circular cross section and part *BC* has a hollow circular cross section. The bar is fixed at both ends and a 150-in-kip axial torque is applied at *B*. Suppose that you want to decrease the weight of the bar by increasing the inside diameter of part *B*. The bar is made of material that will safely support a pure shear stress of 10 ksi. Based on this criterion, what is the largest safe value of the inside diameter?

4.128 The aluminum alloy bar has a circular cross section with 20-mm diameter and a shear modulus of 28 GPa. What is the magnitude of the maximum shear stress in the bar due to the uniformly distributed torque?

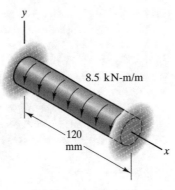

Problems 4.128–4.129

4.129 The aluminum alloy bar has a circular cross section with 20-mm diameter and a shear modulus of 28 GPa. What is the magnitude of the bar's angle of twist (in degrees) relative to the walls at $x = 60$ mm?

Chapter 5
Internal Forces and Moments in Beams

A *beam* is a slender structural member. (The word originally meant either a structural member or a *tree* in the Germanic language that became modern English, because the beams used in constructing buildings and ships were hewn from trees. The word for a tree in modern German is still *baum*.) Beams are the most common structural elements and make up the supporting structures of cars, aircraft, and buildings. We now begin the task of determining the states of stress and strain in beams by analyzing their internal forces and moments.

5.1   Axial Force, Shear Force, and Bending Moment

Consider the beam subjected to an external load and reactions in Fig. 5.1. To determine the internal forces and moments within the beam, in Fig. 5.2(a) we cut the beam by a plane perpendicular to the beam's axis and isolate part of it. The isolated part cannot be in equilibrium unless it is subjected to some system of forces and moments at the plane where it joins the other part of the beam. We know from statics that any system of forces and moments can be represented by an equivalent system consisting of a force acting at a given point and a couple. If the system of external loads and reactions on a beam is two-dimensional, we can represent the internal forces and moments by an equivalent system consisting of two components of force and a couple as shown in Fig. 5.2(b). The *axial force P* is parallel to the beam's axis. The force component V normal to the beam's axis is called the *shear force*, and the couple M is called the *bending moment*.

The directions of the axial force, shear force, and bending moment in Fig. 5.2(b) are the established definitions of the positive directions of these quantities. A positive axial force P subjects the beam to tension. A positive shear force V tends to rotate the longitudinal axis of the beam clockwise (Fig. 5.3(a)). Bending moments are defined to be positive when they tend to bend the axis of the beam in the positive y-axis direction (Fig. 5.3(b)).

© Springer Nature Switzerland AG 2020
A. Bedford, K. M. Liechti, *Mechanics of Materials*,
https://doi.org/10.1007/978-3-030-22082-2_5

Fig. 5.1 Beam subjected to
a load and reactions

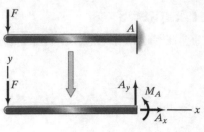

Fig. 5.2 (a) Isolating part
of the beam. (b) Axial force,
shear force, and bending
moment

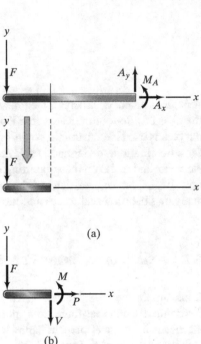

In Chaps. 6 and 9, we show that knowledge of the internal forces and moment in a
beam is essential for evaluating the states of stress and deformations resulting from a
given system of loads. Determining the internal forces and moment at a particular
cross section of a beam typically involves three steps:

1. Draw the free-body diagram of the entire beam and determine the reactions at its
   supports.
2. Cut the beam where you wish to determine the internal forces and moment and
   draw the free-body diagram of one of the resulting parts. You can choose the part
   with the simplest free-body diagram. If your cut divides a distributed load, don't
   represent the distributed load by an equivalent force until *after* you have obtained
   your free-body diagram.
3. Use the equilibrium equations to determine P, V, and M.

Fig. 5.3 (**a**) Positive shear forces tend to rotate the axis of the beam clockwise. (**b**) Positive bending moments tend to bend the axis of the beam in the positive y-axis direction

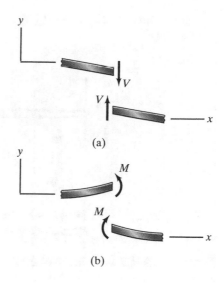

Study Questions

1. What are the axial force, shear force, and bending moment?
2. What are the established definitions of the positive directions of the axial force, shear force, and bending moment?

Example 5.1 Determining P, V, and M at a Given Cross Section
For the beam in Fig. 5.4, determine the internal forces and moment at C.

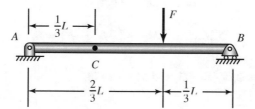

Fig. 5.4

Strategy
We must determine the reactions exerted on the beam by its supports. We can then obtain a free-body diagram by cutting the beam by a plane at C and use equilibrium to determine the internal forces and moment.

Solution
Determine the external forces and moments The first step is to draw the free-body diagram of the entire beam and determine the reactions at its supports. We simply show the results of this step in Fig. (a).

Draw the free-body diagram of part of the beam We cut the beam at C (Fig. (a)) and draw the free-body diagram of the left part, including the internal forces and moment P_C, V_C, and M_C in their defined positive directions (Fig. (b)).

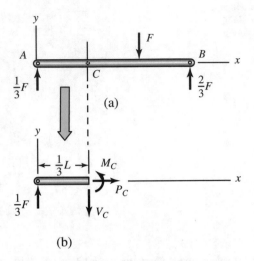

(a) Free-body diagram of the beam. (b) Free-body diagram of the part of the beam to the left of the plane through point C

Apply the equilibrium equations From the equilibrium equations

$$\Sigma F_x = P_C = 0,$$

$$\Sigma F_y = \frac{1}{3}F - V_C = 0,$$

$$\Sigma M_{\text{point } C} = M_C - \left(\frac{1}{3}L\right)\left(\frac{1}{3}F\right) = 0,$$

we obtain $P_C = 0$, $V_C = F/3$, and $M_C = LF/9$.

Discussion

We can check these results by determining them from the free-body diagram of the part of the beam to the right of point C (Fig. (c)). The equilibrium equations are

$$\Sigma F_x = -P_C = 0,$$

$$\Sigma F_y = V_C - F + \frac{2}{3}F = 0,$$

$$\Sigma M_{\text{point } C} = -M_C - \left(\frac{1}{3}L\right)(F) + \left(\frac{2}{3}L\right)\left(\frac{2}{3}F\right) = 0,$$

which confirm that $P_C = 0$, $V_C = F/3$, and $M_C = LF/9$.

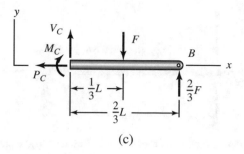

(c)

(c) Free-body diagram of the part of the beam to the right of the plane through point C

Example 5.2 Determining P, V, and M at Given Cross Sections
For the beam in Fig. 5.5, determine the internal forces and moment at B
and at C.

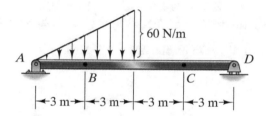

Fig. 5.5

Strategy
We must first determine the reactions at the supports. By cutting the beam at
B and using equilibrium, we can determine the internal forces and moment at
B. Then we can apply the same procedure to point C.

Solution
Determine the external forces and moments   We draw the free-body
diagram of the beam and represent the distributed load by an equivalent
force in Fig. (a). The equilibrium equations are

$$\Sigma F_x = A_x = 0,$$
$$\Sigma F_y = A_y - 180\,\text{N} + D = 0,$$
$$\Sigma M_{\text{point}\,A} = (12\,\text{m})D - (4\,\text{m})(180\,\text{N}) = 0.$$

Solving them, we obtain $A_x = 0$, $A_y = 120\,\text{N}$, and $D = 60\,\text{N}$.

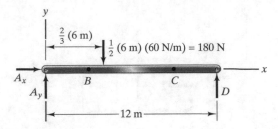

(a) Free-body diagram of the entire beam with the distributed load represented by an equivalent force

Draw the free-body diagram of part of the beam   We cut the beam at B, obtaining the free-body diagram in Fig. (b). Because point B is at the midpoint of the triangular distributed load, the value of the distributed load at B is 30 N/m. By representing the distributed load in Fig. (b) by an equivalent force, we obtain the free-body diagram in Fig. (c). From the equilibrium equations

$$\Sigma F_x = P_B = 0,$$
$$\Sigma F_y = 120\ \text{N} - 45\ \text{N} - V_B = 0,$$
$$\Sigma M_{\text{point }B} = M_B + (1\ \text{m})(45\ \text{N}) - (3\ \text{m})(120\ \text{N}) = 0,$$

we obtain $P_B = 0$, $V_B = 75$ N, and $M_B = 315$ N-m.

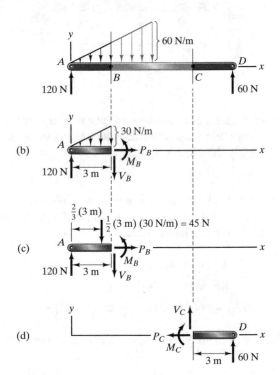

(b, c) Free-body diagrams of the part of the beam to the left of point B. (d) Free-body diagram of the part of the beam to the right of point C

To determine the internal forces and moment at C, we obtain the simplest free-body diagram by isolating the part of the beam to the right of C (Fig. (d)). From the equilibrium equations

$$\Sigma F_x = -P_C = 0,$$
$$\Sigma F_y = V_C + 60 \text{ N} = 0,$$
$$\Sigma M_{\text{point }C} = -M_C + (3 \text{ m})(60 \text{ N}) = 0,$$

we obtain $P_C = 0$, $V_C = -60$ N, and $M_C = 180$ N-m.

Discussion

If you attempt to determine the internal forces and moment at B by cutting the free-body diagram in Fig. (a) at B, you do *not* obtain correct results. (You can confirm that the resulting free-body diagram of the part of the beam to the left of B gives $P_B = 0$, $V_B = 120$ N, and $M_B = 360$ N-m.) The reason is that you do not account properly for the effect of the distributed load on your free-body diagram. You must wait until *after* you have obtained the free-body diagram of part of the beam before representing distributed loads by equivalent forces.

Review Problems

5.1 (a) Draw the free-body diagram of the beam by isolating it from its supports. (b) Determine the reactions at the supports. (c) Determine the internal forces and moment at C by drawing the free-body diagram of the part of the beam to the left of C.

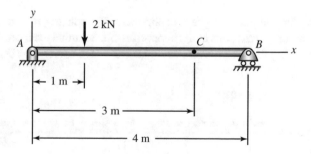

Problems 5.1–5.2

5.2 Determine the internal forces and moment at point C by drawing the free-body diagram of the part of the beam to the right of C.

5.3 (a) Draw the free-body diagram of the beam by isolating it from its support. (b) Determine the reactions at the support. (c) Determine the internal forces and moment at B by drawing the free-body diagram of the part of the beam to the left of B.

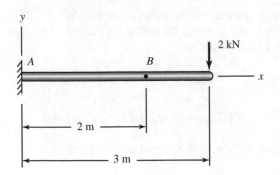

Problems 5.3–5.4

5.4 Determine the internal forces and moment at point B by drawing the free-body diagram of the part of the beam to the right of B.

5.5 Determine the internal forces and moment at B **(a)** if $x = 250$ mm; **(b)** if $x = 750$ mm.

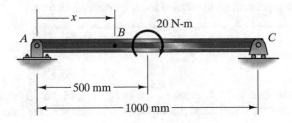

Problem 5.5

5.6 (a) Draw the free-body diagram of the beam by isolating it from its supports. (b) Determine the reactions at the supports. (c) Determine the internal forces and moment at C by drawing the free-body diagram of the part of the beam to the left of C.

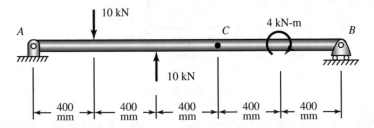

Problems 5.6–5.7

5.7 Determine the internal forces and moment at point C by drawing the free-body diagram of the part of the beam to the right of C.

5.8 Determine the internal forces and moment at A for each loading.

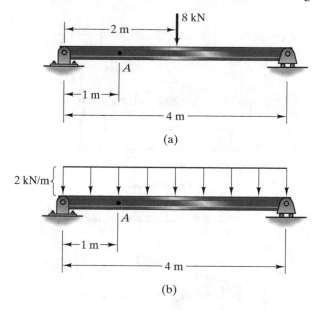

(a)

(b)

Problem 5.8

5.9 Model the ladder rung as a simply supported (pin-supported) beam and assume that the 200-lb load exerted by the person's shoe is uniformly distributed. Determine the internal forces and moment at A.

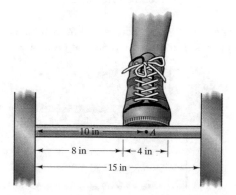

Problems 5.9–5.10

5.10 Model the ladder rung as a simply supported (pin-supported) beam and assume that the 200-lb load exerted by the person's shoe is uniformly distributed. Determine the internal forces and moment at A if the person shifts his foot so that the distance from A to the left edge of the rung is 8 in.

5.11 Show that the internal forces and moment at A are given by

$$P_A = 0, V_A = \left(\frac{L}{6} - \frac{b^2}{2L}\right)w_0, M_A = \left(\frac{bL}{6} - \frac{b^3}{6L}\right)w_0.$$

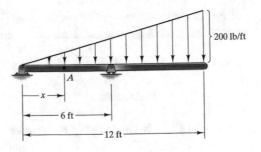

Problem 5.11

5.12 If $x = 3$ ft, what are the internal forces and moment at A?

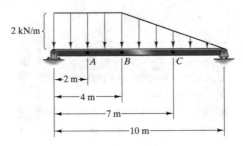

Problems 5.12–5.13

5.13 If $x = 9$ ft, what are the internal forces and moment at A?
5.14 Determine the internal forces and moment at point A.

Problems 5.14–5.16

5.15 Determine the internal forces and moment at point B.
5.16 Determine the internal forces and moment at point C.
5.17 Determine the internal forces and moment at point A of the horizontal member of the frame.

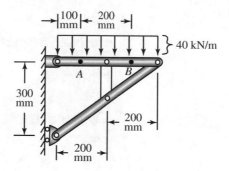

Problems 5.17–5.18

5.18 Determine the internal forces and moment at point B of the horizontal member of the frame.

5.19 Determine the internal forces and moment at point A of the frame.

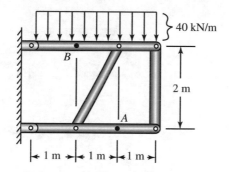

Problems 5.19–5.20

5.20 Determine the internal forces and moment at point B of the frame.

5.2 Shear Force and Bending Moment Diagrams

To determine whether a beam will support a given set of loads, the structural designer must know the state of stress throughout the beam. To evaluate the state of stress, the internal forces and moment must be determined throughout the beam's length. Here we show how the values of P, V, and M can be determined as functions of position along a beam's axis and introduce shear force and bending moment diagrams.

Consider the simply-supported beam in Fig. 5.6(a). Instead of cutting the beam at a specific cross section to determine the internal forces and moment, we cut it at

Fig. 5.6 (a) Beam loaded by a force F and its free-body diagram. (b) Cutting the beam at an arbitrary position between A and B. (c) Cutting the beam at an arbitrary position between B and C

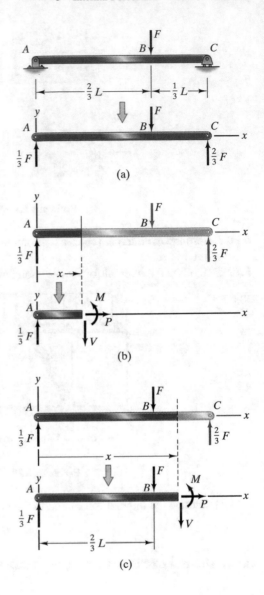

an arbitrary position x between A and B (Fig. 5.6(b)). Applying the equilibrium equations to this free-body diagram, we obtain

$$\left.\begin{array}{l} P = 0 \\[4pt] V = \dfrac{1}{3}F \\[6pt] M = \dfrac{1}{3}Fx \end{array}\right\} \quad 0 < x < \dfrac{2}{3}L.$$

To determine the internal forces and moment for values of x greater than $\frac{2}{3}L$, we obtain a free-body diagram by cutting the beam at an arbitrary position x between B and C (Fig. 5.6(c)). The results are

$$
\left.
\begin{aligned}
P &= 0 \\
V &= -\frac{2}{3}F \\
M &= \frac{2}{3}F(L - x)
\end{aligned}
\right\} \quad \frac{2}{3}L < x < L.
$$

Shear force and bending moment diagrams are simply graphs of V and M, respectively, as functions of x (Fig. 5.7). They permit us to see the changes in the shear force and bending moment that occur along the beam's length as well as their maximum positive and negative values.

Thus, we can determine the distributions of the internal forces and moment in a beam by considering a plane at an arbitrary distance x from the end of the beam and solving for P, V, and M as functions of x. Depending on the complexity of the loading of the beam, it may be necessary to draw several free-body diagrams to determine the distributions over the entire length of the beam. The resulting equations allow us to determine the maximum positive and negative values of the shear force and bending moment and also draw the shear force and bending moment diagrams.

Fig. 5.7 Shear force and bending moment diagrams indicating the maximum positive and negative values of V and M

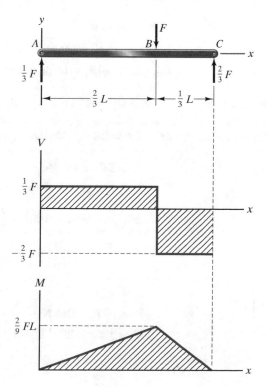

Example 5.3 Shear Force and Bending Moment Diagrams
Draw the shear force and bending moment diagrams for the beam in Fig. 5.8.

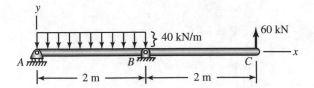

Fig. 5.8

Strategy
By cutting the beam at an arbitrary position between A and B, we can obtain a free-body diagram with which to determine the internal forces and moment for $0 < x < 2$ m. Then by cutting the beam at an arbitrary position between B and C, we can determine the internal forces and moment for $2 < x < 4$ m.

Solution
We begin by drawing the free-body diagram of the entire beam and representing the distributed force by an equivalent force (Fig. (a)). From the equilibrium equations

$$\Sigma F_x = B_x = 0,$$
$$\Sigma F_y = A + B_y - 80 \text{ kN} + 60 \text{ kN} = 0,$$
$$\Sigma M_{\text{point } A} = (2 \text{ m})B_y - (1 \text{ m})(80 \text{ kN}) + (4 \text{ m})(60 \text{ kN}) = 0,$$

we obtain the reactions $A = 100$ kN, $B_x = 0$, and $B_y = -80$ kN. In Fig. (b) we obtain a free-body diagram by cutting the beam at an arbitrary position between A and B. From the equilibrium equations.

$$\Sigma F_x = P = 0,$$
$$\Sigma F_y = 100 - 40x - V = 0,$$
$$\Sigma M_{\text{right end}} = M - 100x + \left(\frac{1}{2}x\right)(40x) = 0,$$

we obtain

$$\left.\begin{array}{l} P = 0 \\ V = 100 - 40x \text{ kN} \\ M = 100x - 20x^2 \text{ kN-m} \end{array}\right\} \quad 0 < x < 2 \text{ m.}$$

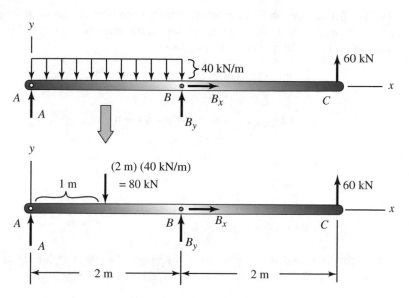

(a) Free-body diagram of the beam representing the distributed load by an equivalent force

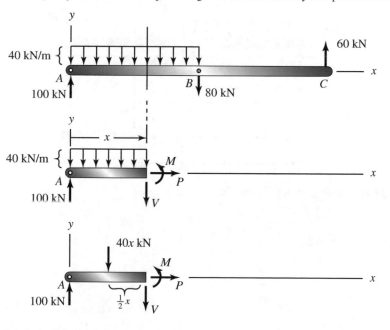

(b) Free-body diagram for $0 < x < 2$ m

In Fig. (c) we cut the beam at an arbitrary position between B and C and draw the free-body diagram of the part of the beam to the right of the cutting plane. From the equilibrium equations

$$\Sigma F_x = -P = 0,$$
$$\Sigma F_y = V + 60 = 0,$$
$$\Sigma M_{\text{left end}} = -M + 60(4 - x) = 0,$$

we obtain

$$\left.\begin{array}{l} P = 0 \\ V = -60 \text{ kN} \\ M = 60(4 - x) \text{ kN-m} \end{array}\right\} \quad 2 < x < 4 \text{ m}.$$

The shear force and bending moment diagrams, obtained by plotting the equations for V and M for the two ranges of x, are shown in Fig. (d).

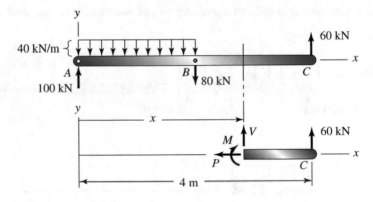

(c) Free-body diagram for $2 < x < 4$ m

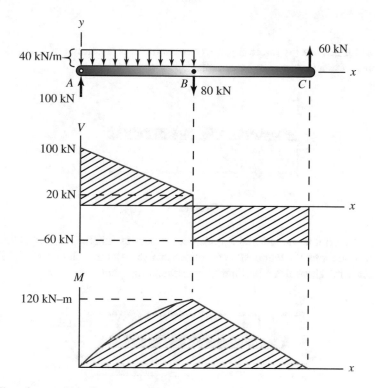

(d) Shear force and bending moment diagrams

Discussion

When you obtain equations for the shear force and bending moment in a beam that apply to different parts of the beam, as we did in this example, there are two conditions you can often use to check your results. (We discuss the bases of these conditions in the next section.) The first one is that *the shear force diagram of a beam is continuous except at points where the beam is subjected to a point force*. The second condition is that *the bending moment diagram of a beam is continuous except at points where the beam is subjected to a point couple*. In this example, the equations we obtained for the bending moment M for $0 < x < 2$ m and for $2 < x < 4$ m must agree at $x = 2$ m. Checking,

$$100(2 \text{ m}) - 20(2 \text{ m})^2 \text{ kN-m} = 60(4 \text{ m} - 2 \text{ m}) \text{ kN-m}:$$
$$120 \text{ kN-m} = 120 \text{ kN-m},$$

we confirm that they agree.

Review Problems

5.21 **(a)** Determine the internal forces and moment as functions of x. **(b)** Draw the shear force and bending moment diagrams.

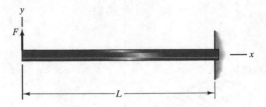

Problem 5.21

5.22 **(a)** Determine the internal forces and moment as functions of x. **(b)** Show that the equations for V and M as functions of x satisfy the equation $V = dM/dx$. **(c)** Draw the shear force and bending moment diagrams.

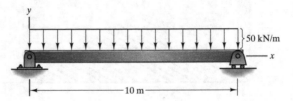

Problems 5.22–5.23

5.23 The beam will safely support a bending moment of 1 MN-m (meganewton-meter) at any cross section. Based on this criterion, what is the maximum safe value of the uniformly distributed load?

5.24 **(a)** Determine the internal forces and moment as functions of x. **(b)** Show that the equations for V and M as functions of x satisfy the equation $V = dM/dx$. **(c)** Determine the maximum bending moment in the beam and the value of x where it occurs.

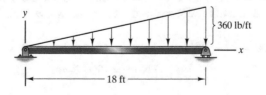

Problem 5.24

5.25 The structural beam is subjected to a uniform distributed load w_0 and reactions $F_0 = w_0 L/2$ and $M_0 = w_0 L^2/12$. Determine the internal forces and moment as functions of x.

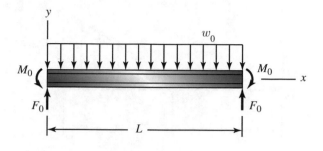

Problem 5.25

5.26 Determine the internal forces and moment as functions of x for $0 < x < L/2$.

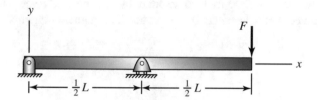

Problems 5.26–5.27

5.27 Determine the internal forces and moment as functions of x for $L/2 < x < L$.
5.28 Determine the internal forces and moment as functions of x for $0 < x < 0.5$ m.
5.29 Determine the internal forces and moment as functions of x for $0.5 < x < 1$ m.
5.30 Determine the internal forces and moment as functions of x for $0 < x < 6$ ft.

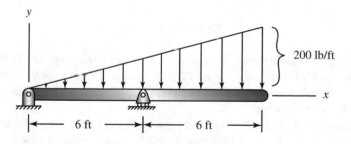

Problems 5.30–5.31

5.31 Determine the internal forces and moment as functions of x for $6 < x < 12$ ft.
5.32 **(a)** Determine the internal forces and moment as functions of x. **(b)** Draw the shear force and bending moment diagrams.

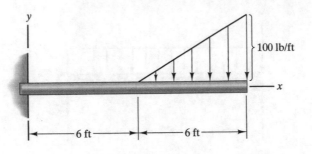

Problem 5.32

5.33∗ Model the ladder rung as a simply supported (pin-supported) beam and assume that the 200-lb load exerted by the person's shoe is uniformly distributed. Draw the shear force and bending moment diagrams for the rung.

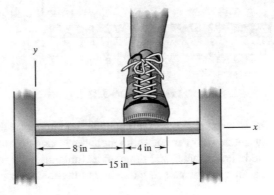

Problems 5.33–5.34

5.34∗ Model the ladder rung as a simply supported (pin-supported) beam and assume that the 200-lb load exerted by the person's shoe is uniformly distributed. What is the maximum bending moment in the ladder rung, and where does it occur?

5.35∗ Assume that the surface on which the beam rests exerts a uniformly distributed load on the beam. Draw the shear force and bending moment diagrams.

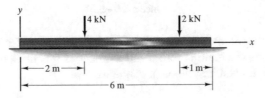

Problem 5.35

5.36 The homogeneous beams AB and CD weigh 600 lb and 500 lb, respectively. Draw the shear force and bending moment diagrams for beam CD. (Remember that the beam's weight is a distributed load.)

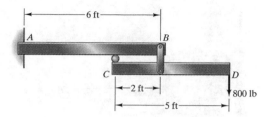

Problems 5.36–5.37

5.37 The homogeneous beams AB and CD weigh 600 lb and 500 lb, respectively. Draw the shear force and bending moment diagrams for beam AB. (Remember that the beam's weight is a distributed load.)

5.38* The load $F = 4650$ lb. Draw the shear force and bending moment diagrams for the beam.

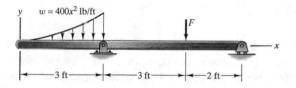

Problems 5.38–5.39

5.39* The load $F = 2150$ lb. What are the maximum positive and negative values of the shear force and bending moment, and at what values of x do they occur?

5.40* Draw the shear force and bending moment diagrams for the beam.

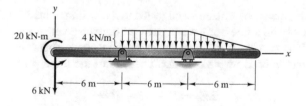

Problems 5.40–5.41

5.41* What are the maximum positive and negative values of the shear force and bending moment, and at what values of x do they occur?

5.42* The lift force on the airplane's wing is given by the distributed load $w = -15$
(1 − 0.04x^2) kN/m, and the weight of the wing is given by the distributed load
$w = 5 − 0.5x$ kN/m. Determine the shear force as a function of x.

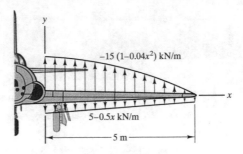

Problem 5.42

5.3   Equations Relating Distributed Load, Shear Force, and Bending Moment

The shear force and bending moment in a beam subjected to a distributed load are
governed by simple differential equations. In this section we derive these equations
and show how they can be used to obtain shear force and bending moment diagrams.
In Chaps. 6 and 9, we show that these equations are also needed for determining the
states of stress and the deflections of beams.

5.3.1   Derivation of the Equations

Suppose that a portion of a beam is subjected to a distributed load w (Fig. 5.9(a)). In
Fig. 5.9(b) we obtain a free-body diagram by cutting the beam at x and at $x + \Delta x$. The
terms ΔP, ΔV, and ΔM are the changes in the axial force, shear force, and bending
moment, respectively, from x to $x + \Delta x$. The sum of the forces in the x direction is

$$\Sigma F_x = P + \Delta P - P = 0.$$

Dividing this equation by Δx and taking the limit as $\Delta x \to 0$, we obtain

$$\frac{dP}{dx} = 0.$$

Fig. 5.9 (a) Portion of a beam subjected to a distributed force w. (b) Obtaining the free-body diagram of an element of the beam

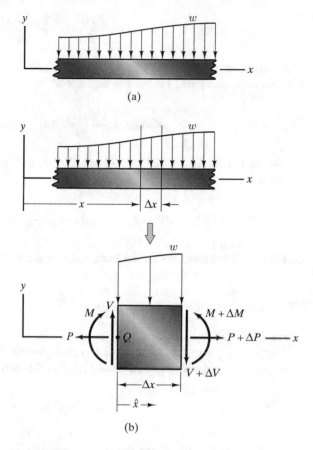

(a)

(b)

This equation simply indicates that the axial force does not depend on x in a portion of a beam that is subjected only to a lateral distributed load. To sum the forces on the free-body diagram in the y direction, we must determine the downward force exerted by the distributed load. In Fig. 5.9(b) we introduce a coordinate $\hat{x}$ that measures distance from the left edge of the free-body diagram. In terms of this coordinate, the downward force exerted on the free-body diagram by the distributed load is

$$\int_0^{\Delta x} w(x+\hat{x})\, d\hat{x}, \tag{5.1}$$

where $w(x+\hat{x})$ denotes the value of w at $x + \hat{x}$. To evaluate this integral, we express $w(x+\hat{x})$ as a Taylor series in terms of $\hat{x}$:

$$w(x + \hat{x}) = w(x) + \frac{dw(x)}{dx}\hat{x} + \frac{1}{2}\frac{d^2w(x)}{dx^2}\hat{x}^2 + \cdots. \tag{5.2}$$

Substituting this expression into Eq. (5.1) and integrating term by term, the downward force exerted by the distributed load is

$$w(x)\,\Delta x + \frac{1}{2}\frac{dw(x)}{dx}(\Delta x)^2 + \cdots.$$

The sum of the forces on the free-body diagram in Fig. 5.9(b) in the y direction is therefore

$$\Sigma F_y = V - V - \Delta V - w(x)\,\Delta x - \frac{1}{2}\frac{dw(x)}{dx}(\Delta x)^2 + \cdots = 0.$$

Dividing by Δx and taking the limit as $\Delta x \to 0$ yields

$$\frac{dV}{dx} = -w, \tag{5.3}$$

where $w = w(x)$.

Our next step is to sum the moments on the free-body diagram in Fig. 5.9(b) about point Q. The clockwise moment about Q due to the distributed load is

$$\int_0^{\Delta x} \hat{x}w(x + \hat{x})\,d\hat{x}.$$

Substituting Eq. (5.2) into this expression and integrating term by term, the moment is

$$\frac{1}{2}w(x)(\Delta x)^2 + \frac{1}{3}\frac{dw(x)}{dx}(\Delta x)^3 + \cdots.$$

The sum of the moments on the free-body diagram about Q is therefore

$$\Sigma M_{\text{point } Q} = M + \Delta M - M - (V + \Delta V)\,\Delta x$$
$$-\frac{1}{2}w(x)(\Delta x)^2 - \frac{1}{3}\frac{dw(x)}{dx}(\Delta x)^3 + \cdots = 0.$$

Dividing this equation by Δx and taking the limit as $\Delta x \to 0$ gives

$$\frac{dM}{dx} = V. \tag{5.4}$$

In principle, Eqs. (5.3) and (5.4) can be used to determine the shear force and bending moment diagrams for a beam. We can integrate Eq. (5.3) to determine V as a function of x and then integrate Eq. (5.4) to determine M as a function of x. However, these equations were derived for a segment of beam subjected only to a distributed load. To apply them for a more general loading, we must also account for the effects of any point forces and couples acting on the beam.

Let us determine what happens to the shear force and bending moment where a beam is subjected to a force F in the positive y direction (Fig. 5.10(a)). By cutting the beam just to the left and just to the right of the force, we obtain the free-body diagram in Fig. 5.10(b), where the subscripts $-$and$+$ denote values to the left and right of the force. Equilibrium requires that

Fig. 5.10 (a) Portion of a beam subjected to a force F in the positive y direction. (b) Obtaining a free-body diagram by cutting the beam to the left and right of F. (c) The shear force diagram undergoes an increase of magnitude F. (d) The bending moment diagram is continuous

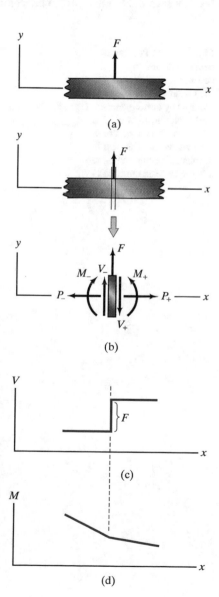

$$V_+ - V_- = F,$$
$$M_+ - M_- = 0.$$

We see that the shear force diagram undergoes an increase of magnitude F (Fig. 5.10(c)), but the bending moment diagram is continuous (Fig. 5.10(d)). The change in the shear force is *positive* if the force F is in the positive y direction.

Now we consider what happens to the shear force and bending moment diagrams where a beam is subjected to a counterclockwise couple C (Fig. 5.11(a)). Cutting the beam just to the left and just to the right of the couple (Fig. 5.11(b)), we determine that

Fig. 5.11 (a) Portion of a beam subjected to a counterclockwise couple C. (b) Obtaining a free-body diagram by cutting the beam to the left and right of C. (c) The shear force diagram is continuous. (d) The bending moment diagram undergoes a decrease of magnitude C

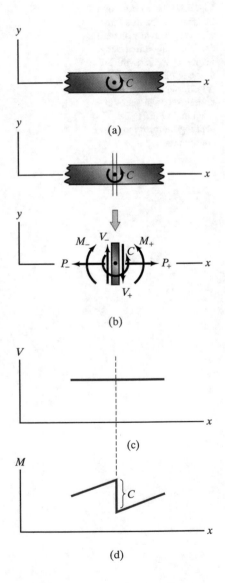

$$V_+ - V_- = 0,$$
$$M_+ - M_- = -C.$$

The shear force diagram is continuous (Fig. 5.11(c)), but the bending moment diagram undergoes a decrease of magnitude C (Fig. 5.11(d)) where a beam is subjected to a counterclockwise couple. The change in the bending moment is *negative* if the couple is in the counterclockwise direction.

5.3.2   *Construction of the Shear Force Diagram*

In a segment of a beam that is subjected only to a distributed load, we have shown that the shear force is related to the distributed load by Eq. (5.3):

$$\frac{dV}{dx} = -w. \tag{5.5}$$

This equation states that the derivative, or slope, of the shear force with respect to x is equal to the negative of the distributed load. Notice that if there is no distributed load ($w = 0$) throughout the segment, the slope is zero, and the shear force is constant. If w is a constant throughout the segment, the slope of the shear force is constant, which means that the shear force diagram for the segment is a straight line. Integrating Eq. (5.5) with respect to x from a position x_A to a position x_B,

$$\int_{x_A}^{x_B} \frac{dV}{dx}\, dx = -\int_{x_A}^{x_B} w\, dx,$$

yields

$$V_B - V_A = -\int_{x_A}^{x_B} w\, dx.$$

The change in the shear force between two positions is equal to the negative of the area defined by the loading curve between those positions (Fig. 5.12):

$$V_B - V_A = -(\text{area defined by the distributed load from } x_A \text{ to } x_B). \tag{5.6}$$

Where a beam is subjected to a point force of magnitude F in the positive y direction, we have shown that the shear force diagram undergoes an increase of magnitude F. Where a beam is subjected to a couple, the shear force diagram is unchanged (continuous).

Fig. 5.12 The change in the shear force is equal to the negative of the area defined by the loading curve

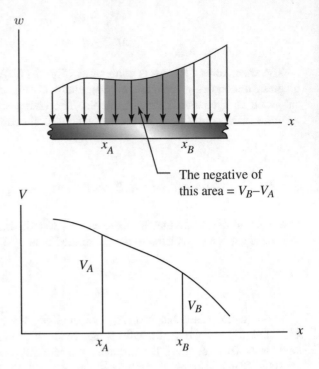

The negative of this area = $V_B - V_A$

Fig. 5.13 Beam loaded by a force F and its free-body diagram

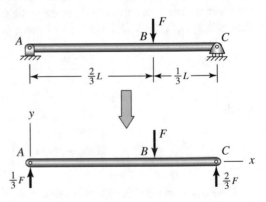

We can demonstrate these results by determining the shear force diagram for the beam in Fig. 5.13. The beam is subjected to a downward force F that results in upward reactions at A and C. Notice that there is no distributed load. Our procedure is to begin at the left end of the beam and construct the diagram from left to right. Figure 5.14(a) shows the increase in the value of V due to the upward reaction at A. Because there is no distributed load, the value of V remains constant between A and B (Fig. 5.14(b)). At B, the value of V decreases due to the downward force

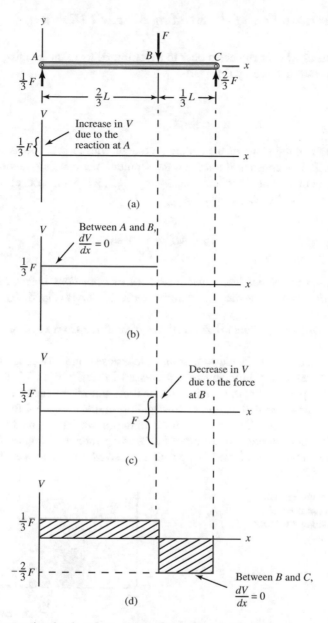

Fig. 5.14 Constructing the shear force diagram for the beam in Fig. 5.13

(Fig. 5.14(c)). The value of V remains constant between B and C, which completes the shear force diagram (Fig. 5.14(d)). Compare Fig. 5.14(d) with the shear force diagram we obtained in Fig. 5.7 by drawing free-body diagrams and applying the equilibrium equations.

5.3.3 *Construction of the Bending Moment Diagram*

In a segment of a beam subjected only to a distributed load, the bending moment is
related to the shear force by Eq. (5.4),

$$\frac{dM}{dx} = V, \tag{5.7}$$

which states that the slope of the bending moment with respect to x is equal to the
shear force. If V is constant throughout the segment, the bending moment diagram
for the segment is a straight line. Integrating Eq. (5.7) with respect to x from a
position x_A to a position x_B yields

$$M_B - M_A = \int_{x_A}^{x_B} V \, dx.$$

*The change in the bending moment between two positions is equal to the area
defined by the shear force diagram between those positions* (Fig. 5.15):

$$M_B - M_A = \text{area defined by the shear force from } x_A \text{ to } x_B. \tag{5.8}$$

Where a beam is subjected to a counterclockwise couple of magnitude C, the
bending moment diagram undergoes a decrease of magnitude C. Where a beam is
subjected to a point force, the bending moment diagram is unchanged.

As an example, we will determine the bending moment diagram for the beam in
Fig. 5.13. We begin with the shear force diagram we have already determined
(Fig. 5.16(a)) and proceed to construct the bending moment diagram from left to
right. The beam is not subjected to a couple at A, so $M_A = 0$. Between A and B, the

Fig. 5.15 The change in the
bending moment is equal to
the area defined by the shear
force diagram

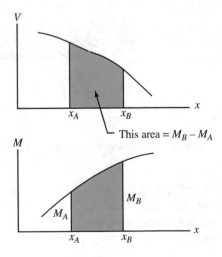

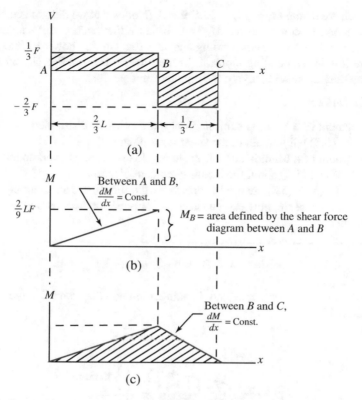

Fig. 5.16 Constructing the bending moment diagram for the beam in Fig. 5.13

slope of the bending moment is constant ($dM/dx = V = F/3$), which tells us that the bending moment diagram between A and B is a straight line (Fig. 5.16(b)). The change in the bending moment from A to B is equal to the area defined by the shear force from A to B:

$$M_B - M_A = \left(\frac{2}{3}L\right)\left(\frac{1}{3}F\right) = \frac{2}{9}LF.$$

Therefore $M_B = 2LF/9$. The slope of the bending moment is also constant between B and C ($dM/dx = V = -2F/3$), so the bending moment diagram between B and C is a straight line. The change in the bending moment from B to C is equal to the area defined by the shear force from B to C,

$$M_C - M_B = \left(\frac{1}{3}L\right)\left(-\frac{2}{3}F\right) = -\frac{2}{9}LF,$$

from which we obtain $M_C = M_B - 2LF/9 = 0$. (Notice that we did not actually need this calculation to conclude that $M_C = 0$, because the beam is not subjected to a couple at C.) The completed bending moment diagram is shown in Fig. 5.16(c). Compare it with the bending moment diagram we obtained in Fig. 5.7 by drawing free-body diagrams and applying the equilibrium equations.

Study Questions

1. If a segment of a beam is subjected only to a constant distributed load w, what does Eq. (5.5) tell you about the shear force diagram?
2. If a segment of a beam is not subject to any loads, what can you conclude from Eqs. (5.5) and (5.7) about the bending moment diagram?
3. When Eqs. (5.5)–(5.8) are used to determine the shear force and bending moment diagrams for a beam, how can you account for point forces and couples acting on the beam?

Example 5.4 Shear Force and Bending Moment Diagrams Using Eqs. (5.5)–(5.8)

Determine the shear force and bending moment diagrams for the beam in Fig. 5.17.

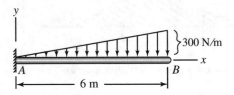

Fig. 5.17

Strategy

We can begin with the free-body diagram of the beam and use Eqs. (5.5) and (5.6) to construct the shear force diagram. Then we can use the shear force diagram and Eqs. (5.7) and (5.8) to construct the bending moment diagram. In determining both the shear force and bending moment diagrams, we must account for the effects of point forces and couples acting on the beam.

Solution

Shear force diagram   The first step is to draw the free-body diagram of the beam and determine the reactions at the built-in support A. Using the results of this step, shown in Fig. (a), we proceed to construct the shear force diagram from left to right. Figure (b) shows the increase in the value of V due to the upward force at A. Between A and B, the distributed load on the beam increases linearly from zero to 300 N/m. Therefore the slope of the shear force diagram decreases linearly from zero to -300 N/m. At B, the shear force must be zero, because no force acts there. With this information, we can sketch the shear force diagram qualitatively (Fig. (c)).

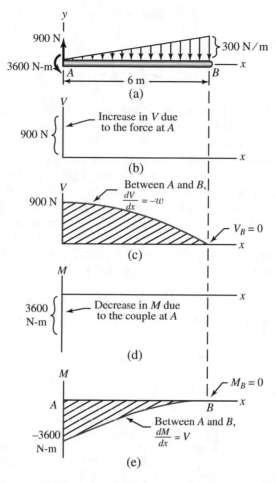

(a) Free-body diagram of the beam. (b, c) Constructing the shear force diagram. (d, e) Constructing the bending moment diagram

We can also obtain an explicit equation for the shear force between A and B by integrating Eq. (5.5). The distributed load as a function of x is $w = (x/6)$ $300 = 50x$ N/m. Writing Eq. (5.5) as

$$dV = -w\,dx = -50x\,dx$$

and integrating,

$$\int_{900}^{V} dV = \int_{0}^{x} -50x\,dx$$

$$[V]_{900}^{V} = [-25x^2]_{0}^{x}$$

$$V - 900 = -25x^2$$

we obtain

$$V = 900 - 25x^2 \text{ N}. \tag{1}$$

Bending moment diagram We construct the bending moment diagram from left to right. Figure (d) shows the initial decrease in the value of M due to the counterclockwise couple at A. Between A and B, the slope of the bending moment diagram is equal to the shear force V. We see from the shear force diagram (Fig. (c)) that at A, the slope of the bending moment diagram has a positive value (900 N). As x increases, the slope begins to decrease, and its rate of decrease grows until the value of the slope reaches zero at B. At B, we know that the value of the bending moment is zero, because no couple acts on the beam at B. Using this information, we can sketch the bending moment diagram qualitatively (Fig. (e)). Notice that its slope decreases from a positive value at A to zero at B, and the rate at which it decreases grows as x increases.

We can obtain an equation for the bending moment between A and B by integrating Eq. (5.7). The shear force as a function of x is given by Eq. (1). Writing Eq. (5.7) as

$$dM = V\,dx = \left(900 - 25x^2\right) dx$$

and integrating,

$$\int_{-3600}^{M} dM = \int_{0}^{x} \left(900 - 25x^2\right) dx$$

$$[M]_{-3600}^{M} = \left[900x - \frac{25}{3}x^3\right]_{0}^{x}$$

$$M + 3600 = 900x - \frac{25}{3}x^3$$

we obtain

$$M = -3600 + 900x - \frac{25}{3}x^3 \text{ N-m}.$$

Discussion

As we demonstrated in this example, Eqs. (5.5)–(5.8) can be applied in two ways. They provide a basis for rapidly obtaining qualitative sketches of shear force and bending moment diagrams. In addition, explicit equations for the diagrams can be obtained by integrating Eqs. (5.5) and (5.7).

Example 5.5 Shear Force and Bending Moment Diagrams Using Eqs. (5.5)–(5.8)

Determine the shear force and bending moment diagrams for the beam in Fig. 5.18.

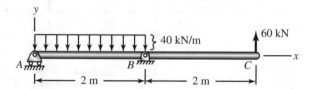

Fig. 5.18

Strategy

Using the same approach as in Example 5.4, we can begin with the free-body diagram of the beam and use Eqs. (5.5) and (5.6) to construct the shear force diagram and then use the shear force diagram and Eqs. (5.7) and (5.8) to construct the bending moment diagram.

Solution

Shear force diagram   We determined the reactions at the supports of this beam in Example 5.3. Using the results, shown in Fig. (a), we proceed to construct the shear force diagram from left to right. Due to the 100-kN upward force at A, $V_A = 100$ kN (Fig. (b)). Between A and B, the distributed load on the beam is constant, which means that the slope of the shear force diagram is constant. Therefore the shear force diagram between A and B is a straight line. We know that the shear force will be discontinuous at B due to the 80-kN force. Let V_B^- and V_B^+ denote the values of V to the left and right of the 80-kN force, respectively. From Eq. (5.6), the change in V from A to B is

$$V_B^- - V_A = -(2 \text{ m})(40 \text{ kN/m}) = -80 \text{ kN},$$

so $V_B^- = V_A - 80 = 20$ kN. The shear force diagram between A and B is shown in Fig. (c).

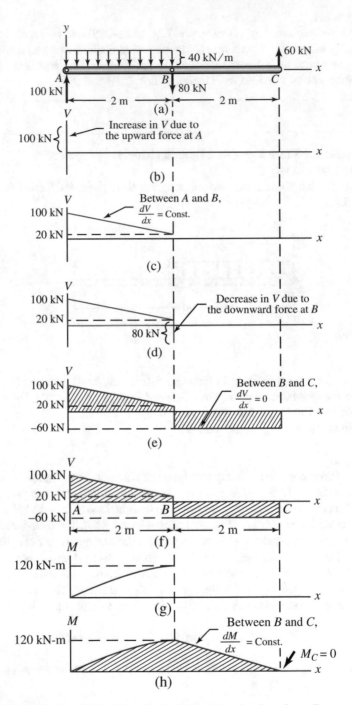

(a) Free-body diagram of the beam. (b–e) Constructing the shear force diagram. (d–h) Constructing the bending moment diagram

We can also determine the shear force between A and B by integrating Eq. (5.5). Writing Eq. (5.5) as

$$dV = w\,dx = -40\,dx$$

and integrating,

$$\int_{100}^{V} dV = \int_{0}^{x} -40\,dx$$
$$[V]_{100}^{V} = [-40x]_{0}^{x}$$
$$V - 100 = -40x$$

we obtain

$$V = 100 - 40x \text{ kN.} \tag{1}$$

Notice that at $x = 2$ m, this equation gives $V_B^- = 20$ kN. The effect of the 80-kN downward force at B is shown in Fig. (d). The value of V to the right of the 80-kN force is

$$V_B^+ = V_B^- - 80 = -60 \text{ kN.}$$

Because there is no distributed load between B and C, the value of V remains constant between B and C, completing the shear force diagram (Fig. (c)).

Bending moment diagram   The beam is not subjected to a couple at A, so $M_A = 0$. Between A and B, the slope of the bending moment diagram equals the shear force. From the shear force diagram (Fig. (f)), we see that the slope is positive between A and B and decreases linearly from A to B. The change in the bending moment between A and B is equal to the area defined by the shear force diagram between A and B,

$$M_B - M_A = (2 \text{ m})(20 \text{ kN}) + \frac{1}{2}(2 \text{ m})(80 \text{ kN}) = 120 \text{ kN-m,}$$

so $M_B = 120$ kN-m. With this information we can sketch the diagram between A and B qualitatively (Fig. (g)). Observe that the slope is positive but decreases from A to B.

We can obtain an equation for the bending moment between A and B by integrating Eq. (5.7). The shear force as a function of x is given by Eq. (1). We write Eq. (5.7) as

$$dM = V\,dx = (100 - 40x)\,dx$$

and integrate:

$$\int_0^M dM = \int_0^x (100 - 40x)\, dx$$

$$[M]_0^M = [100 - 40x]_0^x$$

$$M = 100x - 20x^2 \text{ kN-m.}$$

Because there is no couple at C, $M_C = 0$. The slope of the bending moment is constant between B and C ($dM/dx = V = -60$ kN), so the bending moment diagram between B and C is a straight line (Fig. (h)).

Discussion

Compare this example with Example 5.3, in which we use free-body diagrams and the equilibrium equations to determine the shear force and bending moment diagrams for this beam and loading.

Review Problems

5.43 Determine V and M as functions of x **(a)** by drawing free-body diagrams and using the equilibrium equations and **(b)** by using Eqs. (5.5)–(5.8).

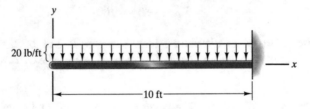

Problem 5.43

5.44 The distributed load acting on the beam is given by the equation

$$w = w_0 \left(1 - \frac{x}{L}\right).$$

Problem 5.44

Determine V and M as functions of x by using Eqs. (5.5)–(5.8).

5.45 Determine V and M as functions of x by using Eqs. (5.5)–(5.8).

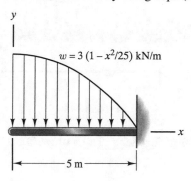

$w = 3\,(1 - x^2/25)$ kN/m

5 m

Problem 5.45

5.46 Determine V and M as functions of x by using Eqs. (5.5)–(5.8).

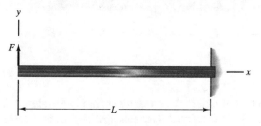

F

L

Problem 5.46

5.47 Determine V and M as functions of x by using Eqs. (5.5)–(5.8).

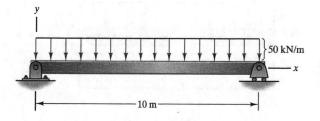

50 kN/m

10 m

Problem 5.47

5.48 Determine V and M as functions of x by using Eqs. (5.5)–(5.8).

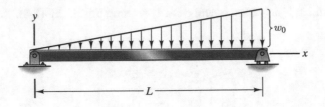

Problem 5.48

5.49 Determine V and M as functions of x by using Eqs. (5.5)–(5.8).

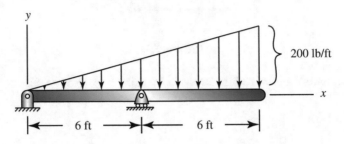

200 lb/ft

6 ft   6 ft

Problem 5.49

5.50 The free-body diagram of the structural beam in Problem 5.25 is shown, where $F_0 = wL/2$ and $M_0 = wL^2/12$. Use Eqs. (5.5)–(5.8) to determine the shear force and bending moment as functions of x.

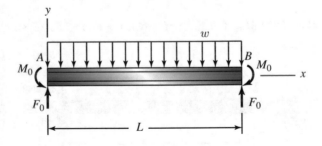

Problem 5.50

5.51 **(a)** Determine V and M as functions of x by using Eqs. (5.5)–(5.8). **(b)** Draw the shear force and bending moment diagrams.

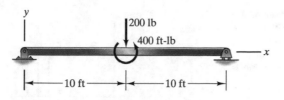

Problem 5.51

5.52 Use Eqs. (5.5)–(5.8) to solve Problem 5.32.
5.53∗ Use Eqs. (5.5)–(5.8) to solve Problem 5.33.
5.54∗ Use Eqs. (5.5)–(5.8) to determine the internal forces and moment as functions of x for the beam in Problem 5.35.
5.55∗ Use Eqs. (5.5)–(5.8) to solve Problem 5.38.
5.56∗ Use Eqs. (5.5)–(5.8) to solve Problem 5.40.

5.4 Singularity Functions

Except for the simplest cases, the methods we have used to determine the shear force and bending moment distributions in a beam have resulted in different sets of equations for different portions of the beam's axis. See Example 5.5, in which we obtained one set of equations for V and M that applied to the left half of the beam and another set of equations that applied to the right half. It would be convenient if we could obtain equations for V and M that apply to an entire beam. This can be achieved by using a class of functions called *singularity functions*.

5.4.1 Delta Function

In Fig. 5.19(a) we define a triangular distributed load w. Its value varies linearly from $w = 0$ at $x = a - \varepsilon$ to $w = 1/\varepsilon$ at $x = a$, and then varies linearly from $w = 1/\varepsilon$ at $x = a$ to $w = 0$ at $x = a + \varepsilon$. The total force exerted by this distributed load (determined by calculating the area under the loading curve) is 1. Thus we have defined a distributed load that exerts a downward force of unit magnitude. In Fig. 5.19(b) we represent this distributed load by an equivalent force by placing a downward force of unit magnitude at the centroid $x = a$ of the loading curve. Figure 5.19(c) is the graph of the integral of the distributed load w from $x = 0$ to x, which is the area under the loading curve from $x = 0$ to x. The value of the integral at $x = a + \varepsilon$ is the total area under the loading curve, which is 1.

Now suppose that we let the parameter ε become progressively smaller. As this occurs, the width of the interval over which w acts becomes smaller and the maximum magnitude of w becomes larger, but *the area under the loading curve,*

Fig. 5.19 (a) A triangular
distributed load. (b) The
distributed load exerts a
force of unit magnitude.
(c) Integral of the distributed
load from 0 to x

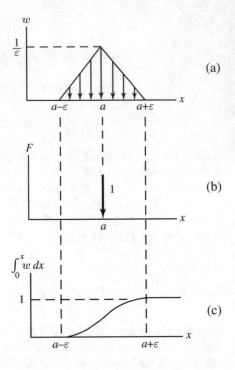

which is the magnitude of the equivalent force exerted by the distributed load,
remains equal to 1. We show what happens to Fig. 5.19 in the limit as $\varepsilon \to 0$ in
Fig. 5.20. The function describing the distributed load w that results from taking the
limit as $\varepsilon \to 0$ is called a *delta function*, which we denote by

$$\langle x - a \rangle^{-1} = \lim_{\varepsilon \to 0} w.$$

(The motivation for this notation will become apparent.) The distributed load
represented by a delta function (Fig. 5.20(a)) exerts a downward force of unit
magnitude (Fig. 5.20(b)). The graph of the integral of w that results from taking
the limit as $\varepsilon \to 0$ is a *step function* (Fig. 5.20(c)), which we denote by $\langle x - a \rangle^0$.
That is,

$$\langle x - a \rangle^0 = \begin{cases} 0 & \text{if } x < a, \\ 1 & \text{if } x \geq a. \end{cases}$$

From these results we see that *the integral of a delta function is a step function*:

$$\int_0^x \langle x - a \rangle^{-1} dx = \langle x - a \rangle^0. \tag{5.9}$$

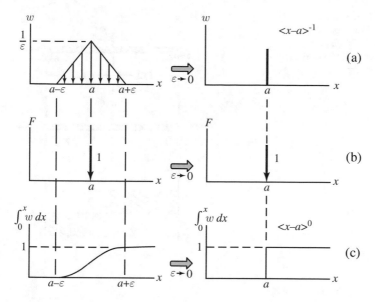

Fig. 5.20 This is what happens to Fig. 5.19 as $\varepsilon \to 0$: (**a**) the triangular distributed load becomes a delta function. (**b**) The delta function exerts a force of unit magnitude. (**c**) The integral of the distributed load becomes a step function

Consider what has happened. We began by defining a distributed load over a finite interval of x that exerted a downward force of unit magnitude. We then let the interval over which the distributed load acted becomes vanishingly small but defined the magnitude of the distributed load so that the force it exerted remained constant. The resulting delta function describes a distributed load that exerts a *downward* force of unit magnitude at the point $x = a$. Furthermore, we have shown that the integral of the delta function is a positive step function of unit magnitude at $x = a$.

What can we do with these arcane results? Figure 5.21(a) shows a beam with pin and roller supports that is subjected to a load F at its midpoint. The free-body diagram showing the reactions exerted by the supports is shown in Fig. 5.21(b), and the shear force diagram of the beam is shown in Fig. 5.21(c). We can determine the distribution of the shear force in this beam in terms of a single equation by using singularity functions. We begin by expressing the forces acting on the free-body diagram of the beam in terms of delta functions. A distributed load $w = \langle x - a \rangle^{-1}$ exerts a *downward* load of unit magnitude at $x = a$, so we define

$$w = -\frac{F}{2}\langle x \rangle^{-1} + F\left\langle x - \frac{L}{2} \right\rangle^{-1} - \frac{F}{2}\langle x - L \rangle^{-1}.$$

Observe that the three terms in this distributed load are an upward force $F/2$ at $x = 0$, a downward force F at $x = L/2$, and an upward force $F/2$ at $x = L$. Now we determine the shear force distribution by integrating Eq. (5.5):

Fig. 5.21 (a) Beam
subjected to a load F.
(b) Free-body diagram
of the beam. (c) The shear
force diagram

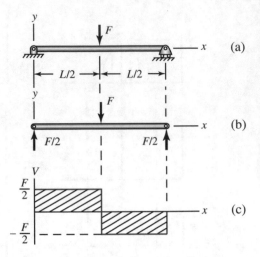

$$V = \int_0^x -w \, dx$$

$$= \int_0^x \frac{F}{2}\langle x\rangle^{-1} dx - \int_0^x F\left\langle x - \frac{L}{2}\right\rangle^{-1} dx + \int_0^x \frac{F}{2}\langle x - L\rangle^{-1} dx$$

$$= \frac{F}{2}\langle x\rangle^0 - F\left\langle x - \frac{L}{2}\right\rangle^0 + \frac{F}{2}\langle x - L\rangle^0,$$

where we have used Eq. (5.9) to integrate the delta functions. We have obtained a
single equation that describes the shear force distribution for the entire beam
(Fig. 5.21(c)). This illustrates how we will use singularity functions, but we need
to define more functions to analyze beams that are subjected to other types of loads.

5.4.2   Dipole

In Fig. 5.22(a) we define a distributed load w whose value is $1/\varepsilon^2$ in the interval
$a - \varepsilon \le x < a$ and $-1/\varepsilon^2$ in the interval $a \le x \le a + \varepsilon$. Calculating the area of the left,
downward-acting part of the distributed load shows that it exerts a downward force
of magnitude $1/\varepsilon$. The right, upward-acting part exerts an upward force of magnitude
$1/\varepsilon$. Representing the left and right parts of the distributed load by their equivalent
forces (Fig. 5.22(b)), we can see that the distributed load w exerts a counterclockwise
couple of unit magnitude. Figure 5.22(c) is the graph of the integral of w from 0 to x.
Notice that it is identical to the function in Fig. 5.19(a).

Now we let the parameter ε become progressively smaller. The width of the
interval over which w acts becomes smaller, and the magnitude of the downward and

Fig. 5.22 (a) A distributed
load. (b) Forces exerted by
the distributed load. (c)
Integral of the distributed
load from 0 to x

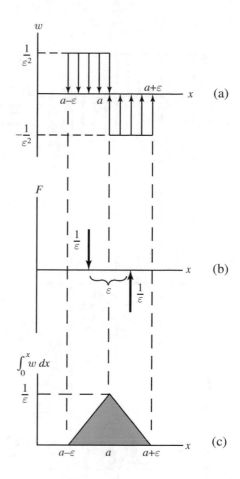

upward forces becomes larger, but *the couple exerted by the distributed load remains equal to* 1. We show what happens to Fig. 5.22 in the limit as $\varepsilon \to 0$ in Fig. 5.23. The function describing the distributed load w that results in the limit as $\varepsilon \to 0$ (Fig. 5.23(a)) is called a *dipole*, which we denote by

$$\langle x - a \rangle^{-2} = \lim_{\varepsilon \to 0} w.$$

The distributed load represented by a dipole exerts a counterclockwise couple of unit magnitude at $x = a$ (Fig. 5.23(b)). Because Fig. 5.22(c) is identical to Fig. 5.19(a), the integral of the distributed load w that results in the limit as $\varepsilon \to 0$ is a delta function (Fig. 5.23c). We see that *the integral of a dipole is a delta function*:

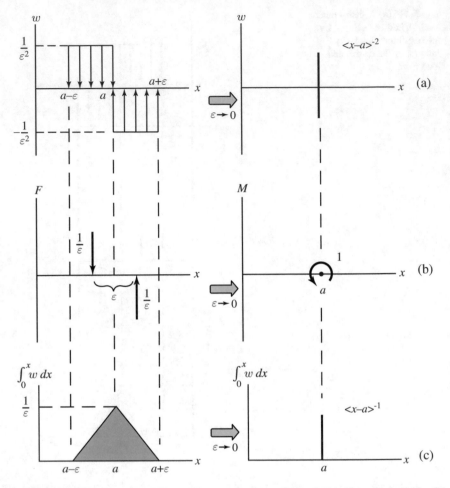

Fig. 5.23 This is what happens to Fig. 5.22 as $\varepsilon \to 0$: (**a**) the distributed load becomes a dipole. (**b**) The dipole exerts a counterclockwise couple of unit magnitude. (**c**) The integral of the distributed load becomes a delta function

$$\int_0^x \langle x - a \rangle^{-2} dx = \langle x - a \rangle^{-1}. \qquad (5.10)$$

We have shown how forces acting on a beam can be expressed as a distributed load in terms of delta functions. In the same way, couples acting on a beam can be expressed as a distributed load in terms of dipoles.

5.4.3   *Macaulay Functions*

We have defined two singularity functions, the delta function and the dipole, which make it possible to express forces and couples acting on beams as distributed loads. The beams we deal with are often subjected to distributed loads as well, and we need singularity functions with which to express them. Many simple distributed loads can be expressed in terms of *Macaulay functions*, which are defined by

$$\langle x - a \rangle^n = \begin{cases} 0 & \text{if } x < a, \\ (x - a)^n & \text{if } x \geq a, \end{cases} \tag{5.11}$$

where $n \geq 0$. The Macaulay functions for $n = 0$ (the step function), $n = 1$ (called the *ramp* function), and $n = 2$ are shown in Fig. 5.24. The integral of the Macaulay function from $x = 0$ to x is

$$\int_0^x \langle x - a \rangle^n dx = \frac{1}{n+1} \langle x - a \rangle^{n+1}. \tag{5.12}$$

The repertoire of singularity functions we have described and their integrals are summarized in Table 5.1. With these results we can express the loads on beams as distributed loads and determine their shear force and bending moment distributions.

Fig. 5.24 Macaulay functions

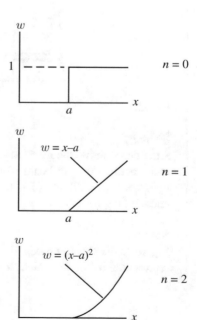

Table 5.1 Singularity functions

Type of function	w	$\int_0^x w\,dx$
Step function	$\langle x - a \rangle^0$	$\langle x - a \rangle$
Delta function	$\langle x - a \rangle^{-1}$	$\langle x - a \rangle^0$
Dipole	$\langle x - a \rangle^{-2}$	$\langle x - a \rangle^{-1}$
Macaulay functions ($n \geq 0$)	$\langle x - a \rangle^n = \begin{cases} 0 & \text{if } x < a, \\ (x-a)^n & \text{if } x \geq a. \end{cases}$	$\frac{1}{n+1}\langle x - a \rangle^{n+1}$

Study Questions

1. If a beam is subjected to a downward force of magnitude F at $x = a$, what singularity function do you use to express it as a distributed load?
2. If a beam is subjected to a counterclockwise couple of magnitude M at $x = a$, what singularity function do you use to express it as a distributed load?
3. If a beam is subjected to a uniformly distributed load of magnitude w_0 from $x = a$ to $x = b$, where $a < b$, how can you represent it as a distributed load using Macaulay functions?

Example 5.6 Shear Force and Bending Moment Using Singularity Functions

A uniformly distributed load acts over part of the cantilever beam in Fig. 5.25. Determine the shear force and bending moment as functions of x.

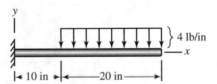

Fig. 5.25

Strategy

After determining the reactions at the fixed support, we will express the loads and reactions on the beam in terms of singularity functions. We can then integrate Eqs. (5.5) and (5.7) to determine the distributions of the shear force and bending moment.

Solution

Figure (a) is the free-body diagram of the beam with the distributed load represented by an equivalent force. From the equilibrium equations

$$\Sigma F_x = A_x = 0,$$

$$\Sigma F_y = A_y - 80 \text{ lb} = 0,$$

$$\Sigma M_{\text{left end}} = M_A - (20 \text{ in})(80 \text{ lb}) = 0,$$

we obtain $A_x = 0$, $A_y = 80$ lb and $M_A = 1600$ in - lb.

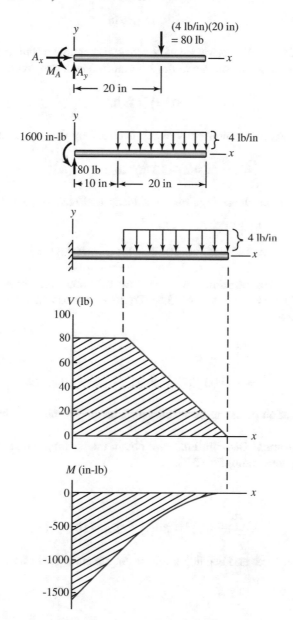

(a) Free-body diagram of the beam with the distributed load represented by an equivalent force. (b) Loads on the beam. (c) Shear force and bending moment diagrams

The reactions are shown on the free-body diagram in Fig. (b). Our objective is to express the reactions and the distributed load in terms of singularity functions using the results in Table 5.1. A delta function $\langle x - a \rangle^{-1}$ exerts a *downward* load of unit magnitude at $x = a$, so we express the 80-lb upward force as

$$-80\langle x \rangle^{-1} \text{ lb/in.}$$

A dipole $\langle x - a \rangle^{-2}$ exerts a counterclockwise couple of unit magnitude at $x = a$, so the contribution due to the 1600 in-lb counterclockwise couple is

$$1600\langle x \rangle^{-2} \text{ lb/in.}$$

We can express the uniformly distributed load by the step function

$$4\langle x - 10 \text{ in} \rangle^{0} \text{ lb/in.}$$

The distributed load describing the loads and reactions on the beam is therefore

$$w = 1600\langle x \rangle^{-2} - 80\langle x \rangle^{-1} + 4\langle x - 10 \text{ in} \rangle^{0} \text{ lb/in.}$$

Shear Force Distribution   We can now determine the shear force as a function of x by integrating Eq. (5.5). To do so, we use the integration rules given in Table 5.1:

$$V = \int_{0}^{x} -w\,dx$$
$$= -1600\langle x \rangle^{-1} + 80\langle x \rangle^{0} - 4\langle x - 10 \text{ in} \rangle \text{ lb.}$$

This equation gives the shear force distribution for the entire beam.

Bending Moment Distribution   We obtain the bending moment as a function of x by integrating Eq. (5.7):

$$M = \int_{0}^{x} V\,dx$$
$$= -1600\langle x \rangle^{0} + 80\langle x \rangle - 2\langle x - 10 \text{ in} \rangle^{2} \text{ in-lb.}$$

The shear force and bending moment diagrams for the beam are shown in Fig. (c).

Discussion
Notice that there is a delta function $\langle x \rangle^{-1}$ in the equation for V. The delta function $\langle x - a \rangle^{-1}$ equals zero everywhere except at $x = a$, where *its value is undefined*. As a result, the value of the equation we obtained for V in this example is undefined at $x = 0$, but is defined for $x > 0$. This is not a new phenomenon—the shear and bending moment distributions we have obtained by other methods also had singularities at points where forces and couples acted, but they did not appear explicitly as one does in this example.

Example 5.7 Shear Force and Bending Moment Using Singularity Functions
The beam in Fig. 5.26 is subjected to a triangular distributed load and a couple. Determine the shear force and bending moment as functions of x.

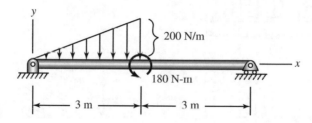

Fig. 5.26

Strategy
We must express the loads and reactions on the beam in terms of singularity functions. We can then integrate Eqs. (5.5) and (5.7) to determine the distributions of the shear force and bending moment.

Solution
The free-body diagram of the beam is shown in Fig. (a) with the distributed load represented by an equivalent force. From the equilibrium equations

$$\Sigma F_x = A_x = 0,$$
$$\Sigma F_y = A_y + B - 300 \text{ N} = 0,$$
$$\Sigma M_{\text{left end}} = (6 \text{ m})B - (2 \text{ m})(300 \text{ N}) + 180 \text{ N-m} = 0,$$

we obtain $A_x = 0$, $A_y = 230$ N, and $B = 70$ N. The free-body diagram of the beam including the reactions is shown in Fig. (b).

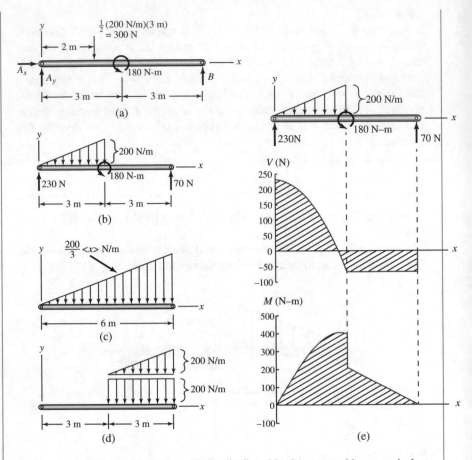

(a) Free-body diagram of the beam with the distributed load represented by an equivalent force. (b) Loads on the beam. (c) Triangular distributed load. (d) Distributed loads that must be subtracted. (e) Shear force and bending moment diagrams

Our objective is to express the loads and reactions as a distributed load in terms of singularity functions using the results in Table 5.1. A delta function $\langle x - a \rangle^{-1}$ exerts a *downward* load of unit magnitude at $x = a$, so the contributions to the distributed load due to the forces at the pin and roller supports are

$$-230\langle x \rangle^{-1} - 70\langle x - 6 \text{ m} \rangle^{-1} \text{ N/m}.$$

A dipole $\langle x - a \rangle^{-2}$ exerts a counterclockwise couple of unit magnitude at $x = a$, so the contribution due to the 180 N-m counterclockwise couple is

$$180\langle x - 3 \text{ m} \rangle^{-2} \text{ N/m}.$$

We can express the distributed load acting from $x = 0$ to $x = 3$ m by the Macaulay function

$$\frac{200}{3}\langle x \rangle \text{ N/m,}$$

but this function describes a distributed load that extends over the beam's entire length (Fig. (c)). We need to subtract the part of this distributed load that acts from $x = 3$ m to $x = 6$ m. To do so, we treat it as the sum of a uniformly distributed load and a triangular load (Fig. (d)). The Macaulay functions we need to *subtract* are

$$200\langle x - 3 \text{ m}\rangle^0 + \frac{200}{3}\langle x - 3 \text{ m}\rangle \text{ N/m.}$$

The total distributed load describing the loads and reactions on the beam is therefore

$$w = -230\langle x \rangle^{-1} + \frac{200}{3}\langle x \rangle + 180\langle x - 3 \text{ m}\rangle^{-2} - 200\langle x - 3 \text{ m}\rangle^0$$
$$- \frac{200}{3}\langle x - 3 \text{ m}\rangle - 70\langle x - 6 \text{ m}\rangle^{-1} \text{ N/m.}$$

Shear Force Distribution   We can determine the shear force as a function of x by integrating Eq. (5.5), using the integration rules summarized in Table 5.1.

$$V = \int_0^x -w \, dx$$
$$= 230\langle x \rangle^0 - \frac{200}{6}\langle x \rangle^2 - 180\langle x - 3 \text{ m}\rangle^{-1} + 200\langle x - 3 \text{ m}\rangle$$
$$+ \frac{200}{6}\langle x - 3 \text{ m}\rangle^2 + 70\langle x - 6 \text{ m}\rangle^0 \text{ N.}$$

This equation gives the shear force distribution for the entire beam.

Bending Moment Distribution   We obtain the bending moment as a function of x by integrating Eq. (5.7):

$$M = \int_0^x V \, dx$$
$$= 230\langle x \rangle - \frac{200}{18}\langle x \rangle^3 - 180\langle x - 3 \text{ m}\rangle^0 + \frac{200}{2}\langle x - 3 \text{ m}\rangle^2$$
$$+ \frac{200}{18}\langle x - 3 \text{ m}\rangle^3 + 70\langle x - 6 \text{ m}\rangle \text{ N-m.}$$

The shear force and bending moment diagrams for the beam are shown in Fig. (e).

Discussion

As this example demonstrates, a variety of distributed loads can be represented by superimposing Macaulay functions.

Review Problems

Problems 5.57–5.66 are to be solved using singularity functions.

5.57 Determine the bending moment in the beam as a function of x.

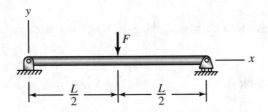

Problem 5.57

5.58 **(a)** Determine the shear force and bending moment in the beam as functions of x. **(b)** Using the results of part (a), draw the shear force and bending moment diagrams.

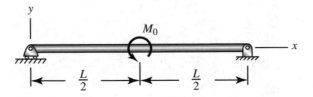

Problem 5.58

5.59 Determine the shear force and bending moment in the beam as functions of x.

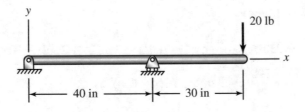

Problem 5.59

5.60 Determine the shear force and bending moment in the beam as functions of x.

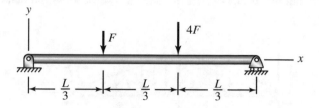

Problem 5.60

5.61 (a) Determine the shear force and bending moment in the beam as functions of x. **(b)** Using the results of part (a), draw the shear force and bending moment diagrams.

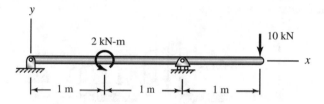

Problem 5.61

5.62 (a) Determine the shear force and bending moment in the beam as functions of x. **(b)** Using the results of part (a), draw the shear force and bending moment diagrams.

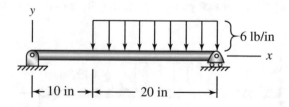

Problem 5.62

5.63 Determine the shear force and bending moment in the beam as functions of x.

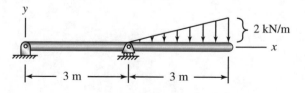

Problem 5.63

5.64 (a) Determine the shear force and bending moment in the beam as functions of *x*. (b) Using the results of part (a), draw the shear force and bending moment diagrams.

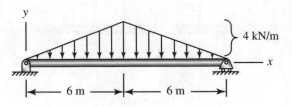

Problem 5.64

5.65 (a) Determine the shear force and bending moment in the beam as functions of *x*. (b) Using the results of part (a), draw the shear force and bending moment diagrams.

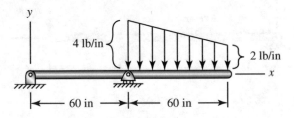

Problem 5.65

5.66 (a) Determine the shear force and bending moment in the beam as functions of *x*. (b) Using the results of part (a), draw the shear force and bending moment diagrams.

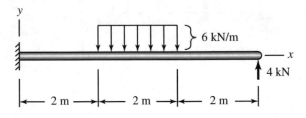

Problem 5.66

Chapter Summary

Axial Force, Shear Force, and Bending Moment

If the system of external loads and reactions on a beam is two-dimensional, the internal forces and moments can be represented by an equivalent system consisting of the *axial force P*, the *shear force V*, and the *bending moment M* (Fig. (a)). The directions of *P*, *V*, and *M* in Fig. (a) are the established positive directions of these quantities. The *shear force and bending moment diagrams* are simply the graphs of *V* and *M*, respectively, as functions of *x*.

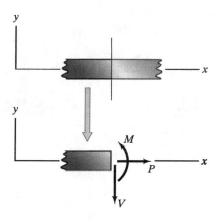

Equations Relating Distributed Load, Shear Force, and Bending Moment

In a segment of a beam that is subjected only to a distributed load (Fig. (b)), the shear force is related to the distributed load by

$$\frac{dV}{dx} = -w. \tag{5.5}$$

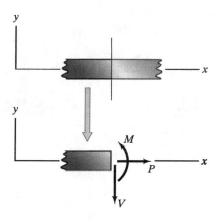

(b)

If there is no distributed load throughout the segment, the shear force is constant. If *w* is a constant throughout the segment, the shear force diagram for the segment is a straight line. The change in the shear force between two positions x_A and x_B is equal to the negative of the area defined by the loading curve between those positions:

$$V_B - V_A = -(\text{area defined by the distributed load from } x_A \text{ to } x_B). \qquad (5.6)$$

Where a beam is subjected only to a point force of magnitude F in the positive y direction, the shear force diagram undergoes an increase of magnitude F (Fig. (c)). Where a beam is subjected only to a couple, the shear force diagram is unchanged (continuous).

In a segment of a beam subjected only to a distributed load, the bending moment is related to the shear force by

$$\frac{dM}{dx} = V. \qquad (5.7)$$

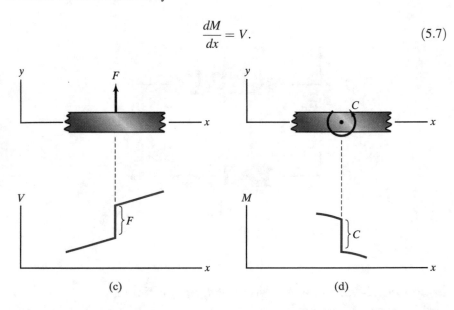

(c)   (d)

The change in the bending moment between two positions x_A and x_B is equal to the area defined by the shear force diagram between those positions:

$$M_B - M_A = \text{area defined by the shear force from } x_A \text{ to } x_B. \qquad (5.8)$$

Where a beam is subjected only to a counterclockwise couple of magnitude C, the bending moment diagram undergoes a decrease of magnitude C (Fig. (d)). Where a beam is subjected only to a point force, the bending moment diagram is unchanged.

Singularity Functions

The loads on a beam can be expressed as a distributed load in terms of the singularity functions summarized in Table 5.1. A delta function exerts a downward force of unit magnitude at $x = a$. A dipole exerts a counterclockwise couple of unit magnitude at

$x = a$. The most common distributed loads can be expressed in terms of Macaulay functions. Once the loads on a beam are expressed as a distributed load in terms of singularity functions, Eq. (5.5) can be integrated to determine the shear force, and Eq. (5.7) can be integrated to determine the bending moment as functions of x.

Review Problems

5.67 (a) Draw the free-body diagram of the beam by isolating it from its supports. (b) Determine the reactions at the supports. (c) Determine the internal forces and moment at C by drawing the free-body diagram of the part of the beam to the left of C.

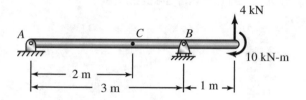

Problems 5.67–5.68

5.68 Determine the internal forces and moment at point C by drawing the free-body diagram of the part of the beam to the right of C.

5.69 If $x = 3$ m, what are the internal forces and moment at C?

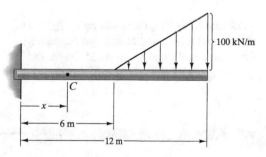

Problems 5.69–5.70

5.70 If $x = 9$ m, what are the internal forces and moment at C?

5.71∗ The loads $F = 200$ N and $C \doteq 800$ N-m. (a) By drawing free-body diagrams and applying the equilibrium equations, determine the internal forces and moment as functions of x. (b) Draw the shear force and bending moment diagrams.

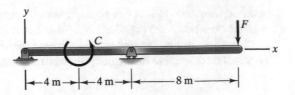

Problems 5.71–5.72

5.72* The beam will safely support shear forces and bending moments of magnitudes 2 kN and 6.5 kN-m, respectively. Based on this criterion, can it safely be subjected to the loads $F = 1$ kN, $C = 1.6$ kN-m?

5.73 The bar BD is rigidly fixed to the beam ABC at B. Draw the shear force and bending moment diagrams for the beam ABC.

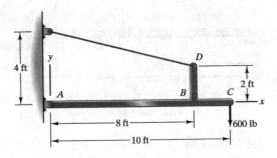

Problem 5.73

5.74* (a) By drawing free-body diagrams and applying the equilibrium equations, determine the internal forces and moment as functions of x for beam ABC. (b) Draw the shear force and bending moment diagrams.

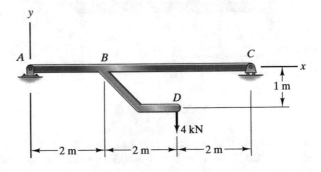

Problem 5.74

5.75 Use singularity functions to solve Problem 5.71.
5.76 Use singularity functions to solve Problem 5.74.

Chapter 6
Stresses in Beams

Because beams are utilized in so many ways and in so many types of structures, determining their states of stress is an important part of structural analysis and design. Here we show that the stresses at a given cross section of a beam can be expressed in terms of the values of the axial force, shear force, and bending moment at that cross section.

6.1 Distribution of the Normal Stress

You can break a small piece of wood by bending it (Fig. 6.1(a)), causing the wood to fracture (Fig. 6.1(b)). You bend the stick by subjecting it to couples at the ends, inducing stresses within the wood that cause it to fail. In the same way, stresses induced by bending moments in the members of a structure can cause the members to become permanently deformed or even lead to collapse of the structure. In this section we analyze the stresses induced in beams by bending moments.

6.1.1 Geometry of Deformation

Let us consider a prismatic beam made of a typical structural material (Fig. 6.2(a)). The x axis of the coordinate system is parallel to the beam's longitudinal axis, and we assume the beam's cross section to be symmetric about the y axis. Suppose that we were to carry out an experiment in which we subject the beam to equal couples M at its ends as shown in Fig. 6.2(b). The resulting deformation of the beam is shown in Fig. 6.3. (The following description of the beam's deformation applies in general only to the part of the beam that is not near the ends where the couples are applied. We will discuss this limitation presently.) Each longitudinal line of the beam that is parallel to the x axis before the couples are applied deforms into a circular arc. Each

© Springer Nature Switzerland AG 2020
A. Bedford, K. M. Liechti, *Mechanics of Materials*,
https://doi.org/10.1007/978-3-030-22082-2_6

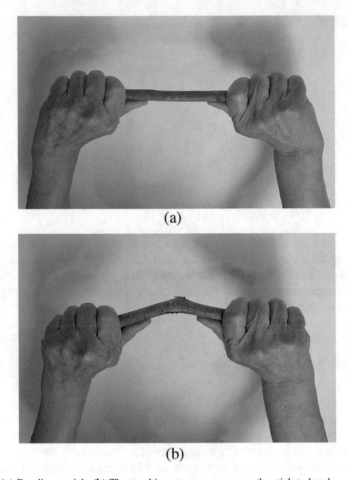

(a)

(b)

Fig. 6.1 (**a**) Bending a stick. (**b**) The resulting stresses can cause the stick to break

Fig. 6.2 (**a**) A prismatic
beam with a symmetrical
cross section. (**b**) Subjecting
the beam to couples at
the ends

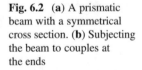

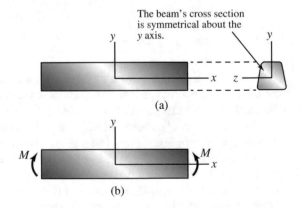

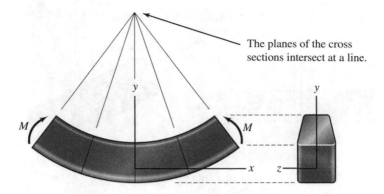

The planes of the cross sections intersect at a line.

Fig. 6.3 Deformation resulting from the applied couples. Each cross section remains plane

cross section remains plane and perpendicular to the beam's curved longitudinal axis. The latter phenomenon is traditionally expressed by the Henry Higgins sounding statement: "Plane sections remain plane." The planes of the cross sections intersect at a line.

Figure 6.4(a) shows a longitudinal plane within the beam that is parallel to the x–z axis before the couples are applied. When the couples are applied, such a plane deforms into a cylindrical surface (Fig. 6.4(b)). Planes near the top of the beam decreases in length when the couples are applied (Fig. 6.4(c)). Planes near the bottom of the beam increase in length. The plane whose length does not change is called the *neutral surface*. The intersection of the neutral surface with a given cross section of the beam is a line called the *neutral axis* of the cross section (Fig. 6.4(d)).

Let us assume that the origin of the coordinate system lies on the neutral axis and consider an element of the beam that is of width dx before the couples are applied (Fig. 6.5). We show this element after the couples are applied in Fig. 6.6, isolating it so that we can analyze its geometry. The radius of the beam's neutral axis is denoted by ρ. Because we assume that the origin lies on the neutral axis, the width of the element at $y = 0$ equals dx after the couples are applied. The width of the element increases below the neutral axis and decreases above it. Let dx' be the width of the element at a distance y from the neutral axis (Fig. 6.6). In terms of the width before and after the deformation, the normal strain in the x direction is

$$\varepsilon_x = \frac{dx' - dx}{dx} = \frac{dx'}{dx} - 1. \tag{6.1}$$

We can express dx and dx' in terms of the radius ρ and the angle $d\theta$ shown in Fig. 6.6:

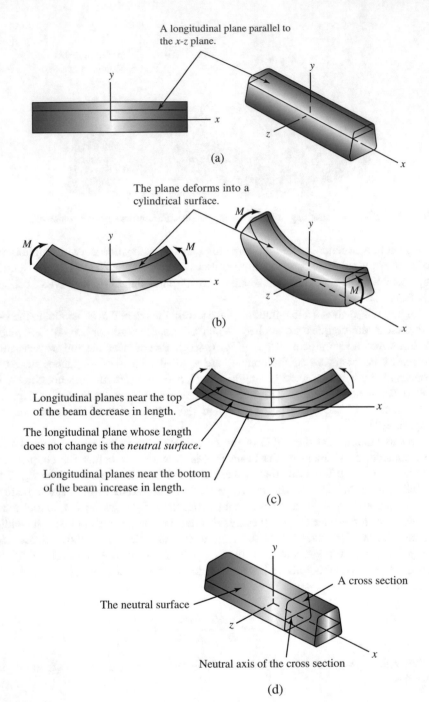

A longitudinal plane parallel to the x-z plane.

(a)

The plane deforms into a cylindrical surface.

(b)

Longitudinal planes near the top of the beam decrease in length.

The longitudinal plane whose length does not change is the *neutral surface*.

Longitudinal planes near the bottom of the beam increase in length.

(c)

The neutral surface

A cross section

Neutral axis of the cross section

(d)

Fig. 6.4 (**a**) A particular plane within the beam. (**b**) Deformation of the plane when the couples are applied. (**c**) Definition of the neutral surface. (**d**) The neutral axis of a particular cross section

Fig. 6.5 Element of the
beam of length dx

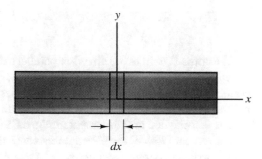

Fig. 6.6 Geometry of the
element after the couples are
applied

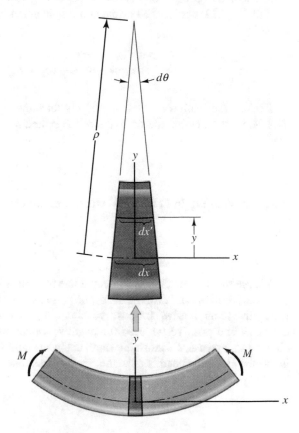

$$dx = \rho \, d\theta,$$
$$dx' = (\rho - y) \, d\theta.$$

Dividing the second equation by the first,

$$\frac{dx'}{dx} = 1 - \frac{y}{\rho},$$

and substituting this result into Eq. (6.1), we obtain

$$\varepsilon_x = -\frac{y}{\rho}. \tag{6.2}$$

The normal strain in the direction parallel to the beam's axis is a linear function of y, which can also be seen from the shape of the element in Fig. 6.6. The negative sign confirms that the width of the element decreases in the positive y direction (above the neutral axis) and increases in the negative y direction (below the neutral axis).

The deformation of the element in Fig. 6.6 implies the presence of normal stresses on the vertical faces of the element that cause the material to be stretched below the neutral axis and compressed above it. Assuming that the material is one for which Eqs. (2.7)–(2.12) apply, the normal strain ε_x is given in terms of the components of stress by

$$\varepsilon_x = \frac{1}{E}\sigma_x - \frac{\nu}{E}\left(\sigma_y + \sigma_z\right). \tag{6.3}$$

Because the beam we are considering is not subjected to loads perpendicular to its axis, let us assume that the normal stresses σ_y and σ_z are zero. Then Eq. (6.3) states that

$$\sigma_x = E\varepsilon_x. \tag{6.4}$$

Substituting Eq. (6.2) into this expression, we obtain

$$\sigma_x = -\frac{Ey}{\rho}. \tag{6.5}$$

We see that the material is subjected to a normal stress σ_x that is a linear function of y. The result of our "effect-cause" analysis is shown in Fig. 6.7. The normal stress is negative (compressive) for positive values of y, causing the width of the element to decrease, and positive (tensile) for negative values of y, causing the width of the element to increase. Because the same analysis applies no matter where the element is located along the beam's axis, the normal stress at every cross section is described by Eq. (6.5).

Fig. 6.7 Normal stresses on the vertical faces of the element

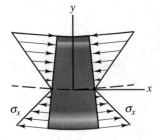

6.1.2 *Relation Between Normal Stress and Bending Moment*

Equations (6.2) and (6.5) indicate that the normal strain ε_x and normal stress σ_x are linear functions of the distance y from the neutral axis and that they are inversely proportional to the radius of curvature ρ. But we have not determined the location of the neutral axis, and we don't know the relationship between ρ and the applied couple M.

In Fig. 6.8 we obtain a free-body diagram by passing a plane perpendicular to the beam's axis. Because the couple M exerts no net force, the horizontal force exerted on the free-body diagram by the distribution of stress σ_x must be zero if the beam is in equilibrium. Letting dA be an element of the beam's cross-sectional area, the horizontal force is

$$\int_A \sigma_x\, dA = 0. \tag{6.6}$$

Substituting Eq. (6.5) into this equation, we find that the horizontal force exerted on the free-body diagram is zero only if

$$\int_A y\, dA = 0. \tag{6.7}$$

From the equation for the y coordinate of the centroid of the beam's cross section,

$$\bar{y} = \frac{\int_A y\, dA}{\int_A dA},$$

we see that Eq. (6.7) implies that the origin of the coordinate system coincides with the centroid. Because we had assumed the origin of the coordinate system to be on the neutral axis, *the centroid of the beam's cross section lies on the neutral axis.*

We have determined the location of the neutral axis from the condition that the sum of the forces on the free-body diagram in Fig. 6.8 must equal to zero. We can relate the radius of curvature ρ of the neutral axis to the couple M, and thus determine the distribution of the normal stress in terms of M, from the condition that the sum of the moments on the free-body diagram must equal to zero. From Fig. 6.9, the moment about the z axis due to the normal stress acting on an element dA of the

Fig. 6.8 Free-body diagram obtained by passing a plane perpendicular to the beam's axis

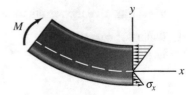

Fig. 6.9 Determining the
moment due to the normal
stress acting on an
element dA

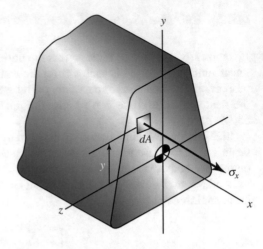

beam's cross section is $-y\sigma_x\,dA$. The total moment about the z axis due to the stress
distribution and the couple M acting on the free-body diagram in Fig. 6.8 is therefore

$$\int_A - y\sigma_x\,dA - M = 0.$$

By substituting Eq. (6.5) into this expression, we can obtain a relation between
M and ρ. We write the resulting equation as

$$\frac{1}{\rho} = \frac{M}{EI}, \tag{6.8}$$

where

$$I = \int_A y^2\,dA$$

is the moment of inertia of the beam's cross-sectional area about the z axis. We now
substitute Eq. (6.8) into Eq. (6.5) to obtain the normal stress distribution in terms
of M:

$$\sigma_x = -\frac{My}{I}. \tag{6.9}$$

We summarize the results for a beam subjected to couples:

1. Radius of curvature:

$$\frac{1}{\rho} = \frac{M}{EI}. \tag{6.10}$$

The sign of ρ indicates the direction of the curvature of the neutral axis. If ρ is positive, the positive y axis is on the concave side of the neutral axis. The product EI is called the *flexural rigidity* of the beam.

2. Distribution of the normal strain:

$$\varepsilon_x = -\frac{y}{\rho} = -\frac{My}{EI}. \qquad (6.11)$$

3. Distribution of the normal stress:

$$\sigma_x = -\frac{My}{I}. \qquad (6.12)$$

6.1.3 Beams Subjected to Arbitrary Loads

The results we have derived apply to a prismatic beam of linear elastic material whose cross section is symmetric about the y axis. Suppose that such a beam could be subjected to couples M at the ends by applying normal stresses given by Eq. (6.12). Then it can be shown that the exact state of stress at each point in the beam consists of the component of normal stress σ_x given by Eq. (6.12) with the other components of stress equal to zero. No matter how the couples M are applied, Saint-Venant's principle (see Sect. 3.1) implies that this state of stress approximates the state of stress in the bar except near the ends.

The internal bending moment in a beam loaded as shown in Fig. 6.10 has the same value M at every cross section. We have seen in Chap. 5 that the internal bending moment in a beam subjected to arbitrary loading is a function of position

Fig. 6.10 Beam subjected to positive couples M

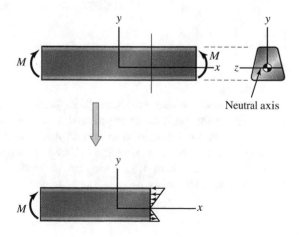

Neutral axis

along the beam's axis. If such a beam is slender (its cross-sectional dimensions are small in comparison with its length), the radius of curvature and distributions of normal strain and normal stress *at a given cross section* can be approximated using Eqs. (6.10)–(6.12). When this is done, M is the value of the bending moment at the given cross section, and ρ is the radius of curvature of the beam's neutral axis in the neighborhood of the cross section.

Study Questions

1. What is the neutral axis of a beam? Where is it located?
2. What is the term I in Eqs. (6.10)–(6.12)?
3. What is the flexural rigidity?
4. If you know the bending moment M at a given cross section of a beam, how can you use Eq. (6.12) to determine the maximum tensile and compressive stresses at that cross section?

Example 6.1 Determining Normal Stresses

The beam in Fig. 6.11 is subjected to couples $M = 4\,\text{kN-m}$. It consists of aluminum alloy with modulus of elasticity $E = 70\,\text{GPa}$. The shape and dimensions of the cross section are shown. Determine the resulting radius of curvature of the beam's neutral axis. What are the maximum tensile and compressive normal stresses, and where do they occur?

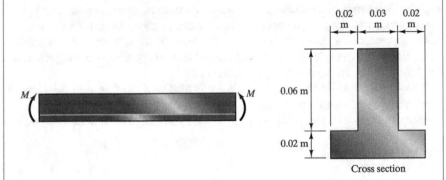

Cross section

Fig. 6.11

Strategy

The radius of curvature is given by Eq. (6.10). To apply that equation, we must determine the position of the neutral axis (the centroid of the beam's cross section lies on the neutral axis) and then use the parallel axis theorem to determine the moment of inertia I. The distribution of normal stress is given by Eq. (6.12), from which we can determine the maximum tensile and compressive stresses.

Solution

We can use any convenient coordinate system to locate the centroid of the cross section (Fig. (a)). Dividing the cross section into rectangles 1 and 2 shown in Fig. (a), the y coordinate of its centroid is

$$\bar{y} = \frac{\bar{y}_1 A_1 + \bar{y}_2 A_2}{A_1 + A_2} = \frac{(0.01 \text{ m})(0.07 \text{ m})(0.02 \text{ m}) + (0.05 \text{ m})(0.03 \text{ m})(0.06 \text{ m})}{(0.07 \text{ m})(0.02 \text{ m}) + (0.03 \text{ m})(0.06 \text{ m})}$$

$$= 0.0325 \text{ m}.$$

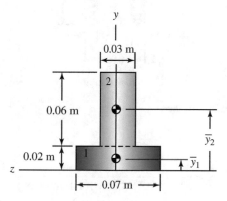

(a) Coordinate system for determining the position of the neutral axis

Placing the origin of the coordinate system on the neutral axis (Fig. (b)), we apply the parallel axis theorem to rectangles 1 and 2 to determine the moment of inertia of the cross section about the z axis:

$$I = I_1 + I_2$$

$$= \left(\frac{1}{12} b_1 h_1^3 + d_1^2 A_1 \right) + \left(\frac{1}{12} b_2 h_2^3 + d_2^2 A_2 \right)$$

$$= \left[\frac{1}{12}(0.07 \text{ m})(0.02 \text{ m})^3 + (0.0225 \text{ m})^2 (0.07 \text{ m})(0.02 \text{ m}) \right]$$

$$+ \left[\frac{1}{12}(0.03 \text{ m})(0.06 \text{ m})^3 + (0.0175 \text{ m})^2 (0.03 \text{ m})(0.06 \text{ m}) \right]$$

$$= 1.85\text{E-}6 \text{ m}^4.$$

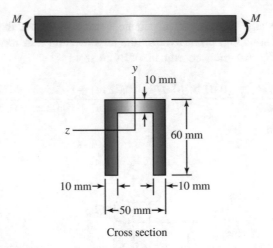

Cross section

(b) Applying the parallel axis theorem

From Eq. (6.10),

$$\frac{1}{\rho} = \frac{M}{EI} = \frac{4000 \text{ N-m}}{(70\text{E9 N/m}^2)(1.85\text{E-6 m}^4)},$$

we obtain $\rho = 32.3$ m.

From Eq. (6.12), the normal stress is

$$\sigma_x = -\frac{My}{I} = -\frac{(4000 \text{ N-m})y}{1.85\text{E-6 m}^4} = -2.17\text{E9}\,y \text{ Pa.}$$

The distribution of normal stress is shown in Fig. (c). The maximum tensile stress occurs at the bottom of the beam:

$$\sigma_x = -(2.17\text{E9})(-0.0325) \text{ Pa} = 70.4 \text{ MPa.}$$

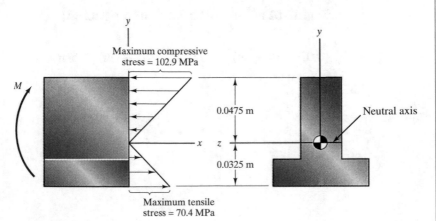

(c) Distribution of the normal stress

The maximum compressive stress occurs at the top:

$$\sigma_x = -(2.17E9)(0.0475) \text{ Pa} = -102.9 \text{ MPa}.$$

Discussion
Notice that the maximum tensile stress is smaller in magnitude than the maximum compressive stress. This occurs in this example because the shape of the cross section causes the neutral axis to be closer to the bottom of the beam, where the maximum tensile stress occurs (Fig. (c)). For this reason, this type of cross section is often used in designing beams made of brittle materials such as concrete, which are relatively weak in tension and strong in compression.

Example 6.2 Determining a Normal Stress
For the beam in Fig. 6.12, determine the normal stress due to bending at point Q.

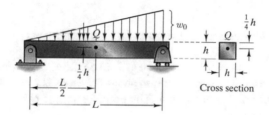

Fig. 6.12

Strategy
We must first determine the value of the bending moment M at the cross section containing point Q. Then we can obtain the normal stress from Eq. (6.12).

Solution
By applying the equilibrium equations to a free-body diagram of the entire beam, we obtain the reactions shown in Fig. (a).

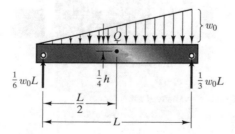

(a) Reactions at the supports

We obtain a free-body diagram by passing a plane through Q (Fig. (b)) and represent the distributed load by an equivalent force (Fig. (c)). From the equilibrium equation

$$\Sigma M_{\text{right end}} = M + \left[\frac{1}{3}\left(\frac{L}{2}\right)\right]\left(\frac{1}{8}w_0L\right) - \left(\frac{L}{2}\right)\left(\frac{1}{6}w_0L\right) = 0,$$

we find that the bending moment at the cross section containing Q is

$$M = \frac{1}{16}w_0L^2.$$

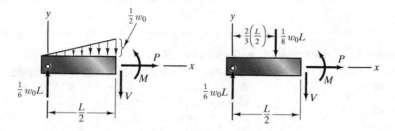

(b) and (c) Free-body diagrams of the part of the beam to the left of Q

Placing the origin of the coordinate system on the neutral axis (Fig. (d)), the moment of inertia of the cross section about the z axis is $I = (1/12)h^4$, and the y coordinate of point Q is $y = h/4$. The normal stress at Q is

$$\sigma_x = -\frac{My}{I} = -\frac{\left(\frac{1}{16}w_0L^2\right)\left(\frac{h}{4}\right)}{\frac{1}{12}h^4} = -\frac{3w_0L^2}{16h^3}.$$

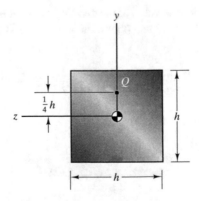

(d) Position of Q relative to the neutral axis

Discussion

We designed this example to demonstrate the meanings of the terms in Eq. (6.12). To determine the normal stress at point Q of the beam, we needed to determine the value of the bending moment M at the cross section containing Q. With the origin of the coordinate system placed on the neutral axis (at the centroid) of the beam's cross section, the term I is the moment of inertia of the cross section about the z axis. Then we obtained the normal stress from Eq. (6.12) by letting y be the vertical coordinate of point Q.

Example 6.3 Determining a Maximum Normal Stress

For the beam in Fig. 6.13, what is the maximum tensile stress due to bending, and where does it occur?

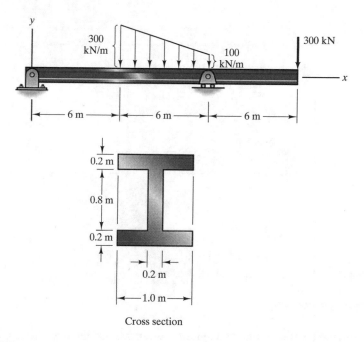

Fig. 6.13

Strategy

The maximum tensile stress will occur at the cross section where the magnitude of the bending moment M is greatest. To locate this cross section and calculate the bending moment, M must be determined as a function of x. We can then determine the maximum tensile stress from Eq. (6.12).

Solution

The distribution of the bending moment for this beam and loading are

$$0 < x < 6\,\text{m}, \qquad M = 200x\,\text{kN-m},$$

$$6 < x < 12\,\text{m}, \qquad M = \frac{50}{9}x^3 - 250x^2 + 2600x - 6600\,\text{kN-m},$$

$$12 < x < 18\,\text{m}, \qquad M = 300x - 5400\,\text{kN-m}.$$

The bending moment diagram is shown in Fig. (a).

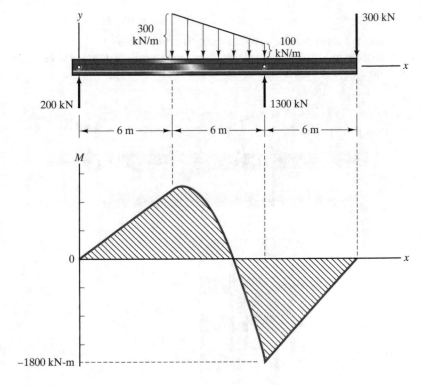

(a) Bending moment diagram

From the bending moment diagram, we can see that the magnitude of M is greatest either where M attains its maximum positive value within the interval $6 < x < 12$ m or at $x = 12$ m. To determine the maximum value of M within the interval $6 < x < 12$ m, we equate the derivative of M with respect to x in that interval to zero:

$$\frac{dM}{dx} = \frac{50}{3}x^2 - 500x + 2600 = 0.$$

Solving this equation, we find that the maximum occurs at $x = 6.69$ m. Substituting this value of x into the equation for M, we obtain $M = 1270$ kN-m. Therefore, the greatest magnitude of the bending moment occurs at $x = 12$ m, where $M = -1800$ kN-m $= -1.8$E6 N-m.

Applying the parallel axis theorem to the cross section (Fig. (b)), the moment of inertia about the z axis is

$$I = I_1 + 2I_2$$
$$= \frac{1}{12}(0.2 \text{ m})(0.8 \text{ m})^3 + 2\left[\frac{1}{12}(1.0 \text{ m})(0.2 \text{ m})^3 + (0.5 \text{ m})^2(1.0 \text{ m})(0.2 \text{ m})\right]$$
$$= 0.110 \text{ m}^4.$$

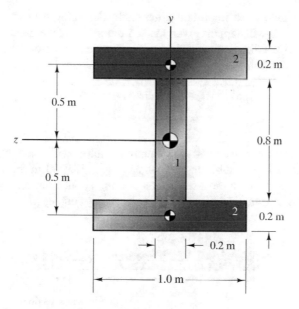

(b) Dividing the cross section into rectangles

From Eq. (6.12), the distribution of the normal stress at $x = 12$ m is

$$\sigma_x = -\frac{My}{I} = \frac{(1.8\text{E6 N-m}) y}{0.110 \text{ m}^4}$$
$$= 16.4\text{E6} y \text{ Pa}.$$

The maximum tensile stress occurs at $y = 0.6$ m (Fig. (c)):

$$\sigma_x = (16.4\text{E6})(0.6) \text{ Pa} = 9.83\text{E6 Pa}.$$

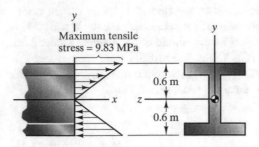

(c) Distribution of normal stress at $x = 12$ m

Discussion
Often the single most important criterion in designing a beam or deciding whether a beam will support given loads is determining the maximum normal stress due to bending, as we did in this example. We discuss the design of beams based on this criterion in the following section.

Problems

6.1 The prismatic beam consists of aluminum alloy with modulus of elasticity $E = 12\text{E}6$ psi. It is subjected to couples $M = 25, 000$ in-lb. (a) What is the moment of inertia I of the beam's cross section about the z axis? (b) What is the radius of curvature of the neutral axis due to the couples M?

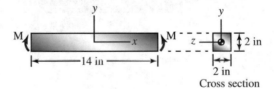

Cross section

Problems 6.1–6.2

6.2 The prismatic beam is subjected to couples $M = 25, 000$ in-lb. (a) What is the normal stress σ_x in the beam at $y = 0.5$ in? (b) What is the maximum tensile stress in the beam?

6.3 The prismatic beam consists of material with modulus of elasticity $E = 70$ GPa. It is subjected to couples $M = 250$ kN-m. (a) What is the resulting radius of curvature of the neutral axis? (b) Determine the maximum tensile stress due to the couples M.

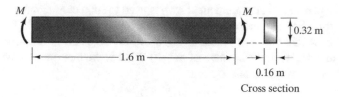

0.32 m

1.6 m

0.16 m

Cross section

Problems 6.3–6.4

6.4 The beam's material will safely support a tensile stress of 180 MPa and a compressive stress of 200 MPa. Based on these criteria, what are the largest couples M to which the beam can be subjected?

6.5 The beam consists of material with modulus of elasticity $E = 14\,E6$ psi and is subjected to couples $M = 150,\,000$ in-lb. **(a)** What is the resulting radius of curvature of the neutral axis? **(b)** Determine the maximum tensile stress in the beam.

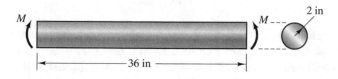

2 in

36 in

Problem 6.5

6.6 A beam with the hollow circular cross section shown consists of material that will safely support a tensile or compressive stress of 30,000 psi. What is the largest bending moment M to which the beam can be subjected?

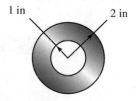

1 in

2 in

Problem 6.6

6.7 The prismatic beam consists of material with modulus of elasticity $E = 190$ GPa. When it is subjected to couples $M = 9000$ N-m, the resulting radius of curvature of the neutral axis is measured and determined to be $\rho = 475$ m. What are the maximum tensile and compressive stresses in the material?

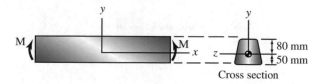

y

y

M

M

x

z

80 mm

50 mm

Cross section

Problem 6.7

6.8   The experimental arrangement shown is used to subject a segment of a beam to a uniform bending moment. The mass of each suspended weight is $m = 15$ kg. **(a)** Show that between B and C the beam is subjected to a uniform bending moment M. What is the value of M? **(b)** What is the maximum tensile stress in the beam between B and C?

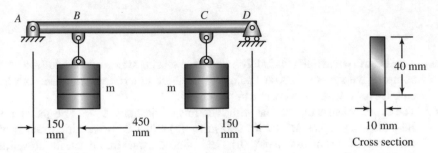

Problem 6.8

6.9∗ The beam consists of material that will safely support a tensile or compressive stress of 350 MPa. Based on this criterion, determine the largest force F the beam will safely support if it has the cross section (a) and if it has the cross section (b). (The two cross sections have approximately the same area.)

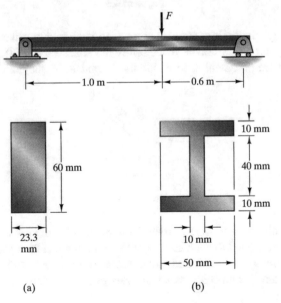

Problems 6.9–6.10

6.10 The beam is subjected to a force $F = 6\,\text{kN}$. Determine the maximum tensile stress due to bending at the cross section midway between the beam's supports if it has the cross section (a) and if it has the cross section (b).

6.11* A prismatic beam has the cross section shown. (a) What is the moment of inertia I of the beam's cross section about the z axis? (b) At a particular cross section, the beam is subjected to the bending moment $M = 2000\,\text{kN-m}$. What are the values of the resulting maximum tensile and compressive stresses?

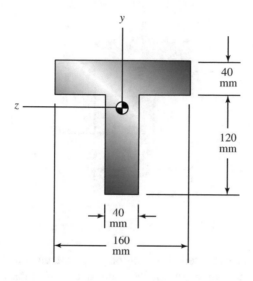

Problem 6.11

6.12* The beam has the cross section shown in Problem 6.11. (a) What is the maximum bending moment in the beam, and where does it occur? (b) Determine the maximum tensile stress in the beam.

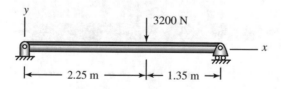

Problem 6.12

6.13 The beam has the cross section shown. The tensile stress at point Q is 4.6 MPa. What is the magnitude w of the distributed load?

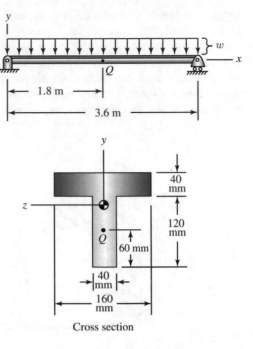

Cross section

Problem 6.13

6.14 The beam is subjected to a uniformly distributed load $w_0 = 300\,\text{lb/in}$. Determine the maximum tensile stress due to bending at $x = 20\,\text{in}$ if the beam has the cross section (a) and if it has the cross section (b). (The two cross sections have approximately the same area.)

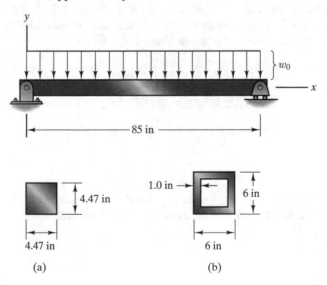

(a)   (b)

Problems 6.14–6.15

6.15 The beam consists of material that will safely support a tensile or compressive stress of 30 ksi. Based on this criterion, determine the largest distributed load w_0 (lb/in) that the beam will safely support if it has the cross section (a) and if it has the cross section (b).

6.16 A bandsaw blade with 2-mm thickness and 20-mm width is wrapped around a pulley with 160-mm radius. The blade is made of steel with modulus of elasticity $E = 200$ GPa. What maximum tensile stress is induced in the blade as a result of being wrapped around the pulley?

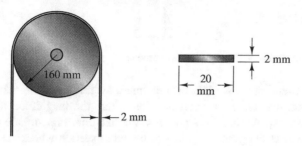

Problem 6.16

6.2   Design Issues

For many applications, the most essential requirement in designing a beam is to ensure that the maximum normal stresses induced by bending moments do not exceed allowable values. In this section we discuss the design of beams *based on this criterion alone*.

6.2.1   Factor of Safety and Section Modulus

The design engineer seeks to ensure that the maximum magnitude of the normal stress in a beam does not exceed the yield stress σ_Y of the material or some chosen fraction of the yield stress, the allowable stress σ_{allow}. The ratio of the yield stress to the allowable stress is the *factor of safety* of the beam:

$$\text{FS} = \frac{\sigma_Y}{\sigma_{\text{allow}}}. \tag{6.13}$$

The distribution of the normal stress in a beam is given by Eq. (6.12):

$$\sigma_x = -\frac{My}{I}. \tag{6.14}$$

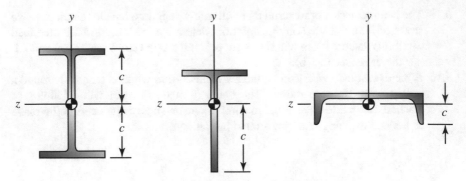

Fig. 6.14 The dimension c for various cross sections

From this equation we see that the maximum magnitude of the normal stress in a prismatic beam occurs at the axial position where the magnitude of the bending moment is greatest. At that axial position, the maximum magnitude of the normal stress will occur at the point or points of the cross section where the magnitude of y is greatest. That is, the magnitude of the normal stress will be greatest at the point or points of the cross section that are farthest from the neutral axis. Let c denote the greatest magnitude of y for a given cross section, as shown in the examples in Fig. 6.14. The ratio of the moment of inertia I of the cross section about the z axis to c,

$$S = \frac{I}{c},\tag{6.15}$$

is called the *section modulus* of the cross section about the z axis. Expressing Eq. (6.14) in terms of the section modulus and the maximum magnitude of the bending moment $|M|_{\mathrm{max}}$, the maximum magnitude of the normal stress in the beam is

$$|\sigma_x|_{\mathrm{max}} = \frac{|M|_{\mathrm{max}}}{S}.\tag{6.16}$$

With these definitions and equations, we can determine the minimum section modulus that a beam consisting of a given material and subjected to specified loads must have in order to achieve a prescribed factor of safety based on the consideration of normal stress alone:

1. Using the yield stress σ_Y of the given material and the prescribed factor of safety FS, solve Eq. (6.13) for the allowable stress σ_{allow}.
2. Given the supports and loads acting on the beam, calculate the distribution of the bending moment, and use it to determine the maximum magnitude of the bending moment $|M|_{\mathrm{max}}$.
3. Assume that the maximum magnitude of the normal stress in the beam $|\sigma_x|_{\mathrm{max}}$ is equal to the allowable stress σ_{allow}, and solve Eq. (6.16) for the section modulus S.

Example 6.4 Factor of Safety of a Beam

The maximum anticipated magnitude of the load on the beam in Fig. 6.15 is $w_0 = 150$ kN/m. The beam's length is $L = 3$ m. The two candidate cross sections (a) and (b) have approximately the same cross-sectional area. The beam is to be made of material with yield stress $\sigma_Y = 700$ MPa. Compare the beam's factor of safety if it has cross section (a) to its factor of safety with cross section (b).

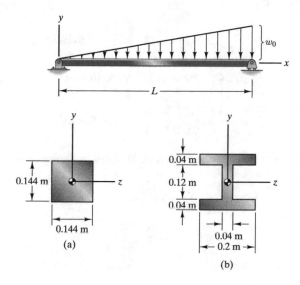

Fig. 6.15

Strategy

We must determine the maximum normal stress resulting from the given load for each cross section. This requires determining the maximum bending moment in the beam and applying Eq. (6.16) for each cross section.

Solution

The distribution of the bending moment in the beam is (see Problem 5.48)

$$M = \frac{1}{6} w_0 \left(Lx - \frac{x^3}{L} \right).$$

To determine where the maximum value of M occurs, we equate the derivative of M with respect to x to zero,

$$\frac{dM}{dx} = \frac{1}{6} w_0 \left(L - \frac{3x^2}{L} \right) = 0,$$

and solve for x, obtaining $x = L/\sqrt{3}$. Substituting this value into the equation for M, the maximum magnitude of the bending moment is

$$|M|_{max} = \frac{w_0 L^2}{9\sqrt{3}}$$

$$= \frac{(150,000 \text{ N/m}) (3 \text{ m})^2}{9\sqrt{3}}$$

$$= 86,600 \text{ N-m}.$$

(a) **Square cross section** The moment of inertia of the square cross section about the z axis is

$$I_x = \frac{1}{12}(0.144 \text{ m})(0.144 \text{ m})^3 = 3.58\text{E-5 m}^4.$$

The value of c is (0.144 m)/2 = 0.072 m, so the section modulus is

$$S = \frac{I}{c} = \frac{3.58\text{E-5 m}^4}{0.072 \text{ m}} = 4.98\text{E-4 m}^3.$$

The maximum magnitude of the normal stress in the beam is

$$|\sigma_x|_{max} = \frac{|M|_{max}}{S}$$

$$= \frac{86,600 \text{ N-m}}{4.98\text{E-4 m}^3}$$

$$= 174 \text{ MPa}.$$

The beam's factor of safety is

$$\text{FS} = \frac{\sigma_Y}{\sigma_{allow}}$$

$$= \frac{700 \text{ MPa}}{174 \text{ MPa}}$$

$$= 4.02.$$

(b) **I-beam cross section** The moment of inertia of the vertical rectangle about the z axis is

$$I_v = \frac{1}{12}(0.04 \text{ m})(0.12 \text{ m})^3 = 0.576\text{E-5 m}^4.$$

Applying the parallel axis theorem, the moment of inertia of each horizontal rectangle about the z axis is

$$I_h = \frac{1}{12}(0.2 \text{ m})(0.04 \text{ m})^3 + (0.06 \text{ m} + 0.02 \text{ m})(0.2 \text{ m})(0.04 \text{ m})$$

$$= 5.227\text{E-5 m}^4.$$

Therefore the moment of inertia of the entire cross section is

$$I = I_v + 2I_h = 11.03\text{E-5 m}^4.$$

The value of c is 0.06 m + 0.04 m = 0.1 m, resulting in the section modulus

$$S = \frac{I}{c} = \frac{11.03\text{E-5 m}^4}{0.1 \text{ m}} = 11.03\text{E-4 m}^3.$$

The maximum magnitude of the normal stress in the beam is

$$\begin{aligned}
|\sigma_x|_{\text{max}} &= \frac{|M|_{\text{max}}}{S} \\
&= \frac{86{,}600 \text{ N-m}}{11.03\text{E-4 m}^3} \\
&= 78.5 \text{ MPa}.
\end{aligned}$$

The beam's factor of safety for this cross section is

$$\begin{aligned}
\text{FS} &= \frac{\sigma_Y}{\sigma_{\text{allow}}} \\
&= \frac{700 \text{ MPa}}{78.5 \text{ MPa}} \\
&= 8.91.
\end{aligned}$$

Discussion
The advantage of a cross section that increases a beam's moment of inertia is apparent in this example. But in particular applications, other factors, such as the availability of stock with a desired cross section, or the cost of manufacturing it, may be overriding considerations.

6.2.2 Standardized Cross Sections

Notice from Eq. (6.14) that the maximum magnitude of the normal stress in a beam is inversely proportional to the moment of inertia I. This explains in large part the designs of many of the beams we observe in use, for example, in highway overpasses and in the frames of buildings. Their cross sections are configured to increase their moments of inertia. Each of the cross sections in Fig. 6.16 has the same cross-sectional area. Below each cross section is the ratio of the moment of inertia I about the z axis to the value of I for the solid square cross section. Clearly, the capacity of a beam to support given loads without failing due to normal stresses can be augmented by tailoring its cross section to increase the moment of inertia.

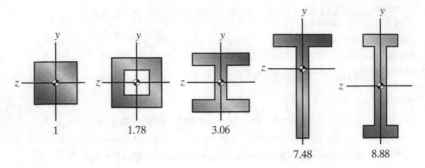

Fig. 6.16 Beam cross sections and the ratio of *I* to the value for a solid square beam with the same cross-sectional area

Designation	Cross-Sectional Area A, mm^2	Dimensions, mm				Moments of Inertia and Section Moduli			
		d	b_f	t_f	t_w	Axis y		Axis z	
						I_y,10^6 mm^4	S_y,10^3 mm^3	I_z,10^6mm^4	S_z,10^3 mm^3
W610X82	10500	599	178	12.8	10.0	12.1	136	565	1890

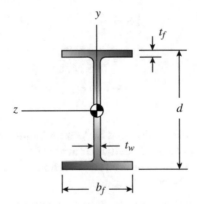

Fig. 6.17 A typical entry from Appendix E

Tables of standardized, commercially available cross sections that have enhanced moments of inertia are published for the convenience of design engineers. Examples of the kinds of information and data that are available are presented in Appendix E, which is abridged with permission from the AISC Shapes Database Version 3.0 of the American Institute of Steel Construction. An entry from Appendix E is shown together with its accompanying figure in Fig. 6.17. The cross section is uniquely identified by its designation W610X82, where W indicates the type of cross section (a wide-flange cross section), 610 is the nominal, or approximate depth *d* in millimeters, and 82 is the mass per unit length in kg/m of a steel beam having this cross section. Notice in Appendix E that the designations of some types of cross sections provide other kinds of information. The tabulated information in Fig. 6.17 includes the cross-sectional area *A*, the exact dimensions of the cross section *d*, b_f, t_f and t_w, and its moments of inertia and section moduli referred to the y and z axes. With this type of information, design engineers can carry out detailed structural design using cross sections that are readily obtainable and achieve desired factors of safety based on the maximum magnitudes of the normal stresses in beams. Additional aspects of design that structural designers must consider in choosing the material and factor of safety of a beam were discussed in Sect. 3.6.

Example 6.5 Beam Design Using Standardized Cross Sections
A beam used in a support structure will consist of ASTM-A572 steel and will be subjected to a maximum bending moment of magnitude $|M|_{max} = 3.8\text{E}6$ in-lb. An S (American Standard) cross section is shown in Fig. 6.18. Choose an S cross section in Appendix E with nominal depth $d = 24$ in so that the beam has a factor of safety of at least 3.

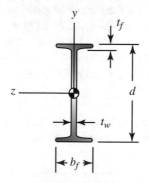

American Standard
cross section

Fig. 6.18

Strategy
Obtaining the yield stress of ASTM-A572 steel from Appendix B and using the given factor of safety, we can solve Eq. (6.13) for the allowable stress. Then by assuming the maximum magnitude of the normal stress in the beam is equal to the allowable stress, we can determine the minimum necessary section modulus from Eq. (6.16) and choose a cross section from Appendix E on that basis.

Solution
Determine the allowable stress   From Appendix B, the yield stress of ASTM-A572 steel is $\sigma_Y = $ 50E3 psi. From Eq. (6.13),

$$FS = \frac{\sigma_Y}{\sigma_{allow}} :$$

$$3 = \frac{50E3 \text{ psi}}{\sigma_{allow}},$$

the allowable stress is $\sigma_{allow} = $ 16.7E3 psi.

Calculate the minimum section modulus   Equating the allowable stress to the maximum magnitude of the normal stress $|\sigma_x|_{max}$ in Eq. (6.16),

$$|\sigma_x|_{max} = \frac{|M|_{max}}{S} :$$

$$16.7E3 \text{ lb/in}^2 = \frac{3.8E6 \text{ in-lb}}{S},$$

we obtain the minimum necessary section modulus:

$$S = 228 \text{ in}^3.$$

Choose a cross section   Of the S cross sections in Appendix E with nominal depths $d = $ 24 in, both S24X106 and S24X121 have section moduli greater than the minimum necessary value. Either of these choices would provide a beam with the desired factor of safety. Using the cross section S24X106 would result in a lighter, more economical beam.

Discussion
This is a reasonably realistic example of the use of standardized cross sections in the design of beams to support prescribed loads. Also see Problems 6.27–6.30.

Problems

6.17 If the beam is made of a material for which the allowable stress in tension and compression is $\sigma_{\text{allow}} = 120\,\text{MPa}$, what is the largest allowable magnitude of the couple M?

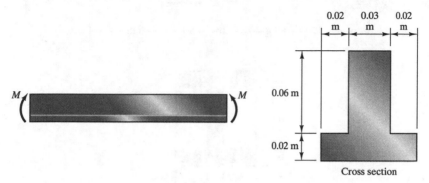

Problem 6.17–6.19

6.18 The beam is made of 7075-T6 aluminum alloy. If it will be subjected to magnitudes of M as large as 10 kN-m, what is the beam's factor of safety? (Assume that the yield stress is the same in tension and compression.)

6.19 The beam is made of a material for which the yield stress in tension is 160 MPa and the yield stress in compression is 200 MPa. If the beam will be subjected to positive values of M as large as 4 kN-m, what is the beam's factor of safety?

6.20 The length of the beam is $L = 9$ ft and it is made of ASTM-A36 structural steel. The maximum anticipated magnitude of the distributed load is $w_0 = 2700\,\text{lb/ft}$. Determine the dimension h so that the beam has a factor of safety FS = 3.

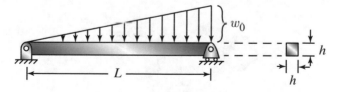

Problem 6.20

6.21 The loads on the beam are the maximum anticipated loads. It is made of wood with yield stress $\sigma_Y = 40\,\text{MPa}$. What is the beam's factor of safety? (See Example 6.3.)

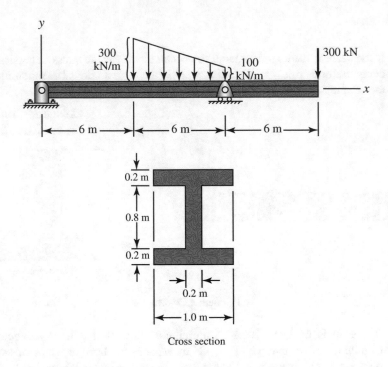

Cross section

Problem 6.21

6.22 A beam made of 7075-T6 aluminum alloy will be subjected to anticipated bending moments as large as 1500 N-m. Determine the beam's factor of safety for two cases: (**a**) it has a solid circular cross section with 20-mm radius; (**b**) it has a hollow circular cross section with 30-mm outer radius and the inner radius chosen so that the beam has the same weight as the beam in case (**a**).

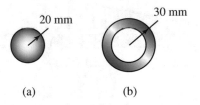

(a)   (b)

Problem 6.22

6.23 The beam is to be made of material with yield stress $\sigma_Y = 700$ MPa and will have a solid circular cross section with radius R. Determine the value of R so that the beam has a factor of safety FS $= 2$. (See Example 6.4.)

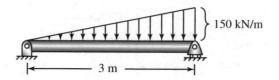

150 kN/m

3 m

Problem 6.23

6.24 The device shown is a playground seesaw. Make a conservative estimate of the maximum weight to which it will be subjected at each end when in use. (Consider contingent situations such as an adult sitting with a child.) Choose a material from Appendix B, and design a cross section for the 4.8-m beam so that it has a factor of safety FS = 4.

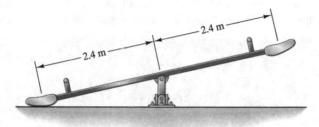

2.4 m

2.4 m

Problem 6.24

6.25* A conservative estimate of the maximum value of the uniformly distributed load on the 8-ft segment of a building's frame is 2000 lb/ft. The beam is to be made of ASTM-A572 steel. Design the I-beam's cross section so that its height is 1 ft and its factor of safety is FS = 4. (Assume the beam has pin supports at the ends.)

Problem 6.25

6.26* The maximum anticipated load on the beam is shown. Choose a material from Appendix B, and design a cross section for the beam so that it has a factor of safety FS = 2.

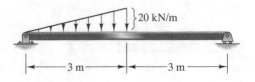

Problem 6.26

6.27 A beam consisting of ASTM-A36 steel (see Appendix B) and having a W150X18 wide-flange cross section (see Appendix E) is to be used in the supporting structure of an architectural dome. The beam will be subjected to an estimated maximum bending moment of magnitude $|M|_{max} = 6$ kN-m. (a) What is the section modulus of the cross section about the z axis? (b) Determine the magnitude of the estimated maximum normal stress to which the beam will be subjected. (c) Determine the beam's factor of safety.

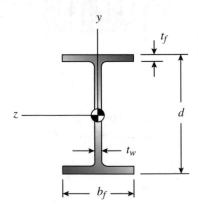

Problems 6.27–6.28 (*Photograph from iStock by B&M Noskowski*)

6.28 A beam consisting of ASTM-A36 steel (see Appendix B) is to be used in the supporting structure of an architectural dome. The beam will be subjected to an estimated maximum bending moment of magnitude $|M|_{max} = 11$ kN-m. Choose a wide-flange cross section in Appendix E with a nominal depth $d = 150$ mm so that the beam will have a factor of safety of at least 4.

6.29 A beam consisting of ASTM-A36 steel (see Appendix B) and having a W6X12 wide-flange cross section (see Appendix E) is to be used in the supporting structure of a pedestrian walkway. The beam will be subjected to an estimated maximum bending moment of magnitude $|M|_{max} = 65, 000$ in-lb. (a) What is the section modulus about the z axis? (b) Determine the magnitude of the estimated maximum normal stress to which the beam will be subjected. (c) Determine the beam's factor of safety.

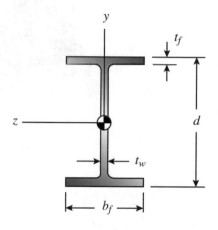

Problems 6.29–6.30

6.30 A beam consisting of ASTM-A36 steel (see Appendix B) and having a W6X12 wide-flange cross section (see Appendix E) is to be used in the supporting structure of a pedestrian walkway. The beam will be subjected to an estimated maximum bending moment of magnitude $|M|_{max} = 90, 000$ in-lb. Choose a wide-flange cross section in Appendix E with a nominal depth $d = 6$ in so that the beam will have a factor of safety of at least 4.

6.3 Composite Beams

If the walls of a box beam are made too thin, the beam can fail by buckling as shown in Fig. 6.19(a). To help prevent buckling of beams with thin walls, a light "filler" material can be used (Fig. 6.19(b)). The filler material does not add significant weight but helps stabilize the walls, resulting in a light, strong beam. This is an example of a *composite beam*, a beam consisting of two or more materials. By taking

Fig. 6.19 (a) A box beam
with thin walls can buckle.
(b) Using a filler material
stabilizes the thin walls

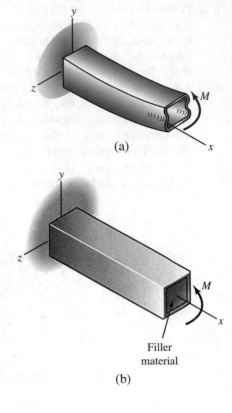

(a)

(b)

Filler
material

Fig. 6.20 Cross section of a
composite beam consisting
of materials A and B

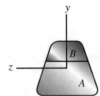

advantage of the different properties of materials, structural engineers can design
composite beams with characteristics that are not possible with a single material. Our
objective is to explain how to determine the normal stress at a given cross section of
a prismatic composite beam consisting of two materials.

6.3.1 Distribution of the Normal Stress

Consider a given cross section of a prismatic composite beam, and let the materials
be denoted A and B (Fig. 6.20). We make no assumption about the geometry of the

cross section except that each of the parts consisting of materials A and B is symmetric about the y axis. Notice that, as in the case of a beam consisting of a single material, we do not know the location of the neutral axis and must determine it as part of our analysis.

If the two materials are bonded together and we retain the fundamental assumption that plane sections remain plane, the geometric arguments leading to Eq. (6.2) hold, and the normal strain in both materials is given by

$$\varepsilon_x = -\frac{y}{\rho},$$

where ρ is the radius of curvature of the neutral axis. Because the materials may have different elastic moduli, we must write Eq. (6.4) for each material, obtaining

$$\sigma_A = E_A \varepsilon_x = -\frac{E_A y}{\rho},$$
$$\sigma_B = E_B \varepsilon_x = -\frac{E_B y}{\rho}. \tag{6.17}$$

Equation (6.6), which expresses the equilibrium requirement that the normal stress distribution exerts no net force, now becomes

$$\int_A \sigma_x \, dA = \int_{A_A} \sigma_A \, dA + \int_{A_B} \sigma_B \, dA = 0,$$

where A_A and A_B denote the cross sections of the materials. Substituting the expressions (6.17) leads to the equation

$$E_A \int_{A_A} y \, dA + E_B \int_{A_B} y \, dA = 0.$$

We can express this equation as

$$E_A A_A \bar{y}_A + E_B A_B \bar{y}_B = 0, \tag{6.18}$$

where $\bar{y}_A$ and $\bar{y}_B$ are the coordinates of the centroids of A_A and A_B relative to the neutral axis. As we demonstrate in Example 6.6, this equation can be used to determine the location of the neutral axis.

The moment about the z axis due to the normal stresses must equal the bending moment M at the given cross section:

$$M = -\int_{A_A} y\sigma_A \, dA - \int_{A_B} y\sigma_B \, dA.$$

Substituting the expressions (6.17) into this equation gives

$$M = \frac{1}{\rho}\left(E_A \int_{A_A} y^2 \, dA + E_B \int_{A_B} y^2 \, dA\right)$$

$$= \frac{1}{\rho}(E_A I_A + E_B I_B),$$

where I_A and I_B are the moments of inertia of A_A and A_B about the z axis. By solving this equation for the radius of curvature ρ and substituting the result into Eq. (6.17), we obtain the distribution of normal stress in the individual materials:

$$\sigma_A = -\frac{My}{I_A + \dfrac{E_B}{E_A}I_B},$$

$$\sigma_B = -\frac{My}{\dfrac{E_A}{E_B}I_A + I_B}. \tag{6.19}$$

Now that we know the stress distributions in the individual materials, we can use them to determine the part of the bending moment M supported by each material:

$$M_A = -\int_{A_A} y\sigma_A \, dA = \frac{I_A}{I_A + \dfrac{E_B}{E_A}I_B}M,$$

$$M_B = -\int_{A_B} y\sigma_B \, dA = \frac{I_B}{\dfrac{E_A}{E_B}I_A + I_B}M. \tag{6.20}$$

Observe that $M_A + M_B = M$.

Example 6.6 Normal Stresses in a Composite Beam
Figure 6.21 is the cross section of a prismatic beam made of steel (material A) and aluminum alloy (material B) with elastic moduli $E_A = 200\,\text{GPa}$ and $E_B = 72\,\text{GPa}$. If $M = 12\,\text{kN-m}$ at a given cross section, determine the distributions of normal stress in the steel and in the aluminum.

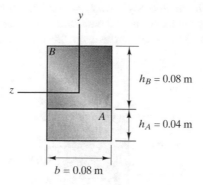

Fig. 6.21

Strategy
We must first determine the position of the neutral axis by using Eq. (6.18).
We can then determine the moments of inertia I_A and I_B of the cross sections of
the two materials about the z axis, and the distributions of the normal stress are
given by Eq. (6.19).

Solution
We begin by placing the coordinate system at an arbitrary position (Fig. (a)).
Equation (6.18) is

$$E_A A_A \bar{y}_A + E_B A_B \bar{y}_B = 0 \ :$$
$$(200 \text{ GPa}) (0.04 \text{ m}) (0.08 \text{ m})\bar{y}_A + (72 \text{ GPa}) (0.08 \text{ m}) (0.08 \text{ m})\bar{y}_B = 0.$$

From Fig. (a) we see that

$$\bar{y}_B - \bar{y}_A = \frac{1}{2}(h_A + h_B) = 0.06 \, \text{m}.$$

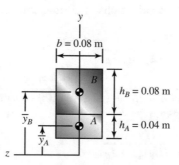

(a) Coordinate system for determining the position of the neutral axis

Solving these two equations yields $\bar{y}_A = -0.0251$ m and $\bar{y}_B = 0.0349$ m. These results tell us the location of the neutral axis. It is 0.0251 m above the centroid of A and 0.0349 m below the centroid of B (Fig. (b)).

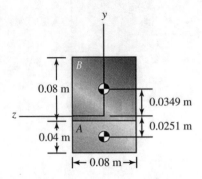

(b) Position of the neutral axis

We can now apply the parallel axis theorem to determine the moments of inertia of the two areas about the z axis:

$$I_A = \frac{1}{12}bh_A^3 + \bar{y}_A^2 A_A$$

$$= \frac{1}{12}(0.08 \text{ m})(0.04 \text{ m})^3 + (-0.0251 \text{ m})^2(0.04 \text{ m})(0.08 \text{ m})$$

$$= 2.45\text{E-}6 \text{ m}^4,$$

$$I_B = \frac{1}{12}bh_B^3 + \bar{y}_B^2 A_B$$

$$= \frac{1}{12}(0.08 \text{ m})(0.08 \text{ m})^3 + (0.0349 \text{ m})^2(0.08 \text{ m})(0.08 \text{ m})$$

$$= 11.2\text{E-}6 \text{ m}^4.$$

From Eq. (6.19), the distributions of normal stress in the two materials are

$$\sigma_A = -\frac{My}{I_A + (E_B/E_A)I_B}$$

$$= -\frac{(12,000 \text{ N-m})y}{2.45\text{E-}6 \text{ m}^4 + (72 \text{ GPa}/200 \text{ GPa})\,11.2\text{E-}6 \text{ m}^4}$$

$$= -1.852y \text{ GPa},$$

$$\sigma_B = -\frac{My}{(E_A/E_B)I_A + I_B}$$

$$= -\frac{(12,000 \text{ N-m})y}{(200 \text{ GPa}/72 \text{ GPa})\,2.45\text{E-}6 \text{ m}^4 + 11.2\text{E-}6 \text{ m}^4}$$

$$= -0.667y \text{ GPa}.$$

The distributions of normal stress are shown in Fig. (c).

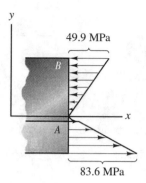

(c) Distribution of the normal stress

Discussion
Notice that the key to determining the distributions of stress in a composite beam is determining the location of the neutral axis and calculating the moments of inertia of the areas A and B about the z axis. Once we had done that in this example, it was straightforward to obtain the distributions of stress in the two materials with Eqs. (6.19).

6.3.2   Transformed Area Method

In Sect. 6.1 we determined the distribution of normal stress in a homogeneous prismatic beam. With the transformed area method, we can obtain the distribution of normal stress in a composite beam by determining the distribution of normal stress in a fictitious homogeneous beam subjected to the same bending moment. The cross section of the fictitious beam is obtained by appropriately changing (transforming) the dimensions of the cross section of the composite beam. Figure 6.22(a) shows the cross section of a prismatic composite beam consisting of materials A and B. Figure 6.22(b) shows the cross section of a fictitious *transformed beam* that *consists entirely of material A*. The area of the cross section of the transformed beam that consisted of material A in the composite beam is unaltered, but the width of the part of the cross section that consisted of material B in the composite beam is scaled (multiplied) by the factor

$$n = \frac{E_B}{E_A}. \tag{6.21}$$

We denote the scaled part of the transformed cross section by T, but keep in mind that it is assumed to consist of the same material as part A. The distribution of normal stress in the transformed beam is given by Eq. (6.12). We will show that *the distribution of normal stress in material A of the composite beam is given by the distribution of normal stress in part A of the transformed beam,*

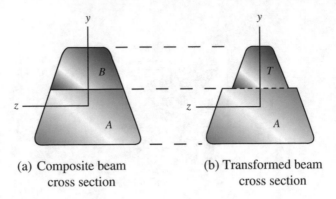

(a) Composite beam
cross section

(b) Transformed beam
cross section

Fig. 6.22 The composite and transformed beam cross sections. The width of part T of the homogeneous transformed beam is equal to the width of part B of the composite beam multiplied by $n = E_B/E_A$

$$\sigma_A = -\frac{My}{I}, \tag{6.22}$$

where I is the moment of inertia of the transformed cross section and y is measured from the neutral axis (centroid) of the transformed cross section. Also, *the distribution of normal stress in material B of the composite beam is obtained by multiplying the distribution of normal stress in part T of the transformed beam by n*:

$$\sigma_B = n\sigma_T = -n\frac{My}{I}. \tag{6.23}$$

To explain why the transformed area method works, suppose that the transformed beam consisting of material A is subjected to couples M (Fig. 6.23(a)). The normal strain is given by Eq. (6.2),

$$\varepsilon_x = -\frac{y}{\rho},$$

where ρ is the radius of curvature of the transformed beam's neutral axis. The distributions of normal stress on parts A and T of the transformed cross section are

$$\sigma_A = E_A\varepsilon_x = -\frac{E_A y}{\rho},$$
$$\sigma_T = E_A\varepsilon_x = -\frac{E_A y}{\rho}. \tag{6.24}$$

The force exerted by normal stress on the free-body diagram in Fig. 6.23(b) must be zero:

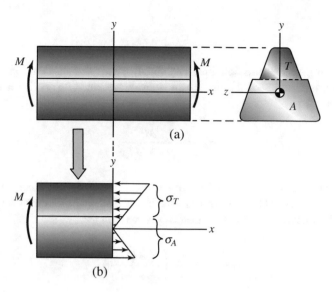

Fig. 6.23 (a) Subjecting the transformed beam to couples. (b) Distribution of the normal stress

$$\int_{A_A} \sigma_A dA + \int_{A_T} \sigma_T dA = 0.$$

Substituting Eq. (6.24) into this equation, we obtain

$$E_A A_A \bar{y}_A + E_A A_T \bar{y}_T = 0. \tag{6.25}$$

Because of the way part B of the composite beam cross section is transformed into part T of the transformed beam cross section, $\bar{y}_T = \bar{y}_B$ and $A_T = nA_B = E_B A_B/E_A$. Therefore we can write Eq. (6.25) as

$$E_A A_A \bar{y}_A + E_B A_B \bar{y}_B = 0.$$

This equation is identical to Eq. (6.18), which proves that *the neutral axis of the transformed beam coincides with the neutral axis of the composite beam*. We next apply Eq. (6.10) to the transformed beam,

$$\frac{1}{\rho} = \frac{M}{E_A I}, \tag{6.26}$$

where ρ is the radius of curvature of the neutral axis and $I = I_A + I_T$ is the moment of inertia of the entire transformed cross section. Because of the way part B of the composite beam cross section is transformed into part T of the transformed beam cross section, $I_T = nI_B = E_B I_B/E_A$. Therefore we can write Eq. (6.26) as

$$\frac{1}{\rho} = \frac{M}{E_A I_A + E_B I_B}.$$

Substituting this expression into Eq. (6.24) yields the distributions of normal stress on parts A and T of the transformed cross section. The result for the normal stress on part A is

$$\sigma_A = -\frac{My}{I_A + \frac{E_B}{E_A}I_B},$$

which is identical to the expression given in Eq. (6.19) for the distribution of the normal stress in material A of the composite beam. The result for the normal stress on part T of the transformed cross section can be written as

$$\sigma_T = -\frac{E_A}{E_B}\left(\frac{My}{\frac{E_A}{E_B}I_A + I_B}\right).$$

The right side of this equation is the product of the term $E_A/E_B = 1/n$ and the expression given in Eq. (6.19) for the distribution of normal stress σ_B in material B of the composite beam. Therefore $\sigma_B = n\sigma_T$, which completes our confirmation of the transformed area method.

Example 6.7 Transformed Area Method
Figure 6.24 is the cross section of a prismatic beam made of steel (material A) and aluminum alloy (material B) with elastic moduli $E_A = 200$ GPa and $E_B = 72$ GPa. If $M = 12$ kN-m at a given cross section, use the transformed area method to determine the distributions of normal stress in the steel and in the aluminum.

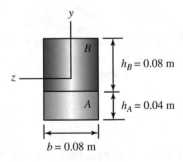

Fig. 6.24

Strategy

After calculating the dimensions of the transformed cross section, we must locate its centroid to determine the location of the neutral axis and calculate its moment of inertia. Then the distributions of the normal stress in parts A and B of the composite beam are given by Eqs. (6.22) and (6.23).

Solution

Determine the dimensions of the transformed cross section. The factor n is

$$n = \frac{E_B}{E_A} = \frac{72 \text{ GPa}}{200 \text{ GPa}} = 0.36.$$

The width of part T of the transformed cross section is the product of n with the width of part B of the composite beam: $(0.36)(0.08 \text{ m}) = 0.0288 \text{ m}$. The transformed cross section is shown in Fig. (a).

Locate the neutral axis and calculate the moment of inertia. Using the coordinate system shown in Fig. (a), we calculate the position of the centroid of the transformed cross section:

$$\bar{y} = \frac{\bar{y}_A A_A + \bar{y}_T A_T}{A_A + A_T}$$

$$= \frac{(0.02 \text{ m})(0.04 \text{ m})(0.08 \text{ m}) + (0.08)(0.08 \text{ m})(0.0288 \text{ m})}{(0.04 \text{ m})(0.08 \text{ m}) + (0.08 \text{ m})(0.0288 \text{ m})} - 0.0451 \text{ m}.$$

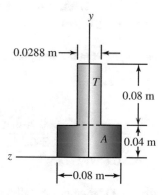

(a) Transformed cross section

Using this result, we place the origin of the coordinate system on the neutral axis of the transformed beam, which is also the location of the neutral axis of the composite beam (Fig. (b)). Applying the parallel axis theorem, the moment of inertia of the transformed cross section is

$$I = \frac{1}{12}(0.08 \text{ m})(0.04 \text{ m})^3 + (0.0251 \text{ m})^2(0.08 \text{ m})(0.04 \text{ m})$$

$$+ \frac{1}{12}(0.0288 \text{ m})(0.08 \text{ m})^3 + (0.0349 \text{ m})^2(0.0288 \text{ m})(0.08 \text{ m})$$

$$= 6.48\text{E-6 m}^4.$$

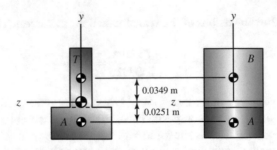

(b) Position of the neutral axis

Determine the distributions of stress. The distribution of stress in part A of the composite beam (the steel) is

$$\sigma_A = -\frac{My}{I}$$

$$= -\frac{(12,000 \text{ N-m})y}{6.48\text{E-6 m}^4}$$

$$= -1.852y \text{ GPa}.$$

The distribution of stress in part B of the composite beam (the aluminum) is

$$\sigma_B = -n\frac{My}{I}$$

$$= -(0.36)\frac{(12,000 \text{ N-m})y}{6.48\text{E-6 m}^4}$$

$$= -0.667y \text{ GPa}.$$

The distributions of stress are shown graphically in Example 6.6.

Discussion
Compare this example to Example 6.6, in which we determine the distributions of normal stress in the steel and in the aluminum using Eqs. (6.18) and (6.19).

Problems

6.31 A composite beam consists of two concentric cylinders. The inner cylinder A has a 1-in radius and is made of wood with modulus of elasticity 1.6E6 psi. The outer cylinder B has a 1-in inner radius and a 1.2-in outer radius and is made of aluminum alloy with modulus of elasticity 10.4E6 psi. If the beam is subjected to a bending moment $M = 2000$ in-lb at a particular cross section, what is the maximum tensile stress in each material?

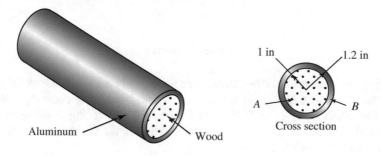

Problem 6.31

6.32 The composite beam is made of steel (material A) and aluminum alloy (material B) with elastic moduli $E_A = 30\text{E}6$ psi and $E_B = 10\text{E}6$ psi. The dimensions $h_A = 3$ in and $h_B = 6$ in. If $M = 90$ in - kip at a given cross section, what is the maximum magnitude of the stress in each material?

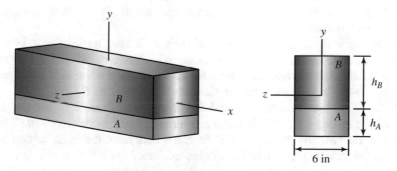

Problems 6.32–6.33

6.33 Suppose that the vertical dimensions of the beam in Problem 6.32 are changed to $h_A = 6$ in and $h_B = 3$ in. If $M = 90$ in - kip at a given cross section, what is the maximum magnitude of the stress in each material?

6.34 The figure shows the cross section of a prismatic beam made of material A with elastic modulus $E_A = 80$ GPa and material B with elastic modulus

$E_B = 16$ GPa. The dimensions $b = 0.10$ m and $h = 0.03$ m. Determine the distance H from the bottom of the cross section to the neutral axis.

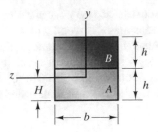

Problems 6.34–6.39

6.35 The figure shows the cross section of a prismatic beam made of material A with elastic modulus $E_A = 80$ GPa and material B with elastic modulus $E_B = 16$ GPa. The dimensions $b = 0.10$ m and $h = 0.03$ m. If the bending moment $M = 720$ N-m at a given cross section, what is the maximum magnitude of the stress in each material?

6.36 The figure shows the cross section of a prismatic beam made of material A with elastic modulus $E_A = 12$E6 psi and material B with elastic modulus $E_B = 3$E6 psi. The dimensions $b = 12$ in and $h = 4$ in. Determine the distance H from the bottom of the cross section to the neutral axis.

6.37 The figure shows the cross section of a prismatic beam made of material A with elastic modulus $E_A = 12$E6 psi and material B with elastic modulus $E_B = 3$E6 psi. The dimensions $b = 12$ in and $h = 4$ in. If the bending moment $M = 8000$ in-lb at a given cross section, how much of the moment is supported by each material?

6.38 The figure shows the cross section of a prismatic beam made of material A with elastic modulus $E_A = 72$ GPa and a material B. The dimensions $b = 0.05$ m and $h = 0.02$ m. If the distance H from the bottom of the cross section to the neutral axis is 0.025 m, what is the elastic modulus of material B?

6.39 The figure shows the cross section of a prismatic beam made of material A with elastic modulus $E_A = 72$ GPa and a material B. The dimensions $b = 0.05$ m and $h = 0.02$ m, and the distance H from the bottom of the cross section to the neutral axis is 0.025 m. If the bending moment $M = 100$ N-m at a given cross section, what is the maximum magnitude of the stress in each material?

6.40 Figure (a) is the cross section of a steel beam with elastic modulus $E = 220$ GPa. Figure (b) is the cross section of a steel box beam (material A) with elastic modulus $E = 220$ GPa whose walls are stabilized by a light filler material (material B) with elastic modulus $E = 6$ GPa. (The cross-sectional area of steel is approximately the same in each cross section.) If the bending moment $M = 400$ N-m, determine the magnitude of the maximum normal stress for each cross section.

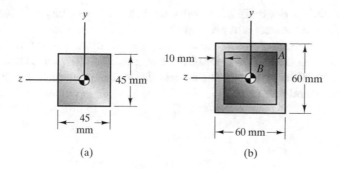

Problem 6.40

6.41 The beam is subjected to a uniformly distributed load $w_0 = 600\,\text{lb/in}$. In case (a), the cross section is square and consists of steel with elastic modulus $E = 30\text{E}6$ psi. In case (b), the cross section consists of a box beam of steel (material A) with elastic modulus $E_A = 30\text{E}6$ psi and a light filler material (material B) with elastic modulus $E_B = 1.2\text{E}6$ psi. (The cross-sectional area of steel is approximately the same in each cross section.) Determine the magnitude of the maximum normal stress at $x = 20$ in for each case.

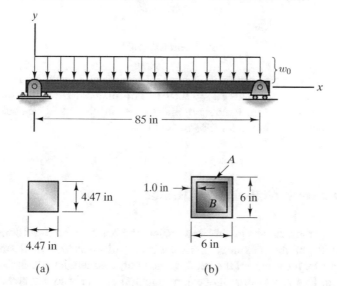

Problems 6.41–6.42

6.42 The beam is subjected to a uniformly distributed load of magnitude w_0. In case (a), the cross section is square and consists of steel with elastic modulus $E = 30\text{E}6$ psi. In case (b), the cross section consists of a box beam of steel (material A) with elastic modulus $E_A = 30\text{E}6$ psi and a light filler material (material B) with elastic modulus $E_B = 1.2\text{E}6$ psi. The yield stress of the steel is

$\sigma_Y = 80$ ksi. For cases (a) and (b), what is the largest value of w_0 for which yielding of the steel will not occur?

6.43 The figure shows the cross section of a prismatic beam made of material A with elastic modulus $E_A = 200$ GPa and material B with elastic modulus $E_B = 120$ GPa. Determine the distance H from the bottom of the cross section to the neutral axis.

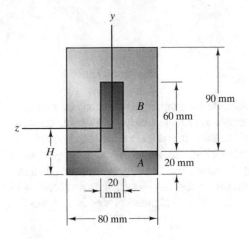

Problems 6.43–6.44

6.44* The figure shows the cross section of a prismatic beam made of material A with elastic modulus $E_A = 200$ GPa and material B with elastic modulus $E_B = 120$ GPa. If the bending moment $M = 400$ N-m at a given cross section, what is the maximum tensile stress in material A?

6.4   Elastic-Perfectly Plastic Beams

If the bending moment at a given cross section of a beam becomes sufficiently large, the magnitude of the maximum normal stress will equal the yield stress of the material. For larger values of the bending moment, the material will undergo plastic deformation. Engineers normally design structural elements so that stresses remain well below the values at which yield occurs. But in some cases beams are intentionally designed to support their loads in a partially yielded state, and in safety and failure analyses of structures, it is often necessary to understand the plastic behavior of beams.

Although more realistic models must usually be used in actual design, we can provide insight into the plastic behavior of beams by assuming that the material is elastic-perfectly plastic (Fig. 6.25). Let M be the bending moment at a given location

Fig. 6.25 Model of an elastic-perfectly plastic material

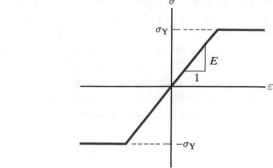

Fig. 6.26 (**a**) Bending moment at a cross section of a rectangular beam. (**b**) For a sufficiently large bending moment, the magnitude of the maximum stress equals the yield stress. (**c**) Stress distribution as the bending moment continues to increase

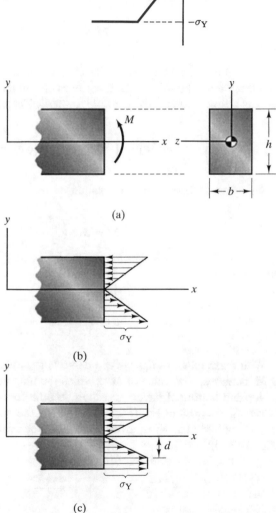

(a)

(b)

(c)

on the axis of a beam with a rectangular cross section (Fig. 6.26(a)). When M is sufficiently large, the magnitudes of the normal stresses at the top and bottom of the cross section are equal to the yield stress (Fig. 6.26(b)). What will the distribution of stress be for a still larger value of M? If we retain the assumption that plane sections remain plane, the distribution of the normal strain ε_x continues to be a linear function of y. But the magnitude of the normal stress cannot exceed the yield stress σ_Y, resulting in the distribution shown in Fig. 6.26(c). The magnitude of the normal stress increases linearly until it reaches the yield stress at some distance d from the neutral axis and then remains constant.

We can calculate the distance d for a given bending moment, and thereby determine the distribution of the normal stress, by equating M to the moment exerted about the z axis by the normal stress:

$$M = -\int_A y\sigma_x \, dA. \tag{6.27}$$

To evaluate the integral, let the element of area dA be a horizontal strip of infinitesimal height: $dA = b\,dy$. For values of y in the range $-d < y < d$, the normal stress is a linear function of y: $\sigma_x = -\sigma_Y(y/d)$. Therefore

$$M = -\int_{-h/2}^{-d} y\sigma_Y b\,dy - \int_{-d}^{d} y\left[-\sigma_Y\left(\frac{y}{d}\right)\right] b\,dy - \int_{d}^{h/2} y(-\sigma_Y) b\,dy. \tag{6.28}$$

We evaluate these integrals, obtaining

$$M = \sigma_Y b\left(\frac{h^2}{4} - \frac{d^2}{3}\right), \tag{6.29}$$

and solve for d:

$$d = \sqrt{3\left(\frac{h^2}{4} - \frac{M}{\sigma_Y b}\right)}. \tag{6.30}$$

With these relationships we can describe the evolution of the stress distribution as M increases. The value of M at which the magnitudes of the normal stresses at the top and bottom of the cross section become equal to the yield stress is obtained by setting $d = h/2$ in Eq. (6.29). This yields the maximum bending moment that can be applied at a given cross section without causing yielding of the material (Fig. 6.27(a)):

$$M = \frac{\sigma_Y b h^2}{6}. \tag{6.31}$$

When M exceeds this value (Fig. 6.27(b)), the portion of the material that has yielded is determined by Eq. (6.30). As M continues to increase, the distance d decreases until all of the material has yielded. The magnitude of the bending moment at which this occurs, obtained by setting $d = 0$ in Eq. (6.29), is called the *ultimate moment* M_U (Fig. 6.27(c)):

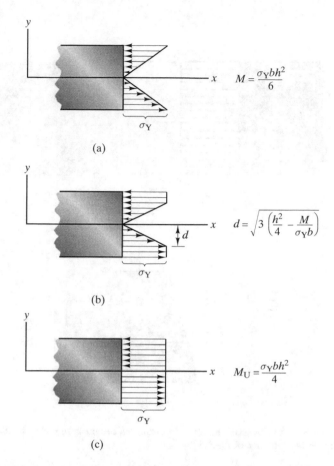

$$M = \frac{\sigma_Y b h^2}{6}$$

(a)

$$d = \sqrt{3\left(\frac{h^2}{4} - \frac{M}{\sigma_Y b}\right)}$$

(b)

$$M_U = \frac{\sigma_Y b h^2}{4}$$

(c)

Fig. 6.27 (**a**) Maximum bending moment that does not cause yielding of the material. (**b**) Stress distribution when the material is partially yielded. (**c**) Bending moment when the material is completely yielded

$$M_U = \frac{\sigma_Y b h^2}{4}. \tag{6.32}$$

When the moment at a given cross section reaches this value, there is no resistance to further bending, and the beam is said to form a *plastic hinge*.

Our analysis has been limited to a beam with a rectangular cross section. For other cross sections, determining the distribution of the normal stress when the material is partially yielded usually requires a numerical solution, but calculating the ultimate moment M_U is straightforward. Let us assume that the material at a given cross section is fully yielded, and let A_T and A_C be the areas of the cross section that are

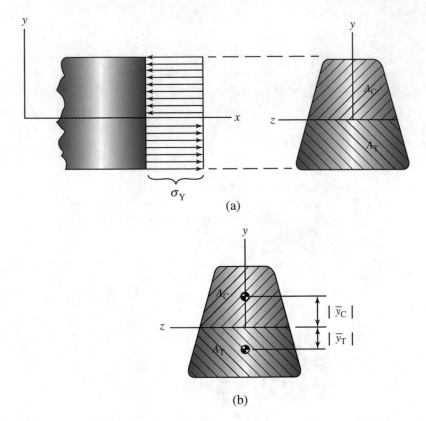

Fig. 6.28 (a) Parts of the cross section subject to tensile and compressive stress. (b) Distances from the neutral axis to the centroids of A_C and A_T

subjected to tensile and compressive stress (Fig. 6.28(a)). From the equilibrium requirement that the distribution of normal stress must exert no net axial force,

$$\int_A \sigma_x \, dA = \int_{A_T} \sigma_Y \, dA + \int_{A_C} (-\sigma_Y) \, dA$$
$$= \sigma_Y (A_T - A_C) = 0,$$

we see that

$$A_T = A_C. \tag{6.33}$$

The areas of the cross section subjected to tensile and compressive stress must be equal. This condition can be used to determine the distribution of the normal stress. The ultimate moment is

$$M_U = -\int_A y\sigma_x\, dA = -\int_{A_T} y\sigma_Y\, dA - \int_{A_C} y(-\sigma_Y)\, dA.$$

We can express this result as

$$M_U = \sigma_Y\left(\bar{y}_C A_C - \bar{y}_T A_T\right), \tag{6.34}$$

where $\bar{y}_T$ and $\bar{y}_C$ are the coordinates of the centroids of A_T and A_C relative to the neutral axis (Fig. 6.28(b)).

Example 6.8 Elastic-Perfectly Plastic Beam
The beam in Fig. 6.29 consists of elastic-perfectly plastic material with yield stress $\sigma_Y = 340$ MPa. (a) Sketch the distribution of normal stress at $x = 3$ m if $w_0 = 70,000$ N/m. (b) If w_0 is increased progressively, at what value will the beam fail by formation of a plastic hinge? Where does the plastic hinge occur?

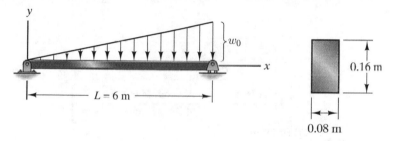

Fig. 6.29

Strategy
(a) With w_0 known, we can determine the bending moment M at $x = 3$ m. Then we can use Eqs. (6.31) and (6.32) to determine whether the material is partially yielded. If there is partial yielding, calculating the distance d from Eq. (6.30) establishes the distribution of normal stress. (b) As w_0 increases, the material will first become completely yielded, and a plastic hinge will form at the cross section where the magnitude of the bending moment is a maximum. The value of the ultimate moment is given by Eq. (6.32).

Solution
(a) Determining the bending moment in the beam as a function of x, we find that

$$M = \frac{w_0}{6}\left(Lx - \frac{x^3}{L}\right). \tag{1}$$

If $w_0 = 70,000$ N/m, the bending moment at $x = 3$ m is

$$M = \frac{70,000 \text{ N/m}}{6}\left[(6\text{ m})(3\text{ m}) - \frac{(3\text{ m})^3}{6\text{ m}}\right] = 157,500 \text{ N-m}.$$

From Eq. (6.31), the maximum moment that will not cause yielding of the material is

$$\frac{\sigma_{Y}bh^2}{6} = \frac{(340\text{E}6 \text{ N/m}^2)(0.08 \text{ m})(0.16 \text{ m})^2}{6} = 116,100 \text{ N-m},$$

and from Eq. (6.32), the ultimate moment when the material is completely yielded is

$$M_{\text{U}} = \frac{\sigma_{Y}bh^2}{4} = \frac{(340\text{E}6 \text{ N/m}^2)(0.08 \text{ m})(0.16 \text{ m})^2}{4} = 174,100 \text{ N-m}. \quad (2)$$

We see that the material is partially yielded. From Eq. (6.30), the distance d is

$$d = \sqrt{3\left(\frac{h^2}{4} - \frac{M}{\sigma_{Y}b}\right)}$$

$$= \sqrt{3\left[\frac{(0.16 \text{ m})^2}{4} - \frac{(157,500 \text{ N-m})}{(340\text{E}6 \text{ N/m}^2)(0.08 \text{ m})}\right]}$$

$$= 0.0428 \text{ m}.$$

Figure (a) shows the distribution of the normal stress.

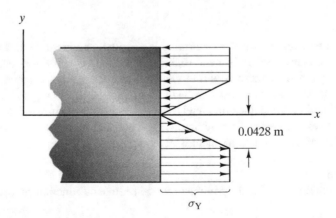

(a) Normal stress at $x = 3$ m

(b) To determine the cross section at which the bending moment is a maximum, we equate the derivative of Eq. (1) to zero:

$$\frac{dM}{dx} = \frac{w_0}{6}\left(L - \frac{3x^2}{L}\right) = 0.$$

Solving this equation, we find that the bending moment is a maximum, and the plastic hinge will occur, at $x = L/\sqrt{3}$. Substituting this value of x into Eq. (1), the value of the maximum bending moment in the beam in terms of w_0 is $M_{max} = w_0 L^2 / 9\sqrt{3}$. The plastic hinge forms when the maximum bending moment is equal to the ultimate moment given by Eq. (2):

$$M_{max} = M_U :$$
$$\frac{w_0 (6\text{ m})^2}{9\sqrt{3}} = 174{,}100 \text{ N-m.}$$

Solving, we obtain $w_0 = 75{,}400$ N/m.

Discussion
Although designers normally seek to prevent yielding in elements of structures, it is often useful to know the ultimate loads they will support. In some cases, for example, in designing structures to survive when subjected to contingent situations such as earthquakes, both the nominal loads to which elements will be subjected in normal use and the ultimate loads they must support are included in design specifications.

Example 6.9 Determining the Ultimate Moment
The beam in Fig. 6.30 is subjected to a moment M. If the material is elastic-perfectly plastic with yield stress $\sigma_Y = 500$ MPa, what is the ultimate moment M_U?

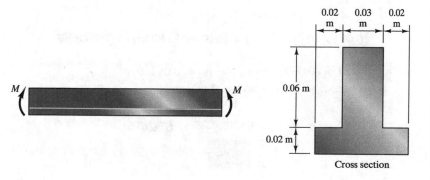

Fig. 6.30

Strategy

The ultimate moment is given by Eq. (6.34) in terms of the areas A_T and A_C of the cross section that are subjected to tensile and compressive stress and the coordinates of their centroids relative to the neutral axis. We must begin by determining the location of the neutral axis from the condition that $A_T = A_C$.

Solution

Let H be the unknown distance from the top of the cross section to the neutral axis (Fig. (a)). In terms of H, the areas A_C and A_T are

$$A_C = (0.03 \text{ m})H,$$
$$A_T = (0.03 \text{ m})(0.06 \text{ m} - H) + (0.07 \text{ m})(0.02 \text{ m}).$$

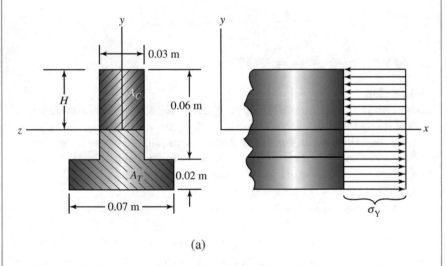

(a)

(a) Distance H to the neutral axis

From the condition

$$A_T = A_C :$$
$$(0.03 \text{ m})(0.06 \text{ m} - H) + (0.07 \text{ m})(0.02 \text{ m}) = (0.03 \text{ m})H,$$

we find that $H = 0.0533 \text{ m}$ and $A_T = A_C = 0.00160 \text{ m}^2$.
The coordinate of the centroid of A_C is

$$\bar{y}_C = \frac{H}{2} = \frac{0.0533 \text{ m}}{2} = 0.0267 \text{ m}.$$

Treating A_T as a composite of two rectangles A_1 and A_2 (Fig. (b)), the coordinate of its centroid is

$$\bar{y}_T = \frac{\bar{y}_1 A_1 + \bar{y}_2 A_2}{A_1 + A_2}$$

$$= \frac{\left[-\left(\frac{0.06\,\text{m} - H}{2}\right)\right](0.03\,\text{m})(0.06\,\text{m} - H) + \left[-(0.06\,\text{m} - H) - \frac{0.02\,\text{m}}{2}\right](0.07\,\text{m})(0.02\,\text{m})}{(0.03\,\text{m})(0.06\,\text{m} - H) + (0.07\,\text{m})(0.02\,\text{m})}$$

$$= -0.0150\,\text{m}$$

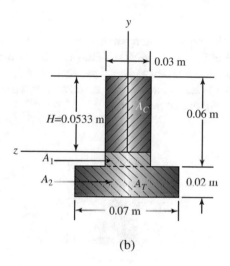

(b)

(b) Determining the centroid of A_T

From Eq. (6.34), the ultimate moment is

$$M_U = \sigma_Y \left(\bar{y}_C A_C - \bar{y}_T A_T \right)$$
$$= (500\text{E6 N/m}^2)[(0.0267\,\text{m})(0.00160\,\text{m}^2) - (-0.0150\,\text{m})(0.00160\,\text{m}^2)]$$
$$= 33.3\,\text{kN-m}.$$

Discussion

We should mention a complication that was glossed over in this section. A beam's neutral axis when it is completely yielded (shown in Figs. (a) and (b) for this example) does not coincide with the centroid of its cross section except for symmetric cases such as rectangular cross sections. As the bending moment increases and the material becomes progressively yielded, the neutral axis moves from the centroid of the cross section to its position when yielding is complete. In order to determine the distribution of the normal stress for a beam such as the one in this example when it is partially yielded, the location of the neutral axis for partial yielding must be determined.

Problems

6.45 A beam with the cross section shown consists of elastic-perfectly plastic material with yield stress $\sigma_Y = 400\,\text{MPa}$. (**a**) What is the maximum bending moment that can be applied at a given cross section without causing yielding of the material? (**b**) What is the ultimate moment that will cause formation of a plastic hinge at a given cross section?

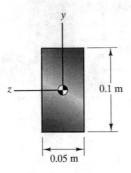

Problems 6.45–6.47

6.46 A beam with the cross section shown consists of elastic-perfectly plastic material with yield stress $\sigma_Y = 400\,\text{MPa}$. Determine the distance d, and sketch the distribution of the normal stress at a given cross section if the bending moment $M = 45,000\,\text{N-m}$.

6.47 A beam with the cross section shown consists of elastic-perfectly plastic material with yield stress $\sigma_Y = 400\,\text{MPa}$. Determine the bending moment M that will cause 50% of the beam's cross section to be yielded.

6.48 A beam with the cross section shown consists of elastic-perfectly plastic material with yield stress $\sigma_Y = 60,000\,\text{psi}$. (**a**) What is the maximum bending moment that can be applied at a given cross section without causing yielding of the material? (**b**) What is the ultimate moment that will cause formation of a plastic hinge at a given cross section?

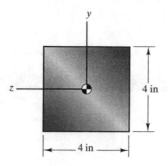

Problems 6.48–6.49

6.49 A beam with the cross section shown consists of elastic-perfectly plastic material with yield stress $\sigma_Y = 60,000$ psi. Determine the distance d, and sketch the distribution of the normal stress at a given cross section if the bending moment $M = 880,000$ in-lb.

6.50 The beam consists of elastic-perfectly plastic material with yield stress $\sigma_Y = 340$ MPa. What is the largest value of w_0 that will not cause yielding of the material?

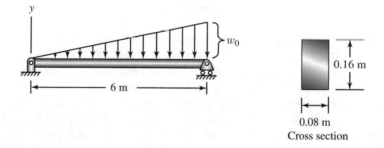

Problems 6.50–6.52

6.51 The beam consists of elastic-perfectly plastic material with yield stress $\sigma_Y = 340$ MPa. Sketch the distribution of normal stress at $x = 2.2$ m if $w_0 = 70,000$ N/m.

6.52 The beam consists of elastic-perfectly plastic material with yield stress $\sigma_Y = 340$ MPa. If $w_0 = 72,000$ N/m, at what axial position x is the distance d a minimum? What is the minimum value of d?

6.53 The beam consists of elastic-perfectly plastic material with yield stress $\sigma_Y = 700$ MPa. Its dimensions are $L = 1.2$ m, $b = 18$ mm, $h = 36$ mm. Determine the distance d, and sketch the distribution of normal stress at $x = 0.4$ m if $w_0 = 22,500$ N/m.

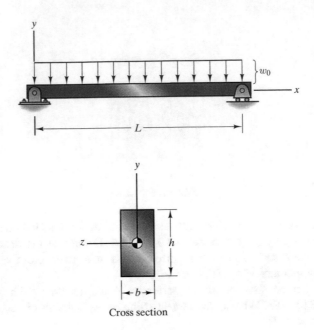

Cross section

Problems 6.53–6.56

6.54 The beam consists of elastic-perfectly plastic material with yield stress $\sigma_Y = 700$ MPa. Its dimensions are $L = 1.2$ m, $b = 18$ mm, $h = 36$ mm. If w_0 is progressively increased, at what value will the beam fail by formation of a plastic hinge? Where does the plastic hinge occur?

6.55 The beam consists of elastic-perfectly plastic material with yield stress $\sigma_Y = 150,000$ psi. Its dimensions are $L = 36$ in, $b = 1/2$ in, $h = 1$ in. Determine the distance d and sketch the distribution of normal stress at $x - 12$ in if $w_0 = 115$ lb/in.

6.56 The beam consists of elastic-perfectly plastic material with yield stress $\sigma_Y = 150,000$ psi. Its dimensions are $L = 36$ in, $b = 1/2$ in, $h = 1$ in. If w_0 is progressively increased, at what value will the beam fail by formation of a plastic hinge? Where does the plastic hinge occur?

6.57 By evaluating the integrals in Eq. (6.28), derive Eq. (6.30) for the distance from the neutral axis at which the magnitude of the normal stress equals the yield stress.

6.58 By applying Eq. (6.34) for the ultimate moment to a rectangular cross section with width b and height h, confirm Eq. (6.32).

6.59 The beam is subjected to a moment M. If the material is elastic-perfectly plastic with yield stress $\sigma_Y = 350$ MPa, what is the ultimate moment M_U?

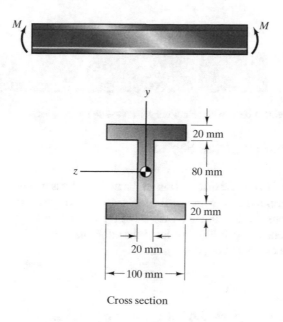

Cross section

Problem 6.59

6.60 The beam is subjected to a moment M. If the material is elastic-perfectly plastic with yield stress $\sigma_Y = 450\,\text{MPa}$, what is the ultimate moment M_U?

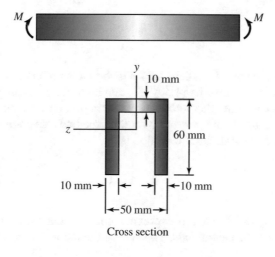

Cross section

Problem 6.60

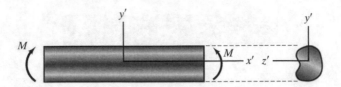

Fig. 6.31 Subjecting a beam with an arbitrary cross section to couples M

6.5   Asymmetric Cross Sections

In Sect. 6.1 we derived the distribution of the normal stress in a beam due to bending under the assumption that the beam's cross section was symmetric about the y axis. Let us now consider a beam with an arbitrary cross section that is subjected to couples at the ends (Fig. 6.31). We label the coordinate axes $x'\,y'\,z'$ and assume that the couples are exerted about the z' axis.

6.5.1   Moment Exerted About a Principal Axis

If we make the same geometric assumptions regarding the beam's deformation that we made in Sect. 6.1, the steps leading to Eq. (6.5) are unchanged, and we conclude that the distribution of normal stress at a given cross section is

$$\sigma_x = -\frac{Ey'}{\rho}, \tag{6.35}$$

where E is the modulus of elasticity and ρ is the radius of curvature of the neutral axis. In Fig. 6.32 we obtain a free-body diagram by passing a plane perpendicular to the beam's axis and show the normal stress acting on an element dA of the cross section. What conditions are necessary for this free-body diagram to be in equilibrium? The force exerted in the x' direction is

$$\int_A \sigma_x\, dA = 0.$$

Substituting Eq. (6.35) into this expression confirms that the centroid of the cross section must lie on the neutral axis. The sum of the moments about the z' axis is

$$-M - \int_A y'\sigma_x\, dA = 0.$$

Just as in Sect. 6.1, substituting Eq. (6.35) into this expression yields the equation for the distribution of the normal stress in terms of the bending moment and the moment of inertia of the cross section about the z' axis:

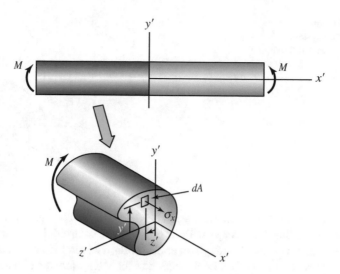

Fig. 6.32 A free-body diagram obtained by passing a plane perpendicular to the beam's axis

$$\sigma_x = -\frac{My'}{I_{z'}}. \tag{6.36}$$

Thus, the two essential results we obtained for a symmetric cross section also apply to an asymmetric cross section: the centroid of the cross section lies on the neutral axis, and the distribution of the normal stress is given by Eq. (6.36). But a third condition is necessary for the free-body diagram in Fig. 6.32 to be in equilibrium. The sum of the moments about the y' axis is

$$\int_A z'\sigma_x \, dA = 0. \tag{6.37}$$

Substituting Eq. (6.35) into this expression, we conclude that

$$\int_A y'z' \, dA = 0. \tag{6.38}$$

This equation is satisfied if the beam's cross section is symmetric about the y' axis. Because our discussion in Sect. 6.1 was limited to such cross sections, it was not necessary to consider this additional equilibrium condition. Equation (6.38) is also satisfied if the cross section is symmetric about the z' axis. More generally, Eq. (6.38) is satisfied, and the distribution of stress is given by Eq. (6.36), only if the z' axis about which the couple M is exerted *is a principal axis of the cross section.* (Principal axes are discussed in Appendix C.)

Fig. 6.33 Orientation of the
$y' z'$ coordinate system
relative to the yz coordinate
system

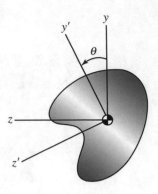

Suppose that we know the moments and product of inertia of a given cross section in terms of a coordinate system yz with its origin at the centroid. In Fig. 6.33 the $y'z'$ coordinate system is rotated through an angle θ relative to the yz coordinate system. The coordinates of a point of the cross section in the $y' z'$ system are given in terms of the coordinates of the point in the yz system by

$$y' = y \cos \theta + z \sin \theta, \tag{6.39}$$

$$z' = -y \sin \theta + z \cos \theta. \tag{6.40}$$

The moments and product of inertia of the cross section in terms of the $y' z'$ system are given in terms of the moments and product of inertia in terms of the yz system by the expressions (see Appendix C.5)

$$I_{y'} = \frac{I_y + I_z}{2} + \frac{I_y - I_z}{2} \cos 2\theta - I_{yz} \sin 2\theta, \tag{6.41}$$

$$I_{z'} = \frac{I_y + I_z}{2} - \frac{I_y - I_z}{2} \cos 2\theta + I_{yz} \sin 2\theta, \tag{6.42}$$

$$I_{y'z'} = \frac{I_y - I_z}{2} \sin 2\theta + I_{yz} \cos 2\theta. \tag{6.43}$$

A value of θ for which y' and z' are principal axes, which we denote by θ_p, satisfies the equation:

$$\tan 2\theta_p = \frac{2I_{yz}}{I_z - I_y}. \tag{6.44}$$

We can determine the orientation of the principal axes by solving this equation for θ_p, and evaluate the moments of inertia about the principal axes from Eqs. (6.41) and (6.42). Then the distribution of the normal stress due to a bending moment M exerted about the z' axis is given by Eq. (6.36).

Fig. 6.34 (**a**) Asymmetric
cross section. (**b**)
Orientation of the
principal axes

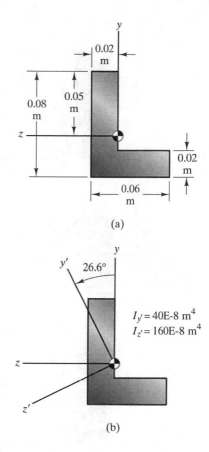

(a)

(b)

For example, consider the asymmetric cross section in Fig. 6.34(a). Its moments
of inertia about the y and z axes are $I_y = 64\,\text{E-8 m}^4$, $I_z = 136\text{E-8 m}^4$, and its product
of inertia is $I_{yz} = 48\text{E-8 m}^4$. From Eq. (6.44),

$$\tan 2\theta_p = \frac{(2)(48\text{E-8})}{(136\text{E-8}) - (64\text{E-8})} = 1.33,$$

we obtain $\theta_p = 26.6°$. Substituting this angle and the values of the moments and
product of inertia into Eqs. (6.41) and (6.42), we obtain $I_{y'} = 40\text{E-8 m}^4$ and $I_{z'} = 160$
E-8 m^4. Figure 6.34(b) shows the orientation of the principal axes and the associated
moments of inertia. Based on this information, we can use Eq. (6.36) to determine
the normal stress due to a bending moment M exerted about the z' axis for each of the
orientations of the beam's cross section shown in Fig. 6.35.

Thus we can use Eq. (6.36) to determine the distribution of normal stress for an
asymmetric cross section when the z' axis about which the bending moment M is
exerted is a principal axis. We next consider the distribution of normal stress when
M is exerted about an arbitrary axis.

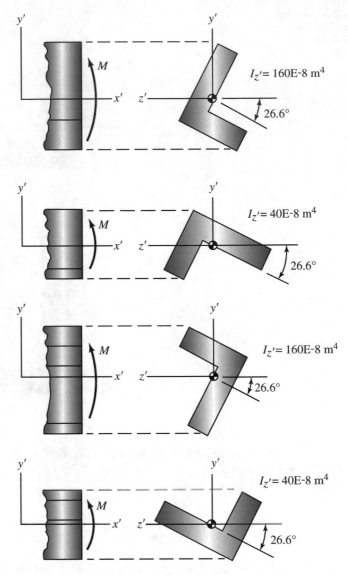

Fig. 6.35 Bending moment M applied about a principal axis of the cross section

6.5.2 Moment Exerted About an Arbitrary Axis

Suppose that the axis about which M acts is not a principal axis of the cross section (Fig. 6.36(a)). By representing the couple M by a vector **M** (Fig. 6.36(b)), we can resolve it into components in terms of a coordinate system $y'\, z'$ that is aligned with the principal axes (Fig. 6.36(c)). Then we can obtain the distribution of the normal stress by superimposing the normal stresses due to the moments about the principal axes:

Fig. 6.36 (a) Moment
M about an arbitrary axis.
(b) Representing the
moment by a vector. (c)
Resolving the vector into
components parallel to the
principal axes

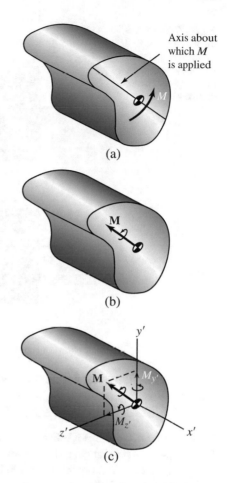

Axis about
which M
is applied

(a)

(b)

(c)

$$\sigma_x = \frac{M_{y'}z'}{I_{y'}} - \frac{M_{z'}y'}{I_{z'}}. \tag{6.45}$$

To understand the signs of the terms in this equation, notice that if the moment component $M_{z'}$ is positive, it results in negative normal stresses for positive values of y' (Fig. 6.37(a)), whereas if the moment component $M_{y'}$ is positive, it results in positive normal stresses for positive values of z' (Fig. 6.37(b)).

By setting $\sigma_x = 0$ in Eq. (6.45), we obtain an equation for a straight line in the $y' - z'$ plane along which the normal stress equals zero. This is the beam's neutral axis:

$$z' = \frac{M_{z'}I_{y'}}{M_{y'}I_{z'}}y'. \tag{6.46}$$

Fig. 6.37 (a) Stress
distribution due to the
moment component $M_{z'}$.
(b) Stress distribution
due to the moment
component $M_{y'}$

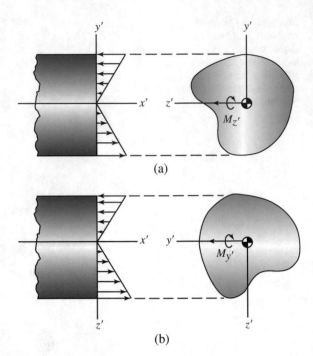

(a)

(b)

Notice that the neutral axis does not coincide with the axis about which the
moment M is applied unless $I_{y'} = I_{z'}$.

Example 6.10 Beam with an Asymmetric Cross Section
At a particular axial position, a beam with the cross section shown in Fig. 6.38
is subjected to a moment $M = 400\,\text{N-m}$ about the z axis. (**a**) Determine the
resulting normal stress at point P. (**b**) Locate the neutral axis.

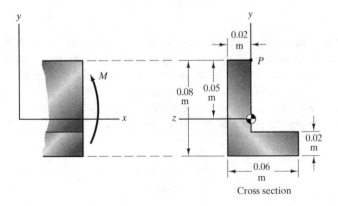

Cross section

Fig. 6.38

Strategy

(a) A $y'z'$ coordinate system aligned with the principal axes of this cross section and the associated moments of inertia are shown in Fig. 6.34(b). If we represent the moment M as a vector and resolve it into components in terms of the $y'z'$ coordinate system, the distribution of the normal stress is given by Eq. (6.45). We can use Eqs. (6.39) and (6.40) to determine the coordinates of point P in terms of the $y'z'$ coordinate system. (b) The neutral axis is described by Eq. (6.46).

Solution

(a) We represent the moment M as a vector $\mathbf{M}$ in Fig. (a). Its components in terms of the $y'z'$ coordinate system are

$$M_{y'} = (400 \text{ N-m}) \sin 26.6^\circ = 179 \text{ N-m},$$
$$M_{z'} = (400 \text{ N-m}) \cos 26.6^\circ = 358 \text{ N-m}.$$

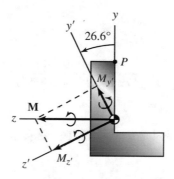

(a) Resolving the moment into components parallel to the principal axes

It was shown in Sect. 6.5 that the moments of inertia about the principal axes are $I_{y'} = 40\text{E-}8 \text{ m}^4$ and $I_{z'} = 160\text{E-}8 \text{ m}^4$. From Eq. (6.45), the distribution of the normal stress is

$$\sigma_x = \frac{M_{y'}z'}{I_{y'}} - \frac{M_{z'}y'}{I_{z'}}$$

$$= \frac{(179 \text{ N-m})z'}{(40\text{E-}8 \text{ m}^4)} - \frac{(358 \text{ N-m})y'}{(160\text{E-}8 \text{ m}^4)} \qquad (1)$$

$$= (447z' - 224y')\text{E6 Pa}.$$

The coordinates of point P in terms of the yz coordinate system are $y = 0.05 \text{ m}$, $z = 0$. Substituting these values into Eqs. (6.39) and (6.40), we obtain

$$y' = y \cos \theta + z \sin \theta$$
$$= (0.05 \text{ m}) \cos 26.6°$$
$$= 0.0447 \text{ m},$$
$$z' = -y \sin \theta + z \cos \theta$$
$$= -(0.05 \text{ m}) \sin 26.6°$$
$$= -0.0224 \text{ m}.$$

Substituting these values into Eq. (1), the normal stress at point P is

$$\sigma_x = [(447)(-0.0224) - (224)(0.0447)] \text{ E6 Pa}$$
$$= -20\text{E6 Pa}.$$

(b) From Eq. (6.46), the equation describing the neutral axis is

$$z' = \frac{M_{z'} I_{y'}}{M_{y'} I_{z'}} y'$$
$$= \frac{(358 \text{ N-m})(40\text{E-8 m}^4)}{(179 \text{ N-m})(160\text{E-8 m}^4)} y'$$
$$= 0.5 y'.$$

The neutral axis is shown in Fig. (b). The angle $\beta = \arctan(0.5) = 26.6°$.

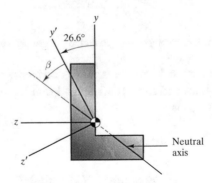

(b) Orientation of the neutral axis

Discussion
Tables of data on standardized cross sections such as the examples in Appendix E provide the locations of the centroids and the orientations of the principal axes. See Problem 6.74.

Problems

6.61 The beam is subjected to a bending moment $M = 400$ N-m about the z' axis, which is a principal axis. The moment of inertia of the cross section about z' is $I_{z'} = 40\text{E-}8$ m^4. Determine the normal stress at point P, which has coordinates $y' = 0.0313$ m, $z' = -0.0269$ m.

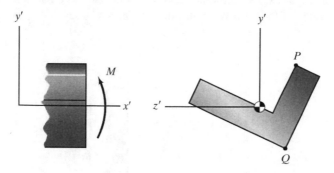

Problems 6.61–6.63

6.62 The beam is subjected to a bending moment $M - 400$ N-m about the z' axis, which is a principal axis. The moment of inertia of the cross section about z' is $I_{z'} = 40\text{E-}8$ m^4. Determine the normal stress at point Q, which has coordinates $y' = -0.0313$ m, $z' = -0.0179$ m.

6.63 Suppose that instead of being applied about the principal axis z' as shown, the bending moment $M = 400$ N-m is applied about the principal axis y' in a direction such that points of the cross section with positive z' coordinates are subjected to tensile stress. The moment of inertia of the cross section about the y' axis is $I_{y'} = 160\text{E-}8$ m^4. Determine the normal stress at point P, which has coordinates $y' = 0.0313$ m, $z' = -0.0269$ m.

6.64 The beam is subjected to a moment $M = 400$ N-m about the z axis. What is the normal stress at the point of the cross section with coordinates $y = 0.05$ m, $z = 0.02$ m? (See Fig. 6.34 and Example 6.10.)

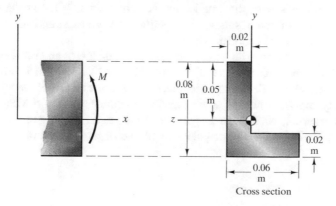

Problems 6.64–6.65

6.65   The beam is subjected to a moment $M = 400$ N-m about the z axis. What is the normal stress at the point of the cross section with coordinates $y = -0.03$ m, $z = -0.04$ m? (See Fig. 6.34 and Example 6.10.)

6.66* In terms of the coordinate system shown, the moments and product of inertia of the cross section are $I_y = I_z = 19.6$ in^4 and $I_{yz} = -11.1$ in^4. A beam with this cross section is subjected to a bending moment $M = 80,000$ in-lb about the principal axis with the larger moment of inertia. What is the maximum magnitude of the resulting distribution of normal stress?

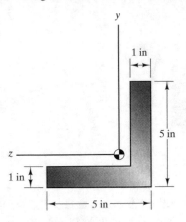

Problems 6.66–6.68

6.67* In terms of the coordinate system shown, the moments and product of inertia of the cross section are $I_y = I_z = 19.6$ in^4 and $I_{yz} = -11.1$ in^4. A beam with this cross section is subjected to a bending moment $M = 80,000$ in-lb about the principal axis with the larger moment of inertia. Determine the magnitude of the normal stress at the point of the cross section with coordinates $y = -1$ in, $z = 2$ in.

6.68* In terms of the coordinate system shown, the moments and product of inertia of the cross section are $I_y = I_z = 19.6$ in^4 and $I_{yz} = -11.1$ in^4. A beam with this cross section is subjected to a moment $M = 80,000$ in-lb about the principal axis with the smaller moment of inertia. What is the maximum magnitude of the resulting distribution of normal stress? Compare your answer to that of Problem 6.66.

6.69   In terms of the coordinate system shown, the moments and product of inertia of the cross section are $I_y = 4.633\text{E-}6$ m^4, $I_z = 7.158\text{E-}6$ m^4, and $I_{yz} = 4.320\text{E-}6$ m^4. A beam with this cross section consists of material that will safely support a normal stress (tensile or compressive) of 5 MPa. Based on this criterion, what is the magnitude of the largest bending moment that can be applied about the principal axis with the larger moment of inertia?

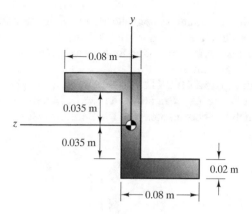

Problems 6.69–6.71

6.70 If the beam described in Problem 6.69 is subjected to a 400-N-m bending moment about the principal axis with the larger moment of inertia, what is the magnitude of the normal stress at the point of the cross section with coordinates $y = 0.045$ m, $z = 0.030$ m?

6.71 In terms of the coordinate system shown, the moments and product of inertia of the cross section are $I_y = 4.633\text{E-}6$ m^4, $I_z = 7.158\text{E-}6$ m^4, and $I_{yz} = 4.320\text{E-}6$ m^4. A beam with this cross section consists of material that will safely support a normal stress (tensile or compressive) of 5 MPa. Based on this criterion, what is the magnitude of the largest bending moment that can be applied about the principal axis with the smaller moment of inertia?

6.72* In terms of the coordinate system shown, the moments and product of inertia of the cross section are $I_y = I_z = 19.6$ in^4 and $I_{yz} = -11.1$ in^4. The beam is subjected to a bending moment $M = 80,000$ in-lb about the z axis. Determine the normal stress at point P.

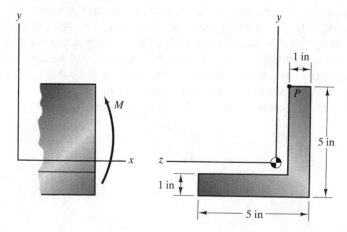

Problems 6.72–6.73

6.73∗ In terms of the coordinate system shown, the moments and product of inertia of the cross section are $I_y = I_z = 19.6 \text{ in}^4$ and $I_{yz} = -11.1 \text{ in}^4$. The beam is subjected to a bending moment about the z axis. Draw a sketch indicating the location of the neutral axis.

6.74∗ A beam with an L152X102X12.7 cross section (see Appendix E) is subjected to a moment $M = 3000$ N-m about the z axis. Determine the normal stress at point P, which has coordinates $y = 101.8$ mm, $z = 24.9$ mm.

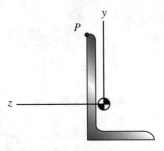

Problem 6.74

6.6   Shear Formula

The internal forces and moments in a beam include the axial force P, shear force V, and bending moment M (Fig. 6.39(a)). In Chap. 3 we discussed the uniform normal stress distribution associated with the axial force (Fig. 6.39(b)). In Sect. 6.1, we discussed the normal stress distribution associated with the bending moment (Fig. 6.39(c)). If the shear force is not zero at a given cross section, there must be a distribution of shear stress on the cross section that exerts a force in the negative y direction equal to V (Fig. 6.40). Here we analyze the shear stresses in beams.

Determining the distribution of the shear stress over a beam's cross section generally requires advanced methods of analysis or the use of a numerical solution. But we can obtain some information about the shear stress by an interesting indirect deduction that leads to a result called the *shear formula*.

From Eq. (5.7), the shear force is related to the bending moment by

$$\frac{dM}{dx} = V, \tag{6.47}$$

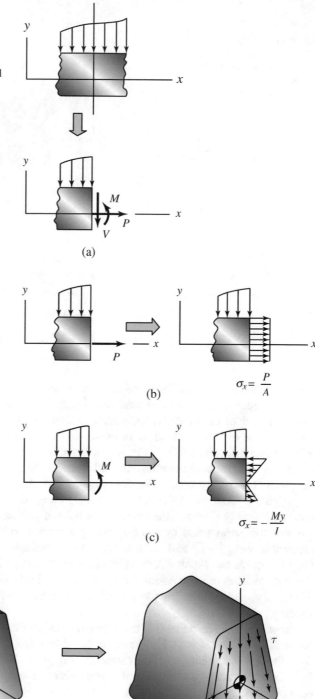

Fig. 6.39 (a) Internal forces and moment. (b) Normal stress distribution associated with the axial load. (c) Normal stress distribution associated with the bending moment

$$\sigma_x = \frac{P}{A}$$

(b)

$$\sigma_x = -\frac{My}{I}$$

(c)

Fig. 6.40 The shear load results from some distribution of shear stress

Fig. 6.41 Element of a
beam of length dx showing
the normal stresses on the
faces

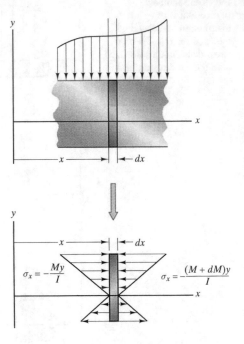

which states that the shear force at a given cross section is equal to the rate of change
of the bending moment with respect to x. Consider a beam whose cross section is
symmetric about the y axis. If we isolate an element of the beam of length dx, the
normal stress distributions on its faces are different if the bending moment varies
with respect to x (Fig. 6.41). Let us pass a horizontal plane through this element at a
position y' relative to the neutral axis and draw the free-body diagram of the part of
the element above the plane (Fig. 6.42). Because of the different normal stresses on
the opposite faces of the element isolated in Fig. 6.42, the element can be in
equilibrium only if shear stress acts on its bottom surface. We denote the average
value of this shear stress by τ_{av}. By passing a second horizontal plane through this
element at $y = y' + dy$ and considering the shear stresses on the resulting element
(Fig. 6.43), we can see that equilibrium requires that the shear stress τ_{av} also act on
the vertical faces of the element. This is illustrated in the oblique view in Fig. 6.44.
This is the shear stress on the beam's cross section whose distribution we are
seeking.

From this analysis we cannot determine the distribution of the shear stress across
the width b of the element. However, we can determine the dependence of τ_{av} on y'
from the free-body diagram of the element in Fig. 6.42. The area acted upon by τ_{av} is
$b\,dx$. Denoting the area of the part of the beam's cross section above $y = y'$ by A', the
sum of the forces on the element is

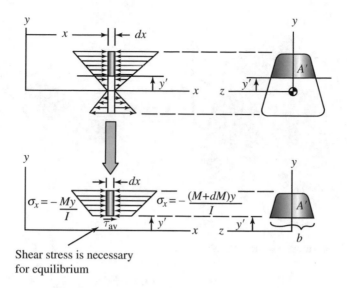

Fig. 6.42   Isolating the part of the element above a horizontal plane at position y'

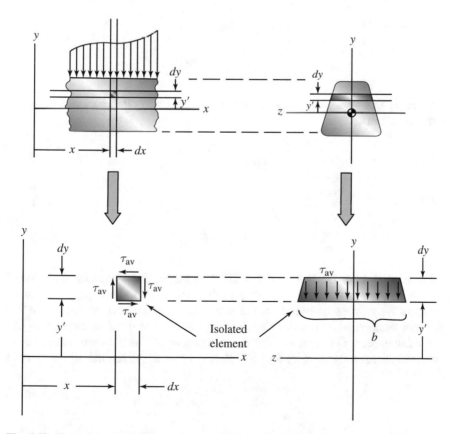

Fig. 6.43   State of shear stress on an element of infinitesimal height dy

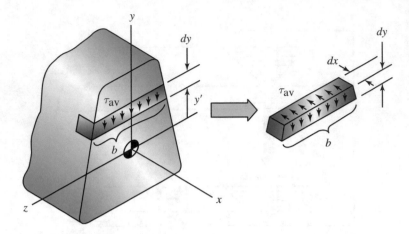

Fig. 6.44 Shear stresses on the element

$$\tau_{av} b\, dx + \int_{A'} \frac{My}{I}\, dA - \int_{A'} \frac{(M+dM)y}{I}\, dA = 0.$$

We solve this equation for τ_{av}, obtaining

$$\tau_{av} = \frac{1}{bI}\frac{dM}{dx}\int_{A'} y\, dA.$$

From Eq. (6.47), $dM/dx = V$, so we obtain an equation for the shear stress in terms of the shear force:

$$\tau_{av} = \frac{VQ}{bI}, \qquad (6.48)$$

where

$$Q = \int_{A'} y\, dA.$$

Equation (6.48) is the shear formula. It determines τ_{av} for a given cross section of a beam at a given position y' relative to the neutral axis. To apply it, we must determine the moment of inertia I of the beam's cross section and the shear force V at the cross section under consideration. We must also determine b and evaluate Q.

We can express Q in terms of the area A' and the position $\bar{y}'$ of the centroid of A' relative to the neutral axis (Fig. 6.45(a)). The definition of the position of the centroid of A' is

Fig. 6.45 Determining Q using (**a**) the area A'; (**b**) the complementary area A''

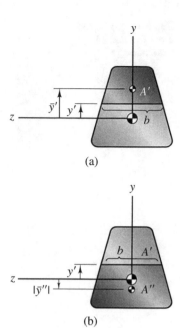

(a)

(b)

$$y' = \frac{\int_{A'} y\, dA}{A'},$$

so Q is given by

$$Q = \bar{y}'A'. \tag{6.49}$$

It is sometimes convenient to express Q in terms of the area complementary to A'. In Fig. 6.45(b) we denote the complementary area by A'' and the position of its centroid by $\bar{y}''$. Because the origin of the coordinate system coincides with position of the centroid of the entire cross section, which we can express as

$$\bar{y} = \frac{\bar{y}'A' + \bar{y}''A''}{A} = 0,$$

we see that $\bar{y}''A'' = -\bar{y}'A'$, so

$$Q = |\bar{y}''|\, A''.$$

Fig. 6.46 (a) Rectangular
cross section. (b)
Determining Q

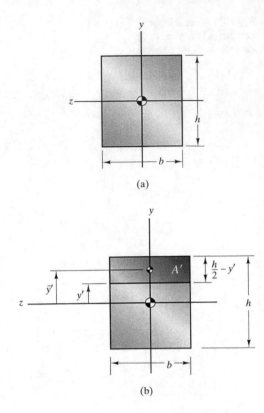

(a)

(b)

6.6.1 *Rectangular Cross Section*

In the particular case of a rectangular cross section (Fig. 6.46(a)), the shear formula
yields an explicit expression for the dependence of the average shear stress on y'.
From Fig. 6.46(b), we see that the area $A' = b(h/2 - y')$ and the position of the
centroid of A' are $\bar{y}' = y' + \frac{1}{2}(h/2 - y')$. Substituting these expressions into
Eq. (6.49), we obtain the term Q as a function of y':

$$Q = \bar{y}'A' = \frac{b}{2}\left[\left(\frac{h}{2}\right)^2 - (y')^2\right].$$

The moment of inertia of the rectangular cross section about the z axis is
$I = \frac{1}{12}bh^3$. From Eq. (6.48), the shear stress is

$$\tau_{av} = \frac{VQ}{bI} = \frac{6V}{bh^3}\left[\left(\frac{h}{2}\right)^2 - (y')^2\right]. \qquad \text{Rectangular cross section} \qquad (6.50)$$

Fig. 6.47 Distribution of τ_{av} on a rectangular cross section

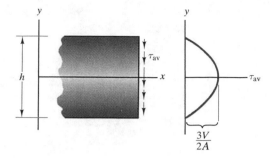

From this equation we see that the average shear stress on a rectangular cross section is a parabolic function of y' (Fig. 6.47). Its value is zero at the bottom of the cross section ($y' = -h/2$). At the neutral axis ($y' = 0$), it reaches its maximum magnitude:

$$(\tau_{av})_{max} = \frac{3V}{2bh} = \frac{3V}{2A}, \quad \text{Rectangular cross section} \quad (6.51)$$

and its value decreases to zero at the top of the cross section ($y' = h/2$).

6.6.2   Shear Stress on an Oblique Element

The shear formula is actually a more general result than our previous derivation indicates. In Fig. 6.42 we passed a horizontal plane through an element of the beam of length dx at a height y' above the neutral axis, obtaining a free-body diagram that allowed us to derive Eqs. (6.48) and (6.49) for the average shear stress on the horizontal element shown in Fig. 6.44. Instead of passing a horizontal plane through the element of length dx, suppose that we pass a plane that is parallel to the x axis but oriented at an arbitrary angle relative to the z axis. The element and the cutting plane are shown in Fig. 6.48(a) with the x axis perpendicular to the page. The derivation of the shear formula from the resulting free-body diagram is identical to our previous derivation, so the average shear stress on the oblique element shown in Fig. 6.48(b) is given by Eqs. (6.48) and (6.49).

Study Questions

1. What is the shear formula?
2. What are the definitions of the terms b and I in Eq. (6.48)?
3. For a given value of y', how can you evaluate the term Q in Eq. (6.48)?
4. If you know the shear force V at a given cross section of a beam with a rectangular cross section, how can you determine the magnitude of τ_{av} at the neutral axis?

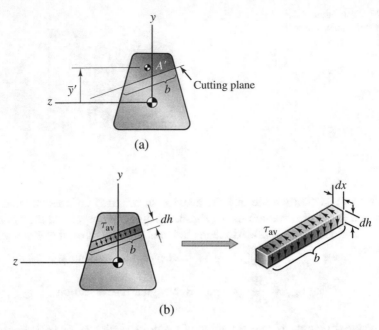

Fig. 6.48 (a) Passing a plane at an arbitrary angle relative to the z axis. The area of the cross section above the plane is A'. (b) The average shear stress on the oblique element is given by the shear formula

Example 6.11 Shear Stresses in a Beam
The beam in Fig. 6.49 is subjected to a uniformly distributed load. For the cross section at $x = 2$ m, determine the average shear stress (**a**) at the neutral axis and (**b**) at $y' = 0.1$ m.

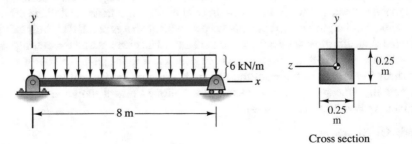

Fig. 6.49

Strategy
We must first determine the shear force V at $x = 2$ m. Then, because the beam has a square cross section, the average shear stress at the neutral axis is given by Eq. (6.51), and the shear stress is given as a function of y' by Eq. (6.50).

Solution

In Fig. (a) we draw a free-body diagram to determine the shear force at $x = 2\,\text{m}$, obtaining $V = 12\,\text{kN}$.

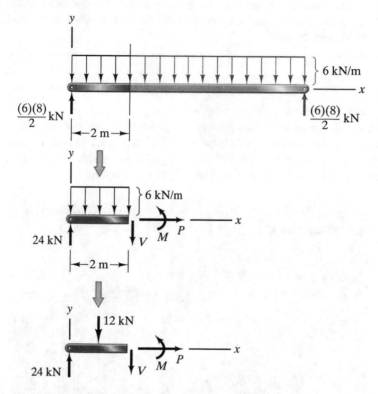

(a) Free-body diagram obtained by passing a plane through the beam at $x = 2\,\text{m}$

(**a**) From Eq. (6.51), the average shear stress at the neutral axis is

$$(\tau_{av})_{y'=0} = \frac{3V}{2A}$$

$$= \frac{(3)\,(12\,\text{kN})}{(2)\,(0.25\,\text{m})\,(0.25\,\text{m})}$$

$$= 288\,\text{kPa}.$$

(**b**) From Eq. (6.50), the average shear stress at $y' = 0.1\,\text{m}$ is

$$\tau_{av} = \frac{6V}{bh^3}\left[\left(\frac{h}{2}\right)^2 - (y')^2\right]$$

$$= \frac{6(12\,\text{kN})}{(0.25\,\text{m})(0.25\,\text{m})^3}\left[\left(\frac{0.25\,\text{m}}{2}\right)^2 - (0.1\,\text{m})^2\right]$$

$$= 104\,\text{kPa}.$$

Discussion
We chose this example to emphasize that Eq. (6.51) applies *only* to a rectangular cross section and yields the average shear stress *only* at the neutral axis. For a rectangular cross section, the average shear stress at other positions relative to the neutral axis is given by Eq. (6.50). For other cross sections, Eq. (6.48) must be used.

Example 6.12 Shear Stresses in a Beam
Beams with the cross section shown in Fig. 6.50 form part of the supporting structure of the subway station escalator. If a beam with this cross section is subjected to a shear force $V = 2$ kN, determine the resulting average shear stress (a) at the neutral axis $y' = 0$ and (b) at $y' = 20$ mm.

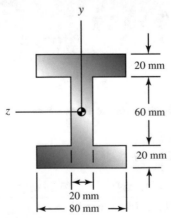

Fig. 6.50 (*Photograph from iStock by Nikada*)

Strategy

The average shear stress is given by the shear formula, Eq. (6.48). To apply it, we must determine the moment of inertia of the beam's cross section about the z axis and use Eq. (6.49) to determine the term Q.

Solution

Dividing the cross section into three rectangles shown in Fig. (a) and applying the parallel axis theorem, the moment of inertia of the cross section is

$$I = \frac{1}{12}(0.02\,\text{m})(0.06\,\text{m})^3 + 2\left[\frac{1}{12}(0.08\,\text{m})(0.02\,\text{m})^3 + (0.04\,\text{m})^2(0.08\,\text{m})(0.02\,\text{m})\right]$$

$$= 5.59\text{E-6}\,\text{m}^4.$$

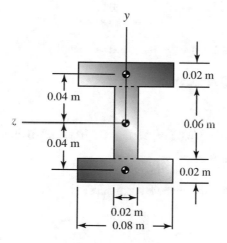

(a) Dividing the cross section into rectangles

(a) The area A' is the part of the cross section above $y' = 0$ (Fig. (b)). Dividing A' into rectangles A'_1 and A'_2 shown in Fig. (b) and noting that the position of the centroid of A' is

$$\bar{y}' = \frac{\bar{y}'_1 A'_1 + \bar{y}'_2 A'_2}{A'_1 + A'_2},$$

we see that the term Q can be written as

$$Q = \bar{y}'A'$$
$$= \bar{y}'(A'_1 + A'_2)$$
$$= \bar{y}'_1 A'_1 + \bar{y}'_2 A'_2.$$

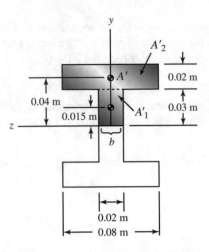

(b) Dividing the area A' above $y' = 0$ into parts A'_1 and A'_2

The value of Q is therefore

$$Q = \bar{y}_1 A'_1 + \bar{y}_2 A'_2$$
$$= (0.015 \text{ m})(0.02 \text{ m})(0.03 \text{ m}) + (0.04 \text{ m})(0.08 \text{ m})(0.02 \text{ m})$$
$$= 73.0\text{E-6 m}^3.$$

From Eq. (6.48), the average shear stress is

$$\tau_{av} = \frac{VQ}{bI}$$
$$= \frac{(2000 \text{ N})(73.0\text{E-6 m}^3)}{(0.02 \text{ m})(5.59\text{E-6 m}^4)}$$
$$= 1.31 \text{ MPa.}$$

(b) The area A' is the part of the cross section above $y' = 0.02$ m (Fig. (c)). Dividing A' into rectangles A'_1 and A'_2 shown in Fig. (c), the value of Q is

$$Q = \bar{y}_1 A'_1 + \bar{y}_2 A'_2$$
$$= (0.025 \text{ m})(0.02 \text{ m})(0.01 \text{ m}) + (0.04 \text{ m})(0.08 \text{ m})(0.02 \text{ m})$$
$$= 69.0\text{E-6 m}^3.$$

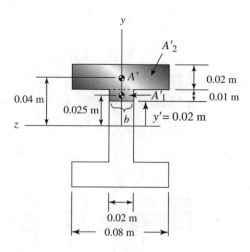

(c) Dividing the area A' above $y' = 0.02$ m into parts A'_1 and A'_2

The average shear stress is

$$\tau_{av} = \frac{VQ}{bI}$$

$$= \frac{(2000\ \text{N})(69.0\text{E-6 m}^3)}{(0.02\ \text{m})(5.59\text{E-6 m}^4)}$$

$$= 1.24\ \text{MPa}.$$

Discussion
Notice how we evaluated Q in this example by dividing the area A' into simple parts. This technique is often useful for a beam whose cross section is not rectangular.

Example 6.13 Shear Stress in a Flange of an I-Beam
A beam with the cross section shown in Fig. 6.51 is subjected to a shear force $V = 2$ kN. Determine the average shear stress τ_{av} on the infinitesimal element indicated.

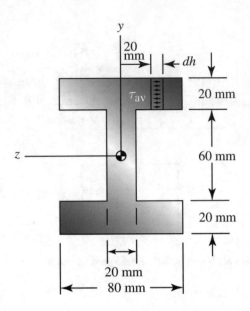

Fig. 6.51

Strategy

We can determine the average shear stress on the element from the shear formula, Eq. (6.48).

Solution

In Example 6.12 we showed that the moment of inertia of the cross section in Fig. 6.51 about the z axis is $I = 5.59\text{E-}6 \text{ m}^4$. The area A' for determining the average shear stress on the infinitesimal element is shown in Fig. (a). The value of Q is

$$
\begin{aligned}
Q &= \bar{y}'A' \\
&= (0.04 \text{ m})(0.02 \text{ m})(0.02 \text{ m}) \\
&= 16.0\text{E-}6 \text{ m}^3.
\end{aligned}
$$

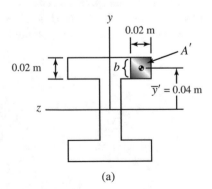

(a)

(a) The area A'. Note the definition of b

From Eq. (6.48), the average shear stress is

$$\tau_{av} = \frac{VQ}{bI}$$
$$= \frac{(2000\ \text{N})(16.0\text{E-6 m}^3)}{(0.02\ \text{m})(5.59\text{E-6 m}^4)}$$
$$= 0.286\ \text{MPa}.$$

Discussion

The flanges of the I-beam in this example are the upper and lower horizontal parts of the cross section. (The vertical part of the cross section that connects the flanges is called the web.) To determine the shear stress in a flange, why did we choose the vertical element in Fig. 6.51? You can gain insight into our choice by determining the average shear stress in the flange acting on a horizontal element located at $y' = 40$ mm.

Problems

6.75 A beam with the cross section shown is subjected to a shear force $V - 8$ kN. What is the average shear stress at the neutral axis ($y' = 0$)?

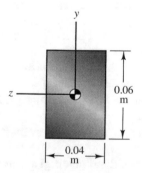

Problems 6.75–6.76

6.76 A beam with the cross section shown is subjected to a shear force $V = 8$ kN. Determine the average shear stress (**a**) at $y' = 0.01$ m; (**b**) at $y' = -0.02$ m.

6.77 The beam is subjected to a uniformly distributed load. Consider the cross section at $x = 3$ m. What is the average shear stress at $y' = 0.05$ m?

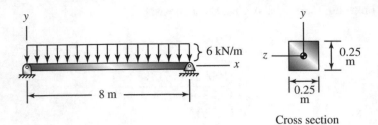

Problems 6.77–6.78

6.78 The beam is subjected to a uniformly distributed load. What is the maximum magnitude of the average shear stress in the beam, and where does it occur?

6.79 For the cross section at $x = 40$ in, determine the average shear stress **(a)** at the neutral axis and **(b)** at $y' = 1.5$ in.

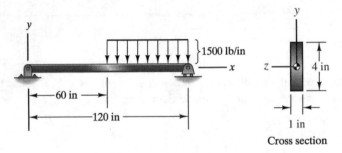

Problems 6.79–6.81

6.80 For the cross section at $x = 80$ in, determine the average shear stress **(a)** at the neutral axis and **(b)** at $y' = 1.5$ in.

6.81 What is the maximum magnitude of the average shear stress in the beam, and where does it occur?

6.82 By integrating the stress distribution given by Eq. (6.50), confirm that the total force exerted on the rectangular cross section by the shear stress is equal to V.

6.83* Prove that the quantity Q defined by Eq. (6.49) is a maximum at the neutral axis ($y' = 0$). Consider an arbitrary cross section that is symmetric about the y axis.

6.84 At a particular axial position, the beam whose cross section is shown is subjected to a shear force $V = 20$ kN. Determine the average shear stress acting on the slanted infinitesimal element.

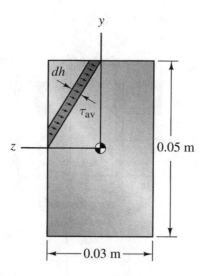

Problem 6.84

6.85 For the cross section at $x = 8$ ft, determine the average shear stress (**a**) at the neutral axis and (**b**) at $y' - 2$ in.

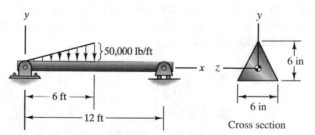

Problems 6.85–6.87

6.86* For the cross section at $x = 8$ ft, determine the value of y' at which the magnitude of the average shear stress is a maximum. (Notice that the maximum magnitude does *not* occur at the neutral axis.) What is the maximum magnitude?

6.87 For the cross section at $x = 4$ ft, determine the average shear stress (**a**) at the neutral axis and (**b**) at $y' = 2$ in.

6.88 At a particular axial position, the beam whose cross section is shown is subjected to a shear force $V = 15$ kN. Determine the average shear stress at the neutral axis $y' = 0$.

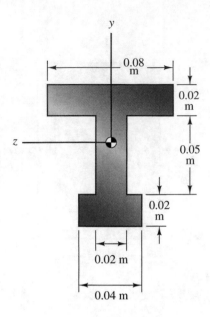

Problem 6.88

6.89 At a particular axial position, the beam whose cross section is shown is subjected to a shear force $V = 15\,\text{kN}$. Determine the average shear stress on the infinitesimal element shown.

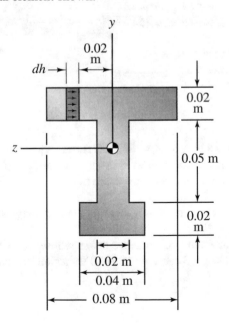

Problem 6.89

6.90 At a particular axial position, the beam whose cross section is shown is subjected to a shear force $V = 40\,\text{kN}$. What is the average shear stress at the neutral axis ($y' = 0$)?

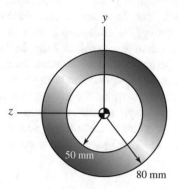

Problems 6.90–6.91

6.91∗ At a particular axial position, the beam whose cross section is shown is subjected to a shear force $V = 40\,\text{kN}$. What is the average shear stress at $y' = 50\,\text{mm}$?

6.92∗ At a particular axial position, the beam whose cross section is shown is subjected to a shear force $V - 40\,\text{kN}$. What is the average shear stress if A' is the area indicated?

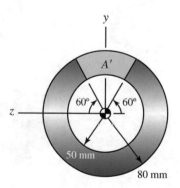

Problem 6.92

6.7 Built-Up Beams

Figure 6.52 illustrates how simple prismatic beams can be joined to obtain beams with enhanced moments of inertia. Such beams, called built-up beams, can be constructed by bonding (gluing) the individual beams to each other or by attaching them with nails or bolts. Here we consider only built-up beams in which the individual beams that are joined consist of the same material.

When the individual beams of a prismatic built-up beam are glued together, the shear formula can be used to determine the average shear stresses the glued joints must support. For example, suppose that the built-up beam in Fig. 6.53(a) is subjected to a shear force V at a particular cross section. Then we can determine the average shear stress on the glued joint at that cross section by using Eq. (6.48), the shear formula

$$\tau_{av} = \frac{VQ}{bI}$$

to determine the average shear stress acting on the plane of the joint (Fig. 6.53(b)), where Q is given by Eq. (6.49):

Fig. 6.52 Examples of built-up beams

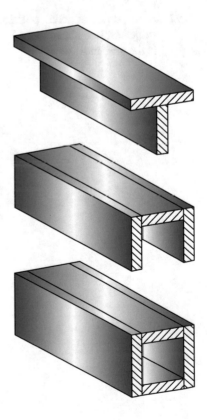

Fig. 6.53 (a) Built-up beam with a glued joint. (b) Applying the shear formula to determine the average shear stress acting on the joint

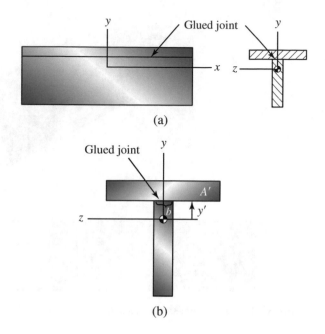

$$Q = \bar{y}'A'.$$

When the individual beams of a built-up beam are connected by nails or bolts equally spaced along the length of the beam, we can use the shear formula to estimate the shear force that each nail or bolt must support. Suppose that the individual beams of the built-up beam shown in Fig. 6.53(a) are connected in this way (Fig. 6.54) and that the beam is subjected to a shear force V at a given cross section. The shear formula, applied as shown in Fig. 6.53(b), determines the average shear stress that the joint between the individual beams would need to support if the beams were bonded over the entire surface of the joint. Multiplying the shear formula by the width b of the joint yields the shear force per unit length of the beam that must be supported at the joint:

$$\text{Shear force per unit length} = \tau_{\text{av}}b = \frac{VQ}{I}.$$

Let n be the number of nails or bolts per unit length of the beam. (Notice that $1/n$ is the distance between the individual nails or bolts.) By dividing the shear force per unit length by the number of nails or bolts per unit length, we obtain the shear force that must be supported by each nail or bolt in the neighborhood of the given cross section:

$$\text{Shear force per nail or bolt} = \frac{VQ}{nI}. \tag{6.52}$$

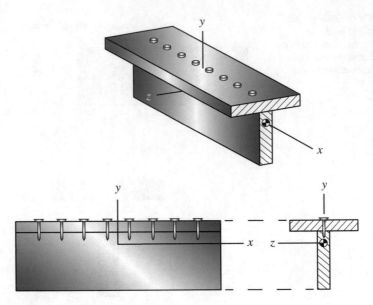

Fig. 6.54 Built-up beam with a nailed joint

Example 6.14 Built-Up Beam with Glued Joints

The beam whose cross section is shown in Fig. 6.55 consists of five planks of wood glued together. At a given axial position, the beam is subjected to a shear force $V = 6000$ lb. What are the magnitudes of the average shear stresses acting on each glued joint?

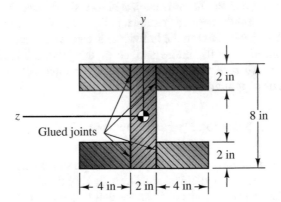

Fig. 6.55

Strategy

The average shear stress is given by the shear formula, Eq. (6.48), where I is the moment of inertia of the entire cross section about the z axis. We must also determine the appropriate values of b and Q.

Solution

We can obtain the moment of inertia of the entire cross section about the z axis by summing the moments of inertia of the planks about the z axis:

$$I = \frac{1}{12}(2 \text{ in})(8 \text{ in})^3 + 4\left[\frac{1}{12}(4 \text{ in})(2 \text{ in})^3 + (3 \text{ in})^2(2 \text{ in})(4 \text{ in})\right] = 384 \text{ in}^4.$$

We can determine the average shear stress acting on the upper-right glued joint by using the area A' and dimension b shown in Fig. (a). The value of Q is

$$\begin{aligned}
Q &= \bar{y}'A' \\
&= (3 \text{ in})(4 \text{ in})(2 \text{ in}) \\
&= 24 \text{ in}^3.
\end{aligned}$$

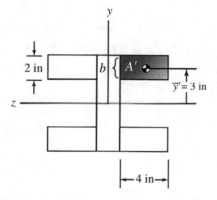

(a) Area A' for determining the average shear stress on a glued joint

The average shear stress is

$$\begin{aligned}
\tau_{av} &= \frac{VQ}{bI} \\
&= \frac{(6000 \text{ lb})(24 \text{ in}^3)}{(2 \text{ in})(384 \text{ in}^4)} \\
&= 188 \text{ psi}.
\end{aligned}$$

Discussion

Notice that we determined the average shear stress on only one of the four glued joints. You can easily show that each joint is subjected to the same shear stress.

Example 6.15 Built-Up Beam with Glued Joints
If the built-up beam in Fig. 6.56 is subjected to a shear force $V = 2500$ N, what average shear stress is exerted on its glued joints?

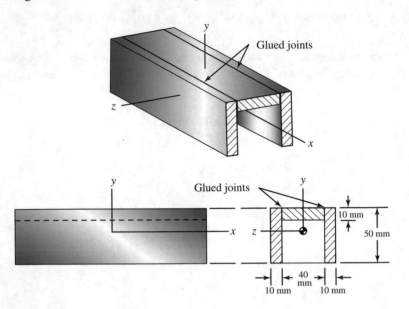

Fig. 6.56

Strategy
We must determine the location of the built-up beam's neutral axis and the moment of inertia of its cross section. Then we can use the shear formula to determine the average shear stress on the glued joints.

Solution
Let A_1 be the area of the horizontal member of the cross section, and let A_2 be the area of each vertical member. In terms of the coordinate system in Fig. (a), the position of the centroid of the built-up beam is

$$\bar{y} = \frac{\bar{y}_1 A_1 + 2\bar{y}_2 A_2}{A_1 + 2A_2}$$

$$= \frac{(0.05 \text{ m} - 0.005 \text{ m})(0.04 \text{ m})(0.01 \text{ m}) + 2(0.025 \text{ m})(0.01 \text{ m})(0.05 \text{ m})}{(0.04 \text{ m})(0.01 \text{ m}) + 2(0.01 \text{ m})(0.05 \text{ m})}$$

$$= 0.0307 \text{ m}.$$

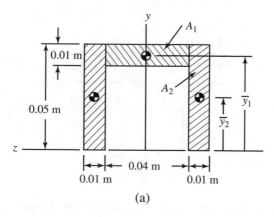

(a)

(a) Coordinate system for determining the position of the neutral axis

In Fig. (b) we place the origin of the coordinate system at the neutral axis (centroid) of the cross section and show the distances between the neutral axis and the centroids of A_1 and A_2. The moment of inertia of the cross section is

$$I = \frac{1}{12}(0.04 \text{ m})(0.01 \text{ m})^3 + (0.0143 \text{ m})^2(0.04 \text{ m})(0.01 \text{ m})$$

$$+2\left[\frac{1}{12}(0.01 \text{ m})(0.05 \text{ m})^3 + (0.00571 \text{ m})^2(0.01 \text{ m})(0.05 \text{ m})\right]$$

$$= 0.326\text{E-6 m}^4.$$

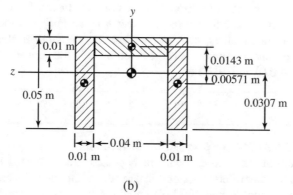

(b)

(b) Determining the moment of inertia about the neutral axis

To determine the average shear stress on the glued joints, we define the area A' as shown in Fig. (c). The term Q is

$$Q = \bar{y}'A'$$
$$= (0.0143 \text{ m})(0.04 \text{ m})(0.01 \text{ m})$$
$$= 5.71\text{E-6 m}^3.$$

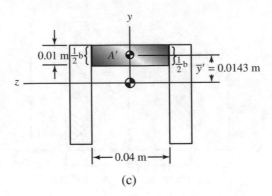

(c)

(c) Area A' for determining the average shear stress on the glued joints. Note the definition of b

Applying the shear formula, the average shear stress on the glued joints is

$$
\begin{aligned}
\tau_{av} &= \frac{VQ}{bI} \\
&= \frac{(2500\ \text{N})(5.71\text{E-6 m}^3)}{(0.02\ \text{m})(0.326\text{E-6 m}^4)} \\
&= 2.19\ \text{MPa}.
\end{aligned}
$$

Discussion
Notice how we defined the distance b in Fig. (c). In the derivation of the shear formula in Sect. 6.6, you can see that b is the length of the boundary of the area A' on which τ_{av} acts. For the area A' we chose in this example, the average shear stress acts on both of the glued joints, so $b = 0.01\ \text{m} + 0.01\ \text{m} = 0.02\ \text{m}$.

Example 6.16 Built-Up Beam with Bolted Joints
The members of the built-up beam in Fig. 6.57 are connected by bolts as shown. The number of bolts per unit length (including the bolts on both sides of the cross section) is $n = 100$ bolts/m. If the beam is subjected to a shear force $V = 2500$ N, what shear force is exerted on each bolt?

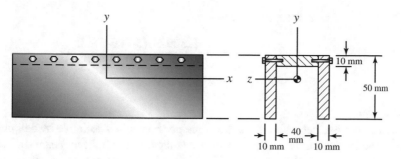

Fig. 6.57

Strategy

We must determine the location of the built-up beam's neutral axis and the moment of inertia of its cross section. Then we can use Eq. (6.52) to determine the shear force on each bolt.

Solution

The location of the neutral axis for this cross section was determined in Example 6.15, and it was shown that the moment of inertia is $I = 0.326$E-6 m^4. To determine the shear force on each bolt, we define the area A' as shown in Fig. (a). The term Q is

$$Q = \bar{y}' A'$$
$$= (0.0143 \text{ m})(0.04 \text{ m})(0.01 \text{ m})$$
$$= 5.71\text{E-6 m}^3.$$

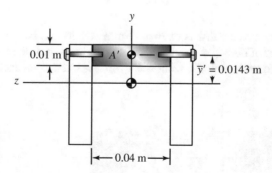

(a) Area A' for determining the shear force per bolt

From Eq. (6.52),

$$\text{Shear force per bolt} = \frac{VQ}{nI}$$

$$= \frac{(2500\ \text{N})(5.71\text{E-6 m}^3)}{(100\ \text{bolts/m})(0.326\text{E-6 m}^4)}$$

$$= 438\ \text{N/bolt}.$$

Discussion
As an exercise, determine the shear force per bolt in this example by defining A' to be the area of one of the vertical members of the built-up beam. In that case only one of the rows of bolts supports the average shear stress, so $n = 50$ bolts/m.

Problems

6.93 The built-up beam consists of two 50 mm × 100 mm beams glued together. If the beam is subjected to a shear force $V = 4$ kN, what average shear stress is exerted on its glued joint?

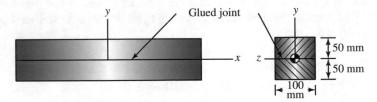

Problem 6.93

6.94 The built-up beam consists of two 50 mm × 100 mm beams bolted together. The number of bolts per unit length is $n = 20$ bolts/m. If the beam is subjected to a shear force $V = 4$ kN, what shear force is exerted on each bolt?

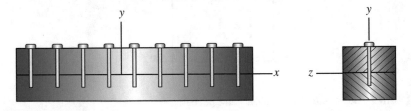

Problem 6.94

6.95 If the built-up beam is subjected to a shear force $V = 36$ kN, what average shear stress is exerted on its glued joint?

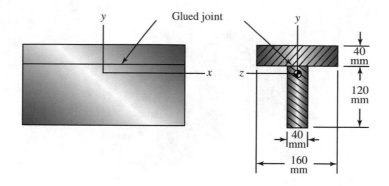

Problem 6.95

6.96 The built-up beam consists of two beams bolted together. The number of bolts per unit length is $n = 25$ bolts/m. If the beam is subjected to a shear force $V = 36$ kN, what shear force is exerted on each bolt?

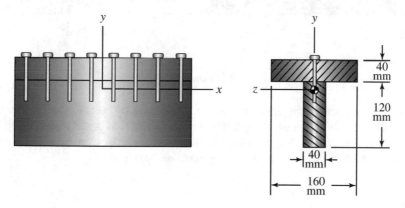

Problem 6.96

6.97 If the built-up beam is subjected to a shear force $V = 4000$ N, what average shear stress is exerted on its glued joints?

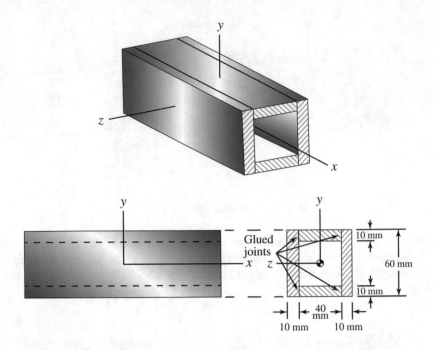

Problem 6.97

6.98 The members of the built-up beam are connected by bolts as shown. The total number of bolts per unit length (including both rows of bolts on each side of the cross section) is 160 bolts/m. If the beam is subjected to a shear force $V = 4000$ N, what shear force is exerted on each bolt?

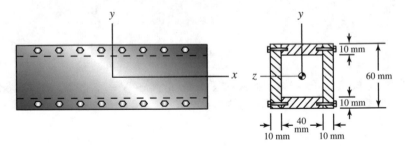

Problem 6.98

6.8 Thin-Walled Cross Sections

In Sect. 6.6 we used the shear formula to determine average values of shear stresses on beam cross sections. Although a numerical solution is generally required to determine the detailed distribution of the shear stress, we can obtain approximate analytical solutions for beams with thin-walled cross sections.

6.8.1 Distribution of the Shear Stress

For such cross sections, the shear stress can be approximated as being parallel to the wall and uniformly distributed across its width. Although proof of this result is beyond our scope, from consideration of an element of the beam wall (Fig. 6.58), it is clear that the component of the shear stress perpendicular to the wall must approach zero as the thickness of the wall decreases, because the wall surfaces are free of stress. We can use the shear formula

$$\tau = \frac{VQ}{bI}$$

to determine the magnitude of the shear stress. Its direction is indicated by the rule of thumb that the shear stress points out of the area A' when V is positive (Fig. 6.59). In this way we can determine the distribution of the shear stress throughout the cross section (Fig. 6.60). The procedure is demonstrated in Examples 6.17 and 6.18.

Fig. 6.58 Shear stress on an element of a thin-walled beam

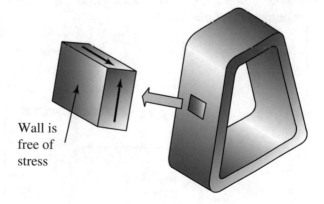

Wall is
free of
stress

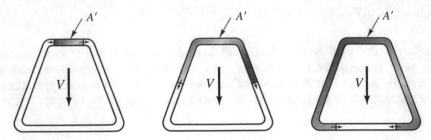

Fig. 6.59 The shear stress points out of A' when V is positive

Fig. 6.60 Distribution of
shear stress

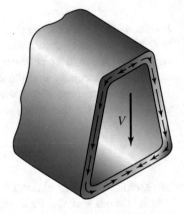

Example 6.17 Beam with a Thin-Walled Cross Section
A beam with the thin-walled cross section in Fig. 6.61 is subjected to a shear
force $V = 5\,\text{kN}$. (The wall thickness is not shown to scale.) Determine the
distribution of the shear stress.

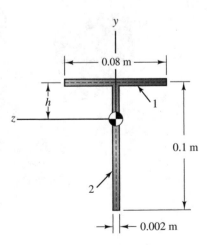

Fig. 6.61

Strategy

We must determine the position of the neutral axis and the moment of inertia I about the z axis. Then we can use the shear formula to determine the distribution of the shear stress.

Solution

The vertical distance h from the midline of the horizontal part of the cross section to the neutral axis is

$$h = \frac{(0)\,(0.08\text{ m})\,(0.002\text{ m}) + (0.0505\text{ m})\,(0.002\text{ m})\,(0.099\text{ m})}{(0.08\text{ m})\,(0.002\text{ m}) + (0.002\text{ m})\,(0.099\text{ m})} = 0.0279\,\text{m}.$$

Denoting the horizontal and vertical parts of the cross section as parts 1 and 2, respectively, the moment of inertia of the cross section about the z axis is

$$
\begin{aligned}
I &= I_1 + I_2 \\
&= \frac{1}{12}(0.08\text{ m})\,(0.002\text{ m})^3 + h^2(0.08\text{ m})\,(0.002\text{ m}) \\
&\quad + \frac{1}{12}(0.002\text{ m})\,(0.099\text{ m})^3 + (0.0505\text{ m} - h)^2(0.002\text{ m})\,(0.099\text{ m}) \\
&= 3.87\text{E-7 m}^4.
\end{aligned}
$$

We will first determine the distribution of shear stress in the horizontal part. Introducing the variable η in Fig. (a) to specify position in the horizontal part, the area $A' = 0.002\eta$ and

$$
\begin{aligned}
Q &= \bar{y}' A' \\
&= h(0.002\eta) \\
&= (5.59\text{E-5})\eta \text{ m}^3.
\end{aligned}
$$

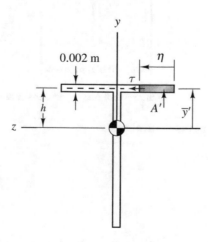

(a) Applying the shear formula to the horizontal part of the cross section

Applying the shear formula, we obtain the shear stress in the right half of the horizontal part as a function of η:

$$\tau = \frac{VQ}{bI}$$

$$= \frac{(5000 \text{ N})\,(5.59\text{E-}5\,\eta \text{ m}^3)}{(0.002 \text{ m})\,(3.87\text{E-}7 \text{ m}^4)} \tag{1}$$

$$= (3.60\text{E}8)\eta \text{ Pa} \quad (\eta \text{ in meters}).$$

We will now determine the distribution of shear stress in the vertical part of the cross section in terms of the distance y' from the neutral axis (Fig. (b)). We can evaluate Q by summing the contributions of the vertical and horizontal parts of A':

$$Q = Q_{\text{horiz}} + Q_{\text{vert}}$$

$$= h(0.08)\,(0.002) + \left[y' + \frac{1}{2}(h - 0.001 - y') \right](h - 0.001 - y')(0.002)$$

$$= 5.19\text{E-}6 - 0.001\,(y')^2 \text{ m}^3.$$

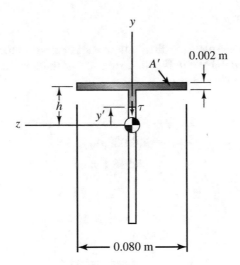

(b) Applying the shear formula to the vertical part of the cross section

The shear stress in the vertical part is

$$\tau = \frac{VQ}{bI}$$

$$= \frac{(5000 \text{ N}) \left[5.19\text{E-}6 - 0.001(y')^2 \text{ m}^3 \right]}{(0.002 \text{ m})(3.87\text{E-}7 \text{ m}^4)} \quad (2)$$

$$= \left[33.5 - 6450(y')^2 \right] \text{MPa} \quad (y' \text{ in meters}).$$

Equations (1) and (2) determine the shear stress throughout the cross section. Figure (c) indicates the direction and magnitude of the distribution.

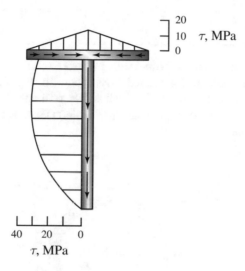

(c) Graph of the stress distribution

Discussion

We pointed out in Sect. 6.6 that the term Q can be expressed in terms of the area A'' that is the complement of the area A' and the position $\bar{y}''$ of its centroid:

$$Q = |\bar{y}''| A''.$$

You can choose whichever area appears simplest to work with. In this example we could have used A'' to determine Q for the vertical web. In terms of y', the area A'' is

$$A'' = (0.1 - h + y')(0.002) \text{ m}^2.$$

The position of its centroid is

$$\bar{y}'' = y' - \frac{1}{2}(0.1 - h + y')$$
$$= -\frac{1}{2}(0.1 - h - y') \text{ m.}$$

Therefore

$$Q = |\bar{y}''|A''$$
$$= \frac{1}{2}(0.1 - h - y')(0.1 - h + y')(0.002)$$
$$= 5.19\text{E-}6 - 0.001(y')^2 \text{ m}^3.$$

Example 6.18 Beam with a Thin-Walled Cross Section
A beam with the circular thin-walled cross section in Fig. 6.62 is subjected to a shear force $V = 20,000$ lb. The radius $R = 6$ in and the wall thickness $t = \frac{1}{4}$ in. (The wall thickness is not shown to scale.) Determine the distribution of the shear stress.

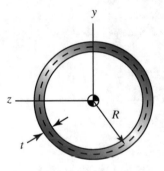

Fig. 6.62

Strategy
We can use the shear formula to determine the distribution of shear stress.

Solution
The moment of inertia of the cross section about the z axis is

$$I = \frac{1}{4}\pi(R + 0.5t)^4 - \frac{1}{4}\pi(R - 0.5t)^4$$
$$= \frac{1}{4}\pi\left[(6.125 \text{ in})^4 - (5.875 \text{ in})^4\right]$$
$$= 170 \text{ in}^4.$$

By using the area A' in Fig. (a), we can determine the shear stress as a function of the angle α.

We can determine the area and centroid of A' by using the results in Appendix D for a circular sector. The area is

$$A' = (R + 0.5t)^2\alpha - (R - 0.5t)^2\alpha$$
$$= \left[(6.125 \text{ in})^2 - (5.875 \text{ in})^2\right]\alpha$$
$$= 3.00\alpha\,\text{in}^2,$$

and the y coordinate of the centroid is

$$\bar{y}' = \frac{\dfrac{2(R+0.5t)\sin\alpha}{3\alpha}(R+0.5t)^2\alpha - \dfrac{2(R-0.5t)\sin\alpha}{3\alpha}(R-0.5t)^2\alpha}{(R+0.5t)^2\alpha - (R-0.5t)^2\alpha}$$
$$= \frac{(R+0.5t)^3 - (R-0.5t)^3}{(R+0.5t)^2 - (R-0.5t)^2}\frac{2\sin\alpha}{3\alpha}$$
$$= \frac{(6.125 \text{ in})^3 - (5.875 \text{ in})^3}{(6.125 \text{ in})^2 - (5.875 \text{ in})^2}\frac{2\sin\alpha}{3\alpha}$$
$$= \frac{6.00\sin\alpha}{\alpha}\,\text{in}.$$

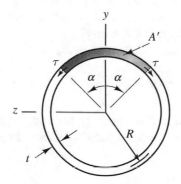

(a) Applying the shear formula

Therefore the term Q is

$$Q = \bar{y}'A' = 18.0\sin\alpha\,\text{in}^3.$$

The distribution of the shear stress is

$$\tau = \frac{VQ}{bI}$$
$$= \frac{(20,000 \text{ lb})\left(18.0\sin\alpha\,\text{in}^3\right)}{\left[(2)\left(\dfrac{1}{4}\text{in}\right)\right](170\,\text{in}^4)}$$
$$= 4240\sin\alpha\,\text{psi}.$$

Figure (b) shows the direction and magnitude of the distribution of the shear stress.

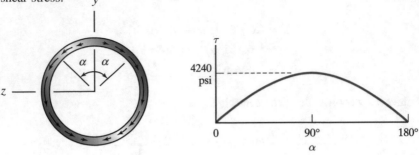

(b) Graph of the stress distribution

Discussion

You know that the cross section in this example is an effective way to support the shear stress resulting from a torque about the axis of the cylinder. Is it an effective way to support a shear load V? Consider a beam that has a solid square cross section whose cross-sectional area is equal to the cross-sectional area of the wall of the cylindrical cross section. Compare the maximum shear stress in the square beam when it is subjected to a shear load $V = 20,000$ lb to the maximum shear stress in this example.

6.8.2   The Shear Center

In Fig. 6.63(a), a lateral force F acts at the end of a thin-walled cantilever beam. At a given cross section, the beam is subjected to a shear force $V = F$ and a bending

Fig. 6.63 (a) Beam subjected to a lateral load. (b) Resulting shear force and bending moment

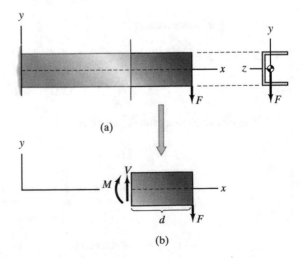

Fig. 6.64 The shear stress exerts a moment about the x axis

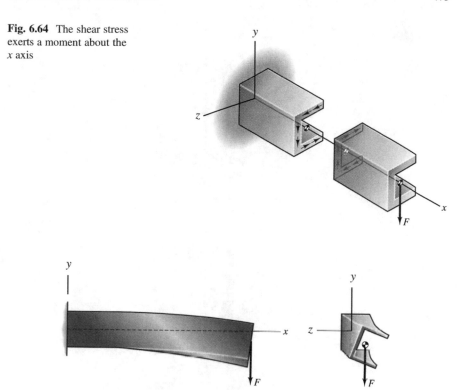

Fig. 6.65 Rotation of the beam's end and distortion of the neutral axis out of the x–y plane

moment $M = -dF$ (Fig. 6.63(b)). Although the beam's cross section is not symmetric about the y axis, the moment M acts about a principal axis of the cross section (see Sect. 6.5). If we make the same geometric assumptions regarding the beam's deformation that we made in Sect. 6.1, the normal stress due to the bending moment is given by the familiar equation $\sigma_x = -My/I$. As a consequence, our derivation of the shear formula applies, and we can use it to determine the distribution of shear stress throughout the beam's cross section as we did in Sect. 6.8.

But there is a shortcoming in this analysis. With the force F applied at the centroid of the cross section as shown in Fig. 6.63(a), the free-body diagram in Fig. 6.63(b) is not in equilibrium. This can be seen in the oblique view in Fig. 6.64, which shows the shear stress on the free-body diagram. The shear stress exerts an unbalanced couple about the neutral axis. If the beam were to be loaded in this way, its end would rotate, and the neutral axis would bend out of the x–y plane, violating our geometric assumptions (Fig. 6.65). This can be prevented by applying a torque T to the end of the beam that balances the moment exerted about the neutral axis by the shear stress (Fig. 6.66(a)). Stated in terms of the shear stress on the left part of Fig. 6.64, *the torque T must equal the moment exerted by the shear stress about the neutral axis* (Fig. 6.66(b)). Alternatively, we can load the beam by the force F alone

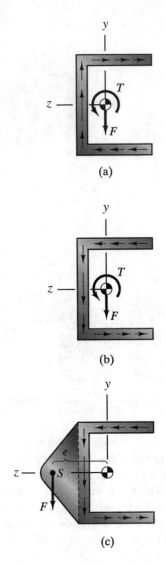

Fig. 6.66 (**a**) Adding a torque T to achieve equilibrium. (**b**) The torque must equal to the moment of the shear stress about the centroid. (**c**) Applying F at the shear center S

by adding a suitable flange to the end and moving F to the left distance e from the centroid such that $eF = T$ (Fig. 6.66(c)). The point S at which F is applied is called the *shear center* of the beam. Loaded in this way, the beam bends in the x–y plane (Fig. 6.67), our geometrical assumptions hold, and the distributions of normal and shear stress can be determined as we have described.

The position of the shear center can be determined by calculating the moment exerted about any given point by the distribution of shear stress on the beam's cross section. The force F must be placed so that it exerts the same moment about that point. We determine shear centers of beams subjected to lateral forces in Examples

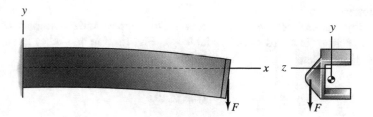

Fig. 6.67 Deformation of the beam when F is applied at the shear center

6.19 and 6.20. In general, whenever a distribution of shear stress is represented by an equivalent force, a point at which the force can be assumed to act is called a shear center for that distribution.

Example 6.19 Shear Center of a Beam
The cantilever beam in Fig. 6.68 is subjected to a lateral force F applied at the shear center. The dimensions of the cross section are $d = 0.06$ m, $h = 0.08$ m and $t = 0.002$ m. (The wall thickness is not shown to scale.) The horizontal distance c from the midline of the vertical part of the cross section to the neutral axis is $c = 0.0180$ m, and the moment of inertia of the cross section about the z axis is $I = 4.696\text{E-}7$ m^4. Determine the distance e from the neutral axis to the shear center S.

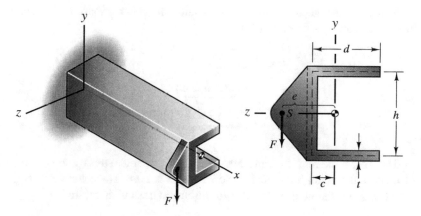

Fig. 6.68

Strategy
We can use the shear formula with $V = F$ to determine the moment exerted by the distribution of shear stress about the neutral axis. The moment eF exerted by the lateral force about the neutral axis must equal to the moment due to the shear stress about the neutral axis. From this condition we can determine e.

Solution

We first determine the shear stress in the upper horizontal part of the cross section. In terms of the variable η in Fig. (a), the area $A' = t\eta$ and the term Q is

$$
\begin{aligned}
Q &= \bar{y}' A' \\
&= \frac{h}{2} t\eta.
\end{aligned}
$$

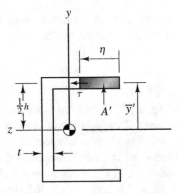

(a) Applying the shear formula to the upper horizontal part

Applying the shear formula, we obtain the shear stress in the upper horizontal part as a function of η:

$$
\begin{aligned}
\tau &= \frac{VQ}{bI} \\
&= \frac{F(h/2)t\eta}{tI} \\
&= \frac{hF}{2I}\eta.
\end{aligned}
$$

The force exerted (toward the left) by the shear stress acting on an element of the upper horizontal part of width $d\eta$ is $\tau t\, d\eta$. The counterclockwise moment about the neutral axis due to the shear stress is therefore

$$
\begin{aligned}
M_{\text{upper part}} &= \int_0^d \frac{h}{2}\tau t\, d\eta \\
&= \int_0^d \frac{h^2 tF}{4I}\eta\, d\eta \\
&= \frac{h^2 d^2 t F}{8I} \\
&= 0.01227F.
\end{aligned}
$$

We leave it as an exercise to show that the same counterclockwise moment is exerted about the neutral axis by the shear stress in the lower horizontal part (see Problem 6.109):

$$M_{\text{lower part}} = 0.01227F.$$

We now determine the shear stress in the vertical part of the cross section in terms of the distance y' from the neutral axis (Fig. (b)). We can evaluate Q by summing the contributions of the vertical and horizontal parts of A':

$$Q = Q_{\text{horiz}} + Q_{\text{vert}}$$
$$= \frac{h}{2}t\left(d + \frac{t}{2}\right) + \left[y' + \frac{1}{2}\left(\frac{h}{2} - \frac{t}{2} - y'\right)\right]t\left(\frac{h}{2} - \frac{t}{2} - y'\right)$$
$$= \left[D - 0.5(y')^2\right]t,$$

where

$$D = \frac{1}{2}hd + \frac{1}{8}(h^2 + t^2).$$

(b) Applying the shear formula to the vertical part

The shear stress in the vertical part is

$$\tau = \frac{VQ}{bI}$$
$$= \frac{F\left[D - 0.5(y')^2\right]t}{tI}$$
$$= \frac{F\left[D - 0.5(y')^2\right]}{I}.$$

The (downward) force exerted by the shear stress acting on an element of the vertical web of height dy' is $\tau t \, dy'$. The counterclockwise moment about the neutral axis due to the shear stress is therefore

$$
\begin{aligned}
M_{\text{vertical part}} &= \int_{-h/2}^{h/2} c\tau t \, dy' \\
&= \int_{-h/2}^{h/2} \frac{ctF}{I}\left[D - 0.5(y')^2\right] dy' \\
&= \frac{ctF}{I}\left(Dh - \frac{h^3}{24}\right) \\
&= 0.01799F.
\end{aligned}
$$

The total moment exerted about the neutral axis by the shear stress is

$$
M_{\text{lower part}} + M_{\text{upper part}} + M_{\text{vertical part}} = 0.0425F.
$$

Equating this to the moment eF exerted by the force F about the neutral axis, we see that the distance from the neutral axis to the shear center is $e = 0.0425$ m.

Discussion
We determined the shear stress in the vertical part and calculated the resulting moment about the neutral axis to demonstrate how to do it, but we didn't actually need to. If we calculate the total moment due to the distribution of shear stress about the point where the z axis intersects the midline of the vertical part, the shear stress in the vertical part exerts no moment. Equating the moment due to the lateral force F about this point to the moment due to the shear stress, we obtain

$$
(e - c)F = M_{\text{lower part}} + M_{\text{upper part}} :
$$
$$
(e - 0.0180)F = 2(0.01227F).
$$

Solving, we again obtain $e = 0.0425$ m.

Example 6.20 Shear Center of a Beam
The cantilever beam in Fig. 6.69 has a semicircular thin-walled cross section and is subjected to a lateral force F applied at the shear center. The radius is $R = 0.06$ m and the wall thickness is $t = 0.002$ m. (The wall thickness is not shown to scale.) The horizontal distance c from the center of the semicircle to the neutral axis is $c = 0.0382$ m, and the moment of inertia of the cross section

about the z axis is $I = 6.79 \times 10^{-7}\,\mathrm{m}^4$. Determine the distance e from the neutral axis to the shear center S.

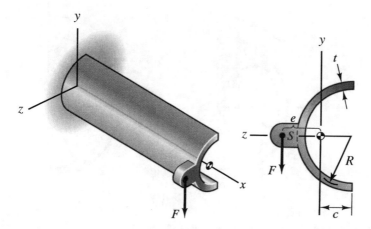

Fig. 6.69

Strategy
We can use the shear formula with $V - F$ to determine the moment exerted by the distribution of shear stress about the center of the semicircular cross section. The moment $(c + e)F$ exerted by the lateral force about the center of the cross section must equal to the moment due to the shear stress.

Solution
We can use the area A' in Fig. (a) to determine the shear stress as a function of the angle θ.

To determine the term Q for A', we will use the areas A_1' and A_2' in Fig. (b) and apply the results in Appendix D for a circular sector.

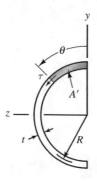

(a) Applying the shear formula

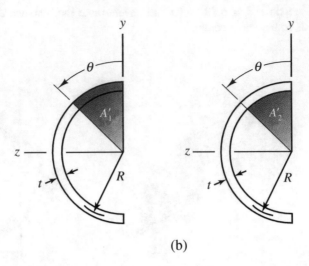

(b)

(b) Areas A_1' and A_2' for determining the value of Q for A'

The areas A_1' and A_2' are

$$A_1' = \frac{1}{2}\theta\left(R + \frac{t}{2}\right)^2, \quad A_2' = \frac{1}{2}\theta\left(R - \frac{t}{2}\right)^2,$$

and the radial distances to their centroids are

$$\bar{r}_1' = \frac{2(R + t/2)\sin\frac{1}{2}\theta}{\frac{3}{2}\theta}, \quad \bar{r}_2' = \frac{2(R - t/2)\sin\frac{1}{2}\theta}{\frac{3}{2}\theta}.$$

Therefore, the value of Q for the area A' is

$$Q = \bar{y}_1 A_1' - \bar{y}_2 A_2' = \left(\bar{r}_1' \cos\frac{1}{2}\theta\right)A_1' - \left(\bar{r}_2' \cos\frac{1}{2}\theta\right)A_2'$$

$$= \frac{2}{3}\left[\left(R + \frac{t}{2}\right)^3 - \left(R - \frac{t}{2}\right)^3\right]\sin\frac{1}{2}\theta\cos\frac{1}{2}\theta.$$

The shear stress as a function of θ is

$$\tau = \frac{VQ}{bI}$$

$$= \frac{2F}{3tI}\left[\left(R + \frac{t}{2}\right)^3 - \left(R - \frac{t}{2}\right)^3\right]\sin\frac{1}{2}\theta\cos\frac{1}{2}\theta. \tag{1}$$

The moment about the center of the semicircular cross section due to the shear stress acting on an element of angular dimension $d\theta$ is $R\tau\,dA = R\tau t\,R\,d\theta$ (Fig. (c)).

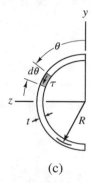

(c)

(c) Element of area for calculating the moment due to the shear stress

The total moment is therefore

$$M = \int_0^\pi R^2 \tau t\,d\theta.$$

By substituting Eq. (1) and integrating, we obtain

$$M = \frac{2FR^2}{3I}\left[\left(R+\frac{t}{2}\right)^3 - \left(R-\frac{t}{2}\right)^3\right].$$

We equate this expression to the moment about the center of the semicircular cross section due to the lateral force F acting at the shear center (Fig. 6.69):

$$(c+e)F = \frac{2FR^2}{3I}\left[\left(R+\frac{t}{2}\right)^3 - \left(R-\frac{t}{2}\right)^3\right].$$

Solving for e, we obtain $e = 0.0382$ m.

Discussion
Why should you care where the shear center is in this example? You may be able to see intuitively that if the force F in Fig. 6.69 is applied at the centroid of the cross section, the end of the beam would move downward and to the right. The beam warps out of the x–y plane, and when the force is sufficiently large, it would fail by buckling, or geometric instability. If the beam is loaded at the shear center, it does not warp and will support substantially larger loads without failing.

Problems

6.99 A beam with the thin-walled cross section shown is subjected to a shear force $V = 5$ kN. What is the shear stress at the neutral axis? (See Example 6.17.)

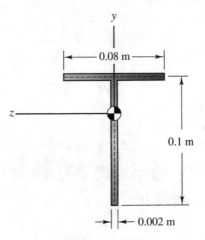

Problems 6.99–6.100

6.100 A beam with the thin-walled cross section shown is subjected to a shear force $V = 5$ kN. Determine the shear stress at $y' = -0.02$ m.

6.101 A beam with the thin-walled cross section shown is subjected to a shear force $V = 20,000$ lb. The radius of the cross section is $R = 6$ in and its wall thickness is $t = \frac{1}{4}$ in. Determine the shear stress at $\alpha = 45°$. (See Example 6.18.)

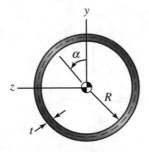

Problems 6.101–6.102

6.102 A beam with the thin-walled cross section shown is subjected to a shear force $V = 20,000$ lb. The radius of the cross section is $R = 8$ in and its wall thickness is $t = \frac{1}{4}$ in. Determine the shear stress at $\alpha = 45°$.

6.103 A beam with the thin-walled cross section shown is subjected to a shear force $V = 2.4\,kN$. (The wall thickness is not shown to scale.) Determine the magnitude of the shear stress in the left vertical part of the cross section as a function of the variable η (eta) shown.

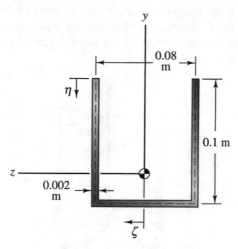

Problems 6.103–6.104

6.104 A beam with the thin-walled cross section shown is subjected to a shear force $V = 2.4\,kN$. (The wall thickness is not shown to scale.) Determine the magnitude of the shear stress in the horizontal part of the cross section as a function of the variable ζ (zeta) shown.

6.105* A beam with the semicircular thin-walled cross section shown is subjected to a shear force $V = 20,000\,lb$. The radius $R = 6\,in.$ and the wall thickness $t = \frac{1}{4}\,in.$ (The wall thickness is not shown to scale.) What is the magnitude of the shear stress at the neutral axis, and at what value of the angle α does it occur?

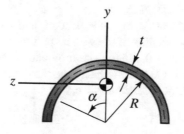

Problems 6.105–6.106

6.106∗ A beam with the semicircular thin-walled cross section shown is subjected to a shear force $V = 20,000$ lb. The radius $R = 6$ in. and the wall thickness $t = \frac{1}{4}$ in. (The wall thickness is not shown to scale.) Draw a graph of the shear stress as a function of the angle α.

6.107∗ A beam with the thin-walled cross section shown is subjected to a shear force $V = 4.5$ kN. (The wall thickness is not shown to scale.) Determine the magnitude of the shear stress in the vertical part of the cross section as a function of the variable η shown.

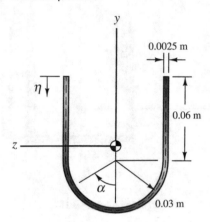

Problems 6.107–6.108

6.108∗ A beam with the thin-walled cross section shown is subjected to a shear force $V = 4.5$ kN. (The wall thickness is not shown to scale.) Determine the shear stress in the circular part of the cross section as a function of the angle α.

6.109 In Example 6.19, show that the counterclockwise moment exerted about the centroid of the beam's cross section by the shear stress in the lower horizontal part of the cross section is

$$M_{\text{lower part}} = 0.0123F \text{ N-m}.$$

6.110 The dimensions $d = 0.50$ m, $h = 1.00$ m, and $t = 0.03$ m. The horizontal distance c from the midline of the vertical part of the thin-walled cross section to the centroid is $c = 0.125$ m. The moment of inertia of the cross section about the z axis is $I = 0.01$ m^4. Determine the distance e from the neutral axis to the shear center S. (See Example 6.19.)

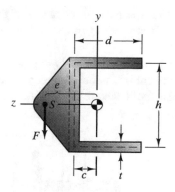

Problems 6.110–6.111

6.111 The dimensions $d = 5$ in, $h = 9$ in, and $t = \frac{1}{8}$ in. The horizontal distance c from the midline of the vertical part of the thin-walled cross section to the centroid is $c = 1.316$ in. The moment of inertia of the cross section about the z axis is $I = 32.91$ in^4. Determine the distance e from the neutral axis to the shear center S. (See Example 6.19.)

6.112 The dimensions $d_L = 0.04$ m, $d_R = 0.08$ m, $h = 0.08$ m, and $t = 0.002$ m. The horizontal distance c from the midline of the vertical part of the thin-walled cross section to the centroid is $c = 0.0151$ m. The moment of inertia of the cross section about the z axis is $I = 8.47\text{E-}7$ m^4. Determine the distance e from the neutral axis to the shear center S.

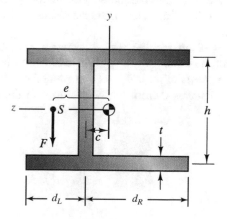

Problems 6.112–6.113

6.113 The dimensions $d_L = 3$ in, $d_R = 8$ in, $h = 8$ in, and $t = \frac{1}{8}$ in. The horizontal distance c from the midline of the vertical part of the thin-walled cross section to the centroid is $c = 1.84$ in. The moment of inertia of the cross section about the z axis is $I = 49.1$ in^4. Determine the distance e from the neutral axis to the shear center S.

6.114 The dimensions $d = 5$ in, $h = 9$ in, and $t = \frac{1}{8}$ in. The horizontal distance c is 4.62 in., and the moment of inertia of the cross section about the z axis is $I = 117$ in^4. Determine the distance e from the neutral axis to the shear center S.

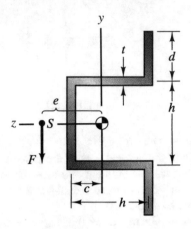

Problems 6.114–6.115

6.115 The dimensions $d = 0.08$ m, $h = 0.12$ m, and $t = 0.004$ m. The horizontal distance c is 0.0646 m, and the moment of inertia of the cross section about the z axis is $I = 1.08$E-5 m^4. Determine the distance e from the neutral axis to the shear center S.

6.116 The radius $R = 0.1$ m and the wall thickness $t = 0.003$ m. The horizontal distance c from the center of the semicircle to the centroid is $c = 0.0637$ m. The moment of inertia of the cross section about the z axis is $I = 4.71$E-6 m^4. Determine the distance e from the neutral axis to the shear center S. (See Example 6.20.)

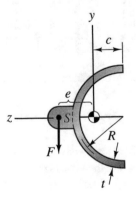

Problems 6.116–6.117

6.117 The radius $R = 4$ in and the wall thickness $t = \frac{1}{16}$ in. The horizontal distance c from the center of the semicircle to the centroid is $c = 2.55$ in. The moment of inertia of the cross section about the z axis is $I = 6.28$ in^4. Determine the distance e from the neutral axis to the shear center S. (See Example 6.20.)

6.118 The radius $R = 4$ in, the dimension $d = 2$ in, and the wall thickness is $t = \frac{1}{8}$ in. The horizontal distance c from the center of the semicircle to the centroid is $c = 1.95$ in. The moment of inertia of the cross section about the z axis is $I = 25.0$ in^4. Determine the distance e from the neutral axis to the shear center S. (See Example 6.20.)

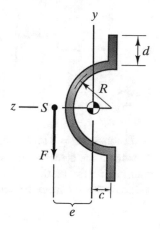

Problems 6.118–6.119

6.119 The radius $R = 0.035$ m, the dimension $d = 0.020$ m, and the wall thickness is $t = 0.003$ m. The horizontal distance c from the center of the semicircle to the centroid is $c = 0.0163$ m. The moment of inertia of the cross section about the z axis is $I = 4.66\text{E-}7$ m^4. Determine the distance e from the neutral axis to the shear center S. (See Example 6.20.)

6.120 The radius $R = 0.035$ m, the dimension $d = 0.020$ m, and the wall thickness is $t = 0.003$ m. The horizontal distance c from the center of the semicircle to the centroid is $c = 0.0137$ m. The moment of inertia of the cross section about the z axis is $I = 3.5\text{E-}7$ m^4. Determine the distance e from the neutral axis to the shear center S. (See Example 6.20.)

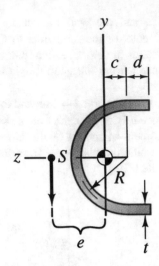

Problems 6.120–6.121

6.121 The radius $R = 4$ in, the dimension $d = 2$ in, and the wall thickness is $t = \frac{1}{8}$ in. The horizontal distance c from the center of the semicircle to the centroid is $c = 1.69$ in. The moment of inertia of the cross section about the z axis is $I = 20.6$ in^4. Determine the distance e from the neutral axis to the shear center S. (See Example 6.20.)

Chapter Summary

Normal Stress

Distribution of the Normal Stress

Consider a slender prismatic beam of isotropic linearly elastic material subjected to arbitrary loads. At the cross section with axial coordinate x (Fig. (a)), the radius of curvature of the beam's neutral axis is given by the equation

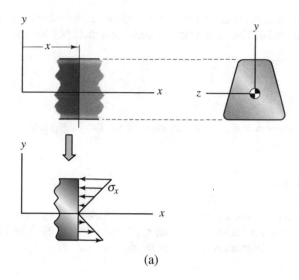

(a)

$$\frac{1}{\rho} = \frac{M}{EI},\qquad(6.10)$$

where I is the moment of inertia of the beam's cross section about the z axis and M is the value of the bending moment at x. If ρ is positive, the positive y axis is on the concave side of the neutral axis. The product EI is the beam's *flexural rigidity*. The distribution of the normal strain is

$$\varepsilon_x = -\frac{y}{\rho} = -\frac{My}{EI},\qquad(6.11)$$

and the distribution of the normal stress is

$$\sigma_x = -\frac{My}{I}.\qquad(6.12)$$

Design Issues

The ratio of the yield stress of the material to a defined allowable stress is the *factor of safety* of a beam:

$$FS = \frac{\sigma_Y}{\sigma_{\text{allow}}}.\qquad(6.13)$$

At a given cross section of a beam, the magnitude of the normal stress is greatest at the point or points of the cross section that are farthest from the neutral axis. Let

c denote the greatest magnitude of y for a given cross section, as shown in the examples in Fig. 6.14. The ratio of the moment of inertia I of the cross section about the z axis to c,

$$S = \frac{I}{c},$$ (6.15)

is called the *section modulus* of the cross section about the z axis.

Composite Beams

Consider a given cross section of a prismatic composite beam consisting of materials A and B that are each symmetric about the y axis (Fig. (b)). The location of the neutral axis can be determined from the relation

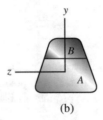

(b)

$$E_A A_A \bar{y}_A + E_B A_B \bar{y}_B = 0,$$ (6.18)

where $\bar{y}_A$ and $\bar{y}_B$ are the coordinates of the centroids of the areas A_A and A_B relative to the neutral axis. The distributions of normal stress in the individual materials are

$$\sigma_A = \frac{My}{I_A + (E_B/E_A)I_B}, \quad \sigma_B = \frac{My}{(E_A/E_B)I_A + I_B}.$$ (6.19)

The *transformed area method* makes it possible to determine the distribution of normal stress in a composite beam by determining the distribution of normal stress in a fictitious homogeneous beam subjected to the same bending moment. Figure (c) shows the cross section of a prismatic composite beam consisting of materials A and B and the cross section of a fictitious *transformed beam* consisting entirely of material A. The area of the cross section of the transformed beam that consisted of material A in the composite beam is unaltered, but the width of the part of the cross section that consisted of material B in the composite beam is scaled (multiplied) by the factor

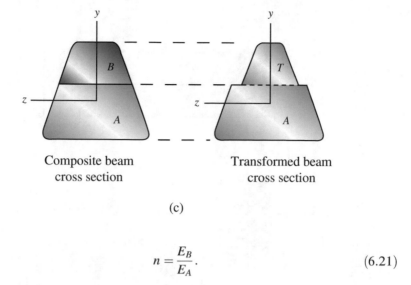

Composite beam
cross section

Transformed beam
cross section

(c)

$$n = \frac{E_B}{E_A}. \qquad (6.21)$$

The scaled part of the transformed cross section is denoted by T. The distribution of normal stress in material A of the composite beam is given by the distribution of normal stress in part A of the transformed beam,

$$\sigma_A = -\frac{My}{I} \qquad (6.22)$$

where I is the moment of inertia of the transformed cross section and y is measured from the neutral axis (centroid) of the transformed cross section. The distribution of normal stress in material B of the composite beam is obtained by multiplying the distribution of normal stress in part T of the transformed beam by n:

$$\sigma_B = n\sigma_T = -n\frac{My}{I}. \qquad (6.23)$$

Elastic-Perfectly Plastic Beams

Let M be the bending moment at a given location of a beam of elastic-perfectly plastic material with a rectangular cross section. When M exceeds the value that causes the maximum normal stress to equal the yield stress, the normal stress increases linearly until it reaches the yield stress at some distance d from the neutral axis and then remains constant (Fig. (d)). The distance d is given by

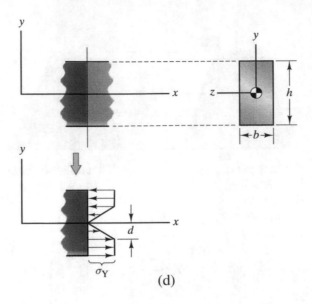

(d)

$$d = \sqrt{3\left(\frac{h^2}{4} - \frac{M}{\sigma_Y b}\right)}. \tag{6.30}$$

The maximum bending moment that can be applied without causing yielding of the material is

$$M = \frac{\sigma_Y b h^2}{6}. \tag{6.31}$$

The magnitude of the bending moment at which all the material is yielded ($d = 0$) is the *ultimate moment*:

$$M_U = \frac{\sigma_Y b h^2}{4}. \tag{6.32}$$

When the moment reaches this value, there is no resistance to further bending, and the beam forms a *plastic hinge*.

For other cross sections, the distribution of the normal stress when the material is completely yielded can be determined from the condition that the areas A_T and A_C that are subjected to tensile and compressive stress are equal. The ultimate moment is

$$M_U = \sigma_Y \left(\bar{y}_C A_C - \bar{y}_T A_T\right), \tag{6.34}$$

where $\bar{y}_T$ and $\bar{y}_C$ are the coordinates of the centroids of A_T and A_C relative to the neutral axis (Fig. (e)).

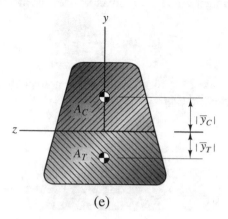

(e)

Asymmetric Cross Sections

The moments and product of inertia of the cross section in Fig. (f) in terms of the $y'\,z'$ system are given in terms of the moments and product of inertia in terms of the yz system by

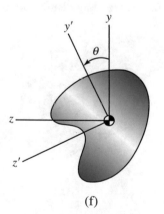

(f)

$$I_{y'} = \frac{I_y + I_z}{2} + \frac{I_y - I_z}{2} \cos 2\theta - I_{yz} \sin 2\theta, \qquad (6.41)$$

$$I_{z'} = \frac{I_y + I_z}{2} - \frac{I_y - I_z}{2} \cos 2\theta + I_{yz} \sin 2\theta, \qquad (6.42)$$

$$I_{y'z'} = \frac{I_y - I_z}{2} \sin 2\theta + I_{yz} \cos 2\theta. \tag{6.43}$$

A value of θ for which y' and z' are principal axes satisfies

$$\tan 2\theta_p = \frac{2I_{yz}}{I_z - I_y}. \tag{6.44}$$

If z' is a principal axis, the distribution of the normal stress due to a bending moment M exerted about z' (Fig. (g)) is

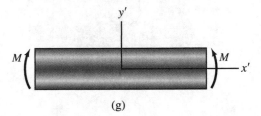

(g)

$$\sigma_x = -\frac{My'}{I_{z'}}. \tag{6.36}$$

If the axis about which M acts is not a principal axis, the vector representing M can be resolved into components in terms of a coordinate system $y'\,z'$ aligned with the principal axes (Fig. (h)). The resulting distribution of normal stress is

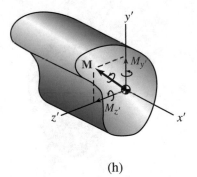

(h)

$$\sigma_x = \frac{M_{y'}z'}{I_{y'}} - \frac{M_{z'}y'}{I_{z'}}. \tag{6.45}$$

The beam's neutral axis is given by

$$z' = \frac{M_{z'}I_{y'}}{M_{y'}I_{z'}}y'. \tag{6.46}$$

Shear Stress

Distribution of the Average Stress

Consider a slender prismatic beam of isotropic linearly elastic material subjected to arbitrary loads. At a given cross section, the average of the component of the shear stress perpendicular to the line of length b in Fig. (i) is given by the *shear formula*

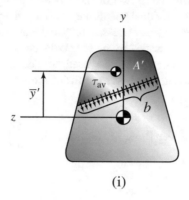

(i)

$$\tau_{av} = \frac{VQ}{bI},\qquad(6.48)$$

where I is the moment of inertia of the beam's cross section about the z axis, V is the value of the shear force at the given cross section, and

$$Q = \bar{y}'A'.\qquad(6.49)$$

For a beam with a rectangular cross section, the average of the shear stress over the horizontal line in Fig. (j) as a function of y' is

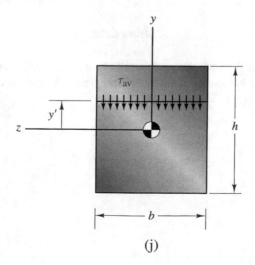

(j)

$$\tau_{\text{av}} = \frac{VQ}{bI} = \frac{6V}{bh^3}\left[\left(\frac{h}{2}\right)^2 - (y')^2\right].$$ (6.50)

Built-Up Beams

Beams constructed by connecting two or more beams along their lengths are called built-up beams. When the individual beams are glued together, the shear formula can be used to determine the average shear stresses the glued joints must support. The individual beams are sometimes connected by nails or bolts equally spaced along the length. In that case the shear formula can be used to determine the shear force that must be supported by each nail or bolt in terms of the number n of nails or bolts per unit length:

$$\text{Shear force per nail or bolt} = \frac{VQ}{nI}.$$ (6.52)

Review Problems

6.122 The beam with the cross section shown is subjected to couples M. The resulting maximum tensile stress in the beam is 3000 psi. Determine M.

Cross section

Problem 6.122

6.123 Assume that the surface on which the beam rests exerts a uniformly distributed load on it. Determine the maximum tensile and compressive stresses due to bending at $x = 3$ m.

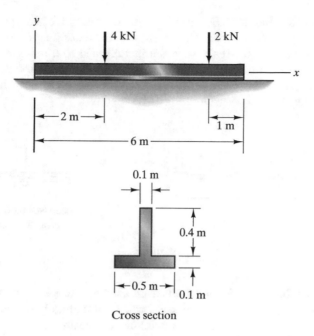

Cross section

Problem 6.123

6.124 For a preliminary analysis of the ladder rung, model it as the simply supported beam AB with a 1.2-kN point force. What is the maximum magnitude of the normal stress in the rung?

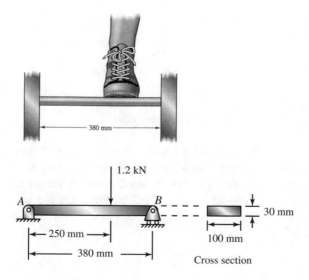

Cross section

Problem 6.124

6.125   The beam has a C200X20.5 American Standard Channel cross section
(see Appendix E) oriented as shown and consists of steel with yield
stress $\sigma_Y = 360$ MPa. The beam is subjected to a uniformly distributed
load $w_0 = 1250$ N/m, which *includes the beam's weight*. (a) Determine the section modulus about the z axis. (b) What is the beam's factor
of safety?

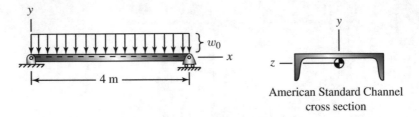

American Standard Channel
cross section

Problem 6.125

6.126   If the C200X20.5 American Standard Channel cross section of the beam in
Problem 6.125 is reoriented as shown, what uniformly distributed load w_0
(which includes the beam's weight) can the beam support with a factor of
safety of 2?

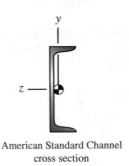

American Standard Channel
cross section

Problem 6.126

6.127∗ The value of the triangular distributed load at $x = 4$ m is $w_0 = 2$ kN/m.
The composite beam consists of two aluminum alloy beams (material A)
with elastic modulus $E_A = 70$ GPa that are bonded to a wood beam (material
B) with elastic modulus $E_B = 12$ GPa. What are the magnitudes of the
maximum normal stresses in the aluminum alloy and in the wood at
$x = 2$ m?

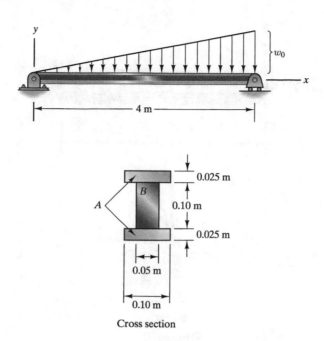

Cross section

Problems 6.127–6.128

6.128* The composite beam consists of two aluminum alloy beams (material A) with elastic modulus $E_A = 70\,\text{GPa}$ that are bonded to a wood beam (material B) with elastic modulus $E_B = 12\,\text{GPa}$. The yield stress of the aluminum alloy is 35 MPa and the yield stress of the wood is 14 MPa. What is the largest value of w_0 for which yielding will not occur in either material?

6.129 The beam consists of elastic-perfectly plastic material with yield stress $\sigma_Y = 250\,\text{MPa}$. The force $F = 80\,\text{kN}$. Determine the distance d and sketch the distribution of normal stress at $x = 2\,\text{m}$.

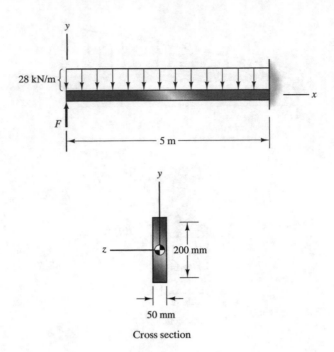

Cross section

Problems 6.129–6.131

6.130 The beam consists of elastic-perfectly plastic material with yield stress $\sigma_Y = 250$ MPa. The force $F = 80$ kN. At what axial position x is the distance d a minimum? What is the minimum value of d?

6.131 The beam consists of elastic-perfectly plastic material with yield stress $\sigma_Y = 250$ MPa. If the force F is progressively increased from its initial value of 80 kN, at what value will the beam fail by formation of a plastic hinge? Where does the plastic hinge occur?

6.132∗ In terms of the coordinate system shown, the moments and product of inertia of the cross section are $I_y = 4.633\text{E-}6$ m^4, $I_z = 7.158\text{E-}6$ m^4, and $I_{yz} = 4.320\text{E-}6$ m^4. The beam is subjected to a moment $M = 200$ N-m about the z axis. Determine the normal stress at point P.

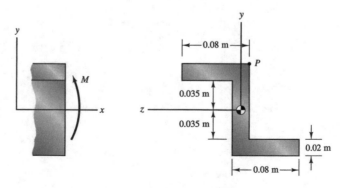

Problems 6.132–6.133

6.133* In terms of the coordinate system shown, the moments and product of inertia of the cross section are $I_y = 4.633\text{E-6 m}^4$, $I_z = 7.158\text{E-6 m}^4$, and $I_{yz} = 4.320\text{E-6 m}^4$. The beam is subjected to a bending moment about the z axis. Draw a sketch indicating the location of the neutral axis.

6.134 The beam is subjected to a distributed load. For the cross section at $x = 0.6\,\text{m}$, determine the average shear stress (**a**) at the neutral axis and (**b**) at $y' = 0.02\,\text{m}$.

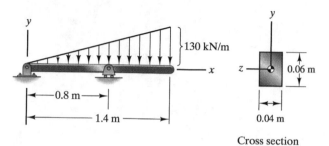

Cross section

Problems 6.134–6.135

6.135 The beam is subjected to a distributed load. For the cross section at $x = 1.0\,\text{m}$, determine the average shear stress (**a**) at the neutral axis and (**b**) at $y' = 0.02\,\text{m}$.

6.136 At a given axial position the beam whose cross section is shown is subjected to a shear force $V = 2400\,\text{lb}$. What is the average shear stress at the neutral axis ($y' - 0$)?

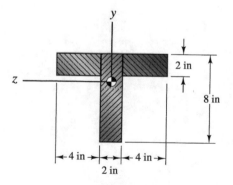

Problems 6.136–6.137

6.137 The beam whose cross section is shown consists of three planks of wood glued together. At a given axial position, it is subjected to a shear force $V = 2400\,\text{lb}$. What are the magnitudes of the average shear stresses acting on each glued joint?

Chapter 7
States of Stress

Internal forces in materials are described in terms of their states of stress. We have described and analyzed the states of stress in bars subjected to axial and torsional loads and in beams subjected to lateral loads. In this chapter we consider arbitrary states of stress and show that the state of stress at a point can be used to determine the normal and shear stresses acting on *any* plane through that point. Because of the importance of maximum values of normal and shear stresses in design, we describe how to determine maximum stresses and the orientations of the planes on which they act. Before beginning, we must review the definition of the state of stress.

7.1   Components of Stress

Consider a particular point p within some object. In Fig. 7.1(a) we introduce a coordinate system and pass a plane through the object that intersects p and is perpendicular to the x axis. In Fig. 7.1(b) we isolate the part of the object to the left of the plane. The part of the object that was (conceptually) removed may exert forces on the part we have isolated at the plane where they were separated. The normal stress at p (the force per unit area in the direction perpendicular to the plane) is the stress component σ_x (Fig. 7.1(b)). The shear stress at p (the force per unit area tangential to the plane) in the direction of the y axis is the stress component τ_{xy}, and the shear stress at p in the direction of the z axis is τ_{xz}. Next, we pass a plane through p that is perpendicular to the y axis (Fig. 7.2(a)) and isolate the part of the object below the plane (Fig. 7.2(b)). The normal stress on this plane is the stress component σ_y, and the shear stresses are τ_{yx} and τ_{yz}. Finally, we pass a plane through p perpendicular to the z axis (Fig. 7.3(a)) and isolate the part of the object behind the plane (Fig. 7.3(b)). The normal stress on this plane is the stress component σ_z, and the shear stresses are τ_{zx} and τ_{zy}. The shear stresses $\tau_{xy} = \tau_{yx}$, $\tau_{yz} = \tau_{zy}$, and

© Springer Nature Switzerland AG 2020
A. Bedford, K. M. Liechti, *Mechanics of Materials*,
https://doi.org/10.1007/978-3-030-22082-2_7

Fig. 7.1 (**a**) Passing a plane perpendicular to the x axis. (**b**) Normal stress and components of the shear stress at p

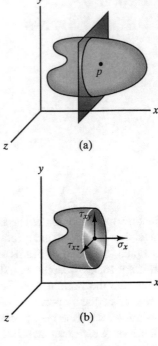

Fig. 7.2 (**a**) Passing a plane perpendicular to the y axis. (**b**) Normal stress and components of the shear stress at p

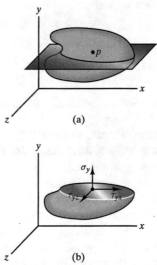

$\tau_{xz} = \tau_{zx}$, so there are six independent components of stress. The components σ_x, σ_y, σ_z, τ_{xy}, τ_{yz}, and τ_{xz} specify the *state of stress* at p in terms of the given coordinate system. It can be represented compactly as a matrix in terms of the components of stress:

Fig. 7.3 (**a**) Passing a plane
perpendicular to the z axis.
(**b**) Normal stress and
components of the shear
stress at p

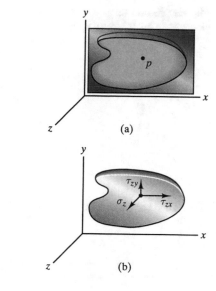

(a)

(b)

Fig. 7.4 Components of
stress on an element
containing p

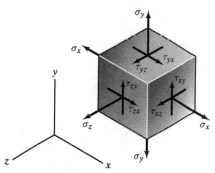

$$\begin{bmatrix} \sigma_x & \tau_{xy} & \tau_{xz} \\ \tau_{yx} & \sigma_y & \tau_{yz} \\ \tau_{zx} & \tau_{zy} & \sigma_z \end{bmatrix} . \tag{7.1}$$

Figure 7.4 shows the components of stress on an element whose faces are
perpendicular to the coordinate axes. If the state of stress is uniform, or *homogeneous*, in a finite neighborhood surrounding the point, the stresses in Fig. 7.4 can be
regarded as the stresses on an element of finite size containing p. Otherwise, they
must be interpreted as the average values of the stress components on an element
containing p.

As an example, consider a prismatic bar subjected to axial loads. If the coordinate
system is oriented with its x axis parallel to the axis of the bar (Fig. 7.5), the only
nonzero stress component is $\sigma_x = P/A$, where A is the bar's cross-sectional area. The
homogeneous state of stress is

Fig. 7.5 Stresses on an
element of a bar subjected to
axial forces

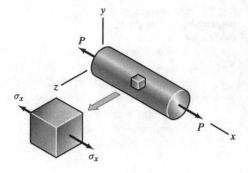

Fig. 7.6 Stresses on an
element of a bar subjected to
torsion

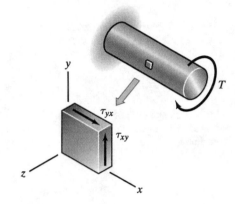

$$
\begin{bmatrix} \sigma_x & \tau_{xy} & \tau_{xz} \\ \tau_{yx} & \sigma_y & \tau_{yz} \\ \tau_{zx} & \tau_{zy} & \sigma_z \end{bmatrix} = \begin{bmatrix} \sigma_x & 0 & 0 \\ 0 & 0 & 0 \\ 0 & 0 & 0 \end{bmatrix}.
$$

In Fig. 7.6 we isolate an infinitesimal element of a cylindrical bar subjected to axial torsion. If we orient the coordinate system so that its x axis is parallel to the axis of the bar and the element lies in the x-y plane, the element is in a state of pure shear stress:

$$
\begin{bmatrix} \sigma_x & \tau_{xy} & \tau_{xz} \\ \tau_{yx} & \sigma_y & \tau_{yz} \\ \tau_{zx} & \tau_{zy} & \sigma_z \end{bmatrix} = \begin{bmatrix} 0 & \tau_{xy} & 0 \\ \tau_{yx} & 0 & 0 \\ 0 & 0 & 0 \end{bmatrix}.
$$

In citing these examples, we have emphasized that the values of the components of stress depend on the orientation of the coordinate system. If the coordinate system is rotated, the orientations of the planes perpendicular to the axes change, and so in general the normal and shear stresses acting on them change. This provides a way to determine the normal and shear stresses acting on different planes through point p. We do this in the following section for the subset of states of stress called plane stress.

Study Questions

1. What are the definitions of the components of stress at a point?
2. How many components of stress must be known to define the state of stress at a point?
3. Suppose that you know the components of stress at a point in terms of a coordinate system xyz. Define a new coordinate system $x'y'z'$ that has the x' axis coincident with the y axis, the y' axis coincident with the z axis, and the z' axis coincident with the x axis. What are the components of stress at the point in terms of the $x'y'z'$ coordinate system?

7.2   Transformations of Plane Stress

The stress at a point is said to be a state of *plane stress* if it is of the form

$$
\begin{bmatrix}
\sigma_x & \tau_{xy} & \tau_{xz} \\
\tau_{yx} & \sigma_y & \tau_{yz} \\
\tau_{zx} & \tau_{zy} & \sigma_z
\end{bmatrix}
=
\begin{bmatrix}
\sigma_x & \tau_{xy} & 0 \\
\tau_{yx} & \sigma_y & 0 \\
0 & 0 & 0
\end{bmatrix} .
\tag{7.2}
$$

That is, the stress components σ_z, τ_{xz}, and τ_{yz} are zero (Fig. 7.7). This does not imply that the three stress components σ_x, σ_y, and τ_{xy} are each necessarily nonzero, but they are the only stress components that may be nonzero in plane stress.

Notice that the states of stress shown in Figs. 7.5 and 7.6 for a bar subjected to axial loads and axial torsion are plane stress. Many important applications in stress analysis result in states of plane stress. In this section we assume that the state of plane stress at a point of a material is known and we wish to know the stresses on

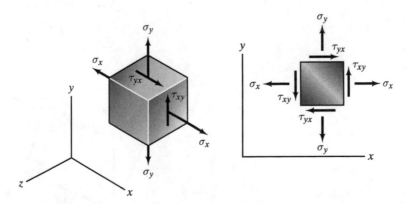

Fig. 7.7 Plane stress

planes other than the three planes perpendicular to the coordinate axes. We also address a crucial question from the standpoint of design. What are the maximum normal and shear stresses, and what are the orientations of the planes on which they act?

7.2.1 Coordinate Transformations

Suppose that we know the state of plane stress at a point p of a material in terms of the coordinate system shown in Fig. 7.8(a), and we want to know the state of stress at p in terms of a coordinate system $x'y'z'$ oriented as shown in Fig. 7.8(b). (The z and z' axes are coincident.) We begin with the element in Fig. 7.8(a) and pass a plane through it as shown in Fig. 7.9. The oblique surface of the resulting free-body diagram is perpendicular to the x' axis, so the normal and shear stresses acting on it are $\sigma_{x'}$ and $\tau_{x'y'}$. We can determine $\sigma_{x'}$ and $\tau_{x'y'}$ by writing the equilibrium equations for this free-body diagram.

Let the area of the oblique surface of the free-body diagram be ΔA. The sum of the forces in the x' direction is

Fig. 7.8 Stress components at p: (**a**) in terms of the xyz coordinate system, (**b**) in terms of the $x'y'z'$ coordinate system

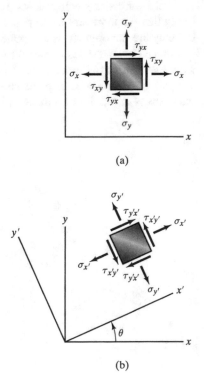

(a)

(b)

Fig. 7.9 Free-body
diagram for determining $\sigma_{x'}$
and $\tau_{x'y'}$.

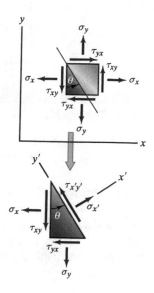

$$\sigma_{x'}\,\Delta A - (\sigma_x\,\Delta A\cos\theta)\cos\theta - (\sigma_y\,\Delta A\sin\theta)\sin\theta$$
$$- (\tau_{xy}\,\Delta A\cos\theta)\sin\theta - (\tau_{yx}\,\Delta A\sin\theta)\cos\theta = 0.$$

Solving for $\sigma_{x'}$, we obtain

$$\sigma_{x'} = \sigma_x\cos^2\theta + \sigma_y\sin^2\theta + 2\tau_{xy}\sin\theta\cos\theta. \tag{7.3}$$

The sum of the forces in the y' direction is

$$\tau_{x'y'}\,\Delta A + (\sigma_x\Delta A\cos\theta)\sin\theta - (\sigma_y\Delta A\sin\theta)\cos\theta$$
$$- (\tau_{xy}\,\Delta A\cos\theta)\cos\theta + (\tau_{yx}\Delta A\sin\theta)\sin\theta = 0.$$

The solution for $\tau_{x'y'}$ is

$$\tau_{x'y'} = -(\sigma_x - \sigma_y)\sin\theta\cos\theta + \tau_{xy}\left(\cos^2\theta - \sin^2\theta\right). \tag{7.4}$$

By using the trigonometric identities

$$2\cos^2\theta = 1 + \cos 2\theta,$$
$$2\sin^2\theta = 1 - \cos 2\theta,$$
$$2\sin\theta\cos\theta = \sin 2\theta,$$
$$\cos^2\theta - \sin^2\theta = \cos 2\theta, \tag{7.5}$$

we can write Eqs. (7.3) and (7.4) in alternative forms that will be useful:

$$\sigma_{x'} = \frac{\sigma_x + \sigma_y}{2} + \frac{\sigma_x - \sigma_y}{2}\cos 2\theta + \tau_{xy}\sin 2\theta, \qquad (7.6)$$

$$\tau_{x'y'} = -\frac{\sigma_x - \sigma_y}{2}\sin 2\theta + \tau_{xy}\cos 2\theta. \qquad (7.7)$$

We can obtain an equation for $\sigma_{y'}$ by setting θ equal to $\theta + 90°$ in the expression for $\sigma_{x'}$. The result is

$$\sigma_{y'} = \frac{\sigma_x + \sigma_y}{2} - \frac{\sigma_x - \sigma_y}{2}\cos 2\theta - \tau_{xy}\sin 2\theta. \qquad (7.8)$$

Given the state of plane stress shown in Fig. 7.8(a), Eqs. (7.6)–(7.8) determine the state of plane stress shown in Fig. 7.8(b) for any value of θ. This means that when we know a state of plane stress at a point p, we can determine the normal and shear stresses on planes through p other than the three planes perpendicular to the coordinate axes. Notice, however, that we can do so only for planes that are parallel to the z axis.

Study Questions

1. What is a state of plane stress?
2. How many components of stress must be known to define a state of plane stress at a point?
3. What is the definition of the angle θ in Eqs. (7.6)–(7.8)?

Example 7.1 Stresses on a Specified Plane
The components of plane stress at point p of the material in Fig. 7.10 are $\sigma_x = 4$ ksi, $\sigma_y = -2$ ksi and $\tau_{xy} = 2$ ksi. What are the normal stress and the magnitude of the shear stress on the plane $\mathcal{P}$ at point p?

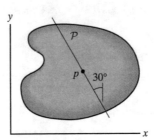

Fig. 7.10

Strategy
If we orient the $x'\,y'$ coordinate system so that the x' axis is perpendicular to the plane $\mathcal{P}$, then $\sigma_{x'}$ and $\tau_{x'y'}$ are the normal and shear stresses on $\mathcal{P}$ (Fig. 7.9). We can use Eqs. (7.6) and (7.7) to determine $\sigma_{x'}$ and $\tau_{x'y'}$.

Solution

The x' axis is perpendicular to the plane $\mathcal{P}$ if $\theta = 30^\circ$ (Fig. (a)).

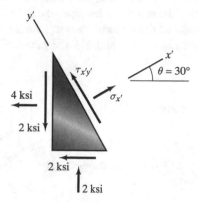

(a) Orienting the $x'y'$ coordinate system so that the x' axis is perpendicular to $\mathcal{P}$

From Eq. (7.6),

$$\sigma_{x'} = \frac{4\,\text{ksi} + (-2\,\text{ksi})}{2} + \frac{4\,\text{ksi} - (-2\,\text{ksi})}{2}\cos 60^\circ + (2\,\text{ksi})\sin 60^\circ$$

$$= 4.23\,\text{ksi},$$

and from Eq. (7.7):

$$\tau_{x'y'} = -\frac{4\,\text{ksi} - (-2\,\text{ksi})}{2}\sin 60^\circ + (2\,\text{ksi})\cos 60^\circ = -1.60\,\text{ksi}.$$

The normal stress on $\mathcal{P}$ at point p is 4.23 ksi, and the magnitude of the shear stress is 1.60 ksi.

Discussion

You need to be familiar with the definitions of the positive directions of the stress components, which are shown for plane stress in Fig. 7.8. Equations (7.6)–(7.8) are derived based on those definitions, and you need to know how to interpret the results you obtain with them. In this example we used Eqs. (7.6) and (7.7) to determine the normal and shear stresses on the plane shown in Fig. (a). In that figure we show the components $\sigma_{x'}$ and $\tau_{x'y'}$ in their defined positive directions. The positive result we obtained for $\sigma_{x'}$ indicates that the normal stress is tensile. The negative result we obtained for $\tau_{x'y'}$ indicates that the shear stress acts in the direction opposite to the direction shown in Fig. (a).

Example 7.2 Stresses as Functions of θ
The state of plane stress at a point p is shown on the left element in Fig. 7.11. **(a)**
Determine the state of plane stress at p acting on the right element in Fig. 7.11 if
$\theta = 60°$. Draw a sketch of the element showing the stresses acting on it. **(b)**
Draw graphs of $\sigma_{x'}$ and $\tau_{x'y'}$ as functions of θ for values of θ from zero to 360°.

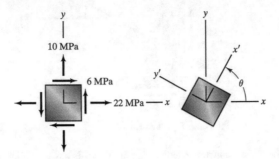

Fig. 7.11

Strategy
The components of plane stress on the left element in Fig. 7.11 are
$\sigma_x = 22$ MPa, $\sigma_y = 10$ MPa and $\tau_{xy} = 6$ MPa. The components of plane stress
on the right element in Fig. 7.11 are given by Eqs. (7.6)–(7.8).

Solution
(a) For $\theta = 60°$, the components of stress are

$$\sigma_{x'} = \frac{22\,\text{MPa} + 10\,\text{MPa}}{2} + \frac{22\,\text{MPa} - 10\,\text{MPa}}{2}\cos 120° + (6\,\text{MPa})\sin 120° = 18.20\,\text{MPa},$$

$$\sigma_{y'} = \frac{22\,\text{MPa} + 10\,\text{MPa}}{2} - \frac{22\,\text{MPa} - 10\,\text{MPa}}{2}\cos 120° - (6\,\text{MPa})\sin 120° = 13.80\,\text{MPa},$$

$$\tau_{x'y'} = -\frac{22\,\text{MPa} - 10\,\text{MPa}}{2}\sin 120° + (6\,\text{MPa})\cos 120° = -8.20\,\text{MPa}.$$

We show the components of stress acting on the element in Fig. (a). Notice
the directions of the shear stresses due to the negative value of $\tau_{x'y'}$.

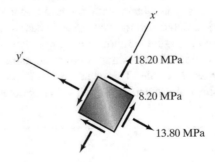

(a) State of stress for $\theta = 60°$.

(b) The stresses $\sigma_{x'}$ and $\tau_{x'y'}$ as functions of θ are

$$\sigma_{x'} = \frac{22\,\text{MPa} + 10\,\text{MPa}}{2} + \frac{22\,\text{MPa} - 10\,\text{MPa}}{2}\cos 2\theta + (6\,\text{MPa})\sin 2\theta$$
$$= 16\,\text{MPa} + (6\,\text{MPa})\cos 2\theta + (6\,\text{MPa})\sin 2\theta,$$

$$\tau_{x'y'} = -\frac{22\,\text{MPa} - 10\,\text{MPa}}{2}\sin 2\theta + (6\,\text{MPa})\cos 2\theta$$
$$= -(6\,\text{MPa})\sin 2\theta + (6\,\text{MPa})\cos 2\theta.$$

The graphs of these expressions are shown in Fig. (b).

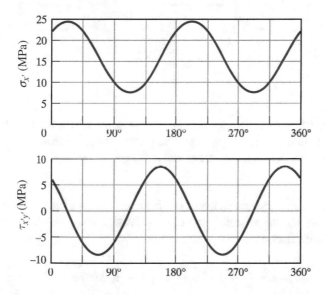

(b) Graphs of $\sigma_{x'}$ and $\tau_{x'y'}$ as functions of θ.

Discussion

The graphs of $\sigma_{x'}$ and $\tau_{x'y'}$ show that the normal stress and shear stress attain maximum and minimum values at particular values of θ. (Not, however, at the same values of θ. Notice that at angles for which the normal stress is a maximum or minimum, the shear stress is zero.) Because the maximum values of the stresses are so important with regard to design, we could use graphs such as these to determine the maximum stresses and the orientations of the planes on which they occur. But in the following sections, we introduce more efficient ways to obtain this information.

Example 7.3 Determining the Orientation of an Element
The state of plane stress at a point p is shown on the left element in Fig. 7.12, and the values of the stresses $\sigma_{x'}$ and $\tau_{x'y'}$ are shown on a rotated element. Determine the normal stress $\sigma_{y'}$ and the angle θ.

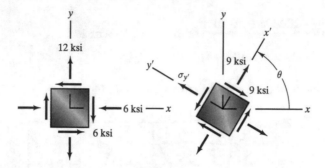

Fig. 7.12

Strategy
Equations (7.6)–(7.8) give the components of stress on the rotated element in terms of θ. Because $\sigma_{x'}$ and $\tau_{x'y'}$ are known, we can solve Eqs. (7.6) and (7.7) for θ, and then determine $\sigma_{y'}$ from Eq. (7.8).

Solution
The components of stress on the left element are $\sigma_x = -6$ ksi, $\tau_{xy} = -6$ ksi and $\sigma_y = 12$ ksi. On the rotated element, $\sigma_{x'} = 9$ ksi and $\tau_{x'y'} = -9$ ksi. Equation (7.6) is (with stresses in ksi)

$$\sigma_{x'} = \frac{\sigma_x + \sigma_y}{2} + \frac{\sigma_x - \sigma_y}{2} \cos 2\theta + \tau_{xy} \sin 2\theta :$$

$$9 = \frac{(-6) + 12}{2} + \frac{(-6) - 12}{2} \cos 2\theta + (-6) \sin 2\theta$$

$$= 3 - 9 \cos 2\theta - 6 \sin 2\theta,$$

and Eq. (7.7) is

$$\tau_{x'y'} = -\frac{\sigma_x - \sigma_y}{2} \sin 2\theta + \tau_{xy} \cos 2\theta :$$

$$-9 = -\frac{(-6) - 12}{2} \sin 2\theta + (-6) \cos 2\theta$$

$$= 9 \sin 2\theta - 6 \cos 2\theta.$$

We can solve these two equations for $\sin 2\theta$ and $\cos 2\theta$. The results are $\sin 2\theta = -1$ and $\cos 2\theta = 0$, from which we obtain $\theta = 135^\circ$. Substituting this result into Eq. (7.8), the stress $\sigma_{y'}$ is

$$\sigma_y = \frac{\sigma_x + \sigma_y}{2} - \frac{\sigma_x - \sigma_y}{2}\cos 2\theta - \tau_{xy}\sin 2\theta$$

$$= \frac{(-6\,\text{ksi}) + 12\,\text{ksi}}{2} - \frac{(-6\,\text{ksi}) - 12\,\text{ksi}}{2}\cos 2(135^\circ) - (-6\,\text{ksi})\sin 2(135^\circ)$$

$$= -3\,\text{ksi}.$$

The stresses are shown on the rotated element in Fig. (a).

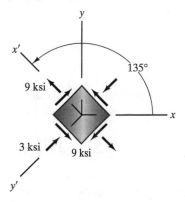

(a) Stresses on the properly oriented element

Discussion
Instead of being asked to determine the components of stress on a rotated element, in this example, they were given, and we were asked to determine the orientation of the element. Our next objective is to determine the orientations of the elements on which the normal and shear stresses have their maximum and minimum values.

7.2.2 Maximum and Minimum Stresses

Given the state of plane stress at a point p, the normal and shear stresses on the plane shown in Fig. 7.13 are given by Eqs. (7.6) and (7.7):

Fig. 7.13 (a) Point p of a material. (b) Normal and shear stresses at p

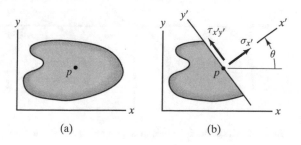

(a)   (b)

$$\sigma_{x'} = \frac{\sigma_x + \sigma_y}{2} + \frac{\sigma_x - \sigma_y}{2} \cos 2\theta + \tau_{xy} \sin 2\theta, \tag{7.9}$$

$$\tau_{x'y'} = -\frac{\sigma_x - \sigma_y}{2} \sin 2\theta + \tau_{xy} \cos 2\theta. \tag{7.10}$$

Thus, we can determine the normal and shear stresses at p for any value of θ. But it is the maximum values of stresses that determine whether a material will fail under a given state of stress. How can we determine them and the orientations of the planes through point p on which they act?

7.2.2.1   Principal Stresses

Let a value of θ for which the normal stress $\sigma_{x'}$ is a maximum or minimum be denoted by θ_p. To determine it, we evaluate the derivative of Eq. (7.9) with respect to 2θ:

$$\frac{d\sigma_{x'}}{d(2\theta)} = -\frac{\sigma_x - \sigma_y}{2} \sin 2\theta + \tau_{xy} \cos 2\theta, \tag{7.11}$$

and equate it to zero, obtaining the equation

$$\tan 2\theta_p = \frac{2\tau_{xy}}{\sigma_x - \sigma_y}. \tag{7.12}$$

When σ_x, σ_y, and τ_{xy} are known, we can solve this equation for $\tan 2\theta_p$, which allows us to determine θ_p. Then substituting θ_p into Eq. (7.9) yields the maximum or minimum value of the normal stress.

Because of the periodic nature of the tangent, Eq. (7.12) yields more than one solution for θ_p. Observe in Fig. 7.14 that if $2\theta_p$ is a solution of Eq. (7.12), so are $2\theta_p + 180°$, $2\theta_p + 2(180°)$, ... This means that the normal stress is a maximum or

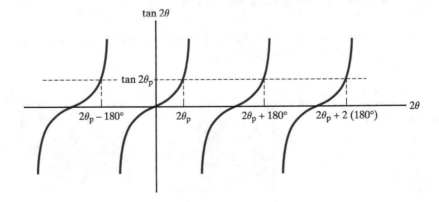

Fig. 7.14 The periodic nature of the tangent gives rise to multiple roots

Fig. 7.15 (a) Angles at which maximum or minimum normal stresses occur. (b) The planes correspond to the faces of a rectangular element

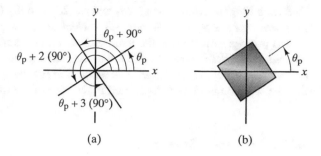

(a)   (b)

Fig. 7.16 Principal stresses and planes on which they act

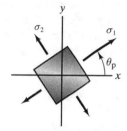

minimum at θ_p, $\theta_p + 90°$, $\theta_p + 2(90°)$, ... (Fig. 7.15(a)). *The planes on which the maximum and minimum normal stresses act correspond to the faces of a rectangular element* (Fig. 7.15(b)).

Recognizing that the maximum and minimum normal stresses act on the faces of a particular rectangular element makes it easy to determine and visualize the planes on which these stresses act. We simply determine one value of θ_p from Eq. (7.12), establishing the orientation of the rectangular element. We can then determine the values of the maximum and minimum stresses, called the *principal stresses* and denoted by σ_1 and σ_2 (Fig. 7.16), from Eq. (7.9).

We can obtain analytical expressions for the values of the principal stresses that are useful when we are not concerned with the planes on which they act. By solving equations

$$\frac{\sin 2\theta_p}{\cos 2\theta_p} = \tan 2\theta_p = \frac{2\tau_{xy}}{\sigma_x - \sigma_y} \tag{7.13}$$

and

$$\sin^2 2\theta_p + \cos^2 2\theta_p = 1 \tag{7.14}$$

for $\sin 2\theta_p$ and $\cos 2\theta_p$ and substituting the results into Eq. (7.9), we obtain

$$\sigma_1, \sigma_2 = \frac{\sigma_x + \sigma_y}{2} \pm \sqrt{\left(\frac{\sigma_x - \sigma_y}{2}\right)^2 + \tau_{xy}^2}. \tag{7.15}$$

Substituting the results for $\sin 2\theta_p$ and $\cos 2\theta_p$ into Eq. (7.10) yields $\tau_{x'y'} = 0$. *The shear stresses are zero on the planes on which the principal stresses act.*

The principal stresses are the maximum and minimum normal stresses acting on planes through point p that are parallel to the z axis. It can be shown that *there are no normal stresses of greater magnitude on any plane through p.* However, we will see that the situation is more complicated in the case of the maximum shear stress.

7.2.2.2   Maximum Shear Stresses

We approach the determination of maximum or minimum shear stresses in the same way we did for normal stresses. Let a value of θ for which the shear stress is a maximum or minimum be denoted by θ_s. Evaluating the derivative of Eq. (7.10) with respect to 2θ and setting it equal to zero, we obtain the equation

$$\tan 2\theta_s = -\frac{\sigma_x - \sigma_y}{2\tau_{xy}}. \tag{7.16}$$

With this equation, we can determine θ_s and substitute it into Eq. (7.10) to determine the maximum or minimum value of the shear stress.

As in the case of the normal stress, if the shear stress is a maximum or minimum at θ_s, it is also a maximum or minimum at $\theta_s + 90°$, $\theta_s + 2\,(90°)$, ... *The planes on which the maximum and minimum shear stresses act also correspond to the faces of a rectangular element.* Because the shear stresses on the faces of a rectangular element are equal in magnitude, the magnitudes of the maximum and minimum shear stresses are equal. Furthermore, the orientation of this rectangular element is related in a simple way to the orientation of the rectangular element on which the principal stresses act. Notice from Eqs. (7.12) and (7.16) that $\tan 2\theta_s$ is the negative inverse of $\tan 2\theta_p$. This implies that the directions defined by the angles $2\theta_s$ and $2\theta_p$ are perpendicular (Fig. 7.17), which means that *the rectangular element on which the maximum and minimum shear stresses act is rotated 45° relative to the rectangular element on which the principal stresses act.* Once we have determined the orientation of the element on which the principal stresses act (Fig. 7.18(a)), we also know the orientation of the element on which the maximum and minimum shear stresses act (Fig. 7.18(b)).

Fig. 7.17 The tangents of the angles defining these two perpendicular directions relative to the x axis are b/a and $-a/b$. One tangent is the negative inverse of the other

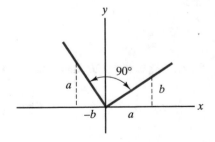

Fig. 7.18 Relationship between the orientation of the elements (**a**) on which the principal stresses act and (**b**) on which the maximum and minimum shear stresses act

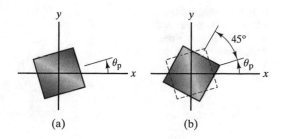

(a) (b)

To obtain an analytical expression for the magnitude of the maximum shear stress, we solve the equations

$$\frac{\sin 2\theta_s}{\cos 2\theta_s} = \tan 2\theta_s = -\frac{\sigma_x - \sigma_y}{2\tau_{xy}} \tag{7.17}$$

and

$$\sin^2 2\theta_s + \cos^2 2\theta_s = 1 \tag{7.18}$$

for $\sin 2\theta_s$ and $\cos 2\theta_s$, and substitute the results into Eq. (7.10). The result is

$$\tau_{max} = \sqrt{\left(\frac{\sigma_x - \sigma_y}{2}\right)^2 + \tau_{xy}^2}. \tag{7.19}$$

This stress is called the *maximum in-plane shear stress*, because it is the greatest shear stress that occurs on any plane parallel to the z axis. However, we will see that greater shear stresses may occur on other planes through p.

To determine the complete state of plane stress on the rectangular element on which the maximum in-plane shear stresses act, we must evaluate the normal stresses. Substituting our results for $\sin 2\theta_s$ and $\cos 2\theta_s$ into Eqs. (7.6) and (7.8), we obtain

$$\sigma_x' = \sigma_y' = \frac{\sigma_x + \sigma_y}{2}.$$

The normal stresses on the element on which the maximum in-plane shear stresses act are equal. We denote this normal stress by σ_s:

$$\sigma_s = \frac{\sigma_x + \sigma_y}{2}. \tag{7.20}$$

Figure 7.19 shows the complete state of stress.

Fig. 7.19 Element on
which the maximum and
minimum in-plane shear
stresses act

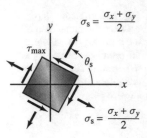

7.2.2.3   The Absolute Maximum Shear Stress

Now let us consider whether shear stresses greater in magnitude than the value given
by Eq. (7.19) can occur on other planes through p. We begin with the element on
which the principal stresses occur and realign the coordinate system with the faces of
the element (Fig. 7.20(a)). In terms of this new coordinate system, the components of
plane stress are $\sigma_x = \sigma_1$, $\sigma_y = \sigma_2$ and $\tau_{xy} = 0$. Substituting these components into
Eq. (7.19), we obtain a new expression for the magnitude of the maximum in-plane
shear stress:

$$\sqrt{\left(\frac{\sigma_x - \sigma_y}{2}\right)^2 + \tau_{xy}^2} = \sqrt{\left(\frac{\sigma_1 - \sigma_2}{2}\right)^2 + 0} = \left|\frac{\sigma_1 - \sigma_2}{2}\right|. \qquad (7.21)$$

This equation is expressed in terms of the principal stresses, but it gives the same
result as Eq. (7.19). We have still considered only planes parallel to the original
z axis. Now, however, let's consider the element on which the principal stresses
occur and reorient the coordinate system as shown in Fig. 7.20(b). In terms of this
coordinate system, $\sigma_x = 0$, $\sigma_y = \sigma_1$ and $\tau_{xy} = 0$. Substituting these components into
Eq. (7.19), we obtain the maximum shear stress on planes parallel to the new z axis:

$$\sqrt{\left(\frac{\sigma_x - \sigma_y}{2}\right)^2 + \tau_{xy}^2} = \sqrt{\left(\frac{0 - \sigma_1}{2}\right)^2 + 0} = \left|\frac{\sigma_1}{2}\right|. \qquad (7.22)$$

Next, we reorient the coordinate system as shown in Fig. 7.20(c). In terms of this
coordinate system, $\sigma_x = \sigma_2$, $\sigma_y = 0$ and $\tau_{xy} = 0$. Substituting these components into
Eq. (7.19), we obtain the maximum shear stress on planes parallel to this z axis:

$$\sqrt{\left(\frac{\sigma_x - \sigma_y}{2}\right)^2 + \tau_{xy}^2} = \sqrt{\left(\frac{\sigma_2 - 0}{2}\right)^2 + 0} = \left|\frac{\sigma_2}{2}\right|. \qquad (7.23)$$

Depending on the values of the principal stresses, Eq. (7.22) and/or Eq. (7.23) can
result in larger values of the magnitude of the maximum shear stress than the
maximum in-plane shear stress. Although we have still considered only a subset of
the possible planes through point p, there are no shear stresses of greater magnitude

Fig. 7.20 Element on which the principal stresses act. (**a**) Realigned coordinate system. (**b**), (**c**) Other orientations of the coordinate system

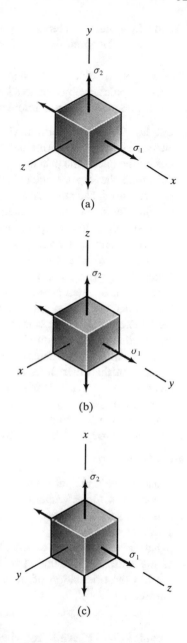

(a)

(b)

(c)

on any plane through p than the largest value given by Eq. (7.21), (7.22) or (7.23). We denote this value, called the *absolute maximum shear stress*, by

$$\tau_{abs} = \text{Max}\left(\left|\frac{\sigma_1}{2}\right|, \left|\frac{\sigma_2}{2}\right|, \left|\frac{\sigma_1 - \sigma_2}{2}\right|\right). \tag{7.24}$$

7.2.2.4   Summary: Determining the Principal Stresses and the Maximum Shear Stress

Here we give a sequence of steps for determining the maximum and minimum stresses at a point p subjected to a known state of plane stress and the orientations of the planes on which the maximum stresses act.

1. Use Eq. (7.12) to determine θ_p, establishing the orientation of the rectangular element on which the principal stresses act.
2. Determine one of the principal stresses by substituting θ_p into Eq. (7.9), and then determine the other principal stress by substituting $\theta_p + 90^\circ$ into Eq. (7.9). Because no shear stresses act on the element on which the principal stresses act, this determines the complete state of stress on the element. Notice that you can determine the principal stresses from Eq. (7.15), but doing so does not tell you which principal stress acts on which faces of the element.
3. Use Eq. (7.16) to determine θ_s, establishing the orientation of the element on which the maximum and minimum in-plane shear stresses act. Alternatively, because this element is rotated 45° relative to the element on which the principal stresses act, you can simply use $\theta_s = \theta_p + 45^\circ$.
4. Determine the shear stress on the element by substituting θ_s into Eq. (7.10). The result tells you the magnitude and direction of the shear stress on one face of the element, which also determines the magnitudes and directions of the shear stresses on the other faces (Fig. 7.19). The normal stress on each face of this element is $(\sigma_x + \sigma_y)/2$, which completes the determination of the state of stress on the element.
5. The absolute maximum shear stress (the magnitude of the largest shear stress on any plane through p) is given by Eq. (7.24).

Study Questions

1. What are the principal stresses?
2. If you know the orientation of the plane on which one of the principal stresses acts, what do you know about the orientation of the plane on which the other principal stress acts?
3. What shear stresses act on the planes on which the principal stresses act?
4. If you know the orientation of the element on which the principal stresses act, what is the orientation of the element on which the maximum in-plane shear stresses act?

Example 7.4 Determining Maximum and Minimum Stresses
The state of plane stress at a point p is shown on the element in Fig. 7.21. Determine the principal stresses and the maximum in-plane shear stress and show them acting on properly oriented elements. Also determine the absolute maximum shear stress.

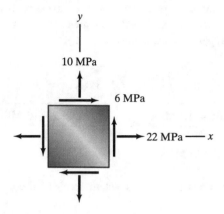

Fig. 7.21

Strategy
The components of plane stress on the element are $\sigma_x = 22$ MPa, $\sigma_y = 10$ MPa and $\tau_{xy} = 6$ MPa. We can follow the steps given in the preceding summary to determine the principal stresses and the maximum shear stress.

Solution
Step 1   From Eq. (7.12),

$$\tan 2\theta_p = \frac{2\tau_{xy}}{\sigma_x - \sigma_y} = \frac{2(6 \text{ MPa})}{22 \text{ MPa} - 10 \text{ MPa}} = 1.$$

Solving this equation, we obtain $\theta_p = 22.5°$. This angle tells us the orientation of the element on which the principal stresses act.

Step 2   We substitute θ_p into Eq. (7.9) to determine the first principal stress:

$$\sigma_1 = \frac{\sigma_x + \sigma_y}{2} + \frac{\sigma_x - \sigma_y}{2} \cos\ 2\theta_p + \tau_{xy} \sin 2\theta_p$$

$$= \frac{22 \text{ MPa} + 10 \text{ MPa}}{2} + \frac{22 \text{ MPa} - 10 \text{ MPa}}{2} \cos 45° + (6 \text{ MPa}) \sin 45°$$

$$= 24.49 \text{ MPa}.$$

We then substitute $\theta_p + 90°$ into Eq. (7.9) to determine the second principal stress:

$$\sigma_2 = \frac{\sigma_x + \sigma_y}{2} + \frac{\sigma_x - \sigma_y}{2} \cos\ 2(\theta_p + 90°) + \tau_{xy} \sin 2(\theta_p + 90°)$$

$$= \frac{22 \text{ MPa} + 10 \text{ MPa}}{2} + \frac{22 \text{ MPa} - 10 \text{ MPa}}{2} \cos 225° + (6 \text{ MPa}) \sin 225°$$

$$= 7.51 \text{ MPa}.$$

The principal stresses are shown on the properly oriented element in Fig. (a). No shear stresses act on this element. We can also determine the values of the principal stresses from Eq. (7.15):

$$\sigma_1, \ \sigma_2 = \frac{\sigma_x + \sigma_y}{2} \ \pm \ \sqrt{\left(\frac{\sigma_x - \sigma_y}{2}\right)^2 + \tau_{xy}^2}$$

$$= \frac{22\,\text{MPa} + 10\,\text{MPa}}{2} \ \pm \ \sqrt{\left(\frac{22\,\text{MPa} - 10\,\text{MPa}}{2}\right)^2 + (6\,\text{MPa})^2}$$

$$= 24.49, \ 7.51\,\text{MPa},$$

but this procedure does not tell us which faces of the element the stresses act on.

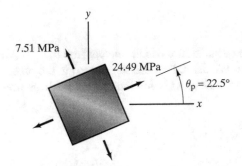

(a) The principal stresses

Step 3   From Eq. (7.16)

$$\tan 2\theta_s = -\frac{\sigma_x - \sigma_y}{2\tau_{xy}} = -\frac{22\,\text{MPa} - 10\,\text{MPa}}{2(6\,\text{MPa})} = -1,$$

from which we obtain $\theta_s = -22.5°$. We can also determine θ_s by using the fact that the element on which the maximum in-plane shear stresses act is rotated $45°$ relative to the element on which the principal stresses act. In this way we obtain $\theta_s = \theta_p + 45° = 67.5°$. This angle differs from the result we obtained using Eq. (7.16) by $90°$, so the resulting orientation of the element is the same. This emphasizes that the angle θ_s can only be determined from Eq. (7.16) within a multiple of $90°$.

Step 4   We substitute θ_s into Eq. (7.10) to determine the maximum in-plane shear stress.

$$\tau_{max} = -\frac{\sigma_x - \sigma_y}{2} \sin 2\theta_s + \tau_{xy} \cos 2\theta_s$$

$$= -\frac{22\,\text{MPa} - 10\,\text{MPa}}{2} \sin\left(-45°\right) + (6\,\text{MPa}) \cos\left(-45°\right)$$

$$= 8.49\,\text{MPa}.$$

The normal stress on each face of this element is

$$\sigma_s = \frac{\sigma_x + \sigma_y}{2} = \frac{22\,\text{MPa} + 10\,\text{MPa}}{2} = 16\,\text{MPa}.$$

The maximum in-plane shear stresses and associated normal stresses are shown on the properly oriented element in Fig. (b).

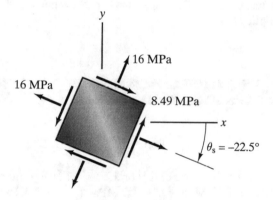

(b) Maximum in-plane shear stresses and associated normal stresses

Step 5 The absolute maximum shear stress is given by the largest of the three values

$$\left|\frac{\sigma_1 - \sigma_2}{2}\right| = \left|\frac{24.49\,\text{MPa} - 7.51\,\text{MPa}}{2}\right| = 8.49\,\text{MPa},$$

$$\left|\frac{\sigma_1}{2}\right| = \left|\frac{24.49\,\text{MPa}}{2}\right| = 12.24\,\text{MPa},$$

$$\left|\frac{\sigma_2}{2}\right| = \left|\frac{7.51\,\text{MPa}}{2}\right| = 3.76\,\text{MPa}.$$

Discussion
This example demonstrates the importance of the concept of the absolute maximum shear stress. For a given state of plane stress, the maximum

in-plane shear stress is the maximum shear stress acting on the plane shown in Fig. 7.13(b) for any value of θ. *But that is only a subset of the planes through the point—planes that are parallel to the z axis.* In this example, the maximum in-plane shear stress is 8.49 MPa, but the absolute maximum shear stress is 12.24 MPa. That means that the maximum shear stress occurs on a plane that is not parallel to the z axis.

Problems

7.1 The prismatic bar has a circular cross section with 100-mm diameter. In a test to determine its elastic properties, the bar is subjected to 4.8-MN (meganewton) axial loads. In terms of the coordinate system shown, what are the six components of the state of stress at a point of the bar that is not near the ends where the forces are applied?

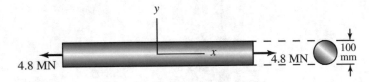

Problem 7.1

7.2 The components of stress at point p of the object in Fig. (a) are $\sigma_x = 2$ MPa, $\sigma_y = -4$ MPa, $\sigma_z = 4$ MPa, $\tau_{xy} = 1.5$ MPa, $\tau_{yz} = -3$ MPa, $\tau_{zx} = -2$ MPa. The plane $\mathcal{P}$ intersects point p and is perpendicular to the y axis. In Fig. (b) the part of the object below the plane is isolated. (**a**) What is the normal stress on the plane $\mathcal{P}$ at p? (**b**) What is the magnitude of the shear stress on the plane $\mathcal{P}$ at p?

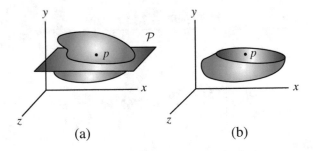

(a)   (b)

Problem 7.2

7.3 In terms of the coordinate system shown in Fig. (a), the components of stress at point p of the object are $\sigma_x = -30$ ksi, $\sigma_y = 65$ ksi, $\sigma_z = 40$ ksi, $\tau_{xy} = 50$ ksi, $\tau_{yz} = 35$ ksi, and $\tau_{zx} = -45$ ksi. The $x'y'z'$ coordinate system shown in Fig. (b) is obtained by rotating the coordinate system in Fig. (a) $90°$ about the y axis. What are the components of stress at p in terms of the $x'y'z'$ coordinate system?

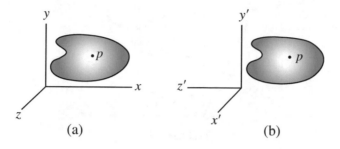

(a)   (b)

Problem 7.3

Strategy: Use the definitions of the components of stress. It isn't necessary to use equations.

7.4 The components of plane stress at point p of a material are shown on the element in Fig. (a).

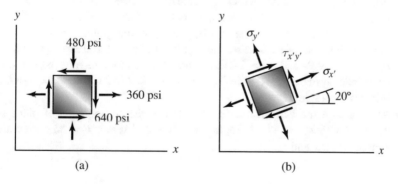

(a)   (b)

Problem 7.4

(a) What are the components of stress σ_x, σ_y, σ_z, τ_{xy}, τ_{yz} and τ_{xz} at p? (b) The element shown in Fig. (b) is located at the same point p. What are the components of stress $\sigma_{x'}$, $\sigma_{y'}$ and $\tau_{x'y'}$ at p?

7.5 The components of plane stress at a point p of a material are $\sigma_x = 20$ MPa, $\sigma_y = 0$, and $\tau_{xy} = 0$. If $\theta = 45°$, what are the components of stress $\sigma_{x'}$, $\sigma_{y'}$ and $\tau_{x'y'}$ at p?

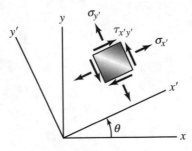

Problems 7.5–7.9

7.6 The components of plane stress at a point p of a material are $\sigma_x = 0$, $\sigma_y = 0$ and $\tau_{xy} = 25$ ksi. If $\theta = 45°$, what are the components of stress $\sigma_{x'}$, $\sigma_{y'}$ and $\tau_{x'y'}$ at p?

7.7 The components of plane stress at a point p of a material are $\sigma_x = -8$ ksi, $\sigma_y = 6$ ksi and $\tau_{xy} = -6$ ksi. If $\theta = 25°$, what are the components of stress $\sigma_{x'}$, $\sigma_{y'}$ and $\tau_{x'y'}$ at p?

7.8 Strain gauges attached to a rocket engine nozzle determine that the components of plane stress $\sigma_{x'} = 66.46$ MPa, $\sigma_{y'} = 82.54$ MPa and $\tau_{x'y'} = 6.75$ MPa at $\theta = 20°$. What are the stresses σ_x, σ_y, and τ_{xy} at that point?

7.9* The components of plane stress at a point p of a material are $\sigma_x = 240$ MPa, $\sigma_y = -120$ MPa and $\tau_{xy} = 240$ MPa, and the components referred to the $x'y'z'$ coordinate system are $\sigma_{x'} = 347$ MPa, $\sigma_{y'} = -227$ MPa, and $\tau_{x'y'} = -87$ MPa. What is the angle θ?

7.10 A point p of the car's frame is subjected to the components of plane stress $\sigma_{x'} = 32$ MPa, $\sigma_{y'} = -16$ MPa, and $\tau_{x'y'} = -24$ MPa. If $\theta = 35°$, what are the stresses σ_x, σ_y, and τ_{xy} at p?

Problems 7.10–7.11 (*Photograph from iStock by mevans*)

7.11 A point p of the car's frame is subjected to the components of plane stress $\sigma_{x'} = 32$ MPa, $\sigma_{y'} = -16$ MPa, and $\tau_{x'y'} = -24$ MPa. If $\theta = -40°$, what are the stresses σ_x, σ_y, and τ_{xy} at p?

7.12 The components of plane stress at point p of the material shown are $\sigma_x = 4$ ksi, $\sigma_y = -2$ ksi and $\tau_{xy} = 2$ ksi. What are the normal stress and the magnitude of the shear stress on the plane $\mathcal{P}$ at point p?

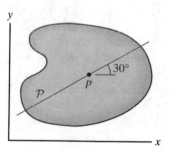

Problem 7.12–7.13

7.13 The components of plane stress at point p of the material shown are $\sigma_x = -10.5$ MPa, $\sigma_y = 60$ MPa, and $\tau_{xy} = -4.5$ MPa. What are the normal stress and the magnitude of the shear stress on the plane $\mathcal{P}$ at point p?

7.14 The architectural column is subjected to an 800-kN compressive axial load. The cross section of the column is shown. Use Eq. (7.6) to determine the normal stresses acting on planes (a), (b), and (c).

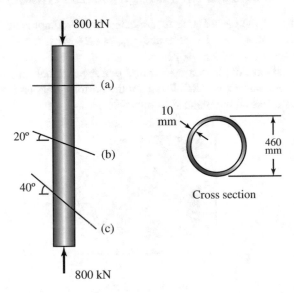

Problems 7.14–7.15

7.15   The architectural column is subjected to an 800-kN compressive axial load. The cross section of the column is shown. Use Eq. (7.7) to determine the magnitudes of the shear stresses acting on planes (a), (b), and (c).

7.16   Determine the stresses σ and τ **(a)** by writing equilibrium equations for the plate shown and **(b)** by using Eqs. (7.6) and (7.7).

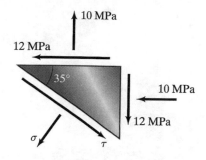

Problem 7.16

7.17 The stress $\tau_{xy} = 14$ MPa and the angle $\theta = 25°$. Determine the components of stress on the right element.

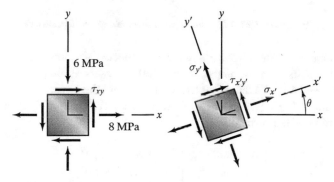

Problems 7.17–7.19

7.18∗ The stresses $\tau_{xy} = 12$ MPa, $\sigma_{x'} = 14$ MPa, and $\sigma_{y'} = -12$ MPa. Determine the stress $\tau_{x'y'}$ and the angle θ.

7.19 The stress $\tau_{x'y'} = 12$ MPa and the angle $\theta = 35°$. Determine $\sigma_{x'}$, $\sigma_{y'}$, and τ_{xy}.

7.20 A point p of an airplane's wing is subjected to plane stress. When $\theta = 55°$, $\sigma_x = 100$ psi, $\sigma_y = -200$ psi, and $\sigma_{x'} = -175$ psi. Determine the stresses τ_{xy} and $\tau_{x'y'}$ at p.

Problems 7.20–7.21 (*Photograph from iStock by sillycruiser*)

7.21* A point p of the airplane's wing is subjected to plane stress. The stress components $\sigma_x = 80$ psi, $\sigma_y = -120$ psi, $\tau_{xy} = -100$ psi, $\sigma_{x'} = -80$ psi, and $\sigma_{y'} = 40$ psi. Determine the stress $\tau_{x'y'}$ and the angle θ.

7.22 Equations (7.6), (7.7), (7.8) apply to plane stress, but they also apply to states of stress of the form

$$
\begin{bmatrix}
\sigma_x & \tau_{xy} & \tau_{xz} \\
\tau_{yx} & \sigma_y & \tau_{yz} \\
\tau_{zx} & \tau_{zy} & \sigma_z
\end{bmatrix}
=
\begin{bmatrix}
\sigma_x & \tau_{xy} & 0 \\
\tau_{yx} & \sigma_y & 0 \\
0 & 0 & \sigma_z
\end{bmatrix}.
$$

That is, the stress components τ_{xz} and τ_{yz} are zero. The state of stress at a point within a fluid at rest is

$$
\begin{bmatrix}
\sigma_x & \tau_{xy} & \tau_{xz} \\
\tau_{yx} & \sigma_y & \tau_{yz} \\
\tau_{zx} & \tau_{zy} & \sigma_z
\end{bmatrix}
=
\begin{bmatrix}
-P & 0 & 0 \\
0 & -P & 0 \\
0 & 0 & -P
\end{bmatrix},
$$

where P is the pressure of the fluid at the point. Use Eqs. (7.6) and (7.7) to show that at a point of a fluid at rest, $\sigma_{x'} = -P$ and $\tau_{x'y'} = 0$ for any value of θ. That is, the normal stress is the negative of the pressure, and the in-plane shear stress is zero for any value of θ.

7.23 By substituting the trigonometric identities (7.5) into Eqs. (7.3) and (7.4), derive Eqs. (7.6) and (7.7).

7.24 The components of plane stress acting on an element of a bar subjected to axial loads are shown. Assuming the stress σ_x to be known, determine the principal stresses and the maximum in-plane shear stress and show them acting on properly oriented elements.

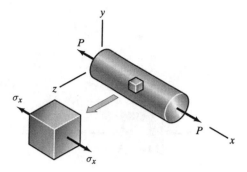

Problem 7.24

7.25 The components of plane stress acting on an element of a bar subjected to torsion are shown. Assuming the stress τ_{xy} to be known, determine the principal stresses and the maximum in-plane shear stress and show them acting on properly oriented elements.

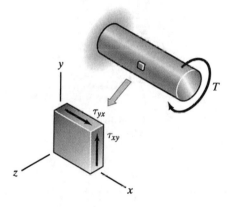

Problem 7.25

For the states of plane stress shown in Problems 7.26–7.29, determine the principal stresses and the maximum in-plane shear stress and show them acting on properly oriented elements.

7.26 Text

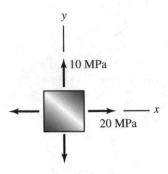

Problem 7.26

7.27 Text

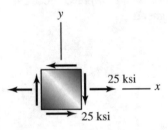

Problem 7.27

7.28 Text

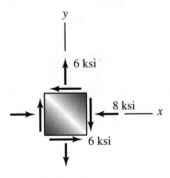

Problem 7.28

7.29 Text

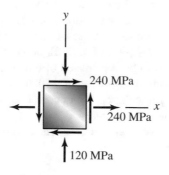

Problem 7.29

7.30 For the state of plane stress $\sigma_x = 20$ MPa, $\sigma_y = 10$ MPa, and $\tau_{xy} = 0$, what is the absolute maximum shear stress?

7.31 For the state of plane stress $\sigma_x = 25$ ksi, $\sigma_y = 0$, and $\tau_{xy} = -25$ ksi, what is the absolute maximum shear stress?

7.32 A point of an antenna's supporting structure is subjected to the state of plane stress $\sigma_x = 40$ MPa, $\sigma_y = -20$ MPa, and $\tau_{xy} = 30$ MPa. Determine the principal stresses and the absolute maximum shear stress.

Problems 7.32–7.33 (*Photograph from iStock by Loren Klein*)

7.33* A point of the antenna's supporting structure is subjected to plane stress. The components $\sigma_y = -20$ MPa and $\tau_{xy} = 30$ MPa. The allowable normal stress of the material (in tension and compression) is 80 MPa. Based on this criterion, what is the allowable range of values of the stress component σ_x?

7.34 By setting θ equal to $\theta + 90°$ in Eq. (7.6), derive Eq. (7.8).

7.35 The connecting rod is to be constructed of a material that can be subjected to an absolute maximum shear stress of 170 MPa. A finite element analysis determines that at a particular operating condition, a point of the rod would be subjected to the state of plane stress $\sigma_x = -160$ MPa, $\sigma_y = 120$ MPa, and $\tau_{xy} = -80$ MPa. Based on this criterion, determine whether the operating condition is acceptable.

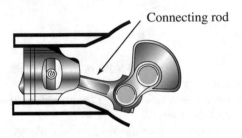

Connecting rod

Problems 7.35–7.36

7.36 The connecting rod is to be constructed of a material that can be subjected to an absolute maximum shear stress of 170 MPa. A finite element analysis determines that at a particular operating condition, a point of the rod would be subjected to the state of plane stress $\sigma_x = -180$ MPa, $\sigma_y = 150$ MPa, $\tau_{xy} = -55$ MPa. Based on this criterion, determine whether the operating condition is acceptable.

7.3 Mohr's Circle for Plane Stress

Mohr's circle is a graphical method for solving Eqs. (7.6)–(7.8). For a given state of plane stress (Fig. 7.22(a)), Mohr's circle determines the components of stress in terms of a coordinate system rotated through a specified angle θ (Fig. 7.22(b)). Why do we discuss this method in an age when computers have made most graphical methods of solution obsolete? The reason is that Mohr's circle allows the solutions to Eqs. (7.6)–(7.8) to be visualized and understood to an extent not possible with other approaches. We first show how to apply Mohr's circle and then explain why it works.

7.3.1 Constructing the Circle

Suppose that we know the components of plane stress σ_x, τ_{xy}, and σ_y in Fig. 7.22(a) and we want to determine the components $\sigma_{x'}$, $\tau_{x'y'}$, and $\sigma_{y'}$ in Fig. 7.22(b) for a given angle θ. Determining this information with Mohr's circle involves four steps:

1. Establish a set of horizontal and vertical axes with normal stress measured along the horizontal axis and shear stress measured along the vertical axis (Fig. 7.23(a)). Positive normal stress is measured to the right and positive shear stress is measured *downward*.
2. Plot two points, point P with coordinates (σ_x, τ_{xy}) and point Q with coordinates $(\sigma_y, -\tau_{xy})$, as shown in Fig. 7.23(b).
3. Draw a straight line connecting points P and Q. Using the intersection of the straight line with the horizontal axis as the center, draw a circle that passes through the two points (Fig. 7.23(c)).
4. Draw a straight line through the center of the circle at an angle 2θ measured counterclockwise from point P (Fig. 7.23(d)). Point P' at which this line intersects the circle has coordinates $(\sigma_{x'}, \tau_{x'y'})$, and point Q' has coordinates $(\sigma_{y'}, -\tau_{x'y'})$.

Fig. 7.22 (a) State of plane stress. (b) Components in terms of a rotated coordinate system

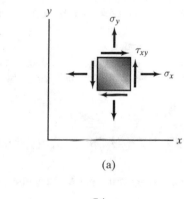

(a)

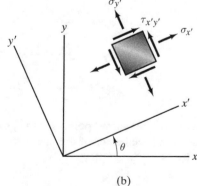

(b)

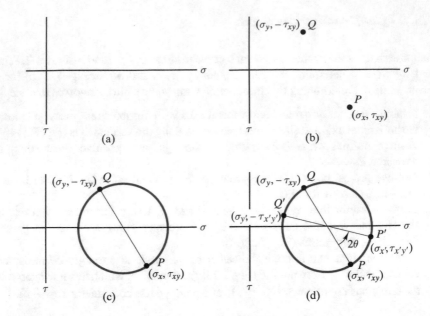

Fig. 7.23 (a) Establishing the axes. Shear stress is positive downward. (b) Plotting points P and Q. (c) Drawing Mohr's circle. The center of the circle is the intersection of the line between points P and Q with the horizontal axis. (d) Determining the stresses

This construction indicates that for any value of the angle θ, the stress components $\sigma_{x'}, \tau_{x'y'}$, and $\sigma_{y'}$ can be determined from the coordinates of two points on Mohr's circle. But we must prove this result.

7.3.2   Why Mohr's Circle Works

We will now prove that Mohr's circle solves Eqs. (7.6)–(7.8):

$$\sigma_{x'} = \frac{\sigma_x + \sigma_y}{2} + \frac{\sigma_x - \sigma_y}{2} \cos 2\theta + \tau_{xy} \sin 2\theta,$$

$$\tau_{x'y'} = -\frac{\sigma_x - \sigma_y}{2} \sin 2\theta + \tau_{xy} \cos 2\theta,$$

$$\sigma_{y'} = \frac{\sigma_x + \sigma_y}{2} - \frac{\sigma_x - \sigma_y}{2} \cos 2\theta - \tau_{xy} \sin 2\theta.$$

Figure 7.24(a) shows points P and Q and Mohr's circle. Notice that the horizontal coordinate of the center of the circle is $(\sigma_x + \sigma_y)/2$, and the radius of the circle R is given by

Fig. 7.24 (a) Dimensions of Mohr's circle. (b) Dimensions including points P' and Q'.

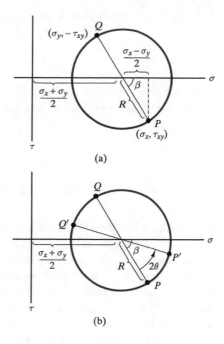

(a)

(b)

$$R = \sqrt{\left(\frac{\sigma_x - \sigma_y}{2}\right)^2 + \left(\tau_{xy}\right)^2}.$$

The sine and cosine of the angle β are

$$\sin \beta = \frac{\tau_{xy}}{R}, \quad \cos \beta = \frac{\sigma_x - \sigma_y}{2R}.$$

From Fig. 7.24(b), the horizontal coordinate of point P' is

$$
\begin{aligned}
\frac{\sigma_x + \sigma_y}{2} + R\cos(\beta - 2\theta) &= \frac{\sigma_x + \sigma_y}{2} + R(\cos\beta\cos 2\theta + \sin\beta\sin 2\theta) \\
&= \frac{\sigma_x + \sigma_y}{2} + \frac{\sigma_x - \sigma_y}{2}\cos 2\theta + \tau_{xy}\sin 2\theta \\
&= \sigma_{x'},
\end{aligned}
$$

and the horizontal coordinate of point Q' is

$$
\begin{aligned}
\frac{\sigma_x + \sigma_y}{2} - R\cos(\beta - 2\theta) &= \frac{\sigma_x + \sigma_y}{2} - R(\cos\beta\cos 2\theta + \sin\beta\sin 2\theta) \\
&= \frac{\sigma_x + \sigma_y}{2} - \frac{\sigma_x - \sigma_y}{2}\cos 2\theta - \tau_{xy}\sin 2\theta \\
&= \sigma_{y'}.
\end{aligned}
$$

The vertical coordinate of point P' is

$$R \sin{(\beta - 2\theta)} = R(-\cos{\beta} \sin{2\theta} + \sin{\beta} \cos{2\theta})$$
$$= -\frac{\sigma_x - \sigma_y}{2} \sin{2\theta} + \tau_{xy} \cos{2\theta}$$
$$= \tau_{x'y'},$$

which implies that the vertical coordinate of point Q' is $-\tau_{x'y'}$. Thus, we have shown that the coordinates of point P' are $(\sigma_{x'}, \tau_{x'y'})$ and the coordinates of point Q' are $(\sigma_{y'}, -\tau_{x'y'})$.

7.3.3 Determining Maximum and Minimum Stresses

Mohr's circle is a map of the stresses $\sigma_{x'}, \tau_{x'y'}$, and $\sigma_{y'}$ for all values of θ, so once the circle is constructed, the values of the maximum and minimum normal stresses and the magnitude of the maximum in-plane shear stress are apparent. *The coordinates of the points where the circle intersects the horizontal axis determine the two principal stresses* (Fig. 7.25). The coordinates of the points at the bottom and top of the circle determine the maximum and minimum in-plane shear stresses, so *the radius of the circle equals the magnitude of the maximum in-plane shear stress*. Notice that Mohr's circle demonstrates very clearly that the shear stresses are zero on the planes on which the principal stresses act, and also that the normal stresses are equal on the planes on which the maximum and minimum in-plane shear stresses act.

We can also use Mohr's circle to determine the orientations of the elements on which the principal stresses and maximum in-plane shear stresses act. If we let point P' coincide with either principal stress (Fig. 7.26(a)), we can measure the angle $2\theta_p$ and thereby determine the orientation of the plane on which that principal stress acts (Fig. 7.26(b)).

We can then let P' coincide with either the maximum or minimum shear stress (Fig. 7.27(a)) and measure the angle $2\theta_s$, determining the orientation of the plane on which that shear stress acts (Fig. 7.27(b)). Notice in the example illustrated in Fig. 7.27 that point P' coincides with the minimum shear stress, so $\tau_{x'y'} = -\tau_{max}$.

Fig. 7.25 Mohr's circle indicates the values of the principal stresses and the magnitude of the maximum in-plane shear stress

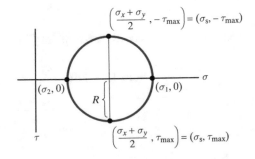

Fig. 7.26 Using Mohr's circle to determine the orientation of the element on which the principal stresses act

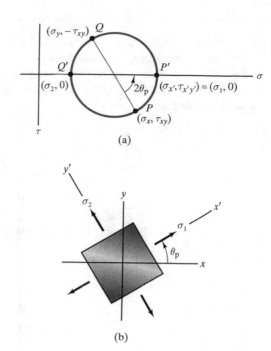

(a)

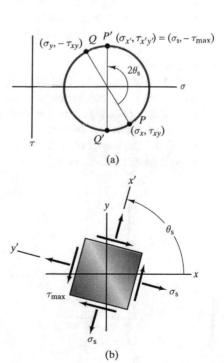

(b)

Fig. 7.27 Using Mohr's circle to determine the orientation of the element on which the maximum in-plane shear stresses act

Example 7.5 Determining Stresses Using Mohr's Circle

The state of plane stress at a point p of an object is shown on the left element in Fig. 7.28. Use Mohr's circle to determine the state of plane stress at p acting on the right element in Fig. 7.28. Draw a sketch of the element showing the stresses acting on it.

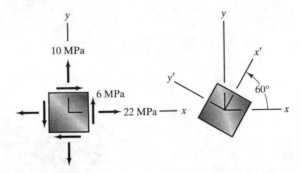

Fig. 7.28

Strategy

The components of plane stress on the left element in Fig. 7.28 are $\sigma_x = 22$ MPa, $\sigma_y = 10$ MPa, and $\tau_{xy} = 6$ MPa. We can follow the steps given for constructing Mohr's circle and determining the stress components $\sigma_{x'}$, $\tau_{x'y'}$, and $\sigma_{y'}$ to determine the stresses on the right element in Fig. 7.28.

Solution

Step 1 Establish a set of horizontal and vertical axes with normal stress measured along the horizontal axis and shear stress measured along the vertical axis (Fig. (a)). Positive shear stress is measured downward. The scales of normal stress and shear stress must be equal and chosen so that the circle will fit on the page but be large enough for reasonable accuracy. (This may require some trial and error.)

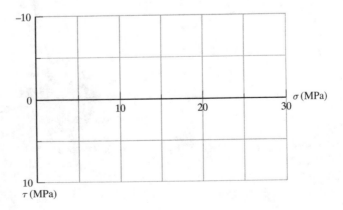

(a) Establishing the axes

Step 2 Plot point P with coordinates $(\sigma_x, \tau_{xy}) = (22, 6)$ and point Q with coordinates $(\sigma_y, -\tau_{xy}) = (10, -6)$, as shown in Fig. (b).

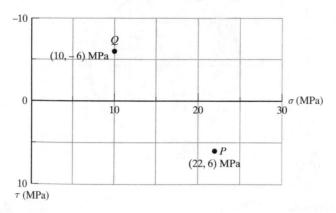

(b) Plotting points P and Q

Step 3 Draw a straight line connecting points P and Q. Using the intersection of the straight line with the horizontal axis as the center, draw a circle that passes through the two points (Fig. (c)).

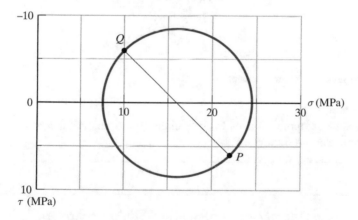

(c) Drawing the circle

Step 4 Draw a straight line through the center of the circle at an angle $2\theta = 120°$ measured counterclockwise from point P (Fig. (d)). Point P' at which this line intersects the circle has coordinates $\left(\sigma_{x'}, \tau_{x'y'}\right)$. The values we estimate from the graph are $\sigma_{x'} = 18.3$ MPa and $\tau_{x'y'} = -8.3$ MPa. Point Q' has coordinates $\left(\sigma_{y'}, -\tau_{x'y'}\right)$, from which we estimate that $\sigma_{y'} = 13.8$ MPa. The stresses are shown in Fig. (e).

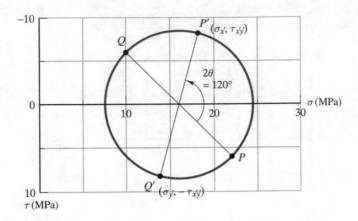

(d) Locating points P' and Q'.

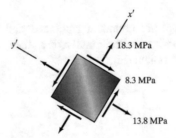

(e) Stresses on the rotated element

Discussion
Compare this application of Mohr's circle to the analytical solution of the same problem in Example 7.2.

Example 7.6 Determining Maximum and Minimum Stresses Using Mohr's Circle
The state of plane stress at a point p of an object is shown on the element in Fig. 7.29. Use Mohr's circle to determine the principal stresses and the maximum in-plane shear stress and show them acting on properly oriented elements.

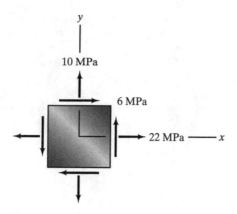

Fig. 7.29

Strategy
The components of plane stress on the element in Fig. 7.29 are $\sigma_x = 22$ MPa, $\sigma_y = 10$ MPa, and $\tau_{xy} = 6$ MPa. Mohr's circle determines the principal stresses and the maximum and minimum in-plane shear stresses. By letting point P' coincide first with one of the principal stresses and then with the maximum or minimum in-plane shear stress, we can determine the orientations of the elements on which these stresses act.

Solution
We first plot points P and Q and draw Mohr's circle (Fig. (a)). Then we let point P' coincide with one of the principal stresses (Fig. (b)). From the circle, we estimate that $\sigma_1 = 24.5$ MPa and $\sigma_2 = 7.5$ MPa. Measuring the angle $2\theta_p$, we estimate that $\theta_p = 22.5°$, which determines the orientation of the plane on which the principal stress σ_1 acts (Fig. (c)).

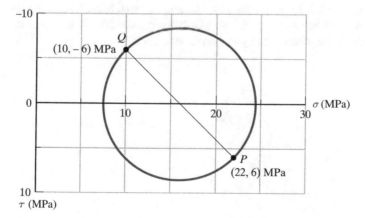

(a) Mohr's circle

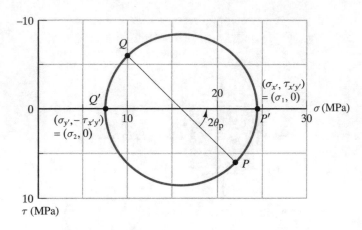

(b) Letting P' coincide with the principal stress σ_1.

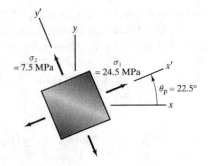

(c) Element on which the principal stresses act

We next let point P' coincide with either the minimum or maximum in-plane shear stress. In this case we choose the minimum stress (Fig. (d)). We estimate that $\tau_{\max} = 8.5$ MPa and the normal stress $\sigma_s = 16$ MPa. Measuring the angle $2\theta_s$, we estimate that $\theta_s = 67.5°$, which determines the orientation of the plane on which the minimum in-plane shear stress acts (Fig. (e)).

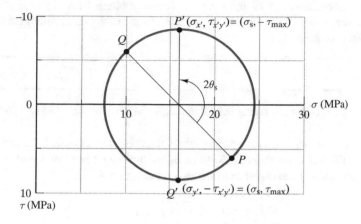

(d) Letting P' coincide with the minimum in-plane shear stress

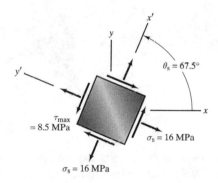

(e) Element on which the maximum and minimum in-plane shear stresses act. Notice that $\tau_{x'y'} = -\tau_{max}$.

Discussion

We have shown that the solutions of Eqs. (7.6)–(7.8) are mapped onto points of Mohr's circle. Conversely, every point of Mohr's circle corresponds to the components of stress on the plane shown in Fig. 7.13(b) for some value of θ. In this example, as soon as we had drawn Mohr's circle in Fig. (a), we could visualize every state of stress on planes oriented as shown in Fig. 7.13(b). In particular, we could immediately see the values of the principal stresses (the points of the circle corresponding to the maximum and minimum values of σ) and the magnitude of the maximum in-plane shear stress (the point of the circle corresponding to the maximum value of τ). This rapid visualization of all the states of stress resulting from a given state of plane stress, especially the extremal values, is one of the most useful attributes of Mohr's circle.

Example 7.7 Determining the Orientation of an Element Using Mohr's Circle

The state of plane stress at a point p of an object is shown on the left element in Fig. 7.30, and the values of the stresses $\sigma_{x'}$ and $\tau_{x'y'}$ are shown on the right element. Determine the normal stress $\sigma_{y'}$ and the angle θ.

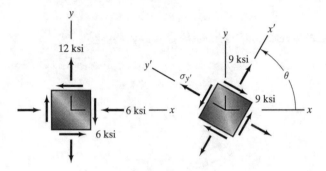

Fig. 7.30

Strategy

The components of stress on the left element are $\sigma_x = -6$ ksi, $\tau_{xy} = -6$ ksi, and $\sigma_y = 12$ ksi. With this information, we can plot points P and Q and draw Mohr's circle. On the rotated element, the stresses $\sigma_{x'} = 9$ ksi and $\tau_{x'y'} = -9$ ksi, which permits us to locate point P'. Then we can measure the angle 2θ and determine the normal stress $\sigma_{y'}$ from the coordinates of point Q'.

Solution

Figure (a) shows points P and Q and Mohr's circle. In Fig. (b) we plot point P' and draw a straight line from P' through the center of the circle. From the coordinates of point Q', we estimate that $\sigma_{y'} = -3$ ksi. Measuring the angle 2θ, we estimate that $\theta = 135°$. The stresses are shown on the properly oriented element in Fig. (c).

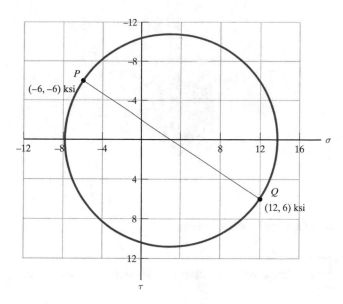

(a) Plotting points P and Q and drawing Mohr's circle

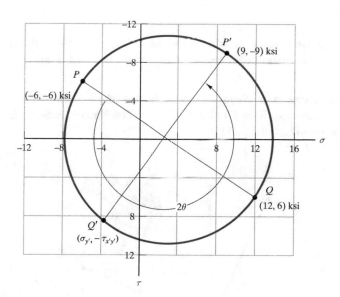

(b) Plotting P'.

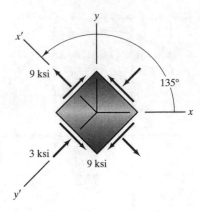

(c) Stresses on the properly oriented element

Discussion
Compare this solution with Example 7.3, in which we solve the same problem analytically.

Problems

We usually express results obtained using Mohr's circle to three significant digits. The accuracy you obtain will depend on the graph paper used and the size of your figures.

7.37 The components of plane stress at a point p of a material are $\sigma_x = 20$ MPa, $\sigma_y = 0$, and $\tau_{xy} = 0$, and the angle $\theta = 45°$. Use Mohr's circle to determine the stresses $\sigma_{x'}, \sigma_{y'}$, and $\tau_{x'y'}$ at p.

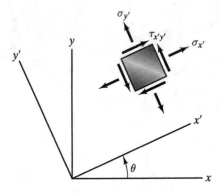

Problems 7.37–7.42

7.38 The components of plane stress at a point p of a material are $\sigma_x = 0$, $\sigma_y = 0$, and $\tau_{xy} = 25$ ksi, and the angle $\theta = 45°$. Use Mohr's circle to determine the stresses $\sigma_{x'}$, $\sigma_{y'}$, and $\tau_{x'y'}$ at p.

7.39 The components of plane stress at a point p of a material are $\sigma_x = 240$ MPa, $\sigma_y = -120$ MPa and $\tau_{xy} = 240$ MPa, and the components referred to the $x'\,y'\,z'$ coordinate system are $\sigma_{x'} = 347$ MPa, $\sigma_{y'} = -227$ MPa, and $\tau_{x'y'} = -87$ MPa. Use Mohr's circle to determine the angle θ.

7.40 Strain gauges attached to a rocket engine nozzle determine that the components of plane stress $\sigma_{x'} = 66.46$ MPa, $\sigma_{y'} = 82.54$ MPa, and $\tau_{x'y'} = 6.75$ MPa at $\theta = 20°$. Use Mohr's circle to determine the stresses σ_x, σ_y, and τ_{xy} at that point.

7.41 The components of plane stress at a point p of a material referred to the $x'\,y'\,z'$ coordinate system are $\sigma_{x'} = -8$ MPa, $\sigma_{y'} = 6$ MPa, and $\tau_{x'y'} = -16$ MPa, and the angle $\theta = 20°$. Use Mohr's circle to determine the stresses σ_x, σ_y, and τ_{xy} at p.

7.42 The components of plane stress at a point p of a drill bit are $\sigma_x = 40$ ksi, $\sigma_y = -30$ ksi, and $\tau_{xy} = 30$ ksi, and the components referred to the $x'\,y'\,z'$ coordinate system are $\sigma_{x'} = 12.5$ ksi, $\sigma_{y'} = -2.5$ ksi, and $\tau_{x'y'} = 45.5$ ksi. (See the figure for Problems 7.37–7.42.) Use Mohr's circle to estimate the angle θ.

Problem 7.42 (*Photograph from iStock by monkeybusinessimages*)

7.43 The components of plane stress at point p of the material shown are $\sigma_x = 4$ ksi, $\sigma_y = -2$ ksi, and $\tau_{xy} = 2$ ksi. Use Mohr's circle to determine the normal stress and the magnitude of the shear stress on the plane $\mathcal{P}$ at point p.

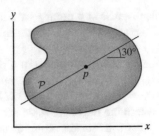

Problems 7.43–7.44

7.44 The components of plane stress at point p of the material shown are $\sigma_x = -10.5$ MPa, $\sigma_y = 6.0$ MPa, and $\tau_{xy} = -4.5$ MPa. Use Mohr's circle to determine the normal stress and the magnitude of the shear stress on the plane $\mathcal{P}$ at point p.

7.45 Determine the stresses σ and τ **(a)** by using Mohr's circle and **(b)** by using Eqs. (7.6) and (7.7).

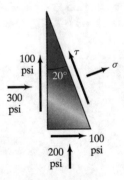

Problems 7.45–7.46

7.46 Solve Problem 7.45 if the 300-psi stress on the element is in tension instead of compression.

7.47 The components of plane stress on an element of an industrial robot are shown. Determine the stresses σ and τ **(a)** by using Mohr's circle and **(b)** by using Eqs. (7.6) and (7.7).

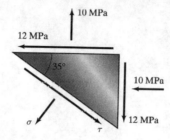

Problem 7.47 (*Photograph from iStock by Thossaphol*)

For the states of plane stress given in Problems 7.48–7.51, use Mohr's circle to determine the principal stresses and the maximum in-plane shear stress and show them acting on properly oriented elements.

7.48 $\sigma_x = 20$ MPa, $\sigma_y = 10$ MPa, $\tau_{xy} = 0$.

7.49 $\sigma_x = 25$ ksi, $\sigma_y = 0$, $\tau_{xy} = -25$ ksi.

7.50 $\sigma_x = -8$ ksi, $\sigma_y = 6$ ksi, $\tau_{xy} = -6$ ksi.

7.51 $\sigma_x = 240$ MPa, $\sigma_y = -120$ MPa, $\tau_{xy} = 240$ MPa.

7.52 A point of the supporting structure of the antenna is subjected to the state of plane stress $\sigma_x = 40$ MPa, $\sigma_y = -20$ MPa, and $\tau_{xy} = 30$ MPa. Use Mohr's circle to determine the principal stresses and the maximum in-plane shear stress at that point.

Problem 7.52 (*Photograph from iStock by sasha85ru*)

7.53 During landing, a point of the supporting strut of an airplane's landing gear will be subjected to the state of plane stress $\sigma_x = -120$ MPa, $\sigma_y = 80$ MPa, and $\tau_{xy} = -50$ MPa. Use Mohr's circle to determine the principal stresses and the maximum in-plane shear stress at that point.

Problem 7.53

7.54 For the state of plane stress $\sigma_x = 8$ ksi, $\sigma_y = 6$ ksi, and $\tau_{xy} = -6$ ksi, use Mohr's circle to determine the absolute maximum shear stress.

> *Strategy*: Use Mohr's circle to determine the principal stresses and then determine the absolute maximum shear stress from Eq. (7.24).

7.4   Maximum and Minimum Stresses in Three Dimensions

Many important applications involve states of stress more general than plane stress. Because of the crucial importance of maximum stresses in design, in this section we explain how to determine the principal stresses and absolute maximum shear stress for a general state of stress and for the subset of three-dimensional states of stress called *triaxial stress*.

7.4.1   General State of Stress

We have seen in our discussion of plane stress that the components of the state of stress depend on the orientation of the coordinate system in which they are

Fig. 7.31 The xyz
coordinate system and a
system $x'\,y'\,z'$ with a
different orientation

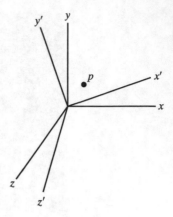

expressed. Suppose that we know the components of stress at a point p in terms of a
particular coordinate system xyz:

$$[I] = \begin{bmatrix} \sigma_x & \tau_{xy} & \tau_{xz} \\ \tau_{yx} & \sigma_y & \tau_{yz} \\ \tau_{zx} & \tau_{zy} & \sigma_z \end{bmatrix}.$$

These components will generally have different values when expressed in terms
of a coordinate system $x'\,y'\,z'$ having a different orientation (Fig. 7.31). For *any* state
of stress, at least one coordinate system $x'\,y'\,z'$ exists for which the state of stress is of
the form

$$\begin{bmatrix} \sigma_{x'} & \tau_{x'y'} & \tau_{x'z'} \\ \tau_{y'x'} & \sigma_{y'} & \tau_{y'z'} \\ \tau_{z'x'} & \tau_{z'y'} & \sigma_{z'} \end{bmatrix} = \begin{bmatrix} \sigma_1 & 0 & 0 \\ 0 & \sigma_2 & 0 \\ 0 & 0 & \sigma_3 \end{bmatrix}.$$

The axes x', y', z' are called *principal axes* and σ_1, σ_2, and σ_3 are the principal
stresses. An infinitesimal element at p that is oriented with the principal axes is
subject to the principal stresses and no shear stress (Fig. 7.32). It can be shown that
the principal stresses are the roots of the cubic equation

$$\sigma^3 - I_1\sigma^2 + I_2\sigma - I_3 = 0, \tag{7.25}$$

where

$$\begin{aligned}
I_1 &= \sigma_x + \sigma_y + \sigma_z, \\
I_2 &= \sigma_x\sigma_y + \sigma_y\sigma_z + \sigma_z\sigma_x - \tau_{xy}^2 - \tau_{yz}^2 - \tau_{zx}^2, \\
I_3 &= \sigma_x\sigma_y\sigma_z - \sigma_x\tau_{yz}^2 - \sigma_y\tau_{zx}^2 - \sigma_z\tau_{xy}^2 + 2\tau_{xy}\tau_{yz}\tau_{zx}.
\end{aligned} \tag{7.26}$$

Fig. 7.32 Stresses on an element oriented with the principal axes

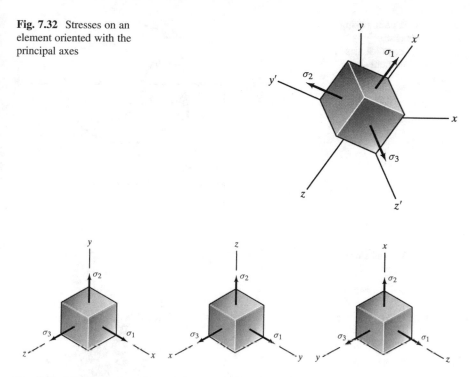

Fig. 7.33 Different orientations of the coordinate system relative to the element on which the principal stresses act

(Although the components of the state of stress depend on the orientation of the coordinate system in which they are expressed, the values of these three coefficients do not. This can be deduced from the fact that the principal stresses, the roots of Eq. (7.25), cannot depend on the orientation of the coordinate system used to evaluate them. For this reason, I_1, I_2, and I_3 are called *stress invariants*.) Thus, for a given state of stress, the principal stresses can be determined by evaluating the coefficients I_1, I_2, and I_3 and solving Eq. (7.25).

The absolute maximum shear stress can be determined by the same approach we applied to plane stress in Sect. 7.2. We begin with the element on which the principal stresses occur and align the coordinate system with the faces of the element in the three ways shown in Fig. 7.33. Applying Eq. (7.19) to each of these orientations, we find that the absolute maximum shear stress is given by

$$\tau_{abs} = \text{Max}\left(\left|\frac{\sigma_1 - \sigma_2}{2}\right|, \left|\frac{\sigma_1 - \sigma_3}{2}\right|, \left|\frac{\sigma_2 - \sigma_3}{2}\right|\right). \tag{7.27}$$

Although this analysis considers only a subset of the possible planes through point p, no shear stresses of greater magnitude act on any plane. In Fig. 7.34 we show that the absolute maximum shear stress determined in this way can be visualized

Fig. 7.34 Superimposing
the Mohr's circles
graphically demonstrates the
absolute maximum shear
stress

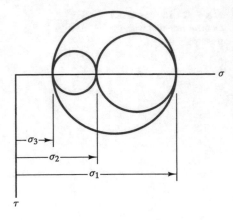

Fig. 7.35 Triaxial stress

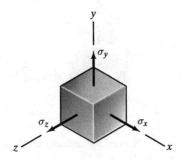

very clearly by superimposing the Mohr's circles obtained from the three orientations of the coordinate system in Fig. 7.33. Notice that if $\sigma_1 > \sigma_2 > \sigma_3$, the absolute
maximum shear stress is $\tau_{abs} = |(\sigma_1 - \sigma_3)/2|$.

7.4.2 *Triaxial Stress*

The state of stress at a point is said to be *triaxial* if it is of the form

$$\begin{bmatrix} \sigma_x & 0 & 0 \\ 0 & \sigma_y & 0 \\ 0 & 0 & \sigma_z \end{bmatrix}. \tag{7.28}$$

The shear stress components τ_{xy}, τ_{xz}, and τ_{yz} are zero (Fig. 7.35). In triaxial stress,
x, y, and z are principal axes, and σ_x, σ_y, and σ_z are the principal stresses. From
Eq. (7.27), the absolute maximum shear stress in triaxial stress is

$$\tau_{abs} = \text{Max}\left(\left|\frac{\sigma_x - \sigma_y}{2}\right|, \left|\frac{\sigma_x - \sigma_z}{2}\right|, \left|\frac{\sigma_y - \sigma_z}{2}\right|\right). \qquad (7.29)$$

Study Questions

1. How can you determine the principal stresses for a general state of stress?
2. What are the principal axes? How can you determine their directions?
3. Why are the coefficients I_1, I_2, and I_3 in Eq. (7.25) called stress invariants?
4. What is triaxial stress? What are the principal stresses for a state of triaxial stress?

Example 7.8 Maximum and Minimum Stresses in Three Dimensions
A point p in the frame of the car in Fig. 7.36 is subjected to the state of stress (in MPa):

$$\begin{bmatrix} \sigma_x & \tau_{xy} & \tau_{xz} \\ \tau_{yx} & \sigma_y & \tau_{yz} \\ \tau_{zx} & \tau_{zy} & \sigma_z \end{bmatrix} = \begin{bmatrix} 4 & 2 & 1 \\ 2 & 2 & 1 \\ 1 & 1 & 3 \end{bmatrix}.$$

Fig. 7.36

Determine the principal stresses and the absolute maximum shear stress at p.

Strategy

Because the state of stress is known, we can use Eq. (7.26) to evaluate the coefficients I_1, I_2, and I_3, then solve Eq. (7.25) to determine the principal stresses. The absolute maximum shear stress is given by Eq. (7.27).

Solution

Principal stresses   Substituting the components of stress in MPa into Eq. (7.26), we obtain $I_1 = 9$, $I_2 = 20$, and $I_3 = 10$, so Eq. (7.25) is

$$\sigma^3 - 9\sigma^2 + 20\sigma - 10 = 0.$$

We can estimate the roots of this cubic equation by drawing a graph of the left side of the equation as a function of σ (Fig. (a)). By using software designed to obtain roots of nonlinear algebraic equations, we obtain $\sigma_1 = 5.895$ MPa, $\sigma_2 = 2.397$ MPa, and $\sigma_3 = 0.708$ MPa.

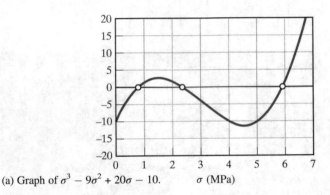

(a) Graph of $\sigma^3 - 9\sigma^2 + 20\sigma - 10$. σ (MPa)

Absolute maximum shear stress   The three arguments in Eq. (7.27) are

$$\left|\frac{\sigma_1 - \sigma_2}{2}\right| = \left|\frac{5.895 \text{ MPa} - 2.397 \text{ MPa}}{2}\right| = 1.749 \text{ MPa},$$

$$\left|\frac{\sigma_1 - \sigma_3}{2}\right| = \left|\frac{5.895 \text{ MPa} - 0.708 \text{ MPa}}{2}\right| = 2.594 \text{ MPa},$$

$$\left|\frac{\sigma_2 - \sigma_3}{2}\right| = \left|\frac{2.397 \text{ MPa} - 0.708 \text{ MPa}}{2}\right| = 0.845 \text{ MPa}.$$

The absolute maximum shear stress is $\tau_{\text{abs}} = 2.594$ MPa.

Discussion

In Sect. 7.2 we described procedures for determining the maximum and minimum normal stresses and the absolute maximum shear stress for a state of plane stress. In this example we determine them for a general state of stress. Why do we devote so much attention to determining maximum stresses? The reason is that they are used in determining whether structural elements will fail under the loads to which they are subjected. We discuss failure criteria in Chap. 12.

Problems

7.55 At a point p, a material is subjected to the state of *plane* stress $\sigma_x = 20$ MPa, $\sigma_y = 10$ MPa, $\tau_{xy} = 0$. Use Eq. (7.25) to determine the principal stresses and use Eq. (7.27) to determine the absolute maximum shear stress. Confirm the absolute maximum shear stress by drawing the superimposed Mohr's circle as shown in Fig. 7.34.

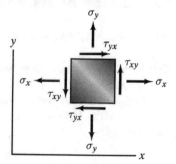

Problems 7.55–7.58

7.56 At a point p, a material is subjected to the state of *plane* stress $\sigma_x = 25$ ksi, $\sigma_y = 0$, $\tau_{xy} = -25$ ksi. Use Eq. (7.25) to determine the principal stresses, and use Eq. (7.27) to determine the absolute maximum shear stress.

7.57 At a point p, a material is subjected to the state of *plane* stress $\sigma_x = 240$ MPa, $\sigma_y = -120$ MPa, $\tau_{xy} = 240$ MPa. Use Eq. (7.25) to determine the principal stresses, and use Eq. (7.27) to determine the absolute maximum shear stress.

7.58 Strain gauges attached to a tunneling drill determine that the components of *plane* stress are $\sigma_x = 67.34$ MPa, $\sigma_y = 82.66$ MPa, $\tau_{xy} = 6.43$ MPa. Use Eq. (7.25) to determine the principal stresses, and use Eq. (7.27) to determine the absolute maximum shear stress.

Problem 7.58 (*Photograph from iStock by PixHouse*)

7.59   Use Eq. (7.25) to determine the principal stresses for an arbitrary state of *plane* stress σ_x, σ_y, τ_{xy} and confirm Eq. (7.15).

7.60   At a point p, a material is subjected to the state of *triaxial* stress $\sigma_x = 240$ MPa, $\sigma_y = -120$ MPa, $\sigma_z = 240$ MPa. Determine the principal stresses and the absolute maximum shear stress.

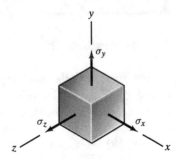

Problems 7.60–7.61

7.61∗   At a point p, a material is subjected to a state of *triaxial* stress. The stresses $\sigma_x = 40$ ksi and $\sigma_z = -20$ ksi. The absolute maximum shear stress is $\tau_{abs} = 60$ ksi. What are the possible values of the stress σ_y?

7.62∗ At a point p, a material is subjected to the state of stress (in ksi):

$$\begin{bmatrix} \sigma_x & \tau_{xy} & \tau_{xz} \\ \tau_{yx} & \sigma_y & \tau_{yz} \\ \tau_{zx} & \tau_{zy} & \sigma_z \end{bmatrix} = \begin{bmatrix} 300 & 150 & -100 \\ 150 & 200 & 100 \\ -100 & 100 & -200 \end{bmatrix}.$$

Determine the principal stresses and the absolute maximum shear stress.

7.63∗ A finite element analysis of a bearing housing indicates that at a point p, the material is subjected to the state of stress (in MPa):

$$\begin{bmatrix} \sigma_x & \tau_{xy} & \tau_{xz} \\ \tau_{yx} & \sigma_y & \tau_{yz} \\ \tau_{zx} & \tau_{zy} & \sigma_z \end{bmatrix} = \begin{bmatrix} 20 & 20 & 0 \\ 20 & -30 & -10 \\ 0 & -10 & 40 \end{bmatrix}.$$

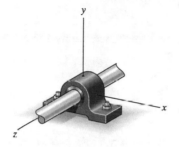

Problem 7.63

Determine the principal stresses and the absolute maximum shear stress.

7.64 The components of *plane* stress at a point p of a material are $\sigma_x = -8$ ksi, $\sigma_y = 6$ ksi, and $\tau_{xy} = -6$ ksi. Use Eqs. (7.6)–(7.8) to determine the components of stress $\sigma_{x'}, \sigma_{y'}$, and $\tau_{x'y'}$ corresponding to a coordinate system $x'y'z'$ oriented at $\theta = 30°$. (**a**) Determine the principal stresses from Eq. (7.25) using the components of stress σ_x, σ_y, and τ_{xy}. (**b**) Determine the principal stresses from Eq. (7.25) using the components of stress $\sigma_{x'}, \sigma_{y'}$, and $\tau_{x'y'}$.

7.65 By using the results from Sect. 7.2, prove for a state of *plane* stress that the coefficient I_1 in Eq. (7.25) does not depend on the orientation of the coordinate system used to evaluate it.

7.66 By using the results from Sect. 7.2, prove for a state of *plane* stress that the coefficient I_2 in Eq. (7.25) does not depend on the orientation of the coordinate system used to evaluate it.

7.5   Application: Bars Subjected to Combined Loads

In previous chapters we have described the states of stress in simple structural
elements subjected to three fundamental types of loading: bars subjected to axial
loads, bars subjected to torsional loads about their longitudinal axes, and beams
subjected to lateral loads. These specific types of loads occur in many applications.
But situations also arise in which structural elements are subjected to combinations
of these loads simultaneously. Here we discuss some important examples of combined loads and show how the techniques developed in this chapter can be used to
analyze the resulting states of stress.

7.5.1   The Fundamental Loads

Before considering combined loads, we briefly review the states of stress for the
individual types of loading.

7.5.1.1   Axially Loaded Bar

Consider a prismatic bar of arbitrary cross section with axial loads acting at the
centroid of its cross section (Fig. 7.37(a)). An element of the bar oriented as shown is
subjected to the homogeneous normal stress $\sigma = P/A$, where A is the bar's cross-
sectional area. In terms of a coordinate system with its x axis parallel to the axis of

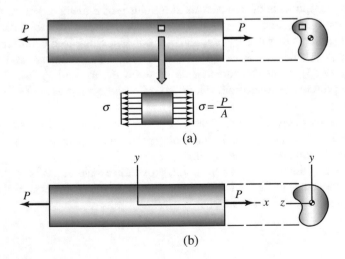

Fig. 7.37 (**a**) Normal stress acting on an element of a bar subjected to axial load. (**b**) Coordinate
system with its x axis coincident with the bar's axis

the bar (Fig. 7.37(b)), each point of the bar not near the ends is subjected to the state of plane stress $\sigma_x = P/A$, $\sigma_y = 0$, and $\tau_{xy} = 0$.

7.5.1.2   Bar Subjected to Torsion

Consider a prismatic bar with circular cross section subjected to torques at its ends (Fig. 7.38(a)). An *infinitesimal* element of the bar, oriented as shown and at radial distance r from the bar's axis, is subjected to the shear stress given by Eq. (4.9): $\tau = Tr/J$, where J is the polar moment of inertia of the bar's circular cross section. In terms of a coordinate system oriented as shown in Fig. 7.38(b), point p of the bar,

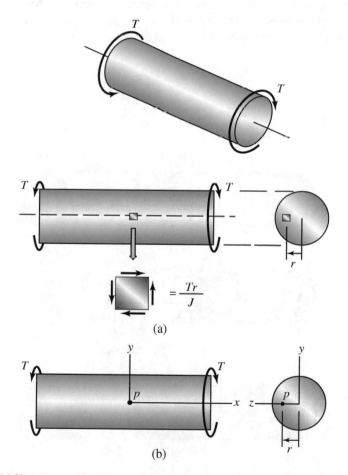

(a)

(b)

Fig. 7.38 (a) Shear stress acting on an element of a bar subjected to torsion. (b) Point p is located at a radial distance r from the bar's axis

which is at a radial distance r from the bar's axis, is subjected to the state of plane stress $\sigma_x = 0$, $\sigma_y = 0$, and $\tau_{xy} = Tr/J$. Observe in Fig. 7.38 that we have chosen the direction of the torque T so that the resulting shear stress τ_{xy} in terms of the coordinate system in Fig. 7.38(b) is positive.

7.5.1.3   Beam Subjected to Bending Moment

In Fig. 7.39(a) a prismatic beam is subjected to couples at its ends. The cross section is symmetric about the y axis. The x axis of the coordinate system coincides with the beam's neutral axis, which is located at the centroid of the cross section. Obtaining a free-body diagram by passing a perpendicular plane $\mathcal{P}$ through the beam at an arbitrary axial position, the distribution of the normal stress on $\mathcal{P}$ is given by Eq. (6.12): $\sigma_x = -My/I$, where I is the moment of inertia of the cross section about the z axis. A point p of the beam that is at vertical position y relative to the neutral axis (Fig. 7.39(b)) is subjected to the state of plane stress $\sigma_x = -My/I$, $\sigma_y = 0$, and $\tau_{xy} = 0$.

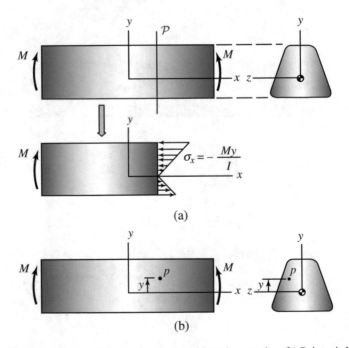

Fig. 7.39 (a) Normal stress distribution in a beam subjected to couples. (b) Point p is located at a height y relative to the neutral axis

7.5.2   *Combined Loads*

We can determine the states of stress that result from combinations of the fundamental types of loading by superimposing the states of stress due to each individual
loading. For example, the bar in Fig. 7.40(a) is subjected to both axial loads and
torques. In terms of the coordinate system shown in Fig. 7.40(b), the state of stress
at any point of the bar that is not near the ends due to the axial loads alone is
$\sigma_x = P/A$, $\sigma_y = 0$, and $\tau_{xy} = 0$, where A is the bar's cross-sectional area. At point
p in Fig. 7.40(b), which is at a radial distance r from the bar's axis, the state of stress
due to the torques alone in terms of the coordinate system shown is $\sigma_x = 0$, $\sigma_y = 0$,
and $\tau_{xy} = Tr/J$, where J is the polar moment of inertia of the bar's circular cross
section. Therefore, the state of stress at point p due to the combined axial and
torsional loads is

$$\sigma_x = \frac{P}{A}, \quad \sigma_y = 0, \quad \tau_{xy} = \frac{Tr}{J}.$$

In Fig. 7.41(a), a prismatic bar with rectangular cross section is fixed at the left
end. The right end is subjected to an axial load P that does not act at the centroid of
the cross section. As a result, the load P exerts a moment about the centroid of the
cross section. From Fig. 7.41(b), we can see that the load exerts a moment about the

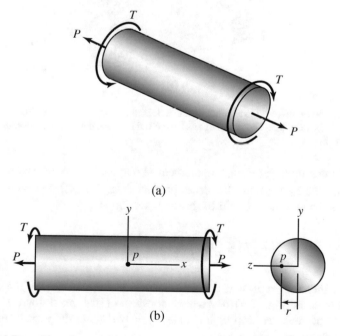

(a)

(b)

Fig. 7.40   (a) Bar subjected to axial loads and torques. (b) Point p is a radial distance r from the
bar's axis

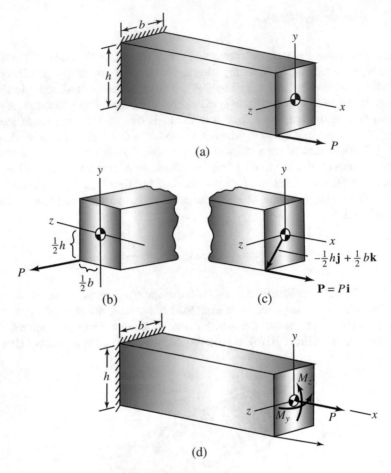

(a)

(b)   (c)

(d)

Fig. 7.41 (a) Beam subjected to an axial load that does not act at the centroid. (b), (c) Determining the moment of the axial load about the centroid. (d) Axial load and couples that are equivalent to the axial load in (a)

y axis of magnitude $\frac{1}{2}bP$ and a moment about the z axis of magnitude $\frac{1}{2}hP$. Confirming this by applying the cross product (Fig. 7.41(c)), the moment due to the axial load about the centroid of the cross section is

$$\left(-\frac{1}{2}h\mathbf{j} + \frac{1}{2}b\mathbf{k}\right) \times P\mathbf{i} = \frac{1}{2}bP\mathbf{j} + \frac{1}{2}hP\mathbf{k}.$$

We can therefore represent the axial load P in Fig. 7.41(a) by the equivalent system shown in Fig. 7.41(d), which consists of the axial load P acting at the centroid of the cross section, a couple $M_y = \frac{1}{2}bP$ about the y axis, and a couple $M_z = \frac{1}{2}hP$ about the z axis.

We can obtain the state of stress at an arbitrary point in the beam by superimposing the individual states of stress resulting from the axial load P acting at the centroid and the couples M_y and M_z. In terms of the coordinate system shown in Fig. 7.41(c), the state of stress due to the axial load P is $\sigma_x = P/A = P/bh$, $\sigma_y = 0$, and $\tau_{xy} = 0$. Applying Eq. (6.12), the state of stress due to the couple M_y is $\sigma_x = M_y z/I_y$, $\sigma_y = 0$, and $\tau_{xy} = 0$, where I_y is the moment of inertia of the cross section about the z axis, and the state of stress due to the couple M_z is $\sigma_x = -M_z y/I_z$, $\sigma_y = 0$, and $\tau_{xy} = 0$, where I_z is the moment of inertia of the cross section about the z axis. (To understand the signs in the expressions for σ_x due to M_y and M_z, notice in Fig. 7.41(c) that M_y results in positive normal stresses when z is positive, whereas M_z results in positive normal stresses when y is negative.) Superimposing these individual states of stress, the state of stress at an arbitrary point in the beam is

$$\sigma_x = \frac{P}{A} + \frac{M_y z}{I_y} - \frac{M_z y}{I_z}, \quad \sigma_y = 0, \quad \tau_{xy} = 0.$$

We need to make two observations about these results. The components of the state of stress at a point depend on the orientation of the coordinate system. Therefore, in order to superimpose states of stress as we have done, *the individual states of stress must be expressed in terms of coordinate systems having the same orientation.* Also, once the state of stress at a point is known, all the techniques developed in this chapter for analyzing states of stress can be applied. Of particular importance for design, the principal stresses and the absolute maximum shear stress can be determined (see Example 7.10).

Example 7.9 Beam Subjected to Combined Loads
The beam in Fig. 7.42 has a rectangular cross section. It is subjected to an axial load $P = 60$ kips and a couple $M_y = 18$ in - kips exerted about the y axis in the direction shown. Determine the state of stress due to the combined loading at point p, which is located at $x = 9$ in, $y = 1.5$ in, $z = 3$ in.

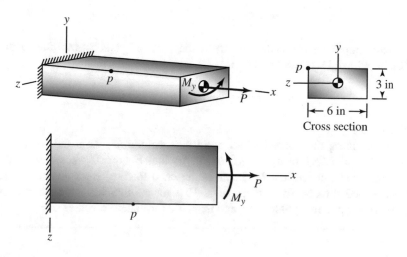

Fig. 7.42

Strategy

We can determine the state of stress at p by superimposing the states of stress due to the axial load and couple.

Solution

The normal stress due to the axial load alone is

$$\sigma_x = \frac{P}{A}$$

$$= \frac{60{,}000 \text{ lb}}{(6 \text{ in})(3 \text{ in})}$$

$$= 3330 \text{ psi},$$

so the state of plane stress due to the axial load alone is

$$\sigma_x = 3330 \text{ psi}, \quad \sigma_y = 0, \quad \tau_{xy} = 0.$$

The distribution of normal stress due to the couple alone is

$$\sigma_x = \frac{M_y z}{I_y}$$

$$= \frac{(18{,}000 \text{ in-lb})(3 \text{ in})}{\frac{1}{12}(3 \text{ in})(6 \text{ in})^3}$$

$$= 1000 \text{ psi},$$

so the state of plane stress at p due to the couple alone is

$$\sigma_x = 1000 \text{ psi}, \quad \sigma_y = 0, \quad \tau_{xy} = 0.$$

Superimposing the states of stress, the state of plane stress due to the combined axial load and couple is

$$\sigma_x = 4330 \text{ psi}, \quad \sigma_y = 0, \quad \tau_{xy} = 0.$$

Discussion

In determining the normal stress due to the couple M_y, how did we know whether $\sigma_x = M_y z / I_y$ or $\sigma_x = -M_y z / I_y$? If you look at how M_y causes the beam to bend, you can see that the beam is stretched for positive values of z and compressed for negative values of z. That means that the normal stress is positive for positive values of z and negative for negative values of z.

Example 7.10 State of Stress in a Propeller Shaft

A helicopter's rotor shaft has the hollow circular cross section shown in Fig. 7.43. The shaft is subjected to an axial load $P = 34,000$ N and a torque $T = 600$ N-m. Determine the principal stresses and the absolute maximum shear stress at a point p on the outer surface of the shaft.

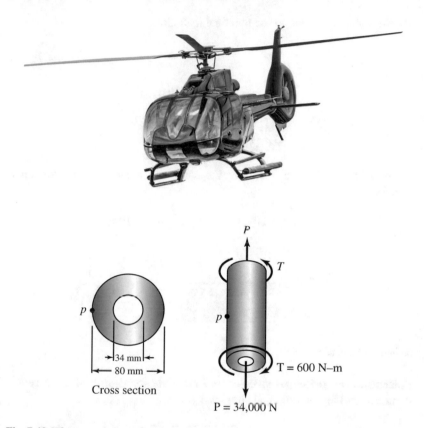

Fig. 7.43 (*Photograph from iStock by filrom*)

Strategy

We will introduce a coordinate system and determine the state of stress at p by superimposing the states of stress due to the axial load and torsion. Then we can calculate the principle stresses and absolute maximum shear stress using results from Sect. 7.2.

Solution

Determine the combined state of stress   In terms of the coordinate system in Fig. (a), the normal stress due to the axial load alone is

$$\sigma_x = \frac{P}{A}$$

$$= \frac{34{,}000 \text{ N}}{\pi\left[(0.040 \text{ m})^2 - (0.017 \text{ m})^2\right]}$$

$$= 8.26 \text{ MPa},$$

and the shear stress due to the torsional load alone is

$$\tau_{xy} = \frac{Tr}{J}$$

$$= \frac{(600 \text{ N-m})(0.040 \text{ m})}{\frac{\pi}{2}\left[(0.040 \text{ m})^4 - (0.017 \text{ m})^4\right]}$$

$$= 6.17 \text{ MPa}.$$

The state of stress at p due to the combined axial load and torsion is therefore

$$\sigma_x = 8.26 \text{ MPa}, \quad \sigma_y = 0, \quad \tau_{xy} = 6.17 \text{ MPa}.$$

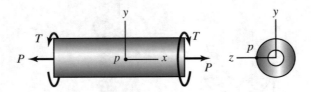

(a) Introducing a coordinate system

Determine the principal stresses and absolute maximum shear stress
From Eq. (7.15), the principal stresses at p are

$$\sigma_1, \sigma_2 = \frac{\sigma_x + \sigma_y}{2} \pm \sqrt{\left(\frac{\sigma_x - \sigma_y}{2}\right)^2 + \tau_{xy}^2}$$

$$= \frac{8.26 \text{ MPa}}{2} \pm \sqrt{\left(\frac{8.26 \text{ MPa}}{2}\right)^2 + (6.17 \text{ MPa})^2}$$

$$= 11.55 \text{ MPa}, -3.30 \text{ MPa}.$$

From Eq. (7.24), the absolute maximum shear stress at p is

$$\tau_{abs} = \text{Max}\left(\left|\frac{\sigma_1}{2}\right|, \left|\frac{\sigma_2}{2}\right|, \left|\frac{\sigma_1 - \sigma_2}{2}\right|\right)$$

$$= \text{Max}\left(\left|\frac{11.55 \text{ MPa}}{2}\right|, \left|\frac{-3.30 \text{ MPa}}{2}\right|, \left|\frac{[11.55 - (-3.30)] \text{ MPa}}{2}\right|\right)$$

$$= 7.42 \text{ MPa}.$$

Discussion

This example is a culmination of the developments we have presented to this point. The propeller shaft must be designed to support the combined axial and torsional loads to which it is subjected. The critical state of stress occurs at the surface of the shaft, where the magnitude of the shear stress due to torsion is greatest. The techniques developed in this chapter make it possible to determine the maximum normal and shear stresses acting on the material of the shaft. We discuss failure criteria in more detail in Chap. 12.

Problems

7.67 The cylindrical bar is subjected to an axial load $P = 600$ kN and a torque $T = 10$ kN-m. Point p is on the surface of the bar. In terms of the coordinate system shown, determine the state of stress at p due to the combined axial load and torque.

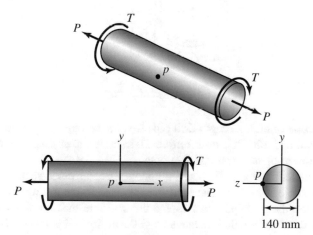

Problems 7.67–7.68

7.68 The cylindrical bar is subjected to an axial load $P = 600$ kN and a torque $T = 10$ kN-m. Point p is on the surface of the bar. Determine the principal stresses at p. (See Example 7.10.)

7.69 The section of high-strength drill pipe is subjected to a 22,000 lb *compressive* axial load and a 36,000 in-lb torque. In terms of the coordinate system shown, determine the state of stress at point p on the pipe's outer surface.

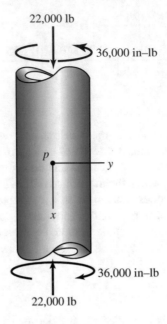

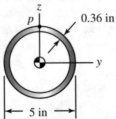

Problems 7.69–7.70

7.70 The section of high-strength drill pipe is subjected to a 22,000 lb *compressive* axial load and a 36,000 in-lb torque. Determine the principal stresses and the maximum in-plane shear stress at point *p*.

7.71 A helicopter's rotor shaft has the hollow circular cross section shown. The shaft is subjected to an axial load $P = 34,000$ N and a torque $T = 600$ N-m. Determine the principal stresses and the absolute maximum shear stress at a point *p* that is at a radial distance $r = 30$ mm from the axis of the shaft. (See Example 7.10.)

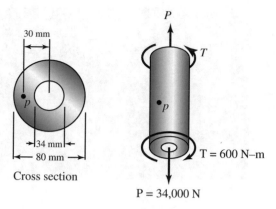

Problems 7.71–7.72

7.72 A helicopter's rotor shaft has the hollow circular cross section shown. The allowable shear stress of the material is 15 MPa. If the shaft is subjected to an axial load $P = 40,000$ N, what is the largest simultaneous torque T to which it can be subjected?

7.73 The cylindrical bar has radius $R = 2$ in. It is subjected to a torque about its axis $T = 12$ in - kip and a couple $M_y = 16$ in - kip about the y axis in the direction shown. In terms of the coordinate system shown, determine the state of stress at point p, which is located at $x = 6$ in, $y = 0$, and $z = 2$ in.

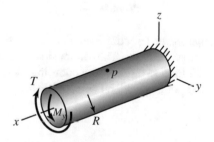

Problems 7.73–7.74

7.74 The cylindrical bar has radius $R = 2$ in. It is subjected to a torque about its axis $T = 12$ in - kip and a couple $M_y = 16$ in - kip about the y axis in the direction shown. Determine the principal stresses and the maximum in-plane shear stress at point p, which is located at $x = 6$ in, $y = 0$, and $z = 2$ in. (See Example 7.10.)

7.75 The C-clamp exerts a 60-lb force on the circular disk. Determine the maximum tensile and compressive stresses in the C-clamp at the cross section shown. The area of the cross section is 0.3 in^2, and its moment of inertia about the z axis is $I_z = 0.03125$ in^4.

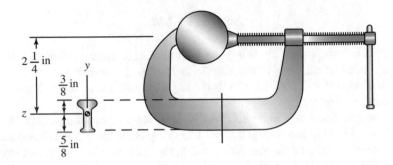

Problem 7.75

7.76* The beam is subjected to an axial load $P = 2$ kN that does not act at the centroid of the cross section. The dimensions of the cross section are $b = 60$ mm, $h = 30$ mm. In terms of the coordinate system shown, determine the state of stress at point p, which has coordinates $y = \frac{1}{2}h, z = \frac{1}{2}b$.

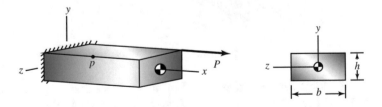

Problem 7.76

7.77 The L-shaped bar BCD is subjected to a downward force $F = 850$ N at D. In terms of the coordinate system shown, what is the state of stress at point p?

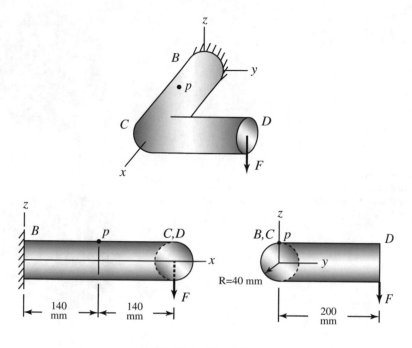

Problems 7.77–7.78

7.78∗ The L-shaped bar BCD is subjected to a downward force F at D. Determine the value of F if the absolute maximum shear stress at p is 3 MPa.

7.6 Application: Pressure Vessels

The critical importance of this subject in engineering has been emphasized by pressure vessel accidents from the early days of steam power to *Apollo 13*. We discuss it in this chapter because pressure vessels provide interesting examples of triaxial states of stress (see Sect. 7.4).

7.6.1 Spherical Vessels

Consider a spherical pressure vessel with radius R and wall thickness t, where $t << R$ (Fig. 7.44). We assume that the vessel contains a gas with uniform pressure p_i and that the outer wall is subjected to a uniform pressure p_o, and we denote the difference between the inner and outer pressures by $p = p_i - p_o$. In applications, p_o is often atmospheric pressure (approximately 10^5 Pa or 14.7 psi).

Fig. 7.44 Spherical
pressure vessel. The wall
thickness is exaggerated

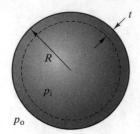

Fig. 7.45 Free-body
diagram of half of the
pressure vessel, including
the enclosed gas

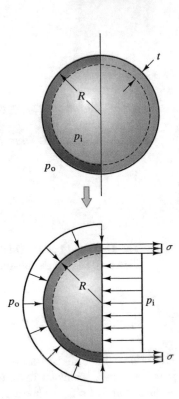

We can approximate the state of stress in a thin-walled spherical pressure vessel if
we assume that the effects of loads other than those exerted by the internal and
external pressures are negligible. In Fig. 7.45 we bisect the vessel by a plane, and
draw a free-body diagram of one-half, *including the gas it contains*. The normal
stress in the wall, which can be approximated as being uniformly distributed for a
thin-walled vessel, is denoted by σ. We want to determine the stress σ by summing
the horizontal forces on the free-body diagram, but what horizontal force is exerted
by the pressure p_o? Suppose that a solid hemisphere of radius R is subjected to a
uniform pressure p_o (Fig. 7.46). No net force is exerted on an object by a uniform
distribution of pressure, so the horizontal force to the right exerted on the hemispherical surface must equal the force to the left exerted on the plane circular face:

Fig. 7.46 Hemisphere
subjected to uniform
pressure p_o.

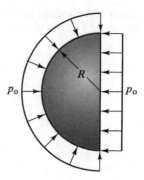

Fig. 7.47 Stresses on an
element at the outer surface

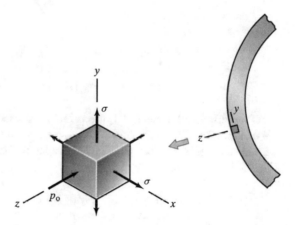

$p_o(\pi R^2)$. Therefore, the sum of the horizontal forces on the free-body diagram in
Fig. 7.45 is

$$p_o\left(\pi R^2\right) + \sigma(2\pi Rt) - p_i\left(\pi R^2\right) = 0.$$

Notice that we use the assumption $t \ll R$ in writing this equation. From this
equation, we obtain the normal stress σ in terms of the dimensions of the vessel and
the pressure difference p:

$$\sigma = \frac{(p_i - p_o)R}{2t} = \frac{pR}{2t}. \tag{7.30}$$

An element isolated from the vessel's outer surface is subjected to the stresses
shown in Fig. 7.47. In terms of the coordinate system shown (the z axis is perpendicular to the wall), the triaxial state of stress is

Fig. 7.48 Stresses on an
element at the inner surface

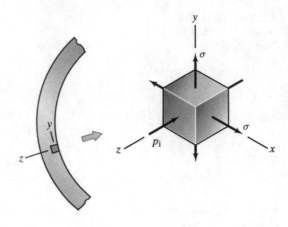

$$\begin{bmatrix} \sigma_x & \tau_{xy} & \tau_{xz} \\ \tau_{yx} & \sigma_y & \tau_{yz} \\ \tau_{zx} & \tau_{zy} & \sigma_z \end{bmatrix} = \begin{bmatrix} \sigma & 0 & 0 \\ 0 & \sigma & 0 \\ 0 & 0 & -p_o \end{bmatrix}. \tag{7.31}$$

As we discussed in Sect. 7.4, the principal stresses for this state of stress are σ, σ, $-p_o$, and from Eq. (7.29) the absolute maximum shear stress is $|(\sigma + p_o)/2|$. The triaxial state of stress on an element isolated from the vessel's inner surface (Fig. 7.48) is

$$\begin{bmatrix} \sigma_x & \tau_{xy} & \tau_{xz} \\ \tau_{yx} & \sigma_y & \tau_{yz} \\ \tau_{zx} & \tau_{zy} & \sigma_z \end{bmatrix} = \begin{bmatrix} \sigma & 0 & 0 \\ 0 & \sigma & 0 \\ 0 & 0 & -p_i \end{bmatrix}. \tag{7.32}$$

In this case, the principal stresses are σ, σ, $-p_i$, and the absolute maximum shear stress is $\tau_{abs} = |(\sigma + p_i)/2|$.

7.6.2   Cylindrical Vessels

We now consider a cylindrical vessel with radius R and wall thickness t (Fig. 7.49). The vessel shown has hemispherical ends, but the results we will obtain for the state of stress in the cylindrical wall do not depend on the shapes of the ends. We again assume that the vessel contains a gas with uniform pressure p_i and that the outer wall is subjected to a uniform pressure p_o.

In Fig. 7.50 we obtain a free-body diagram by passing a plane through the cylindrical wall perpendicular to its axis. The normal stress in the wall is denoted by σ. The horizontal forces on this free-body diagram are identical to those on the free-body diagram of the spherical vessel in Fig. 7.45, so the normal stress σ is given by Eq. (7.30):

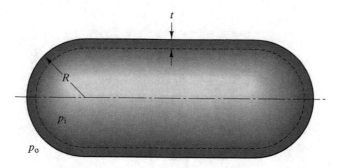

Fig. 7.49 Cylindrical pressure vessel

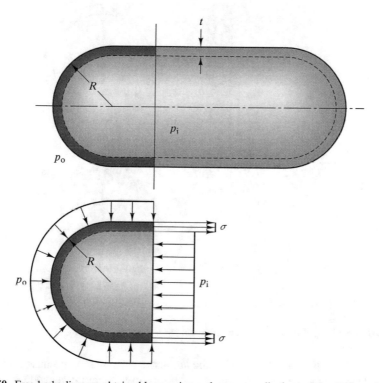

Fig. 7.50 Free-body diagram obtained by passing a plane perpendicular to the cylinder axis

$$\sigma = \frac{(p_i - p_o)R}{2t} = \frac{pR}{2t}. \tag{7.33}$$

But the state of stress in a cylindrical vessel is slightly more complicated than that in a spherical vessel. In Fig. 7.51, we first isolate a "slice" of the cylindrical wall of length Δx, including the gas it contains, and then obtain a free-body diagram by bisecting the slice by a plane parallel to the cylinder axis. The normal stress σ_h on the

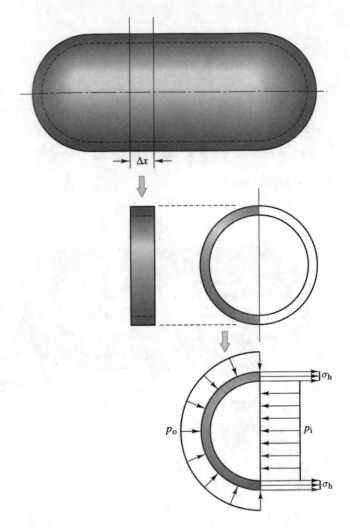

Fig. 7.51 Free-body diagram for determining the hoop stress σ_h.

resulting free-body diagram is called the *hoop stress* (so named because it plays the same role as the tensile stresses in the metal hoops used to reinforce wooden barrels). The planes on which σ_h acts are perpendicular to planes on which the normal stress σ acts (Fig. 7.52).

The sum of the horizontal forces on the free-body diagram in Fig. 7.51 is

$$p_o(2R \ \Delta x) + \sigma_h(2t \ \Delta x) - p_i(2R \ \Delta x) = 0.$$

Solving for the hoop stress, we obtain

Fig. 7.52 Planes on which the normal stresses σ and σ_h act

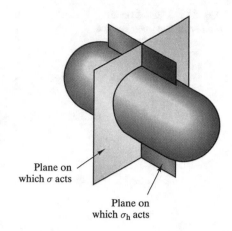

Plane on which σ acts

Plane on which σ_h acts

Fig. 7.53 Stresses on an element at the outer surface

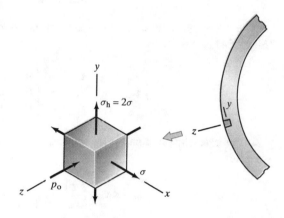

$$\sigma_h = \frac{(p_i - p_o)R}{t} = \frac{pR}{t}. \qquad (7.34)$$

Observe that $\sigma_h = 2\sigma$, which is an important factor in favor of spherical pressure vessels.

An element isolated from the outer surface of the cylindrical wall is subjected to the stresses shown in Fig. 7.53. In terms of the coordinate system shown (the z axis is perpendicular to the wall and the x axis is parallel to the axis of the cylinder), the triaxial state of stress is

$$\begin{bmatrix} \sigma_x & \tau_{xy} & \tau_{xz} \\ \tau_{yx} & \sigma_y & \tau_{yz} \\ \tau_{zx} & \tau_{zy} & \sigma_z \end{bmatrix} = \begin{bmatrix} \sigma & 0 & 0 \\ 0 & \sigma_h & 0 \\ 0 & 0 & -p_o \end{bmatrix}. \qquad (7.35)$$

The principal stresses are σ_h, σ, $-p_o$, and from Eq. (7.29), the absolute maximum shear stress is the largest of the three values

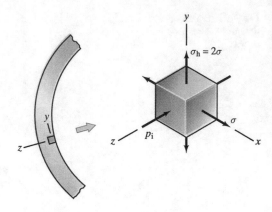

Fig. 7.54 Stresses on an element at the inner surface

$$\tau_{abs} = \text{Max}\left(\left|\frac{\sigma}{2}\right|, \left|\frac{\sigma + p_o}{2}\right|, \left|\frac{\sigma_h + p_o}{2}\right|\right). \tag{7.36}$$

The triaxial state of stress on an element isolated from the vessel's inner surface (Fig. 7.54) is

$$\begin{bmatrix} \sigma_x & \tau_{xy} & \tau_{xz} \\ \tau_{yx} & \sigma_y & \tau_{yz} \\ \tau_{zx} & \tau_{zy} & \sigma_z \end{bmatrix} = \begin{bmatrix} \sigma & 0 & 0 \\ 0 & \sigma_h & 0 \\ 0 & 0 & -p_i \end{bmatrix}. \tag{7.37}$$

In this case, the principal stresses are σ_h, σ, $-p_i$, and the absolute maximum shear stress is

$$\tau_{abs} = \text{Max}\left(\left|\frac{\sigma}{2}\right|, \left|\frac{\sigma + p_i}{2}\right|, \left|\frac{\sigma_h + p_i}{2}\right|\right). \tag{7.38}$$

7.6.3   Design Issues

The one aspect of pressure vessel design we will consider is the most important: making sure that the stresses in the vessel walls do not exceed those the material will support. We have determined the principal stresses and absolute maximum shear stresses in the walls of spherical and cylindrical vessels. The criterion for design we use in the example and problems is to ensure that the absolute maximum shear stress in the material does not exceed a shear yield stress τ_Y, or some specified allowable shear stress τ_{allow}. This is called the *Tresca criterion*, and we discuss it further in Chap. 12. The factor of safety is now defined by

$$FS = \frac{\tau_Y}{\tau_{allow}}. \tag{7.39}$$

Because the maximum shear stress in a tensile test is one-half the applied normal stress, we will assume that the shear yield stress τ_Y is given in terms of the normal yield stress σ_Y tabulated in Appendix B by the relation

$$\tau_Y = \frac{1}{2}\sigma_Y. \tag{7.40}$$

Of course, the values in Appendix B are merely representative, and in actual design, the yield stress must be determined for the specific material being used.

Study Questions

1. What are the principal stresses at a point of the inner surface of a spherical pressure vessel?
2. What is the hoop stress?
3. From the standpoint of stresses, why is a spherical pressure vessel superior to a cylindrical one?
4. What is the Tresca criterion? How can it be used in pressure vessel design?

Example 7.11 Design of a Cylindrical Pressure Vessel
A cylindrical pressure vessel with 2-m radius and hemispherical ends is to be designed to support an internal pressure as large as $p_i = 8E5$ Pa (8 atmospheres) with the outer pressure equal to atmospheric pressure $p_o = 1E5$ Pa. It is to be constructed of ASTM-A514 steel. Determine the vessel's wall thickness so that it has a factor of safety FS $= 4$.

Strategy
We can determine the normal yield stress σ_Y for ASTM-A514 steel from Appendix B and then use Eqs. (7.40) and (7.39) to determine the allowable value of the absolute maximum shear stress τ_{allow}. Comparing the terms (7.36) and (7.38), and remembering that $\sigma_h = 2\sigma$, it is clear that the absolute maximum shear stress is $(\sigma_h + p_i)/2$. By equating this expression to τ_{allow} and using Eq. (7.34), we can determine the necessary wall thickness.

Solution
From Appendix B, the normal yield stress σ_Y for ASTM-A514 steel is 700 MPa, so from Eq. (7.40), $\tau_Y = 350$ MPa. Then from Eq. (7.39), the allowable value of the absolute maximum shear stress is

$$\tau_{allow} = \frac{\tau_Y}{FS} = \frac{350 \text{ MPa}}{4} = 87.5 \text{ MPa}.$$

We equate the absolute maximum shear stress to the allowable value:

$$\frac{\sigma_h + p_i}{2} = \tau_{\text{allow}}.$$

Substituting Eq. (7.34) into this expression and solving for the wall thickness, we obtain

$$t = \frac{(p_i - p_o)R}{2\tau_{\text{allow}} - p_i}$$

$$= \frac{(7\text{E}5 \text{ Pa}) (2 \text{ m})}{2 (87.5\text{E}6 \text{ Pa}) - 8\text{E}5 \text{ Pa}}$$

$$= 0.00804 \text{ m}.$$

The necessary wall thickness is 8.04 mm.

Discussion
Factors of safety are not theoretical abstractions. They can and should be confirmed by testing whenever possible. In this example, that would be done by testing the vessel (with appropriate safety precautions) to a gauge pressure that subjected the material to an absolute maximum shear stress of 350 MPa, four times the value to which the vessel is designed. Confirm that if atmospheric pressure is $p_o = 1\text{E}5$ Pa, that would require subjecting the vessel to an internal pressure $p_i = 2.9\text{E}6$ Pa.

Problems

7.79 The spherical pressure vessel has a 2.5-m radius and a 5-mm wall thickness. It contains a gas with pressure $p_i = 6\text{E}5$ Pa, and the outer wall is subjected to atmospheric pressure $p_o - 1\text{E}5$ Pa. Determine the maximum normal stress in the vessel wall.

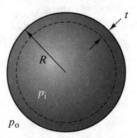

Problems 7.79–7.81

7.80 The spherical pressure vessel has a 24-in radius and a $\frac{1}{64}$-in wall thickness. It contains a gas with pressure $p_i = 200$ psi, and the outer wall is subjected to atmospheric pressure $p_o = 14.7$ psi. Determine the maximum normal stress and the absolute maximum shear stress at the vessel's inner surface.

7.81 The spherical pressure vessel has a 24-in radius and a $\frac{1}{64}$-in wall thickness. It consists of material with a yield shear stress $\tau_Y = 100$ ksi. If the vessel is designed to contain gas with a maximum pressure $p_i = 150$ psi and the outer wall is subjected to atmospheric pressure $p_o = 14.7$ psi, what is the factor of safety?

7.82 The cylindrical pressure vessel with hemispherical ends has a 2.5-m radius and a 5-mm wall thickness. It contains a gas with pressure $p_i = 6E5$ Pa, and the outer wall is subjected to atmospheric pressure $p_o = 1E5$ Pa. Determine the maximum normal stress in the vessel wall. Compare your answer to the answer to Problem 7.79.

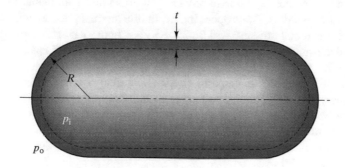

Problems 7.82–7.86

7.83 A cylindrical pressure vessel with hemispherical ends has a 2-m radius. It is to be designed to support an internal pressure as large as $p_i = 8E5$ Pa (8 atmospheres) with the outer pressure equal to atmospheric pressure $p_o = 1E5$ Pa. If you don't want to vessel wall to be subjected to a normal stress greater than 140 MPa, what is the minimum necessary wall thickness?

7.84 A cylindrical pressure vessel has a 600-mm radius and an 8-mm wall thickness. It contains a gas with pressure $p_i = 3E5$ Pa, and the outer wall is subjected to atmospheric pressure $p_o = 1E5$ Pa. Determine the maximum normal stress and the absolute maximum shear stress at the inner surface of the vessel's cylindrical wall.

7.85 A cylindrical pressure vessel used for natural gas storage has a 6-ft radius and a $\frac{1}{2}$-in wall thickness. It contains gas with pressure $p_i = 80$ psi, and the outer wall is subjected to atmospheric pressure $p_o = 14.7$ psi. Determine the maximum normal stress and the absolute maximum shear stress at the inner surface of the vessel's cylindrical wall.

7.86 A cylindrical pressure vessel with hemispherical ends has a 2-m radius. The vessel is to be constructed of ASTM-A514 steel. It is to be designed to support an internal pressure as large as $p_i = 8E5$ Pa (8 atmospheres) with the outer pressure equal to atmospheric pressure $p_o = $ 1E5 Pa. Determine the vessel's wall thickness so that the factor of safety is FS = 3.

7.87 Suppose that you are designing a spherical pressure vessel with a 200-mm radius to be used in a fuel cell to provide power in a satellite. The maximum internal pressure will be 16 MPa, and the external pressure will be negligible. Choose an aluminum alloy from Appendix B, and determine the wall thickness to obtain a safety factor FS = 1.5.

7.88 If design constraints require that the pressure vessel in Problem 7.87 be a cylindrical vessel with 150-mm radius, hemispherical ends, and the same internal volume as the spherical vessel, determine the wall thickness needed to obtain a safety factor FS = 1.5.

7.89 Suppose that you are making preliminary design calculations for a deep submersible vehicle that is to have a cylindrical hull with hemispherical ends and a 1.5-m radius. At its operating depth, the internal pressure will be 1E5 Pa, and the external pressure will be 330E5 Pa. Choose a steel from Appendix B, and determine the necessary thickness of the hull so that it has a safety factor FS = 2.

7.90 Suppose that you are making preliminary design calculations for a deep submersible vehicle that is to have a cylindrical hull with hemispherical ends and a 1.5-m radius. At its operating depth, the internal pressure will be 1E5 Pa, and the external pressure will be 330E5 Pa. Choose an aluminum alloy from Appendix B, and determine the necessary thickness of the hull so that it has a safety factor FS = 2.

7.7   Tetrahedron Argument

We have seen that the normal and shear stresses acting on a plane through a point p of a material depend in general on the orientation of the plane. We have derived equations and also a graphical solution, Mohr's circle, that allow determination of the normal and shear stresses, but only for a subset of planes through p and for the special case of plane stress. We conclude this chapter by showing how the normal and shear stresses can be determined on any plane through p and for a general state of stress.

7.7.1   Determining the Traction

We consider a point p of a sample of material and a tetrahedral element containing p (Fig. 7.55). Three faces of the tetrahedron are perpendicular to the coordinate axes. The fourth face is perpendicular to a unit vector **n** which specifies its orientation. Let the area of the face perpendicular to **n** be ΔA, and let the areas of the

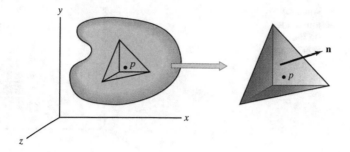

Fig. 7.55 Point p of a material and a tetrahedron containing p

Fig. 7.56 Areas of the faces of the tetrahedron. The angle between the vector **n** and its component n_x is θ_x.

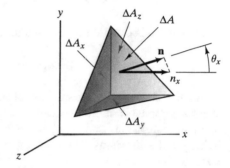

faces perpendicular to the x, y, and z axes be ΔA_x, ΔA_y, and ΔA_z (Fig. 7.56). Denoting the angle between the vector **n** and the x axis by θ_x, the x component of **n** is $n_x = \cos\theta_x$. Because **n** is perpendicular to the surface with area ΔA and the x axis is perpendicular to the surface with area ΔA_x, the angle between these two surfaces is θ_x. The area ΔA_x is the projection onto the y-z plane of the area ΔA:

$$\Delta A_x = \Delta A \cos\theta_x$$
$$= \Delta A\, n_x. \tag{7.41}$$

Applying the same argument to the surface A_y perpendicular to the y axis and the surface A_z perpendicular to the z axis yields the relations

$$\Delta A_y = \Delta A\, n_y, \quad \Delta A_z = \Delta A\, n_z. \tag{7.42}$$

Let **t** be the average value of the traction acting on the surface with area ΔA. Figure 7.57 shows the x component of **t** and also the average values of those stress components on the surfaces ΔA_x, ΔA_y, and ΔA_z which exert forces in the x direction. The sum of the forces on the tetrahedron in the x direction is

Fig. 7.57 Component of
the average traction and the
average stresses on the
tetrahedron that exert forces
in the x direction

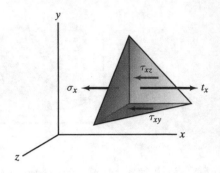

$$t_x \, \Delta A - \sigma_x \, \Delta A_x - \tau_{xy} \, \Delta A_y - \tau_{xz} \, \Delta A_z = 0.$$

Dividing this equation by ΔA and using Eqs. (7.41) and (7.42), we obtain

$$t_x - \sigma_x n_x - \tau_{xy} n_y - \tau_{xz} n_z = 0.$$

As the size of the tetrahedron decreases, the average traction and average
stress components in this equation approach their values at point p. Corresponding expressions can be obtained by summing the forces on the tetrahedron
in the y and z directions. The resulting equations for the components of the traction
t are

$$\begin{aligned}
t_x &= \sigma_x n_x + \tau_{xy} n_y + \tau_{xz} n_z, \\
t_y &= \tau_{yx} n_x + \sigma_y n_y + \tau_{yz} n_z, \\
t_z &= \tau_{zx} n_x + \tau_{zy} n_y + \sigma_z n_z,
\end{aligned} \tag{7.43}$$

which we can write as the matrix equation

$$\begin{bmatrix} t_x \\ t_y \\ t_z \end{bmatrix} = \begin{bmatrix} \sigma_x & \tau_{xy} & \tau_{xz} \\ \tau_{yx} & \sigma_y & \tau_{yz} \\ \tau_{zx} & \tau_{zy} & \sigma_z \end{bmatrix} \begin{bmatrix} n_x \\ n_y \\ n_z \end{bmatrix}. \tag{7.44}$$

Although we have assumed the material to be in equilibrium in deriving these
equations, they hold even when the material is not in equilibrium. If we know the
state of stress at a point p, Eqs. (7.43) or (7.44) determine the components of the
traction **t** at p acting on the plane through p whose orientation is defined by the unit
vector **n** (Fig. 7.58).

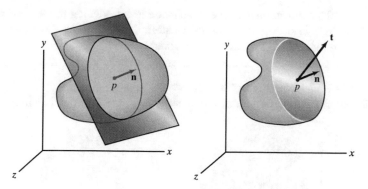

Fig. 7.58 Traction **t** on a plane whose orientation is specified by a unit vector **n**

Fig. 7.59 Normal and shear
stresses

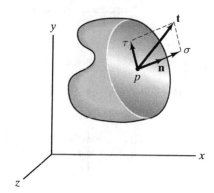

7.7.2 *Determining the Normal and Shear Stresses*

Once the components of the traction **t** are known, we can easily determine the normal
stress and the magnitude of the shear stress. The normal and shear stresses at p are
the components of the traction **t** normal and parallel to the given plane (Fig. 7.59).
The normal stress can be determined from the relation

$$\sigma = \mathbf{t} \cdot \mathbf{n} = t_x n_x + t_y n_y + t_z n_z. \tag{7.45}$$

The vector **t** is the sum of its vector component $\sigma \mathbf{n}$ normal to the plane and its
vector component parallel to the plane, so the magnitude of the shear stress can be
determined by evaluating the magnitude of the parallel vector component $\mathbf{t} - \sigma \mathbf{n}$:

$$|\tau| = |\mathbf{t} - \sigma \mathbf{n}|. \tag{7.46}$$

These results confirm the statement we made in introducing the state of stress. If the state of stress is known at a point of a material, the normal and shear stresses on any plane through that point can be determined.

Example 7.12 Determining Stresses on a Plane
The state of stress at point p of the material in Fig. 7.60(a) is (in MPa)

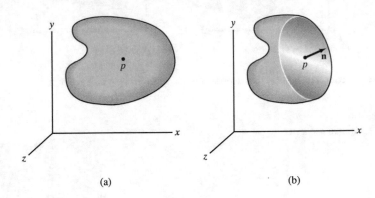

(a)   (b)

Fig. 7.60

$$\begin{bmatrix} \sigma_x & \tau_{xy} & \tau_{xz} \\ \tau_{yx} & \sigma_y & \tau_{yz} \\ \tau_{zx} & \tau_{zy} & \sigma_z \end{bmatrix} = \begin{bmatrix} 4 & -2 & 3 \\ -2 & 2 & 0 \\ 3 & 0 & -2 \end{bmatrix}.$$

(a) Determine the traction $\mathbf{t}$ at p acting on the plane in Fig. 7.60(b) if $\mathbf{n} = 0.818\mathbf{i} + 0.545\mathbf{j} + 0.181\mathbf{k}$. (b) Determine the normal stress and the magnitude of the shear stress at p acting on the plane in Fig. 7.60(b).

Strategy
(a) The components of the traction are given in terms of the components of $\mathbf{n}$ by Eqs. (7.43) or (7.44). (b) The normal stress and the magnitude of the shear stress are given by Eqs. (7.45) and (7.46).

Solution
(a) The stress components are $\sigma_x = 4$ MPa, $\sigma_y = 2$ MPa, $\sigma_z = -2$ MPa, $\tau_{xy} = -2$ MPa, $\tau_{xz} = 3$ MPa, and $\tau_{yz} = 0$. From Eqs. (7.43), the components of the traction are

$$t_x = \sigma_x n_x + \tau_{xy} n_y + \tau_{xz} n_z$$
$$= (4\text{ MPa})(0.818) + (-2\text{ MPa})(0.545) + (3\text{ MPa})(0.181)$$
$$= 2.725\text{ MPa},$$

$$t_y = \tau_{yx} n_x + \sigma_y n_y + \tau_{yz} n_z$$
$$= (-2\text{ MPa})(0.818) + (2\text{ MPa})(0.545) + (0)(0.181)$$
$$= -0.546\text{ MPa},$$

$$t_z = \tau_{zx} n_x + \tau_{zy} n_y + \sigma_z n_z$$
$$= (3\text{ MPa})(0.818) + (0)(0.545) + (-2\text{ MPa})(0.181)$$
$$= 2.092\text{ MPa}.$$

The traction is $\mathbf{t} = 2.725\mathbf{i} - 0.546\mathbf{j} + 2.092\mathbf{k}$ MPa.
(b) From Eq. (7.45), the normal stress is

$$\sigma = \mathbf{t} \cdot \mathbf{n} = t_x n_x + t_y n_y + t_z n_z$$
$$= (2.725\text{ MPa})(0.818) + (-0.546\text{ MPa})(0.545) + (2.092\text{ MPa})(0.181)$$
$$= 2.31\text{ MPa}.$$

The component of $\mathbf{t}$ parallel to the plane is

$$\mathbf{t} - \sigma\mathbf{n} = 2.725\mathbf{i} - 0.546\mathbf{j} + 2.092\mathbf{k} - (2.310)(0.818\mathbf{i} + 0.545\mathbf{j} + 0.181\mathbf{k})$$
$$= 0.835\mathbf{i} - 1.805\mathbf{j} + 1.674\mathbf{k}\,(\text{MPa}),$$

so the magnitude of the shear stress is

$$|\tau| = |\mathbf{t} - \sigma\mathbf{n}|$$
$$= \sqrt{(0.835)^2 + (-1.805)^2 + (1.674)^2}\text{ MPa}$$
$$= 2.60\text{ MPa}$$

Discussion

In part (a) we used the components of stress and the components of the unit vector $\mathbf{n}$ to determine the components of the traction vector $\mathbf{t}$. In mathematical language, the components of stress *transformed* the components of $\mathbf{n}$ into the components of $\mathbf{t}$. Furthermore, the transformation is *linear*. The components of $\mathbf{t}$ are linear functions of the components of $\mathbf{n}$. We say that the components of the state of stress at a point provide a *linear transformation of the vector* $\mathbf{n}$ *into the vector* $\mathbf{t}$. A linear transformation of a vector into a vector is called a *second-order tensor*. For this reason, the state of stress at a point is referred to as the *stress tensor*. In the next chapter, we discuss the state of strain at a point. It is also a second-order tensor and is referred to as the *strain tensor*.

Example 7.13 Applying Eq. (7.44) to Plane Stress
If the material in Fig. 7.61 is subjected to the state of plane stress

$$
\begin{bmatrix}
\sigma_x & \tau_{xy} & 0 \\
\tau_{yx} & \sigma_y & 0 \\
0 & 0 & 0
\end{bmatrix}
$$

at point p, the shear stress $\tau_{x'y'}$ is given by Eq. (7.5):

$$
\tau_{x'y'} = -(\sigma_x - \sigma_y) \sin\theta \cos\theta + \tau_{xy}(\cos^2\theta - \sin^2\theta).
$$

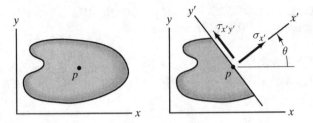

Fig. 7.61

Derive this equation by using Eq. (7.44).

Strategy
With Eq. (7.44) we can obtain the components of the traction on the plane
shown in the right part of Fig. 7.61. Then we can determine the stress $\tau_{x'y'}$ by
calculating the component of the traction parallel to the plane.

Solution
The unit vector normal to the plane (Fig. (a)) has components $\mathbf{n} = \cos\theta\,\mathbf{i} + \sin\theta\,\mathbf{j}$.
Therefore, the components of the traction on the plane are

$$
\begin{bmatrix}
t_x \\
t_y \\
t_z
\end{bmatrix}
=
\begin{bmatrix}
\sigma_x & \tau_{xy} & 0 \\
\tau_{yx} & \sigma_y & 0 \\
0 & 0 & 0
\end{bmatrix}
\begin{bmatrix}
\cos\theta \\
\sin\theta \\
0
\end{bmatrix}
$$

$$
=
\begin{bmatrix}
\sigma_x \cos\theta + \tau_{xy} \sin\theta \\
\tau_{xy} \cos\theta + \sigma_y \sin\theta \\
0
\end{bmatrix}.
$$

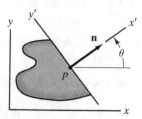

(a) Unit vector **n** normal to the plane

[Equations (7.43) can be used to obtain this result.] The components of the traction are shown in Fig. (b). Summing the components of t_x and t_y parallel to the plane, we obtain the shear stress:

$$\tau_{x'y'} = t_y \cos\theta - t_x \sin\theta$$
$$= (\tau_{xy}\cos\theta + \sigma_y\sin\theta)\cos\theta - (\sigma_x\cos\theta + \tau_{xy}\sin\theta)\sin\theta$$
$$= -(\sigma_x - \sigma_y)\sin\theta\cos\theta + \tau_{xy}(\cos^2\theta - \sin^2\theta).$$

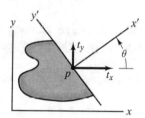

(b) The x and y components of the traction

Discussion

This example demonstrates the connection between this section and our analysis of plane stress in Sect. 7.2. For the special case of plane stress, we used Eq. (7.44) to obtain Eq. (7.4) for the shear stress τ'_{xy}. Also see Problem 7.104.

Problems

7.91 The state of stress at a point p of a material (in MPa) is

$$\begin{bmatrix} \sigma_x & \tau_{xy} & \tau_{xz} \\ \tau_{yx} & \sigma_y & \tau_{yz} \\ \tau_{zx} & \tau_{zy} & \sigma_z \end{bmatrix} = \begin{bmatrix} 8 & 0 & 0 \\ 0 & 4 & 0 \\ 0 & 0 & 6 \end{bmatrix}.$$

Determine the traction **t** acting on the plane whose orientation is specified by the unit vector **n** = −0.857**i** + 0.429**j** + 0.286**k**.

Strategy: The components of the traction are given in terms of the components of **n** by Eqs. (7.43) or (7.44).

7.92 The state of stress at a point p of a material (in MPa) is

$$\begin{bmatrix} \sigma_x & \tau_{xy} & \tau_{xz} \\ \tau_{yx} & \sigma_y & \tau_{yz} \\ \tau_{zx} & \tau_{zy} & \sigma_z \end{bmatrix} = \begin{bmatrix} 8 & 0 & 0 \\ 0 & 4 & 0 \\ 0 & 0 & 6 \end{bmatrix}.$$

Determine the normal stress σ and the magnitude of the shear stress $|\tau|$ acting on the plane whose orientation is specified by the unit vector **n** = − 0.857**i** + 0.429**j** + 0.286**k**.

7.93 The state of stress at a point p of a material (in ksi) is

$$\begin{bmatrix} \sigma_x & \tau_{xy} & \tau_{xz} \\ \tau_{yx} & \sigma_y & \tau_{yz} \\ \tau_{zx} & \tau_{zy} & \sigma_z \end{bmatrix} = \begin{bmatrix} -30 & 20 & 0 \\ 20 & 45 & -20 \\ 0 & -20 & 60 \end{bmatrix}.$$

Determine the normal stress σ and the magnitude of the shear stress $|\tau|$ acting on the plane whose orientation is specified by the unit vector **n** = 0.381**i** − 0.889**j** − 0.254**k**.

7.94 The state of stress at point p of the material shown (in ksi) is

$$\begin{bmatrix} \sigma_x & \tau_{xy} & \tau_{xz} \\ \tau_{yx} & \sigma_y & \tau_{yz} \\ \tau_{zx} & \tau_{zy} & \sigma_z \end{bmatrix} = \begin{bmatrix} 2 & 2 & 1 \\ 2 & -3 & -1 \\ 1 & -1 & 4 \end{bmatrix},$$

and the unit vector **n** = 0.667**i** + 0.333**j** + 0.667**k**. Determine the traction **t** on the plane shown.

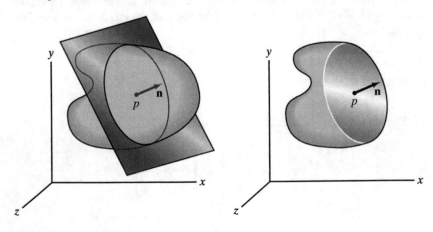

Problems 7.94–7.96

7.95 The state of stress at point p of the material shown (in ksi) is

$$\begin{bmatrix} \sigma_x & \tau_{xy} & \tau_{xz} \\ \tau_{yx} & \sigma_y & \tau_{yz} \\ \tau_{zx} & \tau_{zy} & \sigma_z \end{bmatrix} = \begin{bmatrix} 2 & 2 & 1 \\ 2 & -3 & -1 \\ 1 & -1 & 4 \end{bmatrix},$$

and the unit vector $\mathbf{n} = 0.667\mathbf{i} + 0.333\mathbf{j} + 0.667\mathbf{k}$. Determine the normal stress and the magnitude of the shear stress on the plane shown.

7.96 The state of stress at point p of the material shown (in GPa) is

$$\begin{bmatrix} \sigma_x & \tau_{xy} & \tau_{xz} \\ \tau_{yx} & \sigma_y & \tau_{yz} \\ \tau_{zx} & \tau_{zy} & \sigma_z \end{bmatrix} = \begin{bmatrix} 6 & -2 & 3 \\ -2 & 2 & 1 \\ 3 & 1 & 4 \end{bmatrix},$$

and the plane shown passes through the three points $(3, 0, 0)$ m, $(0, 6, 0)$ m, and $(0, 0, 2)$ m. Determine the normal stress and the magnitude of the shear stress acting on the plane. (You must determine the unit vector $\mathbf{n}$.)

7.97 The state of stress at a point p of a material (in MPa) is

$$\begin{bmatrix} \sigma_x & \tau_{xy} & \tau_{xz} \\ \tau_{yx} & \sigma_y & \tau_{yz} \\ \tau_{zx} & \tau_{zy} & \sigma_z \end{bmatrix} = \begin{bmatrix} 20 & -15 & 30 \\ -15 & -40 & 25 \\ 30 & 25 & \sigma_z \end{bmatrix}.$$

The normal stress on the plane whose orientation is specified by the unit vector $\mathbf{n} = -0.857\mathbf{i} + 0.429\mathbf{j} + 0.286\mathbf{k}$ is $\sigma = 12.6$ MPa. What is the stress component σ_z?

7.98 The state of stress at a point p of a material (in MPa) is

$$\begin{bmatrix} \sigma_x & \tau_{xy} & \tau_{xz} \\ \tau_{yx} & \sigma_y & \tau_{yz} \\ \tau_{zx} & \tau_{zy} & \sigma_z \end{bmatrix} - \begin{bmatrix} 20 & -15 & 30 \\ -15 & -40 & 25 \\ 30 & 25 & \sigma_z \end{bmatrix}.$$

The normal stress on the plane whose orientation is specified by the unit vector $\mathbf{n} = -0.857\mathbf{i} + 0.429\mathbf{j} + 0.286\mathbf{k}$ is $\sigma = 12.6$ MPa. What is the magnitude of the shear stress on the specified plane?

7.99 The state of stress at a point in a liquid or gas at rest is

$$\begin{bmatrix} \sigma_x & \tau_{xy} & \tau_{xz} \\ \tau_{yx} & \sigma_y & \tau_{yz} \\ \tau_{zx} & \tau_{zy} & \sigma_z \end{bmatrix} = \begin{bmatrix} -P & 0 & 0 \\ 0 & -P & 0 \\ 0 & 0 & -P \end{bmatrix},$$

where P is the pressure. Show that the normal stress is $\sigma = -P$ and the shear stress is zero on any plane through the point.

7.100 For *any* state of stress at a point p, at least one coordinate system xyz exists for which the state of stress is of the form

$$
\begin{bmatrix}
\sigma_x & \tau_{xy} & \tau_{xz} \\
\tau_{yx} & \sigma_y & \tau_{yz} \\
\tau_{zx} & \tau_{zy} & \sigma_z
\end{bmatrix}
=
\begin{bmatrix}
\sigma_1 & 0 & 0 \\
0 & \sigma_2 & 0 \\
0 & 0 & \sigma_3
\end{bmatrix},
$$

where σ_1, σ_2, and σ_3 are the principal stresses (see Sect. 7.4). Use Eqs. (7.43), (7.45), and (7.46) to determine the normal stress and the magnitude of the shear stress on a plane through p that is perpendicular to the x axis. (Your answers will be in terms of the principal stresses.)

7.101 For *any* state of stress at a point p, at least one coordinate system xyz exists for which the state of stress is of the form

$$
\begin{bmatrix}
\sigma_x & \tau_{xy} & \tau_{xz} \\
\tau_{yx} & \sigma_y & \tau_{yz} \\
\tau_{zx} & \tau_{zy} & \sigma_z
\end{bmatrix}
=
\begin{bmatrix}
\sigma_1 & 0 & 0 \\
0 & \sigma_2 & 0 \\
0 & 0 & \sigma_3
\end{bmatrix},
$$

where σ_1, σ_2, and σ_3 are the principal stresses (see Sect. 7.4). Determine the normal stress on a plane through p whose orientation is specified by the unit vector $(1/\sqrt{3})\,(\mathbf{i}+\mathbf{j}+\mathbf{k})$.

7.102 For the state of plane stress shown, use Eqs. (7.45) and (7.46) to determine the normal stress σ and the magnitude of the shear stress τ.

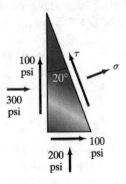

Problem 7.102

7.103 For the state of plane stress shown, use Eqs. (7.45) and (7.46) to determine the normal stress σ and the magnitude of the shear stress τ.

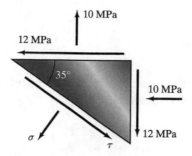

Problem 7.103

7.104 If the material in Fig. (a) is subjected to the state of plane stress

$$\begin{bmatrix} \sigma_x & \tau_{xy} & 0 \\ \tau_{yx} & \sigma_y & 0 \\ 0 & 0 & 0 \end{bmatrix}$$

at point p, the normal stress $\sigma_{x'}$ in Fig. (b) is given by Eq. (7.4):

$$\sigma_{x'} = \sigma_x \cos^2 \theta + \sigma_y \sin^2 \theta + 2\tau_{xy} \sin \theta \cos \theta.$$

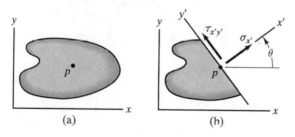

Problem 7.104

Derive this equation by using Eqs. (7.43) and (7.45). (See Example 7.13.)

7.105 If the material in Fig. (a) is subjected to the state of plane stress

$$\begin{bmatrix} \sigma_x & \tau_{xy} & 0 \\ \tau_{yx} & \sigma_y & 0 \\ 0 & 0 & 0 \end{bmatrix}$$

at point p, the shear stress $\tau_{x'y'}$ in Fig. (b) is given by Eq. (7.4):

$$\tau_{x'y'} = -(\sigma_x - \sigma_y) \sin \theta \cos \theta + \tau_{xy} (\cos^2 \theta - \sin^2 \theta).$$

Derive this equation by using Eqs. (7.43).

Chapter Summary

Components of Stress

In terms of a given coordinate system, the *state of stress* at a point p of a material is specified by the *components of stress*:

$$\begin{bmatrix} \sigma_x & \tau_{xy} & \tau_{xz} \\ \tau_{yx} & \sigma_y & \tau_{yz} \\ \tau_{zx} & \tau_{zy} & \sigma_z \end{bmatrix}. \tag{7.1}$$

The directions of these stresses, and the orientations of the planes on which they act, are shown in Fig. (a). The shear stresses $\tau_{xy} = \tau_{yx}$, $\tau_{yz} = \tau_{zy}$ and $\tau_{xz} = \tau_{zx}$.

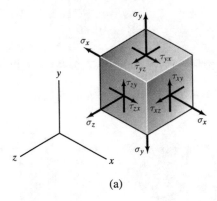

(a)

Transformations of Plane Stress

The stress at a point is said to be a state of *plane stress* if it is of the form

$$\begin{bmatrix} \sigma_x & \tau_{xy} & 0 \\ \tau_{yx} & \sigma_y & 0 \\ 0 & 0 & 0 \end{bmatrix}. \tag{7.3}$$

In terms of a coordinate system $x'\,y'\,z'$ oriented as shown in Fig. (b), the components of stress are

$$\sigma_{x'} = \frac{\sigma_x + \sigma_y}{2} + \frac{\sigma_x - \sigma_y}{2}\cos 2\theta + \tau_{xy}\sin 2\theta, \tag{7.6}$$

$$\tau_{x'y'} = -\frac{\sigma_x - \sigma_y}{2}\sin 2\theta + \tau_{xy}\cos 2\theta, \tag{7.7}$$

$$\sigma_{y'} = \frac{\sigma_x + \sigma_y}{2} - \frac{\sigma_x - \sigma_y}{2}\cos 2\theta - \tau_{xy}\sin 2\theta. \tag{7.8}$$

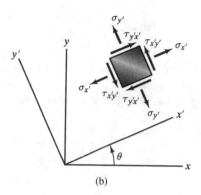

(b)

Maximum and Minimum Stresses in Plane Stress

The orientation of the element on which the principal stresses act (Fig. (c)) is determined from the equation

$$\tan 2\theta_p = \frac{2\tau_{xy}}{\sigma_x - \sigma_y}. \tag{7.12}$$

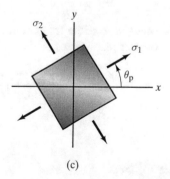

(c)

The values of the principal stresses can be obtained by substituting θ_p into Eqs. (7.6) and (7.8). Their values can also be determined from the equation

$$\sigma_1, \sigma_2 = \frac{\sigma_x + \sigma_y}{2} \pm \sqrt{\left(\frac{\sigma_x - \sigma_y}{2}\right)^2 + \tau_{xy}^2}, \qquad (7.15)$$

although this equation does not indicate the planes on which they act.

The orientation of the element on which the maximum in-plane shear stresses act (Fig. (d)) can be determined from the equation

$$\tan 2\theta_s = -\frac{\sigma_x - \sigma_y}{2\tau_{xy}}, \qquad (7.16)$$

or by using the relation $\theta_s = \theta_p + 45°$. The normal stresses σ_s acting on this element are shown in Fig. (d). The value of the maximum in-plane shear stress can be obtained by substituting θ_s into Eq. (7.7). Its value can also be determined from the equation

$$\tau_{max} = \sqrt{\left(\frac{\sigma_x - \sigma_y}{2}\right)^2 + \tau_{xy}^2}, \qquad (7.19)$$

although this equation does not indicate the direction of the stress on the element.

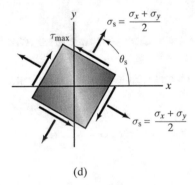

(d)

The *absolute maximum shear stress* (the maximum shear stress on any plane through p) resulting from a state of plane stress is

$$\tau_{abs} = \text{Max}\left(\left|\frac{\sigma_1}{2}\right|, \left|\frac{\sigma_2}{2}\right|, \left|\frac{\sigma_1 - \sigma_2}{2}\right|\right). \qquad (7.24)$$

Mohr's Circle for Plane Stress

Given a state of plane stress σ_x, τ_{xy}, and σ_y, establish a set of horizontal and vertical axes with normal stress measured to the right along the horizontal axis and shear stress measured downward along the vertical axis. Plot two points, point P with coordinates (σ_x, τ_{xy}) and point Q with coordinates $(\sigma_y, -\tau_{xy})$. Draw a straight line connecting points P and Q. Using the intersection of the straight line with the horizontal axis as the center, draw a circle that passes through the two points (Fig. (e)). Draw a straight line through the center of the circle at an angle 2θ measured counterclockwise from point P. Point P' at which this line intersects the circle has coordinates $(\sigma_{x'}, \tau_{x'y'})$, and point Q' has coordinates $(\sigma_{y'}, -\tau_{x'y'})$, as shown in Fig. (f).

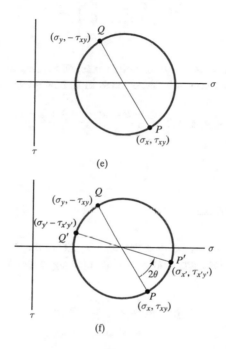

(e)

(f)

Principal Stresses in Three Dimensions

For a general state of stress, the principal stresses are the roots of the cubic equation

$$\sigma^3 - I_1\sigma^2 + I_2\sigma - I_3 = 0, \tag{7.25}$$

where the *stress invariants* are

$$
\begin{aligned}
I_1 &= \sigma_x + \sigma_y + \sigma_z, \\
I_2 &= \sigma_x\sigma_y + \sigma_y\sigma_z + \sigma_z\sigma_x - \tau_{xy}^2 - \tau_{yz}^2 - \tau_{zx}^2, \\
I_3 &= \sigma_x\sigma_y\sigma_z - \sigma_x\tau_{yz}^2 - \sigma_y\tau_{xz}^2 - \sigma_z\tau_{xy}^2 + 2\tau_{xy}\tau_{yz}\tau_{zx}.
\end{aligned}
\tag{7.26}
$$

The absolute maximum shear stress resulting from a general state of stress is

$$\tau_{abs} = \text{Max}\left(\left|\frac{\sigma_1 - \sigma_2}{2}\right|, \left|\frac{\sigma_1 - \sigma_3}{2}\right|, \left|\frac{\sigma_2 - \sigma_3}{2}\right|\right). \tag{7.27}$$

The state of stress at a point is said to be *triaxial* if it is of the form

$$\begin{bmatrix} \sigma_x & 0 & 0 \\ 0 & \sigma_y & 0 \\ 0 & 0 & \sigma_z \end{bmatrix}. \tag{7.28}$$

In triaxial stress, x, y, and z are principal axes, and σ_x, σ_y, and σ_z are the principal stresses. The absolute maximum shear stress in triaxial stress is

$$\tau_{abs} = \text{Max}\left(\left|\frac{\sigma_x - \sigma_y}{2}\right|, \left|\frac{\sigma_x - \sigma_z}{2}\right|, \left|\frac{\sigma_y - \sigma_z}{2}\right|\right). \tag{7.29}$$

Tetrahedron Argument

Given the state of stress at a point p, the components of the traction $\mathbf{t}$ at p acting on a plane through p whose orientation is defined by a perpendicular unit vector $\mathbf{n}$ (Fig. (g)) are

$$\begin{aligned} t_x &= \sigma_x n_x + \tau_{xy} n_y + \tau_{xz} n_z, \\ t_y &= \tau_{yx} n_x + \sigma_y n_y + \tau_{yz} n_z, \\ t_z &= \tau_{zx} n_x + \tau_{zy} n_y + \sigma_z n_z. \end{aligned} \tag{7.43}$$

The normal stress at p is

$$\sigma = \mathbf{t} \cdot \mathbf{n} = t_x n_x + t_y n_y + t_z n_z, \tag{7.45}$$

and the magnitude of the shear stress is

$$|\tau| = |\mathbf{t} - \sigma\mathbf{n}|. \tag{7.46}$$

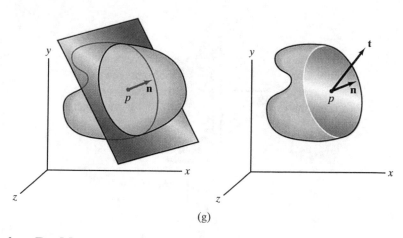

(g)

Review Problems

7.106* The components of plane stress at a point p of a drill bit are $\sigma_x = 40$ ksi, $\sigma_y = -30$ ksi, and $\tau_{xy} = 30$ ksi, and the components referred to the $x'\,y'\,z'$ coordinate system are $\sigma_{x'} = 12.5$ ksi, $\sigma_{y'} = -2.5$ ksi, and $\tau_{x'y'} = 45.5$ ksi. What is the angle θ?

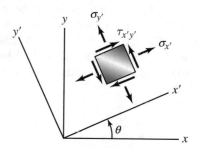

Problems 7.106–7.107

7.107 The components of plane stress at a point p of the drill bit referred to the $x'\,y'\,z'$ coordinate system are $\sigma_{x'} = -8$ MPa, $\sigma_{y'} = 6$ MPa, and $\tau_{x'y'} = -16$ MPa. If $\theta = 20°$, what are the stresses σ_x, σ_y, and τ_{xy} at point p?

7.108 Determine the stresses σ and τ **(a)** by writing equilibrium equations for the element shown and **(b)** by using Eqs. (7.6) and (7.7).

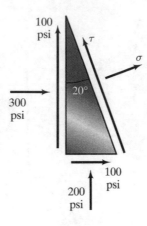

Problems 7.108–7.109

7.109 Solve Problem 7.108 if the 300-psi stress on the element is in tension instead of compression.

7.110 For the state of plane stress $\sigma_x = 8$ ksi, $\sigma_y = 6$ ksi, and $\tau_{xy} = -6$ ksi, what is the absolute maximum shear stress?

7.111 For the state of plane stress $\sigma_x = 240$ MPa, $\sigma_y = -120$ MPa, and $\tau_{xy} = 240$ MPa, what is the absolute maximum shear stress?

7.112 The components of plane stress acting on an element of a bar subjected to axial loads are shown. The stress $\sigma_x = 4000$ psi. Consider a plane through the bar whose normal is $20°$ relative to the x axis. Use Mohr's circle to determine the normal stress and the magnitude of the shear stress on the plane.

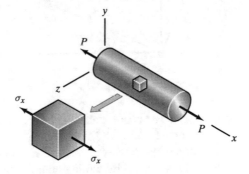

Problem 7.112

7.113 The components of plane stress acting on an element of a bar subjected to torsion are shown. The stress $\tau_{xy} = 4000$ psi. Use Mohr's circle to determine the principal stresses to which the material is subjected.

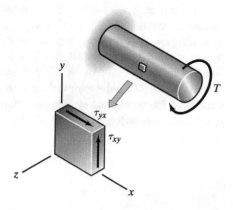

Problem 7.113

7.114 A machine element is subjected to the state of *triaxial* stress $\sigma_x = 300$ MPa, $\sigma_y = -200$ MPa, $\sigma_z = -100$ MPa. What is the absolute maximum shear stress?

7.115* At a point p, a material is subjected to the state of stress (in MPa):

$$\begin{bmatrix} \sigma_x & \tau_{xy} & \tau_{xz} \\ \tau_{yx} & \sigma_y & \tau_{yz} \\ \tau_{zx} & \tau_{zy} & \sigma_z \end{bmatrix} = \begin{bmatrix} 4 & 2 & 1 \\ 2 & -2 & 1 \\ 1 & 1 & -3 \end{bmatrix}.$$

Determine the principal stresses and the absolute maximum shear stress.

7.116 The cylindrical bar has radius $R = 20$ mm. It is subjected to a torque about its axis $T = 60$ N-m and a couple $M_y = 85$ N-m about the y axis in the direction shown. In terms of the coordinate system shown, determine the state of stress at point p, which is located at $x = 60$ mm, $y = 0$, $z = 20$ mm.

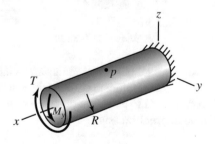

Problems 7.116–7.117

7.117 The cylindrical bar has radius $R = 20$ mm. It is subjected to a torque about its axis $T = 60$ N-m and a couple $M_y = 85$ N-m about the y axis in the direction shown. Determine the principal stresses and the maximum in-plane shear stress at point p, which is located at $x = 60$ mm, $y = 0$, $z = 20$ mm.

7.118 The beam is subjected to an axial load $P = 1200$ lb that does not act at the centroid of the cross section. The dimensions of the cross section are $b = 4$ in and $h = 2$ in. In terms of the coordinate system shown, determine the state of stress at point p, which has coordinates $y = \frac{1}{2}h$ and $z = \frac{1}{4}b$.

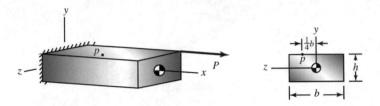

Problem 7.118

7.119 The spherical pressure vessel has a 1-m radius and a 0.002-m wall thickness. It contains a gas with pressure $p_i = 1.8E5$ Pa, and the outer wall is subjected to atmospheric pressure $p_o = 1E5$ Pa. Determine the maximum normal stress and the absolute maximum shear stress at the vessel's inner surface.

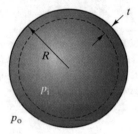

Problems 7.119–7.120

7.120 The spherical pressure vessel has a 1-m radius and a 0.002-m wall thickness. The allowable value of the absolute maximum shear stress is $\tau_{allow} = 14$ MPa. If the outer wall is subjected to atmospheric pressure $p_o = 1E5$ Pa, what is the maximum allowable internal pressure?

Chapter 8
States of Strain and Stress-Strain Relations

In this chapter we review the state of strain at a point of an object and show that it specifies the deformation of the material in the neighborhood of the point. We then discuss the relationship between the state of stress and the state of strain for an elastic material subject to small strains.

8.1 Components of Strain

We have introduced two types of strain, normal strain and shear strain. The normal strain ε is a measure of the change in length of a material line element whose length is dL in a reference state. If the length of the line element in a deformed state is dL' (Fig. 8.1), the normal strain is defined to be the change in length divided by its original length:

$$\varepsilon = \frac{dL' - dL}{dL}.$$

The value of the normal strain at a point depends in general on the direction of the line element. We say that ε is the normal strain in the direction of dL.

The shear strain γ is a measure of the change in the angle between two line elements dL_1 and dL_2 that are perpendicular in a reference state. The angle between the elements in a deformed state is defined to be $(\pi/2) - \gamma$ (Fig. 8.2). The value of γ at a point depends in general on the directions of the two elements. We say that γ is the shear strain referred to the directions of the elements dL_1 and dL_2.

In terms of the coordinate system shown in Fig. 8.3(a), the six components of strain are:

ε_x The normal strain determined with the element dL parallel to the x axis (Fig. 8.3(b))

© Springer Nature Switzerland AG 2020
A. Bedford, K. M. Liechti, *Mechanics of Materials*,
https://doi.org/10.1007/978-3-030-22082-2_8

Fig. 8.1 Material line
element in the reference
and deformed states

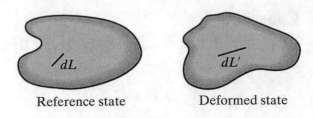

Reference state   Deformed state

Fig. 8.2 Line elements that
are perpendicular in the
reference state. The decrease
in the angle between them
in the deformed state is the
shear strain

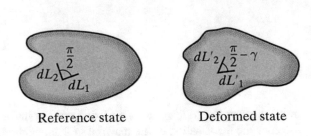

Reference state   Deformed state

Fig. 8.3 (**a**) Introducing a
coordinate system. (**b**) Line
element for determining the
value of ε_x at p. (**c**) Line
elements for determining the
value of γ_{xy} at p

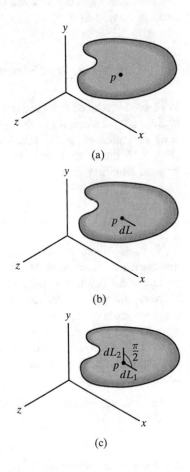

ε_y The normal strain determined with the element dL parallel to the y axis

ε_z The normal strain determined with the element dL parallel to the z axis

γ_{xy} The shear strain determined with the elements dL_1 and dL_2 in the positive x and y directions (Fig. 8.3(c))

γ_{yz} The shear strain determined with the elements dL_1 and dL_2 in the positive y and z directions

γ_{xz} The shear strain determined with the elements dL_1 and dL_2 in the positive x and z directions.

If these six components of strain are known at a point, the normal strain in any direction and the shear strain referred to any two perpendicular directions can be determined. (This result holds only when the components of strain are *small*, meaning that they are sufficiently small that products of the components are negligible in comparison to the components themselves.) For this reason these components are called the *state of strain* at the point. We say that the strain is *homogeneous* in a region if the state of strain is the same at each point. In analogy to the state of stress, we can represent the state of strain as the matrix

$$\begin{bmatrix} \varepsilon_x & \gamma_{xy} & \gamma_{xz} \\ \gamma_{yx} & \varepsilon_y & \gamma_{yz} \\ \gamma_{zx} & \gamma_{zy} & \varepsilon_z \end{bmatrix}, \tag{8.1}$$

where $\gamma_{yx} = \gamma_{xy}$, $\gamma_{zx} = \gamma_{xz}$ and $\gamma_{zy} = \gamma_{yz}$.

The state of strain determines the change in volume of a material due to a deformation involving small strains. Consider an element of material in the reference state with dimensions dx_0, dy_0, dz_0 (Fig. 8.4). The volume of the element is $dV_0 = dx_0\, dy_0\, dz_0$. In the deformed state, the lengths of the edges of the element are $dx = (1 + \varepsilon_x)\, dx_0$, $dy = (1 + \varepsilon_y)\, dy_0$ and $dz = (1 + \varepsilon_z)\, dz_0$ (Fig. 8.4). Its volume in the deformed state is

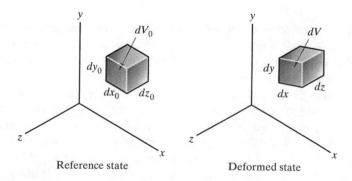

Reference state   Deformed state

Fig. 8.4 Element of material in the reference and deformed states

$$dV = dx\,dy\,dz = (1 + \varepsilon_x)\,(1 + \varepsilon_y)\,(1 + \varepsilon_z)\,dx_0\,dy_0\,dz_0.$$

Neglecting products of the strains, we can write this result as

$$dV = \left(1 + \varepsilon_x + \varepsilon_y + \varepsilon_z\right) dV_0. \tag{8.2}$$

(When products of strains are negligible, the components of shear strain do not affect the change in volume of the material.) The change in volume of the material per unit volume, denoted by e, is called the *dilatation*:

$$e = \frac{dV - dV_0}{dV_0} = \varepsilon_x + \varepsilon_y + \varepsilon_z. \tag{8.3}$$

Study Questions

1. What are the definitions of the components of strain at a point?
2. What does it mean when the strain is said to be homogeneous in a region?
3. Consider an element of material that has volume dV_0 in a reference state. If the material is subjected to a known state of strain, how can you determine the change in volume of the element?

8.2   Transformations of Plane Strain

For a state of *plane strain*, defined by

$$\begin{bmatrix} \varepsilon_x & \gamma_{xy} & 0 \\ \gamma_{yx} & \varepsilon_y & 0 \\ 0 & 0 & 0 \end{bmatrix}, \tag{8.4}$$

we can derive transformation equations equivalent to the equations for plane stress developed in Sect. 5.2. Suppose that we know the state of plane strain at a point p of a material in terms of the xyz coordinate system shown in Fig. 8.5 and we want to know the components of strain $\varepsilon_{x'}, \gamma_{x'y'}$ and $\varepsilon_{y'}$.

To determine the normal strain $\varepsilon_{x'}$, we begin with an infinitesimal element of material at p which in the reference state has the triangular shape shown in Fig. 8.6(a). We denote the infinitesimal length of the hypotenuse by dL_0. In the deformed state (Fig. 8.6(b)), we know the lengths of the sides in terms of the normal strains ε_x, ε_y, and $\varepsilon_{x'}$, and we can express the angle between the sides that were perpendicular in the reference state in terms of γ_{xy}. By analyzing this triangle, we can determine $\varepsilon_{x'}$ in terms of the strains ε_x, ε_y, and γ_{xy}. Applying the law of cosines to the deformed element gives

Fig. 8.5 Coordinate
systems xyz and $x'y'z'$. The
z and z' axes are coincident

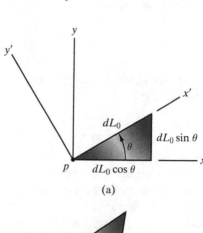

Fig. 8.6 (**a**) Triangular
element in the reference
state. (**b**) Element in the
deformed state

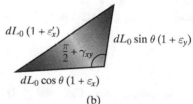

(a)

(b)

$$dL_0^2(1 + \varepsilon_{x'})^2 = dL_0^2 \sin^2\theta(1 + \varepsilon_y)^2 + dL_0^2 \cos^2 0(1 + \varepsilon_x)^2$$
$$- 2dL_0^2 \sin\theta \cos\theta(1 + \varepsilon_x)(1 + \varepsilon_y) \cos(\pi/2 + \gamma_{xy}). \tag{8.5}$$

We apply the identity

$$\cos(\pi/2 + \gamma_{xy}) = \cos(\pi/2)\cos\gamma_{xy} - \sin(\pi/2)\sin\gamma_{xy}$$
$$= -\sin\gamma_{xy},$$

and because strains are assumed to be small, we can approximate $\sin\gamma_{xy}$ by γ_{xy}, obtaining

$$\cos(\pi/2 + \gamma_{xy}) = -\gamma_{xy}.$$

Substituting this result into Eq. (8.5) and neglecting products of strains, it becomes

$$\varepsilon_{x'} = \varepsilon_x \cos^2\theta + \varepsilon_y \sin^2\theta + \gamma_{xy} \sin\theta \cos\theta. \tag{8.6}$$

By using the identities

$$2\cos^2\theta = 1 + \cos 2\theta,$$
$$2\sin^2\theta = 1 - \cos 2\theta,$$
$$2\sin\theta \cos\theta = \sin 2\theta,$$

we can write Eq. (8.6) in the form

$$\varepsilon_{x'} = \frac{\varepsilon_x + \varepsilon_y}{2} + \frac{\varepsilon_x - \varepsilon_y}{2} \cos 2\theta + \frac{\gamma_{xy}}{2} \sin 2\theta. \tag{8.7}$$

We can obtain an equation for $\varepsilon_{y'}$ by setting θ equal to $\theta + 90°$ in this expression. The result is

$$\varepsilon_{y'} = \frac{\varepsilon_x + \varepsilon_y}{2} - \frac{\varepsilon_x - \varepsilon_y}{2} \cos 2\theta - \frac{\gamma_{xy}}{2} \sin 2\theta. \tag{8.8}$$

We determine the shear strain $\gamma_{x'y'}$ by using Eq. (8.7). Instead of using this equation to express the normal strain $\varepsilon_{x'}$ in terms of the strains ε_x, ε_y, and γ_{xy}, we can reverse its role and use it to express the normal strain ε_x in terms of the strains $\varepsilon_{x'}$, $\varepsilon_{y'}$ and $\gamma_{x'y'}$ simply by replacing θ by $-\theta$. The result is

$$\varepsilon_x = \frac{\varepsilon_{x'} + \varepsilon_{y'}}{2} + \frac{\varepsilon_{x'} - \varepsilon_{y'}}{2} \cos 2\theta - \frac{\gamma_{x'y'}}{2} \sin 2\theta.$$

Substituting Eqs. (8.7) and (8.8) into this equation, we can solve for $\gamma_{x'y'}$ in terms of ε_x, ε_y, and γ_{xy}:

$$\frac{\gamma_{x'y'}}{2} = -\frac{\varepsilon_x - \varepsilon_y}{2} \sin 2\theta + \frac{\gamma_{xy}}{2} \cos 2\theta. \tag{8.9}$$

Compare Eqs. (8.7), (8.8), and (8.9) to the transformation equations for plane stress, Eqs. (7.6), (7.7), and (7.8). They are identical in form, with the normal stress replaced by the normal strain and the shear stress replaced by one-half the shear strain. The state of stress and the state of strain with the shear strains γ_{xy}, γ_{yz}, and γ_{xz} replaced by $\gamma_{xy}/2$, $\gamma_{yz}/2$, and $\gamma_{xz}/2$ are both quantities called *tensors*. Although a complete discussion of tensors is beyond our scope, these examples demonstrate how components of tensors transform between coordinate systems. From our present point of view, the similarities of these equations mean that the analysis of strains follows the same path we used for stresses, and the results will therefore be very familiar.

Although we have derived the strain transformation equations under the assumption of a state of plane strain, we will show in Sect. 8.4 that for many materials they also apply to the components of strain resulting from a state of plane stress. The strain component ε_z is not generally zero in plane stress, but that does not affect the derivations of Eqs. (8.7), (8.8), and (8.9).

Study Questions

1. What is a state of plane strain?
2. Do Eqs. (8.7)–(8.9) hold for any values of the components of plane strain ε_x, ε_y, and γ_{xy}? Explain.
3. What is the definition of the term $\gamma_{x'y'}$ in Eq. (8.9)?
4. Equations (8.7)–(8.9) hold for a state of plain strain. Do they also apply when ε_x, ε_y, and γ_{xy} are components of strain resulting from a state of plane stress?

Example 8.1 Transformation of Strain Components
The components of plane strain at point p of the material shown in Fig. 8.7 are $\varepsilon_x = 0.003$, $\varepsilon_y = 0.001$, and $\gamma_{xy} = -0.006$. Determine the components of plane strain in terms of the $x'y'$ coordinate system.

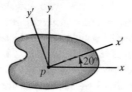

Fig. 8.7

Strategy
We can use Eqs. (8.7)–(8.9) to determine the strain components $\varepsilon_{x'}$, $\varepsilon_{y'}$ and $\gamma_{x'y'}$.

Solution
The components of plane strain in terms of the $x'y'$ coordinate system are

$$\varepsilon_{x'} = \frac{\varepsilon_x + \varepsilon_y}{2} + \frac{\varepsilon_x - \varepsilon_y}{2}\cos 2\theta + \frac{\gamma_{xy}}{2}\sin 2\theta$$

$$= \frac{0.003 + 0.001}{2} + \frac{0.003 - 0.001}{2}\cos 2(20°) + \frac{-0.006}{2}\sin 2(20°)$$

$$= 0.00084,$$

$$\varepsilon_{y'} = \frac{\varepsilon_x + \varepsilon_y}{2} - \frac{\varepsilon_x - \varepsilon_y}{2}\cos 2\theta - \frac{\gamma_{xy}}{2}\sin 2\theta$$

$$= \frac{0.003 + 0.001}{2} - \frac{0.003 - 0.001}{2}\cos 2(20°) - \frac{-0.006}{2}\sin 2(20°)$$

$$= 0.00316,$$

$$\gamma_{x'y'} = 2\left(-\frac{\varepsilon_x - \varepsilon_y}{2}\sin 2\theta + \frac{\gamma_{xy}}{2}\cos 2\theta\right)$$

$$= -(0.003 - 0.001)\sin 2(20°) + (-0.006)\cos 2(20°)$$

$$= -0.00588.$$

Discussion

While calculations of this kind appear rather dry, remember that the results have physical meanings. If a material is subjected to the state of strain given in this example at a point p, our results determine the strain of the material (how much it stretches or contracts) in the x' and y' directions. They also determine the change in the angle between line elements aligned in the x' and y' directions. Thus they provide additional information about how the material deforms.

8.2.1 Strain Gauge Rosette

Before continuing our discussion of the analysis of strains, we will describe an interesting and important application of the strain transformation equations. The term *strain gauge* refers to an instrument for measuring strains. The type we are concerned with here, called a *resistance strain gauge*, is based on the observation that the electrical resistance of a wire varies when the wire is subjected to axial strain (Fig. 8.8). Once the relationship between the strain of a given wire and its electrical resistance has been established experimentally (a procedure called *calibration*), the axial strain of the wire can be determined by measuring its resistance, which can be done very accurately.

To use the calibrated wire as a strain gauge, it is bonded to the surface of an unloaded specimen. When the specimen is loaded, the strain of the specimen in the direction of the wire is determined by measuring the wire's resistance. (The wire must be sufficiently thin that the force the strained wire exerts on the specimen is negligible.) The wire is typically arranged in the pattern shown in Fig. 8.9 to minimize the size of the gauge and so measure the strain within a relatively small neighborhood of a point.

Normal strain can be measured by the type of strain gauge we have described, but how can shear strain be measured? This can be done in a clever way by a *strain gauge rosette*, which consists of three strain gauges measuring normal strain in three directions, as shown in Fig. 8.11(a). [The term *rosette* is said to derive from the resemblance between strain gauge rosettes mounted on colored felt and a small cloth ornament called a rosette that was worn on hats around the time of the French Revolution (Fig. 8.10).] Small gauges placed close together are typically used so that they measure the strains in a relatively small neighborhood of a chosen point. We introduce a coordinate system and denote the directions of the strain gauges by θ_a, θ_b, and θ_c (Fig. 8.11(b)). Using Eq. (8.6) to express the normal strains in the directions of the three strain gauges in terms of the strain components ε_x, ε_y, and γ_{xy}, we obtain

Fig. 8.8 The electrical resistance R of a wire is a function of its axial strain ε

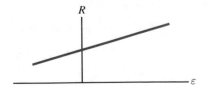

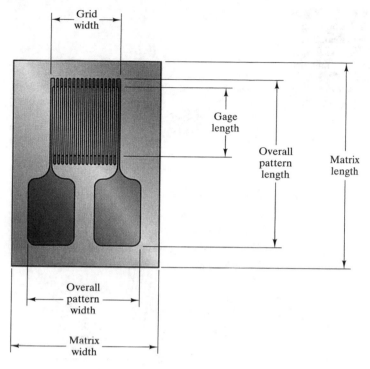

Fig. 8.9 Typical resistance strain gauge

Fig. 8.10 The original
rosette (*Image from iStock
by tatadonets*)

$$\varepsilon_a = \varepsilon_x \cos^2 \theta_a + \varepsilon_y \sin^2 \theta_a + \gamma_{xy} \sin \theta_a \cos \theta_a,$$
$$\varepsilon_b = \varepsilon_x \cos^2 \theta_b + \varepsilon_y \sin^2 \theta_b + \gamma_{xy} \sin \theta_b \cos \theta_b, \qquad (8.10)$$
$$\varepsilon_c = \varepsilon_x \cos^2 \theta_c + \varepsilon_y \sin^2 \theta_c + \gamma_{xy} \sin \theta_c \cos \theta_c.$$

By measuring the normal strains ε_a, ε_b, and ε_c, this system of equations can be
solved for the strain components ε_x, ε_y, and γ_{xy}, thus determining the shear strain.

Fig. 8.11 (a) Strain gauge
rosette. (b) Introducing a
coordinate system. The
angles θ_a, θ_b, and θ_c specify
the directions of the three
strain gauges

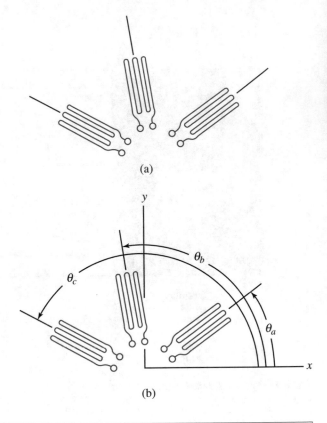

(a)

(b)

Example 8.2 Strain Gauge Rosette
The strains measured by a strain gauge rosette oriented as shown in Fig. 8.12
are $\varepsilon_a = 0.004$, $\varepsilon_b = -0.003$, and $\varepsilon_c = 0.002$. Determine the components of
strain ε_x, ε_y, and γ_{xy}.

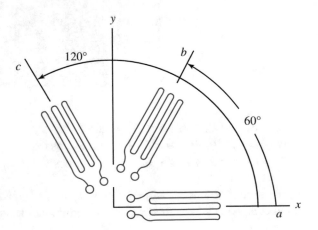

Fig. 8.12

Strategy

The angles specifying the directions of the strain gauges relative to the x axis are $\theta_a = 0$, $\theta_b = 60°$, and $\theta_c = 120°$. By substituting these values and the values of the strains $\varepsilon_a, \varepsilon_b$, and ε_c into Eq. (8.10), we can solve for ε_x, ε_y, and γ_{xy}.

Solution

Equation (8.10) is

$$0.004 = \varepsilon_x,$$
$$-0.003 = \varepsilon_x \cos^2 60° + \varepsilon_y \sin^2 60° + \gamma_{xy} \sin 60° \cos 60°,$$
$$0.002 = \varepsilon_x \cos^2 120° + \varepsilon_y \sin^2 120° + \gamma_{xy} \sin 120° \cos 120°.$$

Solving, we obtain $\varepsilon_x = 0.00400$, $\varepsilon_y = -0.00200$, and $\gamma_{xy} = -0.00577$.

Discussion

Have you thought about how the components of stress at a point of a structural element could be measured? As we demonstrate in this example, a strain gauge rosette can be used to determine the components of plane strain at a point. In Sect. 2.2 we showed that for typical structural materials, the state of stress at a point can be determined if the state of strain is known.

8.2.2 Maximum and Minimum Strains

Given a state of plane strain at a point p, Eqs. (8.7), (8.8), and (8.9) determine the strain components $\varepsilon_{x'}, \varepsilon_{y'}$, and $\gamma_{x'y'}$ corresponding to the coordinate system shown in Fig. 8.5 for any value of θ. In this section we consider the following questions: For what values of θ are the normal strain $\varepsilon_{x'}$ and shear strain $\gamma_{x'y'}$ a maximum or minimum, and what are their values? In other words, what are the maximum and minimum normal and shear strains in the x-y plane?

8.2.2.1 Principal Strains

Let a value of θ for which $\varepsilon_{x'}$ is a maximum or minimum be denoted by θ_p. By evaluating the derivative of Eq. (8.7) with respect to 2θ and setting it equal to zero, we obtain the equation

$$\tan 2\theta_p = \frac{\gamma_{xy}}{\varepsilon_x - \varepsilon_y}. \tag{8.11}$$

When ε_x, ε_y, and γ_{xy} are known, we can solve this equation for θ_p and substitute it into Eq. (8.7) to determine the maximum or minimum value of $\varepsilon_{x'}$. Just as in the case of the maximum and minimum normal stresses, if $2\theta_p$ is a solution of Eq. (8.11), then so is $2\theta_p + 180°$. This means that the normal strain is a maximum or minimum in the direction θ_p and also in the direction $\theta_p + 90°$, which means that *the maximum and minimum normal strains occur in the x' and y' axis directions*. Once we have determined θ_p, the maximum and minimum normal strains in the x-y plane, called the *principal strains* and denoted ε_1 and ε_2, can be determined from Eqs. (8.7) and (8.8). *There are no normal strains of greater magnitude in any direction.*

To obtain analytical expressions for the values of the principal strains, we solve the equations

$$\frac{\sin 2\theta_p}{\cos 2\theta_p} = \tan 2\theta_p = \frac{\gamma_{xy}}{\varepsilon_x - \varepsilon_y} \tag{8.12}$$

and

$$\sin^2 2\theta_p + \cos^2 2\theta_p = 1 \tag{8.13}$$

for $\sin 2\theta_p$ and $\cos 2\theta_p$ and substitute the results into Eq. (8.7), obtaining

$$\varepsilon_1, \varepsilon_2 = \frac{\varepsilon_x + \varepsilon_y}{2} \pm \sqrt{\left(\frac{\varepsilon_x - \varepsilon_y}{2}\right)^2 + \left(\frac{\gamma_{xy}}{2}\right)^2}. \tag{8.14}$$

By substituting the expressions for $\sin 2\theta_p$ and $\cos 2\theta_p$ into Eq. (8.9), we find that *the value of the shear strain $\gamma_{x'y'}$ at $\theta = \theta_p$ is zero*. The physical interpretation of this result is interesting: An infinitesimal square element oriented as shown in Fig. 8.13(a) is subjected to the principal strains in the x' and y' directions and undergoes no shear strain. The element is rectangular in the deformed state (Fig. 8.13(b)).

8.2.2.2   Maximum Shear Strains

Let a value of θ for which $\gamma_{x'y'}$ is a maximum or minimum be denoted by θ_s. Evaluating the derivative of Eq. (8.9) with respect to 2θ and setting it equal to zero, we obtain the equation

$$\tan 2\theta_s = -\frac{\varepsilon_x - \varepsilon_y}{\gamma_{xy}}. \tag{8.15}$$

With this equation we can determine θ_s and substitute it into Eq. (8.9) to obtain the maximum or minimum value of $\gamma_{x'y'}$. If θ_s is a solution of Eq. (8.15), then so are $\theta_s + 90°$, $\theta_s + 2(90°)$, ... As we illustrate in Fig. 8.14, *the maximum and minimum shear strains describe the shear strain of a rectangular element*. Furthermore, because $\tan 2\theta_s$ is the negative inverse of $\tan 2\theta_p$, this element is rotated $45°$ relative to the element which is subjected to the maximum and minimum normal strains.

Fig. 8.13 (a) Reference state of a square element aligned with the directions of the principal strains. (b) The deformed element undergoes no shear strain (The strains are exaggerated. The largest principal strain may occur in either the x' or the y' direction)

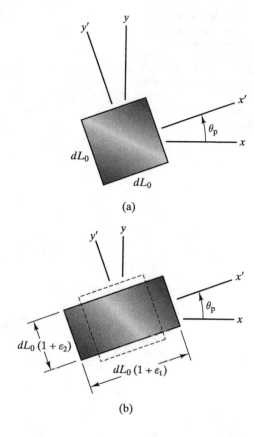

We can demonstrate why this is the case by superimposing these two elements (Fig. 8.15).

To obtain an analytical expression for the maximum magnitude of the shear strain, we solve the equations

$$\frac{\sin 2\theta_s}{\cos 2\theta_s} = \tan 2\theta_s = -\frac{\varepsilon_x - \varepsilon_y}{\gamma_{xy}} \tag{8.16}$$

and

$$\sin^2 2\theta_s + \cos^2 2\theta_s = 1 \tag{8.17}$$

for $\sin 2\theta_s$ and $\cos 2\theta_s$ and substitute the results into Eq. (8.9). The result is

$$\gamma_{max} = \sqrt{(\varepsilon_x - \varepsilon_y)^2 + \gamma_{xy}^2}. \tag{8.18}$$

Fig. 8.14 (a) Reference
state of the square element
that is subjected to the
maximum and minimum
shear strains. (b) Deformed
element (The shear strain is
exaggerated. The strain $\gamma_{x'y'}$
may be either positive or
negative)

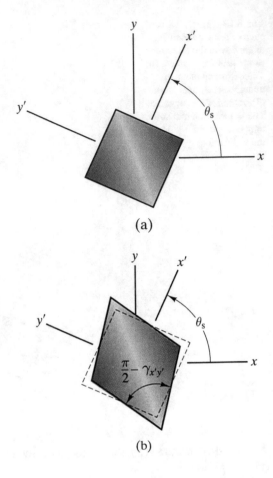

(a)

(b)

This is called the *maximum in-plane shear strain*, because it is the maximum
shear strain in the x-y plane.

By substituting our results for sin $2\theta_s$ and cos $2\theta_s$ into Eqs. (8.7) and (8.8), we
obtain

$$\varepsilon_{x'} = \varepsilon_{y'} = \frac{\varepsilon_x + \varepsilon_y}{2}.$$

The element that is subjected to the maximum and minimum shear strains
(Fig. 8.14) is subjected to equal normal strains in the x' and y' directions.

We can obtain expressions for the absolute maximum shear strain the same way
we determined the absolute maximum shear stress. We begin with the element that is
subjected to the principal strains (Fig. 8.13) and realign the coordinate system with
the faces of the element (Fig. 8.16(a)). In terms of this new coordinate system, the
components of plane strain are $\varepsilon_x = \varepsilon_1$, $\varepsilon_y = \varepsilon_2$, and $\gamma_{xy} = 0$. Substituting these

Fig. 8.15 Element subjected to the greatest shear strain superimposed onto the element subjected to the principal strains

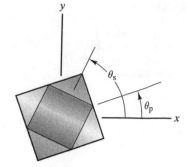

Reference state

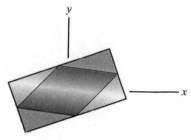

Deformed state

components into Eq. (8.18), we obtain a different expression for the magnitude of the maximum in-plane shear strain:

$$\sqrt{\left(\varepsilon_x - \varepsilon_y\right)^2 + \gamma_{xy}^2} = \sqrt{\left(\varepsilon_1 - \varepsilon_2\right)^2 + 0} = |\varepsilon_1 - \varepsilon_2|. \tag{8.19}$$

This equation is expressed in terms of the principal strains, but it gives the same result as Eq. (8.18). Now we reorient the coordinate system as shown in Fig. 8.16(b). In terms of this coordinate system, $\varepsilon_x = 0$, $\varepsilon_y = \varepsilon_1$, and $\gamma_{xy} = 0$. Substituting these components into Eq. (8.18), we obtain

$$\sqrt{\left(\varepsilon_x - \varepsilon_y\right)^2 + \gamma_{xy}^2} = \sqrt{\left(0 - \varepsilon_1\right)^2 + 0} = |\varepsilon_1|. \tag{8.20}$$

Next, we reorient the coordinate system as shown in Fig. 8.16(c). In terms of this coordinate system, $\varepsilon_x = 0$, $\varepsilon_y = \varepsilon_2$, and $\gamma_{xy} = 0$. Substituting these components into Eq. (8.18), we obtain

$$\sqrt{\left(\varepsilon_x - \varepsilon_y\right)^2 + \gamma_{xy}^2} = \sqrt{\left(0 - \varepsilon_2\right)^2 + 0} = |\varepsilon_2|. \tag{8.21}$$

Fig. 8.16 Element
subjected to the principal
strains. (**a**) Realigned
coordinate system. (**b, c**)
Other orientations of the
coordinate system

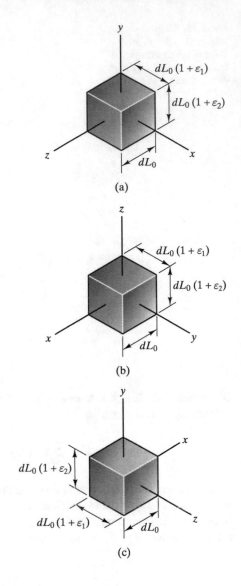

(a)

(b)

(c)

Depending on the values of the principal strains, Eq. (8.20) and/or Eq. (8.21) can result in larger values of the magnitude of the maximum shear strain than the maximum in-plane shear strain. The absolute maximum shear strain is the largest value given by Eqs. (8.19), (8.20), or (8.21):

$$\gamma_{\text{abs}} = \text{Max}(|\varepsilon_1|, |\varepsilon_2|, |\varepsilon_1 - \varepsilon_2|). \tag{8.22}$$

8.2.2.3 Summary: Determining the Principal Strains and the Maximum Shear Strain

Here we give a sequence of steps for determining the principal strains and the maximum shear strain for a given state of plane strain.

1. Use Eq. (8.11) to determine θ_p, establishing the orientation of the $x'y'$ coordinate system that is aligned with the directions of the principal strains.
2. Determine ε_1 and ε_2 and the directions in which they act by substituting θ_p into Eqs. (8.7) and (8.8). You can also determine the values of ε_1 and ε_2 from Eq. (8.14), but this equation does not tell you which principal strain acts in the x' direction and which one acts in the y' direction.
3. Use Eq. (8.15) to determine θ_s, establishing directions of the x' and y' axes for which $\gamma_{x'y'}$ is a maximum or minimum. Alternatively, you can simply use $\theta_s = \theta_p + 45°$.
4. Determine $\gamma_{x'y'}$ by substituting θ_s into Eq. (8.9). The magnitude of this shear strain is the maximum in-plane shear strain. You can also determine the maximum in-plane shear strain from Eq. (8.18).
5. The absolute maximum shear strain is given by Eq. (8.22).

Study Questions

1. What are the principal strains?
2. If you know the direction of one of the principal strains, what do you know about the direction of the other principal strain?
3. What is the shear strain of the element that is subjected to the principal strains?
4. If you know the orientation of the element that is subjected to the principal strains, what do you know about the orientation of the element that is subjected to the maximum in-plane shear strain?

Example 8.3 Determining Maximum and Minimum Strains
The state of plane strain at a point p is $\varepsilon_x = 0.003$, $\varepsilon_y = 0.001$, and $\gamma_{xy} = -0.006$. Determine the principal strains and the maximum in-plane shear strain, and show the orientations of the elements subjected to these strains. Also determine the absolute maximum shear strain.

Strategy
We can follow the steps given in the preceeding summary to determine the principal strains and the maximum shear strain.

Solution
Step 1 From Eq. (8.11)

$$\tan 2\theta_p = \frac{\gamma_{xy}}{\varepsilon_x - \varepsilon_y} = \frac{-0.006}{0.003 - 0.001} = -3.0.$$

Solving this equation, we obtain $\theta_p = -35.78°$. This angle tells us the orientation of the $x'\,y'$ coordinate system aligned with the principal strains.

Step 2   We substitute θ_p into Eqs. (8.7) and (8.8) to determine the principal strains.

$$\varepsilon_{x'} = \frac{\varepsilon_x + \varepsilon_y}{2} + \frac{\varepsilon_x - \varepsilon_y}{2}\cos 2\theta_p + \frac{\gamma_{xy}}{2}\sin 2\theta_p$$

$$= \frac{0.003 + 0.001}{2} + \frac{0.003 - 0.001}{2}\cos 2\left(-35.78°\right) + \frac{-0.006}{2}\sin 2\left(-35.78°\right)$$

$$= 0.00516,$$

$$\varepsilon_{y'} = \frac{\varepsilon_x + \varepsilon_y}{2} - \frac{\varepsilon_x - \varepsilon_y}{2}\cos 2\theta_p - \frac{\gamma_{xy}}{2}\sin 2\theta_p$$

$$= \frac{0.003 + 0.001}{2} - \frac{0.003 - 0.001}{2}\cos 2\left(-35.78°\right) - \frac{-0.006}{2}\sin 2\left(-35.78°\right)$$

$$= -0.00116.$$

The principal strains are $\varepsilon_1 = 0.00516$ and $\varepsilon_2 = -0.00116$. They are shown on the properly oriented element in Fig. (a). This element is subjected to no shear strain.

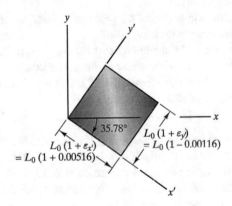

(a) The principal strains. L_0 is the dimension of the square element in the reference state

Step 3   From Eq. (8.15)

$$\tan\ 2\theta_s = -\frac{\varepsilon_x - \varepsilon_y}{\gamma_{xy}} = -\frac{0.003 - 0.001}{-0.006} = 0.333,$$

from which we obtain $\theta_s = 9.22°$.

Step 4   We substitute θ_s into Eq. (8.9) to determine the maximum in-plane shear strain.

$$\frac{\gamma_{x'y'}}{2} = -\frac{\varepsilon_x - \varepsilon_y}{2}\sin 2\theta_s + \frac{\gamma_{xy}}{2}\cos 2\theta_s$$

$$= -\frac{0.003 - 0.001}{2}\sin 2(9.22°) + \frac{-0.006}{2}\cos 2(9.22°)$$

$$= -0.00316.$$

We see that $\gamma_{x'y'} = -0.00632$. The maximum in-plane shear strain is shown on the properly oriented element in Fig. (b).

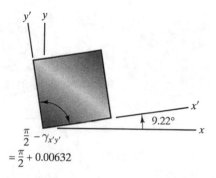

(b) Maximum in-plane shear strain. Notice that $\gamma_{x'y'}$ is negative.

Step 5   The absolute maximum shear strain is the largest of the three values

$$|\varepsilon_1 - \varepsilon_2| = |0.00516 - (-0.00116)| = 0.00632,$$
$$|\varepsilon_1| = |0.00516| = 0.00516,$$
$$|\varepsilon_2| = |-0.00116| = 0.00116.$$

In this example the absolute maximum shear strain equals the magnitude of the maximum in-plane shear strain, 0.00632.

Discussion
You should keep reminding yourself of the physical meanings of results such as those in this example. Suppose that before the given state of strain was applied, you could draw a small square at point p ("small" means small enough so that the state of strain would be approximately homogeneous throughout the square) that had the orientation shown in Fig. (a). When the state of strain was applied, that square would be deformed into a rectangle (because it is subjected to no shear strain), and the lengths of its sides would be given by the results shown in Fig. (a).

Problems

8.1 Perpendicular line elements of lengths dL_1 and dL_2 at a point p of a reference state of a material are parallel to the y and z axes, respectively. In a deformed state, the lengths of the elements are $dL_1' = 1.002dL_1$ and $dL_2' = 1.003dL_2$, and the angle between them is $90.15°$. With this information, which components of the state of strain at p can you determine, and what are their values?

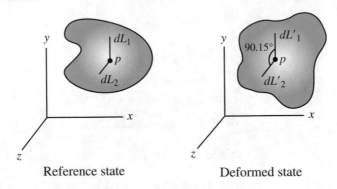

Problem 8.1

8.2 Perpendicular line elements of lengths dL_1 and dL_2 at a point p of a reference state of a material are parallel to the x and z axes, respectively. In a deformed state, the lengths of the elements are $dL_1' = 0.996dL_1$ and $dL_2' = 1.0015dL_2$, and the angle between them is $89.86°$. With this information, which components of the state of strain at p can you determine, and what are their values?

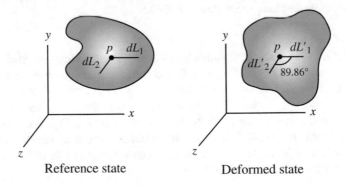

Problem 8.2

8.3 An element of a material is shown before, and after it is subjected to a state of homogeneous plane strain. The x, y coordinates of points P and Q in the deformed state are $(200.2, 0.4)$ mm and $(-0.2, 199.9)$ mm, respectively. What are the values of the normal strains ε_x and ε_y in the deformed state?

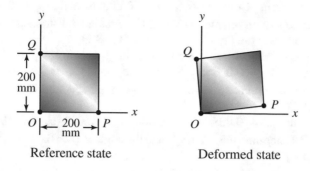

Reference state   Deformed state

Problems 8.3–8.4

8.4 An element of a material is shown before, and after it is subjected to a state of homogeneous plane strain. The x, y coordinates of points P and Q in the deformed state are $(200.2, 0.4)$ mm and $(-0.2, 199.9)$ mm, respectively. What is the value of the shear strain γ_{xy} in the deformed state?

8.5 The components of plane strain at point p are $\varepsilon_x = 0.003$, $\varepsilon_y = 0$, and $\gamma_{xy} = 0$. If $\theta = 45°$, what are the strains $\varepsilon_{x'}, \varepsilon_{y'}$, and $\gamma_{x'y'}$ at point p? (See the figure for Problems 8.5–8.11.)

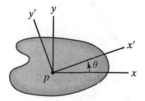

Problems 8.5–8.11

8.6* The components of plane strain at a point of a machine tool during a milling operation are $\varepsilon_x = 0.00400$, $\varepsilon_y = -0.00300$, and $\gamma_{xy} = 0.00600$, and the components referred to the $x'y'z'$ coordinate system are $\varepsilon_{x'} = 0.00125, \varepsilon_{y'} = -0.00025$, and $\gamma_{x'y'} = 0.00910$. What is the angle θ? (See the figure for Problems 8.5–8.11.)

8.7* The components of plane strain at point p are $\varepsilon_x = 0.0024$, $\varepsilon_y = -0.0012$, and $\gamma_{xy} = 0.0048$. The normal strains $\varepsilon_{x'} = 0.00347$ and $\varepsilon_{y'} = -0.00227$. Determine $\gamma_{x'y'}$ and the angle θ. (See the figure for Problems 8.5–8.11.)

8.8 At a point of a loading crane, measurements indicate that the components of plane strain $\varepsilon_{x'} = 0.00665$, $\varepsilon_{y'} = 0.00825$, and $\gamma_{x'y'} = 0.00135$ for a coordinate system oriented at $\theta = 20°$. What are the strains ε_x, ε_y, and γ_{xy} at that point? (See the figure for Problems 8.5–8.11.)

8.9 The components of plane strain at point p referred to the $x'y'$ coordinate system are $\varepsilon_{x'} = 0.0066$, $\varepsilon_{y'} = -0.0086$, and $\gamma_{x'y'} = 0.0028$. If $\theta = 20°$, what are the strains ε_x, ε_y, and γ_{xy} at point p? (See the figure for Problems 8.5–8.11.)

8.10* The strains $\varepsilon_x = 0.008$, $\varepsilon_y = -0.006$, $\gamma_{xy} = 0.024$, $\varepsilon_{x'} = 0.014$, and $\varepsilon_{y'} = -0.012$. Determine the strain $\gamma_{x'y'}$ and the angle θ. (See the figure for Problems 8.5–8.11.)

8.11 The strains $\varepsilon_x = 0.008$, $\varepsilon_y = -0.006$, and $\gamma_{x'y'} = 0.024$, and the angle $\theta = 35°$. Determine $\varepsilon_{x'}$, $\varepsilon_{y'}$, and γ_{xy}. (See the figure for Problems 8.5–8.11.)

8.12 Points P and Q are 1 mm apart in the reference state of a material. If the material is subjected to the homogeneous state of plane strain $\varepsilon_x = 0.003$, $\varepsilon_y = -0.002$, and $\gamma_{xy} = -0.006$, what is the distance between points P and Q in the deformed material?

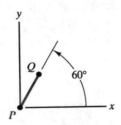

Problem 8.12

8.13 Two points P and Q of the pliers are 2 mm apart when the pliers are unstressed. Under particular loads, the material containing these points is subjected to a homogeneous state of plane stress and the strain components $\varepsilon_x = 0.008$, $\varepsilon_y = 0.002$, and $\gamma_{xy} = -0.003$. What is the distance between points P and Q?

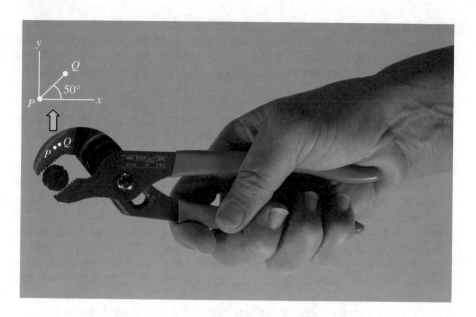

Problems 8.13–8.14

8.14 The points P and Q of the pliers are 2 mm apart when the pliers are unstressed. Under particular loads, they are 1.992 mm apart. The material is in plane stress. If $\varepsilon_x = -0.0088$ and $\varepsilon_y = 0.0024$, what is γ_{xy}?

8.15 The material containing points O, P, and Q is subjected to the state of homogeneous plane strain $\varepsilon_x = 0.00665$, $\varepsilon_y = 0.00825$, and $\gamma_{xy} = 0.00135$. (a) What are the distances between points O and Q and between points O and P in the deformed material? (b) What is the angle between the lines OQ and OP in the deformed material? (See the figure for Problems 8.15–8.19.)

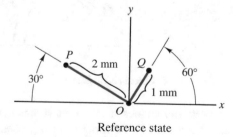

Reference state

Problems 8.15–8.19

8.16 After the material containing points O, P, and Q is subjected to a homogeneous state of strain, points O and Q are 1.0045 mm apart, points O and P are 2.0032 mm apart, and the angle between the lines OQ and OP is 89.426°. What are the strain components ε_x, ε_y, and γ_{xy}? (See the figure for Problems 8.15–8.19.)

8.17* If the material containing points O, P, and Q is subjected to a homogeneous state of plane strain $\varepsilon_x = 0.006$, $\varepsilon_y = 0.002$, and $\gamma_{xy} = -0.004$, what is the distance between points P and Q in the deformed material? (See the figure for Problems 8.15–8.19.)

8.18 If the material containing points O, P, and Q is subjected to a homogeneous state of plane strain $\varepsilon_x = 0.016$, $\varepsilon_y = -0.004$, and $\gamma_{xy} = 0.008$, what is the angle between the lines OQ and OP in the deformed material? (See the figure for Problems 8.15–8.19.)

8.19* After the material containing points O, P, and Q is subjected to a homogeneous state of strain, points O and Q are 1.002 mm apart, points O and P are 1.998 mm apart, and points P and Q are 2.242 mm apart. What are the strain components ε_x, ε_y, and γ_{xy}? (See the figure for Problems 8.15–8.19.)

8.20 The strain gauge rosette shown is mounted on an object that is subjected to the components of strain $\varepsilon_x = 0.0045$, $\varepsilon_y = 0.0016$, and $\gamma_{xy} = -0.002$. What normal strains would be measured by the strain gauges a, b, and c?

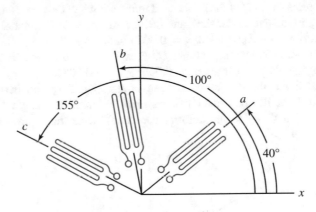

Problems 8.20–8.21

8.21 The strain gauge rosette shown measures normal strains $\varepsilon_a = 0$, $\varepsilon_b = -0.0048$, and $\varepsilon_c = 0.0072$. What are the strain components ε_x, ε_y, and γ_{xy}?

8.22 A bar is subjected to axial forces. The strains measured by a strain gauge rosette oriented as shown are $\varepsilon_a = 0.003$, $\varepsilon_b = 0.001$, and $\varepsilon_c = -0.001$. What are the strain components ε_x, ε_y, and γ_{xy}?

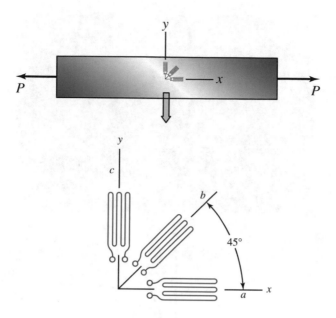

Problem 8.22

8.23 The strains measured by a strain gauge rosette mounted on the bicycle brake are $\varepsilon_a = 0.00220$, $\varepsilon_b = -0.00100$, and $\varepsilon_c = -0.00360$. Determine the strains ε_x, ε_y, and γ_{xy}.

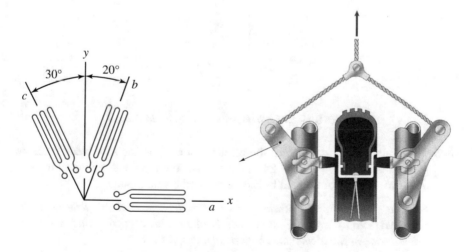

Problem 8.23

8.24 A strain gauge mounted on the blade of a saw measures the normal strains $\varepsilon_a = 0.0080$, $\varepsilon_b = 0.0024$, and $\varepsilon_c = -0.0030$. What is the normal strain in the direction of the line d?

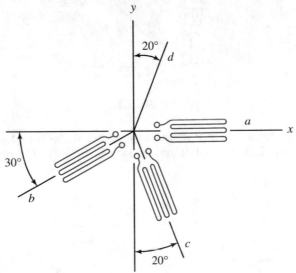

Problems 8.24–8.25 (*Photograph from iStock by Bilanol*)

8.25 The strain gauges a and b of the rosette mounted on the saw blade measure $\varepsilon_a = 0.0045$ and $\varepsilon_b = -0.0024$. If the normal strain in the direction of the line d is zero, what strain would be measured by the strain gauge c?

For the states of plane strain given in Problems 8.26–8.29, determine the principal strains and the maximum in-plane shear strain, and show the orientations of the elements subjected to these strains.

8.26 $\varepsilon_x = 0.002$, $\varepsilon_y = 0.001$, $\gamma_{xy} = 0$.
8.27 $\varepsilon_x = 0.0025$, $\varepsilon_y = 0$, $\gamma_{xy} = -0.0050$.
8.28 $\varepsilon_x = -0.008$, $\varepsilon_y = 0.006$, $\gamma_{xy} = -0.012$.
8.29 $\varepsilon_x = 0.0024$, $\varepsilon_y = -0.0012$, $\gamma_{xy} = 0.0024$.

8.30 A point p of the bearing's housing is subjected to a state of plane stress, and the strain components $\varepsilon_x = -0.0024$, $\varepsilon_y = 0.0044$, and $\gamma_{xy} = -0.0030$. Determine the principal strains and the magnitude of the maximum in-plane shear strain.

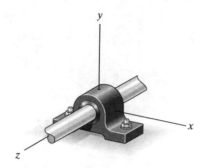

Problem 8.30

8.31 For the state of plane strain $\varepsilon_x = 0.002$, $\varepsilon_y = 0.001$, and $\gamma_{xy} = 0$, what is the absolute maximum shear strain?

8.32 For the state of plane strain $\varepsilon_x - 0.0024$, $\varepsilon_y - 0.0012$, and $\gamma_{xy} - 0.0024$, what is the absolute maximum shear strain?

8.33 A point of the MacPherson strut suspension is subjected to the state of plane strain $\varepsilon_x = -0.0088$, $\varepsilon_y = 0.0024$, and $\gamma_{xy} = -0.0036$. Determine the principal strains and the absolute maximum shear strain.

Problem 8.33

8.34 By setting θ equal to $\theta + 90°$ in Eq. (8.7), derive Eq. (8.8).

8.35∗ Consider an infinitesimal element of material that in the reference state has the triangular shape shown in Fig. (a). In the deformed state (Fig. (b)), the lengths of the sides are known in terms of the normal strains ε_x, $\varepsilon_{x'}$, and $\varepsilon_{y'}$, and the angle between the sides that were perpendicular in the reference state can be expressed in terms of $\gamma_{x'y'}$. By analyzing the deformed triangle, show that

$$\varepsilon_x = \varepsilon_{x'} \cos^2\theta + \varepsilon_{y'} \sin^2\theta - \gamma_{x'y'} \sin\theta \cos\theta.$$

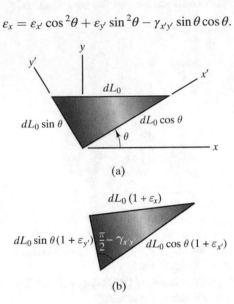

(a)

(b)

Problem 8.35

(In doing so, assume that products of the strains are negligible.)

8.36∗ A circular line in the x-y plane of circumference $C = 2\pi R$ is drawn in the reference state of a material. If the material is then subjected to a homogeneous strain ε_x and the other components of strain are zero, show that the length of the deformed line is $C(1 + 0.5\varepsilon_x)$.

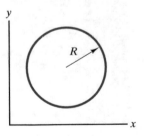

Problem 8.36

8.3 Mohr's Circle for Plane Strain

Because the equations for transforming strains between coordinate systems are so similar to the corresponding equations for stresses, we can apply Mohr's circle to strains in the same way we applied it to stresses. Recall that Eqs. (8.7), (8.8), and (8.9)

$$
\begin{aligned}
\varepsilon_{x'} &= \frac{\varepsilon_x + \varepsilon_y}{2} + \frac{\varepsilon_x - \varepsilon_y}{2} \cos 2\theta + \frac{\gamma_{xy}}{2} \sin 2\theta, \\
\varepsilon_{y'} &= \frac{\varepsilon_x + \varepsilon_y}{2} - \frac{\varepsilon_x - \varepsilon_y}{2} \cos 2\theta - \frac{\gamma_{xy}}{2} \sin 2\theta, \\
\frac{\gamma_{x'y'}}{2} &= -\frac{\varepsilon_x - \varepsilon_y}{2} \sin 2\theta + \frac{\gamma_{xy}}{2} \cos 2\theta,
\end{aligned}
\tag{8.23}
$$

are identical to Eqs. (7.6), (7.7), and (7.8) for plane stress, with the normal stresses replaced by the normal strains and the shear stress replaced by one-half the shear strain. Accounting for the factor of $\frac{1}{2}$ multiplying the shear strain, we can construct Mohr's circle for strain in exactly the same way we did for stresses.

8.3.1 Constructing the Circle

Suppose that we know the components of plane strain ε_x, ε_y, and γ_{xy} at a point p. Mohr's circle allows us to solve graphically for the components $\varepsilon_{x'}$, $\varepsilon_{y'}$, and $\gamma_{x'y'}$ for a given angle θ. This involves four steps:

1. Establish a set of horizontal and vertical axes with normal strain measured along the horizontal axis and one-half the shear strain measured along the vertical axis (Fig. 8.17(a)). Positive normal strain is measured to the right and positive shear strain is measured *downward*.
2. Plot two points, point P with coordinates $(\varepsilon_x, \ \gamma_{xy}/2)$ and point Q with coordinates $(\varepsilon_y, -\gamma_{xy}/2)$, as shown in Fig. 8.17(b).
3. Draw a straight line connecting points P and Q. Using the intersection of the straight line with the horizontal axis as the center, draw a circle that passes through the two points (Fig. 8.17(c)).
4. Draw a straight line through the center of the circle at an angle 2θ measured counterclockwise from point P (Fig. 8.17(d)). The point P' at which this line intersects the circle has coordinates $(\varepsilon_{x'}, \gamma_{x'y'}/2)$, and the point Q' has coordinates $(\varepsilon_{y'}, -\gamma_{x'y'}/2)$.

8.3.2 Determining Maximum and Minimum Strains

Once we have constructed Mohr's circle, we can immediately see the values of the maximum and minimum normal strains and the magnitude of the maximum in-plane

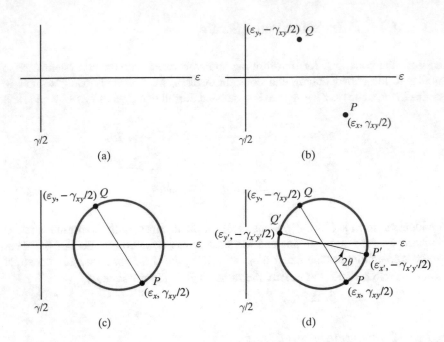

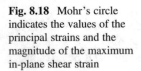

Fig. 8.17 (a) Establishing the axes. Shear strain is positive downward. (b) Plotting points P and Q. (c) Drawing Mohr's circle. The center of the circle is the intersection of the line between points P and Q with the horizontal axis. (d) Determining the strains

Fig. 8.18 Mohr's circle indicates the values of the principal strains and the magnitude of the maximum in-plane shear strain

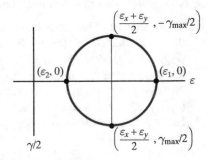

shear strain. The coordinates of the points where the circle intersects the horizontal axis determine the two principal strains (Fig. 8.18). The coordinates of the points at the bottom and top of the circle determine the maximum and minimum in-plane shear strains, so the radius of the circle equals one-half the magnitude of the maximum in-plane shear strain.

We can also use Mohr's circle to determine the orientations of the elements subjected to the principal strains and the maximum in-plane shear strain. By letting the point P' coincide with either principal strain (Fig. 8.19(a)), we can measure the angle $2\theta_p$ and thereby determine the orientation of the x' and y' directions in which the principal strains occur (Fig. 8.19(b)).

Fig. 8.19 Using Mohr's circle to determine the orientation of the element subjected to the principal strains

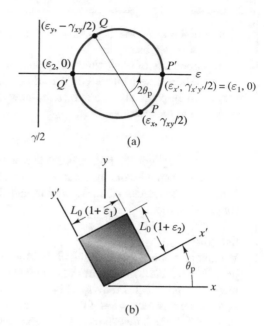

(a)

(b)

Fig. 8.20 Using Mohr's circle to determine the orientation of the element subjected to the maximum in-plane shear strain

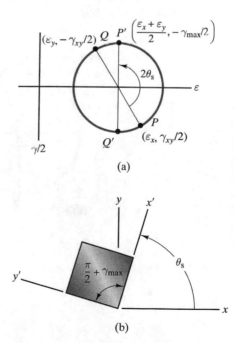

(a)

(b)

We can then let P' coincide with either the maximum or minimum shear strain (Fig. 8.20(a)) and measure the angle $2\theta_s$, determining the orientation of the x' - y' axes associated with that shear strain (Fig. 8.20(b)). Notice that in the example illustrated in Fig. 8.20, point P' coincides with the minimum shear strain, so $\gamma_{x'y'} = -\gamma_{max}$.

Example 8.4 Determining Maximum and Minimum Strains Using Mohr's Circle

The components of plane strain at a point p are $\varepsilon_x = 0.003$, $\varepsilon_y = 0.001$, and $\gamma_{xy} = -0.006$. Use Mohr's circle to determine the principal strains and the maximum in-plane shear strain, and show the orientations of the elements subjected to these strains.

Strategy

By letting the point P' of Mohr's circle coincide first with one of the principal strains and then with the maximum or minimum in-plane shear strain, we can determine the values of these strains and the orientations of the elements subjected to them.

Solution

We first plot points P and Q and draw Mohr's circle (Fig. (a)). Then we let the point P' coincide with one of the principal strains. In this case we choose the minimum principal strain ε_2 (Fig. (b)). From the circle we estimate that $\varepsilon_1 = 0.0052$ and $\varepsilon_2 = -0.0012$. Measuring the angle $2\theta_p$, we estimate that $\theta_p = 54.5°$, which determines the orientation of the element subjected to the principal strains (Fig. (c)).

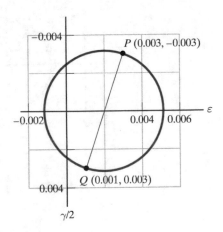

(a) Mohr's circle

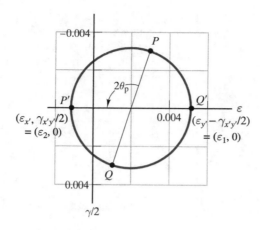

(b) Letting P' coincide with the principal strain ε_2

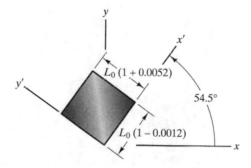

(c) Element subjected to the principal strains

Letting the point P' coincide with the minimum in-plane shear strain (Fig. (d)), we estimate that $\gamma_{\max} = 0.0062$. Measuring the angle $2\theta_s$, we estimate that $\theta_s = 9.5°$, which determines the orientation of the element subjected to the maximum and minimum in-plane shear strains (Fig. (e)).

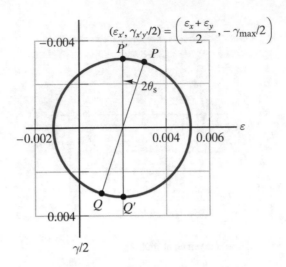

(d) Letting P' coincide with the minimum in-plane shear strain

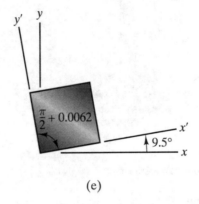

(e)

(e) Element subjected to the maximum and minimum in-plane shear strains. Notice that $\gamma_{x'y'} = -\gamma_{max}$

Discussion

Compare this solution to Example 8.3, in which we began with the same state of strain and analytically determined the principal strains and maximum in-plane shear strain.

Problems

We usually express results obtained using Mohr's circle to three significant digits. The accuracy of the results you obtain will depend of the graph paper used and the size of your figures.

8.37 The components of plane strain at point p are $\varepsilon_x = 0.003$, $\varepsilon_y = 0$, and $\gamma_{xy} = 0$, and the angle $\theta = 45°$. Use Mohr's circle to determine the strains $\varepsilon_{x'}, \varepsilon_{y'}$, and $\gamma_{x'y'}$ at point p. (See the figure for Problems 8.37–8.44.)

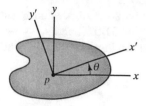

Problems 8.37–8.44

8.38 The components of plane strain at point p are $\varepsilon_x = 0$, $\varepsilon_y = 0$, and $\gamma_{xy} = 0.004$, and the angle $\theta = 45°$. Use Mohr's circle to determine the strains $\varepsilon_{x'}, \varepsilon_{y'}$, and $\gamma_{x'y'}$ at point p. (See the figure for Problems 8.37–8.44.)

8.39 The components of plane strain at point p are $\varepsilon_x = -0.0024$, $\varepsilon_y = 0.0012$, and $\gamma_{xy} = -0.0012$, and the angle $\theta = 25°$. Use Mohr's circle to determine the strains $\varepsilon_{x'}, \varepsilon_{y'}$, and $\gamma_{x'y'}$ at point p. (See the figure for Problems 8.37–8.44.)

8.40 The components of plane strain at point p are $\varepsilon_x = \varepsilon_y = \varepsilon_0$, $\gamma_{xy} = 0$. Use Mohr's circle to show that $\varepsilon_{x'} = \varepsilon_0$ and $\gamma_{x'y'} = 0$ for any value of θ. (See the figure for Problems 8.37–8.44.)

8.41 The components of plane strain at a point of a loading crane are $\varepsilon_x = 0.00400$, $\varepsilon_y = -0.00300$, and $\gamma_{xy} = 0.00600$, and the components referred to the $x'y'z'$ coordinate system are $\varepsilon_{x'} = 0.00125$, $\varepsilon_{y'} = -0.00025$, and $\gamma_{x'y'} = 0.00910$. What is the angle θ?(See the figure for Problems 8.37–8.44.)

8.42 The components of plane strain at point p are $\varepsilon_x = 0.0024$, $\varepsilon_y = -0.0012$, and $\gamma_{xy} = 0.0048$. The normal strains $\varepsilon_{x'} = 0.00347$ and $\varepsilon_{y'} = -0.00227$. Use Mohr's circle to determine $\gamma_{x'y'}$ and the angle θ. (See the figure for Problems 8.37–8.44.)

8.43 Strain gauges attached to a rocket nozzle determine that the components of plane strain $\varepsilon_{x'} = 0.00727$, $\varepsilon_{y'} = 0.00763$, and $\gamma_{x'y'} = 0.00207$ at $\theta = 40°$. Use Mohr's circle to determine the strains ε_x, ε_y, and γ_{xy} at that point. (See the figure for Problems 8.37–8.44.)

8.44 The components of plane strain at point p referred to the $x'y'$ coordinate system are $\varepsilon_{x'} = 0.0066$, $\varepsilon_{y'} = -0.0086$, and $\gamma_{x'y'} = 0.0028$, and the angle $\theta = 20°$. Use Mohr's circle to determine the strains ε_x, ε_y, and γ_{xy} at point p. (See the figure for Problems 8.37–8.44.)

8.45 A point of a material is subjected to the state of plane strain $\varepsilon_x = 0.008$, $\varepsilon_y = -0.004$, and $\gamma_{xy} = -0.006$. Use Mohr's circle to determine the principal strains and the maximum in-plane shear strain.

For the states of plane strain given in Problems 8.46–8.49, use Mohr's circle to determine the principal strains, the maximum in-plane shear strain, and the orientations of the elements subjected to these strains.

8.46 $\varepsilon_x = 0.002$, $\varepsilon_y = 0.001$, $\gamma_{xy} = 0$.

8.47 $\varepsilon_x = 0.0025$, $\varepsilon_y = 0$, $\gamma_{xy} = -0.0050$.

8.48 $\varepsilon_x = -0.008$, $\varepsilon_y = 0.006$, $\gamma_{xy} = -0.012$.

8.49 $\varepsilon_x = 0.0024$, $\varepsilon_y = -0.0012$, $\gamma_{xy} = 0.0024$.

8.50 A point of a MacPherson strut suspension is subjected to the state of plane strain $\varepsilon_x = -0.0088$, $\varepsilon_y = 0.0024$, and $\gamma_{xy} = -0.0036$. Use Mohr's circle to determine the principal strains and the magnitude of the maximum in-plane shear strain.

Problem 8.50

8.4   Stress-Strain Relations

In Sect. 2.2 we stated the equations that relate the components of stress to the components of strain for the model of material behavior called isotropic linear elasticity (Eqs. (2.7)–(2.12)). In this section we discuss why these equations have the forms they do and explain the concept of material isotropy.

We have discussed the state of stress, which is related to the internal forces at a material point, and the state of strain, which is related to the deformation in the neighborhood of a material point. We now want to address the question of whether there is a relationship between them. Consider the example of a crystalline material, such as iron, that can be modeled as a lattice of atoms connected by bonds that behave like springs. When such a material is deformed, the distances between atoms change, and the "springs" become stretched or compressed, altering the forces they exert. Because the internal forces near a point will depend only on the changes in the distances between atoms near that point, it is reasonable to postulate that the state of stress depends only on the state of strain. This is a very simple conceptual model of material behavior, and in general the state of stress at a point of a material can depend on the temperature as well as the history of the state of strain. A material for which the state of stress at a point is a single-valued function of the current state of strain at that point is said to be *elastic*. Some materials, including the most common structural materials subjected to limited ranges of temperature and stress, can be modeled adequately by assuming them to be elastic. Our objective is to derive the equations relating the state of stress in an elastic material to its state of strain for elastic materials subjected to small strains.

8.4.1 Linear Elastic Materials

An elastic material is defined to be one for which the state of stress at a point is a function only of the current state of strain at that point. We can therefore express each component of stress as a function of the components of strain:

$$\sigma_x = \sigma_x\left(\varepsilon_x, \varepsilon_y, \varepsilon_z, \gamma_{xy}, \gamma_{yz}, \gamma_{xz}\right),$$
$$\sigma_y = \sigma_y\left(\varepsilon_x, \varepsilon_y, \varepsilon_z, \gamma_{xy}, \gamma_{yz}, \gamma_{xz}\right),$$
$$\sigma_z = \sigma_z\left(\varepsilon_x, \varepsilon_y, \varepsilon_z, \gamma_{xy}, \gamma_{yz}, \gamma_{xz}\right),$$
$$\tau_{xy} = \tau_{xy}\left(\varepsilon_x, \varepsilon_y, \varepsilon_z, \gamma_{xy}, \gamma_{yz}, \gamma_{xz}\right),$$
$$\tau_{yz} = \tau_{yz}\left(\varepsilon_x, \varepsilon_y, \varepsilon_z, \gamma_{xy}, \gamma_{yz}, \gamma_{xz}\right),$$
$$\tau_{xz} = \tau_{xz}\left(\varepsilon_x, \varepsilon_y, \varepsilon_z, \gamma_{xy}, \gamma_{yz}, \gamma_{xz}\right).$$

These *stress-strain relations*, which determine the state of stress in a given material in terms of its state of strain, are examples of what are called *constitutive equations*. They are functions that depend on the constitution, or physical nature, of a material. Let us express the equation for σ_x as a power series in terms of the components of strain

$$\sigma_x = a_{10} + a_{11}\varepsilon_x + a_{12}\varepsilon_y + a_{13}\varepsilon_z + a_{14}\gamma_{xy} + a_{15}\gamma_{yz} + a_{16}\gamma_{xz}$$
$$+ a_{17}\varepsilon_x^2 + a_{18}\varepsilon_x\varepsilon_y + \cdots,$$

where the coefficients $a_{10}, a_{11}, \ldots$ are constants. If we assume that the stress σ_x is zero when the components of strain are zero, the coefficient $a_{10} = 0$. If we also assume that the components of strain are sufficiently small that products of the components are negligible in comparison to the components themselves, we obtain

$$\sigma_x = a_{11}\varepsilon_x + a_{12}\varepsilon_y + a_{13}\varepsilon_z + a_{14}\gamma_{xy} + a_{15}\gamma_{yz} + a_{16}\gamma_{xz}.$$

Expressing each component of stress in this way, we obtain the equations

$$\sigma_x = a_{11}\varepsilon_x + a_{12}\varepsilon_y + a_{13}\varepsilon_z + a_{14}\gamma_{xy} + a_{15}\gamma_{yz} + a_{16}\gamma_{xz},$$
$$\sigma_y = a_{21}\varepsilon_x + a_{22}\varepsilon_y + a_{23}\varepsilon_z + a_{24}\gamma_{xy} + a_{25}\gamma_{yz} + a_{26}\gamma_{xz},$$
$$\sigma_z = a_{31}\varepsilon_x + a_{32}\varepsilon_y + a_{33}\varepsilon_z + a_{34}\gamma_{xy} + a_{35}\gamma_{yz} + a_{36}\gamma_{xz},$$
$$\tau_{xy} = a_{41}\varepsilon_x + a_{42}\varepsilon_y + a_{43}\varepsilon_z + a_{44}\gamma_{xy} + a_{45}\gamma_{yz} + a_{46}\gamma_{xz}, \tag{8.24}$$
$$\tau_{yz} = a_{51}\varepsilon_x + a_{52}\varepsilon_y + a_{53}\varepsilon_z + a_{54}\gamma_{xy} + a_{55}\gamma_{yz} + a_{56}\gamma_{xz},$$
$$\tau_{xz} = a_{61}\varepsilon_x + a_{62}\varepsilon_y + a_{63}\varepsilon_z + a_{64}\gamma_{xy} + a_{65}\gamma_{yz} + a_{66}\gamma_{xz}.$$

An elastic material that can be modeled by these stress-strain relations is said to be *linear elastic*. The components of stress are linear functions of the components of strain. To model a given material, the 36 constants a_{11}, a_{12}, ... must be known. At this juncture we are apparently faced with the daunting prospect of performing 36 independent experiments to determine the stress-strain relations of a single material. But as we describe in the following section, far fewer than 36 constants must be known to characterize many linear elastic materials.

8.4.2   Isotropic Materials

The simplest way to explain what is meant by an isotropic material is to consider a familiar material that is not isotropic—wood. If we apply a normal stress σ to opposite faces of a cube of wood and measure the resulting normal strain ε, we obtain one result if the grain of the wood is parallel to the direction in which the stress is applied (Fig. 8.21a), and a different result if the grain of the wood is perpendicular to the direction in which the stress is applied (Fig. 8.21b). The behavior of the wood depends on the direction of its grain. The behavior of a material that is not isotropic depends on the orientation of the material. The behavior of a material that is *isotropic* (which, roughly translated from Greek, means "the same in all directions") does not depend on the orientation of the material. Its stress-strain relations are the same for any orientation of the material. Many materials used in engineering are approximately isotropic, although the use of intentionally created

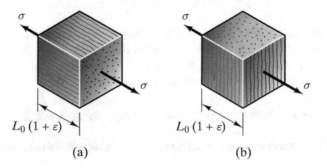

(a)   (b)

Fig. 8.21 Stretching a cube of wood: (**a**) parallel to the grain; (**b**) perpendicular to the grain

anisotropic materials, such as fiber-reinforced and layered composite materials, is increasing.

Another way to state the definition of an isotropic material is that *the stress-strain relations are the same for any orientation of the coordinate system relative to the material.* In other words, instead of requiring the material properties to be the same for different orientations of the material, we require the material properties to be the same for different orientations of the frame of reference relative to the material. Using this definition, we now investigate how material isotropy affects the stress-strain relations of a linear elastic material.

8.4.2.1 Isotropic Stress-Strain Relations

If we regard the stress-strain equations (8.24) as six linear algebraic equations for the components of strain in terms of the components of stress, we can in principle invert them to obtain linear equations for the components of strain in terms of the components of stress. We write the resulting equations as

$$
\begin{aligned}
\varepsilon_x &= b_{11}\sigma_x + b_{12}\sigma_y + b_{13}\sigma_z + b_{14}\tau_{xy} + b_{15}\tau_{yz} + b_{16}\tau_{xz}, \\
\varepsilon_y &= b_{21}\sigma_x + b_{22}\sigma_y + b_{23}\sigma_z + b_{24}\tau_{xy} + b_{25}\tau_{yz} + b_{26}\tau_{xz}, \\
\varepsilon_z &= b_{31}\sigma_x + b_{32}\sigma_y + b_{33}\sigma_z + b_{34}\tau_{xy} + b_{35}\tau_{yz} + b_{36}\tau_{xz}, \\
\gamma_{xy} &= b_{41}\sigma_x + b_{42}\sigma_y + b_{43}\sigma_z + b_{44}\tau_{xy} + b_{45}\tau_{yz} + b_{46}\tau_{xz}, \\
\gamma_{yz} &= b_{51}\sigma_x + b_{52}\sigma_y + b_{53}\sigma_z + b_{54}\tau_{xy} + b_{55}\tau_{yz} + b_{56}\tau_{xz}, \\
\gamma_{xz} &= b_{61}\sigma_x + b_{62}\sigma_y + b_{63}\sigma_z + b_{64}\tau_{xy} + b_{65}\tau_{yz} + b_{66}\tau_{xz}.
\end{aligned}
\tag{8.25}
$$

Isotropy places severe restrictions on the possible values of the constants b_{11}, b_{12}, ... For example, suppose that we subject an isotropic material to the normal stress σ shown in Fig. 8.22. In terms of the coordinate system shown, the only nonzero stress component is $\sigma_x = \sigma$. From Eq. (8.25), the strain components ε_x and ε_y are

$$
\begin{aligned}
\varepsilon_x &= b_{11}\sigma, \\
\varepsilon_y &= b_{21}\sigma.
\end{aligned}
$$

Fig. 8.22 Subjecting an isotropic material to a normal stress

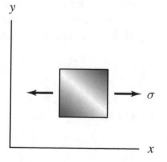

Fig. 8.23 The same state of
stress with a new coordinate
system

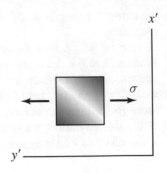

We now introduce a different coordinate system obtained by rotating the original
coordinate system 90° about its z axis (Fig. 8.23). Because the material is isotropic,
the stress-strain relations are given by Eq. (8.25) expressed in terms of the $x'y'z'$
coordinate system with the same coefficients $b_{11}, b_{12}, \ldots$. The only nonzero stress
component is $\sigma_{y'} = \sigma$, so from Eq. (8.25) expressed in terms of the $x'y'z'$ system, we
obtain

$$\varepsilon_{x'} = b_{12}\sigma,$$
$$\varepsilon_{y'} = b_{22}\sigma.$$

Because the x and y' axes of the two coordinate systems are parallel, the normal
strains ε_x and $\varepsilon_{y'}$ are equal.

$$\begin{aligned} \varepsilon_x &= \varepsilon_{y'} : \\ b_{11}\sigma &= b_{22}\sigma. \end{aligned} \tag{8.26}$$

The y and x' axes are parallel, so ε_y and $\varepsilon_{x'}$ are also equal.

$$\begin{aligned} \varepsilon_y &= \varepsilon_{x'} : \\ b_{21}\sigma &= b_{12}\sigma. \end{aligned} \tag{8.27}$$

From Eqs. (8.26) and (8.27), we see that for an isotropic material, the constants
$b_{11} = b_{22}$ and $b_{12} = b_{21}$.

Next, we subject the cube to the shear stress τ shown in Fig. 8.24(a). In terms of
the coordinate system shown, the only nonzero stress component is $\tau_{xy} = \tau$. From
Eq. (8.25), the strain component ε_x is

$$\varepsilon_x = b_{14}\tau.$$

We now introduce an $x'y'z'$ coordinate system obtained by rotating the original
coordinate system 180° about its y axis (Fig. 8.24(b)). The only nonzero stress

Fig. 8.24 Subjecting an isotropic material to a shear stress

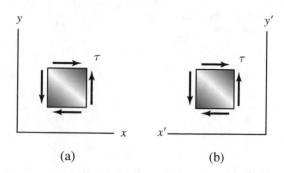

(a)   (b)

component is $\tau_{x'y'} = -\tau$, so from Eq. (8.25) expressed in terms of the $x'y'z'$ system, we obtain

$$\varepsilon_{x'} = -b_{14}\tau.$$

Because the x and x' axes of the two coordinate systems are parallel, the normal strains ε_x and $\varepsilon_{x'}$ are equal.

$$\varepsilon_x = \varepsilon_{x'} :$$
$$b_{14}\tau = -b_{14}\tau.$$

From this equation we conclude that for an isotropic material, the constant $b_{14} = 0$.

Isotropy—the assumption that the stress-strain relations for a material are identical for any orientation of the coordinate system—has forced us to conclude that the coefficients in Eq. (8.25) cannot have any values but must satisfy the restrictions $b_{11} = b_{22}, b_{12} = b_{21}$, and $b_{14} = 0$. The number of independent constants in Eq. (8.25) is reduced from 36 to 33. By continuing with arguments of the kind we used to determine these three restrictions, it can be shown that Eq. (8.25) must be of the forms

$$\varepsilon_x = b_{11}\sigma_x + b_{12}\sigma_y + b_{12}\sigma_z, \tag{8.28}$$

$$\varepsilon_y = b_{12}\sigma_x + b_{11}\sigma_y + b_{12}\sigma_z, \tag{8.29}$$

$$\varepsilon_z = b_{12}\sigma_x + b_{12}\sigma_y + b_{11}\sigma_z, \tag{8.30}$$

$$\gamma_{xy} = b_{44}\tau_{xy}, \tag{8.31}$$

$$\gamma_{yz} = b_{44}\tau_{yz}, \tag{8.32}$$

$$\gamma_{xz} = b_{44}\tau_{xz}. \tag{8.33}$$

The number of coefficients is reduced from 36 to 3. Furthermore, we can express these three coefficients in familiar terms. If we subject an isotropic material to a

Fig. 8.25 Applying a
normal stress σ_x

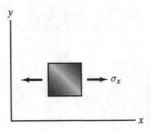

normal stress σ_x and the other components of stress are zero (Fig. 8.25), the ratio of
σ_x to the resulting normal strain ε_x is the modulus of elasticity E of the material:

$$\frac{\sigma_x}{\varepsilon_x} = E. \tag{8.34}$$

Applying Eq. (8.28) to this state of stress, we obtain

$$\varepsilon_x = b_{11}\sigma_x. \tag{8.35}$$

By comparing Eqs. (8.34) and (8.35), we determine the constant b_{11} in terms of E:

$$b_{11} = \frac{1}{E}. \tag{8.36}$$

For the state of stress shown in Fig. 8.25, the negative of the ratio of the lateral
strain (either ε_y or ε_z) to the axial strain ε_x is Poisson's ratio v of the material:

$$-\frac{\varepsilon_y}{\varepsilon_x} = v. \tag{8.37}$$

Applying Eq. (8.29) to this state of stress gives

$$\varepsilon_y = b_{12}\sigma_x. \tag{8.38}$$

Dividing this equation by Eq. (8.35), we obtain

$$\frac{\varepsilon_y}{\varepsilon_x} = \frac{b_{12}}{b_{11}},$$

and by comparing this equation to Eq. (8.37), we determine the constant b_{12} in terms
of E and v:

$$b_{12} = -vb_{11} = -\frac{v}{E}. \tag{8.39}$$

Fig. 8.26 Applying a shear stress τ_{xy}

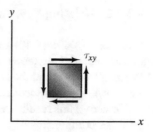

Now we subject the isotropic material to a shear stress τ_{xy} with the other components of stress equal to zero (Fig. 8.26). The ratio of τ_{xy} to the resulting shear strain γ_{xy} is the shear modulus G of the material:

$$\frac{\tau_{xy}}{\gamma_{xy}} = G. \tag{8.40}$$

From Eq. (8.31),

$$\gamma_{xy} = b_{44}\tau_{xy}, \tag{8.41}$$

and by comparing Eqs. (8.40) and (8.41), we determine the constant b_{44} in terms of G:

$$b_{44} = \frac{1}{G}. \tag{8.42}$$

Substituting the expressions (8.36), (8.39), and (8.42) into Eqs. (8.28)–(8.33), we obtain the stress-strain relations for an isotropic linear elastic material in forms in which they are commonly presented:

$$\varepsilon_x = \frac{1}{E}\sigma_x - \frac{\nu}{E}(\sigma_y + \sigma_z), \tag{8.43}$$

$$\varepsilon_y = \frac{1}{E}\sigma_y - \frac{\nu}{E}(\sigma_x + \sigma_z), \tag{8.44}$$

$$\varepsilon_z = \frac{1}{E}\sigma_z - \frac{\nu}{E}(\sigma_x + \sigma_y), \tag{8.45}$$

$$\gamma_{xy} = \frac{1}{G}\tau_{xy}, \tag{8.46}$$

$$\gamma_{yz} = \frac{1}{G}\tau_{yz}, \tag{8.47}$$

$$\gamma_{xz} = \frac{1}{G}\tau_{xz}. \tag{8.48}$$

8.4.2.2   Relating E, v, and G

Equations (8.43)–(8.48) express the stress-strain relations for an isotropic linear elastic material in terms of three coefficients, the modulus of elasticity E, the Poisson's ratio v, and the shear modulus G. *Such a material is actually characterized by only two independent constants*, because G can be expressed in terms of E and v.

To derive this result, we subject an isotropic material to a shear stress $\tau_{xy} = \tau$ and let other components of stress be zero (Fig. 8.27(a)). From Eqs. (8.43)–(8.48), the only nonzero component of strain is $\gamma_{xy} = \tau/G$. Now let us consider an $x'y'z'$ coordinate system obtained by rotating the xyz coordinate system $45°$ about the z axis (Fig. 8.27(b)). Because we know the state of stress, we can use Eqs. (5.7) and (5.9) to determine the stress components $\sigma_{x'}$ and $\sigma_{y'}$:

$$\sigma_{x'} = \tau \sin 2(45°) = \tau,$$
$$\sigma_{y'} = -\tau \sin 2(45°) = -\tau. \tag{8.49}$$

Notice that the stress components $\sigma_{z'} = \sigma_z = 0$. We also know the state of strain, so we can use Eq. (8.7) to determine the strain component $\varepsilon_{x'}$:

$$\varepsilon_{x'} = \frac{\gamma_{xy}}{2} \sin 2(45°) = \frac{1}{2G}\tau.$$

Because the material is isotropic, the strain component $\varepsilon_{x'}$ and the stress components $\sigma_{x'}, \sigma_{y'}$, and $\sigma_{z'}$ must satisfy Eq. (8.43).

$$\varepsilon_{x'} = \frac{1}{E}\sigma_{x'} - \frac{v}{E}(\sigma_{y'} + \sigma_{z'}) :$$
$$\frac{1}{2G}\tau = \frac{1}{E}\tau + \frac{v}{E}\tau.$$

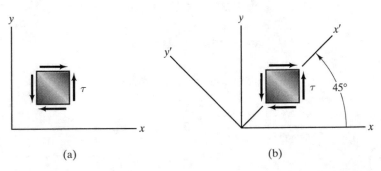

(a)   (b)

Fig. 8.27 (a) Applying a shear stress $\tau_{xy} = \tau$. (b) Introducing a rotated coordinate system $x'y'z'$

Solving this equation for G, we obtain the shear modulus in terms of the modulus of elasticity and Poisson's ratio:

$$G = \frac{E}{2(1+v)}.$$ (8.50)

8.4.2.3   Lamé Constants

Equations (8.43)–(8.48) give the components of strain in terms of the components of stress for an isotropic linear elastic material. When the components of stress are expressed in terms of the components of strain, they can be written in the forms

$$\sigma_x = (\lambda + 2\mu)\varepsilon_x + \lambda(\varepsilon_y + \varepsilon_z),$$ (8.51)

$$\sigma_y = (\lambda + 2\mu)\varepsilon_y + \lambda(\varepsilon_x + \varepsilon_z),$$ (8.52)

$$\sigma_z - (\lambda + 2\mu)\varepsilon_z + \lambda(\varepsilon_x + \varepsilon_y),$$ (8.53)

$$\tau_{xy} = \mu\gamma_{xy},$$ (8.54)

$$\tau_{yz} = \mu\gamma_{yz},$$ (8.55)

$$\tau_{xz} = \mu\gamma_{xz},$$ (8.56)

where λ and μ are called the *Lamé constants*. The constant $\mu = G$, and the constant λ is given in terms of the modulus of elasticity and Poisson's ratio by

$$\lambda = \frac{vE}{(1+v)(1-2v)}.$$ (8.57)

8.4.2.4   Bulk Modulus

Consider an element of material with volume dV_0 in the reference state (Fig. 8.28(a)). We subject the element to a pressure P, so that $\sigma_x = -P$, $\sigma_y = -P$, and $\sigma_z = -P$ (Fig. 8.28(b)). Let the volume of the deformed element be dV. The *bulk modulus K* of the material is defined to be the ratio of the normal stress $-P$ to the dilatation $e = (dV - dV_0)/dV_0$:

$$K = \frac{-P}{e}.$$ (8.58)

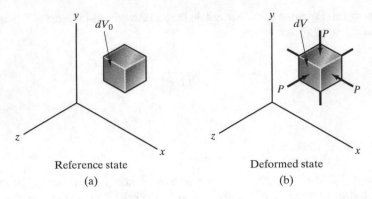

Reference state   Deformed state
(a)   (b)

Fig. 8.28 Element of material in the reference state and when subjected to a pressure p

From Eq. (8.3), the dilatation of a material subjected to small strains is

$$e = \varepsilon_x + \varepsilon_y + \varepsilon_z,$$

and from Eqs. (8.43)–(8.45), the normal strains of the element are

$$\varepsilon_x = \varepsilon_y = \varepsilon_z = \frac{1 - 2v}{E}(-P).$$

Using these relationships, we can express the bulk modulus in the form

$$K = \frac{E}{3(1 - 2v)}. \tag{8.59}$$

The bulk modulus is sometimes used instead of the modulus of elasticity E or the Lamé constant λ in expressing the stress-strain relations of an isotropic elastic material.

Study Questions

1. What is an elastic material? What is a linear elastic material?
2. We proved that for an *isotropic* linear elastic material, the coefficients in Eq. (8.25) must satisfy the restrictions $b_{11} = b_{22}$, $b_{12} = b_{21}$, and $b_{14} = 0$. Why doesn't the proof we used show that these restrictions must be satisfied by any linear elastic material?
3. How many independent coefficients appear in the stress-strain relations for an isotropic linear elastic material?
4. What is the definition of the bulk modulus?

Example 8.5 Determining Elastic Constants

The cylindrical bar in Fig. 8.29 consists of an isotropic linear elastic material and is subjected to axial loads. As a result, the bar is subjected to a normal stress $\sigma_x = 420$ MPa, and the other components of stress are zero. By measuring the changes in the bar's length and diameter, it is determined that the axial strain is $\varepsilon_x = 0.006$ and the lateral strain is $\varepsilon_y = \varepsilon_z = -0.002$. Determine (**a**) the modulus of elasticity, Poisson's ratio, and shear modulus of the material, (**b**) the Lamé constants of the material, and (**c**) the bulk modulus of the material.

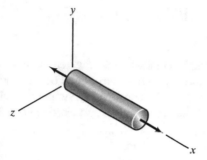

Fig. 8.29

Strategy

(**a**) We know the state of stress and the strain components ε_x and ε_y, so we can solve Eqs. (8.43) and (8.44) for the modulus of elasticity E and the Poisson's ratio v. We can then use Eq. (8.50) to determine the shear modulus G.
(**b**) The Lamé constant $\mu = G$, and λ is given in terms of E and v by Eq. (8.57). (**c**) The bulk modulus is given in terms of E and v by Eq. (8.59).

Solution

(**a**) From Eq. (8.43)

$$\varepsilon_x = \frac{1}{E}\sigma_x - \frac{v}{E}(\sigma_y + \sigma_z):$$
$$0.006 = \frac{1}{E}(420\text{E6 Pa}),$$

we obtain $E = 70.0$ GPa. Then, from Eq. (8.44)

$$\varepsilon_y = \frac{1}{E}\sigma_y - \frac{v}{E}(\sigma_x + \sigma_z):$$
$$-0.002 = \frac{-v}{70.0\text{E9 Pa}}(420\text{E6 Pa}),$$

we obtain $v = 0.333$. From Eq. (8.50), the shear modulus is

$$G = \frac{E}{2(1+v)} = \frac{70.0\,\text{GPa}}{(2)\,(1+0.333)} = 26.3 \ \text{GPa}.$$

(b) The Lamé constant $\mu = G = 26.3$ GPa. From Eq. (8.57)

$$\lambda = \frac{vE}{(1+v)(1-2v)} = \frac{(0.333)\,(70.0\,\text{GPa})}{(1+0.333)\,[1-2(0.333)]} = 52.5\,\text{GPa}.$$

(c) From Eq. (8.59), the bulk modulus is

$$K = \frac{E}{3(1-2v)} = \frac{70.0\,\text{GPa}}{3[1-2(0.333)]} = 70.0\,\text{GPa}.$$

Discussion
This important example demonstrates why tensile tests of materials are so common. You may already have observed one being conducted or may do so in future courses. By a single, conceptually simple test, all of the coefficients that characterize the behavior of an isotropic elastic material can be determined.

Problems

8.51 The state of stress at a point p in a material with modulus of elasticity $E = 28$ GPa and Poisson's ratio $v = 0.3$ is (in MPa)

$$\begin{bmatrix} \sigma_x & \tau_{xy} & \tau_{xz} \\ \tau_{yx} & \sigma_y & \tau_{yz} \\ \tau_{zx} & \tau_{zy} & \sigma_z \end{bmatrix} = \begin{bmatrix} 250 & 60 & -80 \\ 60 & 300 & 40 \\ -80 & 40 & 150 \end{bmatrix}.$$

What is the state of strain at p?

8.52 The state of strain at a point p in a material with modulus of elasticity $E = 28$ GPa and Poisson's ratio $v = 0.3$ is

$$\begin{bmatrix} \varepsilon_x & \gamma_{xy} & \gamma_{xz} \\ \gamma_{yx} & \varepsilon_y & \gamma_{yz} \\ \gamma_{zx} & \gamma_{zy} & \varepsilon_z \end{bmatrix} = \begin{bmatrix} 220 & 180 & 100 \\ 180 & -150 & -120 \\ 100 & -120 & -100 \end{bmatrix} \times 10^{-5}.$$

What is the state of stress at p?

8.53 The state of stress at a point p of a nickel pipe in a gaseous diffusion centrifuge is (in ksi)

$$\begin{bmatrix} \sigma_x & \tau_{xy} & \tau_{xz} \\ \tau_{yx} & \sigma_y & \tau_{yz} \\ \tau_{zx} & \tau_{zy} & \sigma_z \end{bmatrix} = \begin{bmatrix} 600 & 30 & 25 \\ 30 & 450 & 40 \\ 25 & 40 & 520 \end{bmatrix}.$$

What is the state of strain at p?

8.54 Show that a state of plane stress at a point p of an isotropic, linear elastic material does not necessarily result in a state of plane strain at p. What condition must the state of plane stress satisfy to result in a state of plane strain?

8.55 The state of plane stress at a point in the machine part made of 2014-T6 aluminum alloy is $\sigma_x = 450$ MPa, $\sigma_y = -300$ MPa, and $\tau_{xy} = 60$ MPa. What is the state of strain at that point?

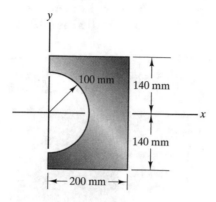

Problem 8.55

8.56 An arm of a robotic actuator made of 7075-T6 aluminum alloy is subjected to a state of plane stress σ_x, σ_y, τ_{xy}. Using a strain gauge rosette, it is determined experimentally that $\varepsilon_x = 0.00350$, $\varepsilon_y = 0.00600$, and $\gamma_{xy} = -0.02400$. What are the components of the state of stress (in ksi)?

Problem 8.56 (*Photograph from iStock by Chesky_W*)

8.57 A material is subjected to a state of plane stress $\sigma_x = 400$ MPa, $\sigma_y = -200$ MPa, $\tau_{xy} = 300$ MPa. Using a strain gauge rosette, it is determined experimentally that $\varepsilon_x = 0.00239$, $\varepsilon_y = -0.00162$, and $\gamma_{xy} = 0.00401$. What are the modulus of elasticity and Poisson's ratio of the material?

8.58 A material is subjected to a state of plane stress $\sigma_x = 400$ MPa, $\sigma_y = -200$ MPa, $\tau_{xy} = 300$ MPa. Using a strain gauge rosette, it is determined experimentally that $\varepsilon_x = 0.00575$, $\varepsilon_y = -0.00400$, and $\gamma_{xy} = 0.00975$. Determine (**a**) the Lamé constants λ and μ; (**b**) the bulk modulus K.

8.59 The state of stress at a point p in a material with modulus of elasticity $E = 15E6$ psi and Poisson's ratio $\nu = 0.33$ is (in ksi)

$$
\begin{bmatrix}
\sigma_x & \tau_{xy} & \tau_{xz} \\
\tau_{yx} & \sigma_y & \tau_{yz} \\
\tau_{zx} & \tau_{zy} & \sigma_z
\end{bmatrix}
=
\begin{bmatrix}
50 & -60 & -40 \\
-60 & 40 & 40 \\
-40 & 40 & -40
\end{bmatrix}.
$$

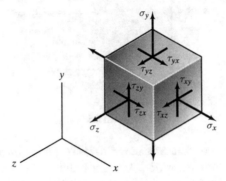

Problem 8.59

What is the state of strain at p?

8.60 The state of strain at a point p in a material with modulus of elasticity $E = 15E6$ psi and Poisson's ratio $\nu = 0.33$ is

$$
\begin{bmatrix}
\varepsilon_x & \gamma_{xy} & \gamma_{xz} \\
\gamma_{yx} & \varepsilon_y & \gamma_{yz} \\
\gamma_{zx} & \gamma_{zy} & \varepsilon_z
\end{bmatrix}
=
\begin{bmatrix}
0.004 & 0.010 & -0.010 \\
0.010 & -0.002 & 0 \\
-0.010 & 0 & -0.003
\end{bmatrix}.
$$

What is the state of stress at p?

8.61 For the material titanium (see Appendix B), determine **(a)** the Lamé constants λ and μ in psi and **(b)** the bulk modulus K in psi.

8.62 The cylindrical bar consists of a material with modulus of elasticity E and Poisson's ratio v and is subjected to axial loads. The resulting state of stress is

$$
\begin{bmatrix}
\sigma_x & \tau_{xy} & \tau_{xz} \\
\tau_{yx} & \sigma_y & \tau_{yz} \\
\tau_{zx} & \tau_{zy} & \sigma_z
\end{bmatrix}
=
\begin{bmatrix}
\sigma_x & 0 & 0 \\
0 & 0 & 0 \\
0 & 0 & 0
\end{bmatrix}.
$$

(a) Determine the state of strain in terms of σ_x, E, and v. **(b)** The length of the unloaded bar is L and its cross-sectional area is A. Determine the volume of the loaded bar in terms of L, A, σ_x, E, and v.

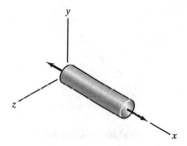

Problems 8.62–8.63

8.63 The cylindrical bar consists of a material with modulus of elasticity E and Poisson's ratio v and is subjected to axial loads that cause a normal stress $\sigma_x = 380$ MPa. The axial and lateral normal strains are measured and determined to be $\varepsilon_x = 0.0020$, $\varepsilon_y = \varepsilon_z = -0.0007$. Determine **(a)** the modulus of elasticity E, Poisson's ratio v, and shear modulus G of the material and **(b)** the Lamé constants λ and μ of the material.

8.64* At a point of a material subjected to a state of plane stress, the strains measured by the strain gauge rosette are $\varepsilon_a = 0.006$, $\varepsilon_b = -0.003$, and $\varepsilon_c = -0.002$. The modulus of elasticity and Poisson's ratio of the material are $E = 30$ GPa and $v = 0.33$. What is the state of stress at the point? (See the figure for Problems 8.64–8.66.)

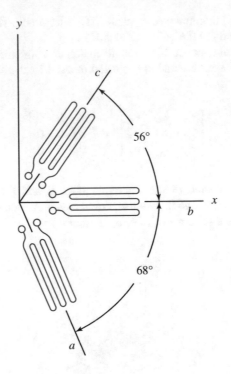

Problems 8.64–8.66

8.65∗ At a point of a steel hydraulic piston that is subjected to a state of plane stress, the strains measured by the strain gauge rosette are $\varepsilon_a = -0.00150$, $\varepsilon_b = 0.00140$, and $\varepsilon_c = 0.00086$. The modulus of elasticity and Poisson's ratio of the material are $E = 200$ GPa and $\nu = 0.33$. What are the principal stresses and the absolute maximum shear stress at the point? (See the figure for Problems 8.64–8.66.)

Problem 8.65

8.66* The strain gauge rosette is mounted on the outer wall of one of an Atlas launch vehicle's nozzles, where the material is in a homogeneous state of plane stress. The strains measured by the rosette are $\varepsilon_a = 0.0053$, $\varepsilon_b = 0.0038$, and $\varepsilon_c = 0.0029$. The modulus of elasticity and Poisson's ratio of the material are $E = 70$ GPa and $v = 0.33$. What are the components of plane stress? (See the figure for Problems 8.64–8.66.)

8.67* An isotropic linear elastic material is subjected to a normal stress $\sigma_x = \sigma$, and the other components of stress are zero. Using Eqs. (8.43)–(8.45) and Eq. (8.51) for this state of stress, using Eq. (8.50), and recalling that $G = \mu$, derive Eq. (8.57):

$$\lambda = \frac{vE}{(1+v)(1-2v)}.$$

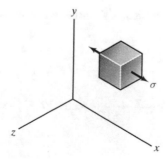

Problem 8.67

Chapter Summary

Components of Strain

In terms of a given coordinate system, the *state of strain* at a point p of a material is defined by the *components of strain*:

$$\begin{bmatrix} \varepsilon_x & \gamma_{xy} & \gamma_{xz} \\ \gamma_{yx} & \varepsilon_y & \gamma_{yz} \\ \gamma_{zx} & \gamma_{zy} & \varepsilon_z \end{bmatrix}. \tag{8.1}$$

The components ε_x, ε_y, and ε_z are the normal strains in the x, y, and z directions. The component $\gamma_{xy} = \gamma_{yx}$ is the shear strain referred to the positive directions of the x and y axes, and the components $\gamma_{yz} = \gamma_{zy}$ and $\gamma_{xz} = \gamma_{zx}$ are defined similarly.

Consider a volume dV_0 of material in a reference state. Its volume in the deformed state is

$$dV = \left(1 + \varepsilon_x + \varepsilon_y + \varepsilon_z\right) dV_0. \tag{8.2}$$

The *dilatation* is the change in volume of the material per unit volume:

$$e = \frac{dV - dV_0}{dV_0} = \varepsilon_x + \varepsilon_y + \varepsilon_z. \tag{8.3}$$

Transformations of Plane Strain

The strain at a point p is said to be a state of *plane strain* if it is of the form

$$\begin{bmatrix} \varepsilon_x & \gamma_{xy} & 0 \\ \gamma_{yx} & \varepsilon_y & 0 \\ 0 & 0 & 0 \end{bmatrix}. \tag{8.4}$$

In terms of a coordinate system $x'y'z'$ oriented as shown in Fig. (a), the components of strain are

$$\varepsilon_{x'} = \frac{\varepsilon_x + \varepsilon_y}{2} + \frac{\varepsilon_x - \varepsilon_y}{2} \cos 2\theta + \frac{\gamma_{xy}}{2} \sin 2\theta, \tag{8.7}$$

$$\varepsilon_{y'} = \frac{\varepsilon_x + \varepsilon_y}{2} - \frac{\varepsilon_x - \varepsilon_y}{2} \cos 2\theta - \frac{\gamma_{xy}}{2} \sin 2\theta, \tag{8.8}$$

$$\frac{\gamma_{x'y'}}{2} = -\frac{\varepsilon_x - \varepsilon_y}{2}\sin 2\theta + \frac{\gamma_{xy}}{2}\cos 2\theta. \tag{8.9}$$

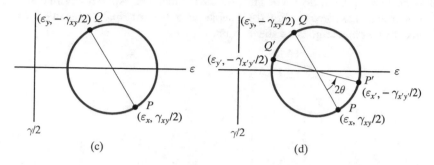

(c)   (d)

For an isotropic linear elastic material, Eqs. (8.7), (8.8), and (8.9) also apply to the components of strain resulting from a state of plane stress.

Strain Gauge Rosette

Suppose that a material is subject to an unknown state of plane strain relative to the x-y coordinate system in Fig. (a). A *strain gauge rosette* measures the normal strain $\varepsilon_{x'}$ in three different directions: θ_a, θ_b, and θ_c. Let the measured strains be ε_a, ε_b, and ε_c. Using Eq. (8.6) to express these strains in terms of the strain components ε_x, ε_y, and γ_{xy} gives

$$\begin{aligned}
\varepsilon_a &= \varepsilon_x \cos^2\theta_a + \varepsilon_y \sin^2\theta_a + \gamma_{xy}\sin\theta_a\cos\theta_a, \\
\varepsilon_b &= \varepsilon_x \cos^2\theta_b + \varepsilon_y \sin^2\theta_b + \gamma_{xy}\sin\theta_b\cos\theta_b, \\
\varepsilon_c &= \varepsilon_x \cos^2\theta_c + \varepsilon_y \sin^2\theta_c + \gamma_{xy}\sin\theta_c\cos\theta_c.
\end{aligned} \tag{8.10}$$

This system of equations can be solved for the strain components ε_x, ε_y, and γ_{xy}.

Maximum and Minimum Strains in Plane Strain

A value of θ for which the normal strain is a maximum or minimum is determined from the equation

$$\tan 2\theta_{\mathrm{p}} = \frac{\gamma_{xy}}{\varepsilon_x - \varepsilon_y}. \tag{8.11}$$

The values of the principal strains can be obtained by substituting θ_{p} into Eqs. (8.7) and (8.8). Their values can also be determined from the equation

$$\varepsilon_1, \varepsilon_2 = \frac{\varepsilon_x + \varepsilon_y}{2} \pm \sqrt{\left(\frac{\varepsilon_x - \varepsilon_y}{2}\right)^2 + \left(\frac{\gamma_{xy}}{2}\right)^2}, \tag{8.14}$$

although this equation does not indicate their directions. An infinitesimal square element oriented as shown in Fig. (b) is subjected to the principal strains in the x' and y' directions and undergoes no shear strain.

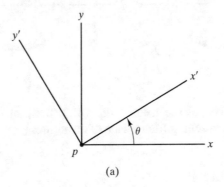

(a)

A value of θ for which the in-plane shear strain is a maximum or minimum is determined from the equation

$$\tan 2\theta_s = -\frac{\varepsilon_x - \varepsilon_y}{\gamma_{xy}}. \tag{8.15}$$

The corresponding shear strain can be obtained by substituting θ_s into Eq. (8.9). The magnitude of the maximum in-plane shear strain can be determined from the equation

$$\gamma_{max} = \sqrt{\left(\varepsilon_x - \varepsilon_y\right)^2 + \gamma_{xy}^2}. \tag{8.18}$$

The *absolute maximum shear strain* is

$$\gamma_{abs} = \text{Max}(|\varepsilon_1|, |\varepsilon_2|, |\varepsilon_1 - \varepsilon_2|). \tag{8.22}$$

Mohr's Circle for Plane Strain

Given a state of plane strain ε_x, ε_y, and γ_{xy}, establish a set of horizontal and vertical axes with normal strain measured to the right along the horizontal axis and shear strain measured downward along the vertical axis. Plot two points, point P with coordinates $(\varepsilon_x, \gamma_{xy}/2)$ and point Q with coordinates $(\varepsilon_x, -\gamma_{xy}/2)$. Draw a straight line connecting points P and Q. Using the intersection of the straight line with the horizontal axis as the center, draw a circle that passes through the two points (Fig. (c)). Draw a straight line through the center of the circle at an angle 2θ measured counterclockwise from point P. Point P' at which this line intersects the circle has coordinates $(\varepsilon_{x'}, \gamma_{x'y'}/2)$, and point Q' has coordinates $(\varepsilon_{y'}, -\gamma_{x'y'}/2)$ (Fig. (d)).

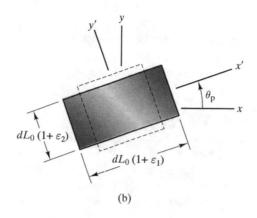

(b)

Stress-Strain Relations

A material for which the state of stress at a point is a single-valued function of the current state of strain at that point is said to be *elastic*. If the function is invariant under changes of orientation of the coordinate system relative to the material, the material is *isotropic*. If the components of strain are linear functions of the components of stress, the material is *linear elastic*. The stress-strain relations for an isotropic linear elastic material are

$$\varepsilon_x = \frac{1}{E}\sigma_x - \frac{\nu}{E}(\sigma_y + \sigma_z), \tag{8.43}$$

$$\varepsilon_y = \frac{1}{E}\sigma_y - \frac{\nu}{E}(\sigma_x + \sigma_z), \tag{8.44}$$

$$\varepsilon_z = \frac{1}{E}\sigma_z - \frac{\nu}{E}(\sigma_x + \sigma_y), \tag{8.45}$$

$$\gamma_{xy} = \frac{1}{G}\tau_{xy}, \tag{8.46}$$

$$\gamma_{yz} = \frac{1}{G}\tau_{yz}, \tag{8.47}$$

$$\gamma_{xz} = \frac{1}{G}\tau_{xz}. \tag{8.48}$$

The shear modulus is related to the modulus of elasticity and Poisson's ratio by

$$G = \frac{E}{2(1+v)}. \tag{8.50}$$

The stress-strain relations can also be expressed as

$$\sigma_x = (\lambda + 2\mu)\varepsilon_x + \lambda(\varepsilon_y + \varepsilon_z), \tag{8.51}$$

$$\sigma_y = (\lambda + 2\mu)\varepsilon_y + \lambda(\varepsilon_x + \varepsilon_z), \tag{8.52}$$

$$\sigma_z = (\lambda + 2\mu)\varepsilon_z + \lambda(\varepsilon_x + \varepsilon_y), \tag{8.53}$$

$$\tau_{xy} = \mu\gamma_{xy}, \tag{8.54}$$

$$\tau_{yz} = \mu\gamma_{yz}, \tag{8.55}$$

$$\tau_{xz} = \mu\gamma_{xz}, \tag{8.56}$$

where λ and μ are the *Lamé constants*. The constant $\mu = G$, and the constant λ is given in terms of the modulus of elasticity and Poisson's ratio by

$$\lambda = \frac{vE}{(1+v)(1-2v)}. \tag{8.57}$$

Let an isotropic linear elastic material be subjected to a pressure P so that $\sigma_x = -P$, $\sigma_y = -P$, and $\sigma_z = -P$. The *bulk modulus* K of the material is the ratio of $-P$ to the dilatation

$$K = \frac{-P}{e}. \tag{8.58}$$

In terms of the modulus of elasticity and Poisson's ratio, the bulk modulus is

$$K = \frac{E}{3(1-2v)}. \tag{8.59}$$

Review Problems

8.68 The components of plane strain at point p are $\varepsilon_x = 0$, $\varepsilon_y = 0$, and $\gamma_{xy} = 0.004$. If $\theta = 45°$, what are the strains $\varepsilon_{x'}, \varepsilon_{y'}$, and $\gamma_{x'y'}$ at point p?

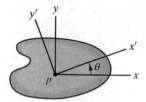

Problems 8.68–8.69

8.69 The components of plane strain at point p are $\varepsilon_x = -0.0024$, $\varepsilon_y = 0.0012$, and $\gamma_{xy} = -0.0012$. If $\theta = 25°$, what are the strains $\varepsilon_{x'}, \varepsilon_{y'}$, and $\gamma_{x'y'}$ at point p?

8.70 A point p of the bearing's housing is subjected to a state of plane stress, and the strain components $\varepsilon_{x'} = 0.0024$, $\varepsilon_{y'} = 0.0044$, and $\gamma_{x'y'} = -0.0030$. If $\theta = 20°$, what are the strains ε_x, ε_y, and γ_{xy} at p?

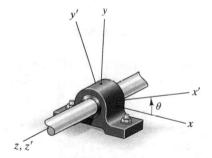

Problems 8.70–8.71

8.71∗ A point p of the bearing's housing is subjected to the state of plane strain $\varepsilon_x = 0.0032$, $\varepsilon_y = -0.0026$, and $\gamma_{xy} = 0.0044$. If $\varepsilon_{x'} = 0.0037$ and $\varepsilon_{y'} = -0.0031$, determine the angle θ and the strain $\gamma_{x'y'}$ at p.

8.72 The strains measured by a strain gauge rosette oriented as shown are $\varepsilon_a = -0.00116$, $\varepsilon_b = -0.00065$, and $\varepsilon_c = 0.00130$. Determine the strains ε_x, ε_y, and γ_{xy}.

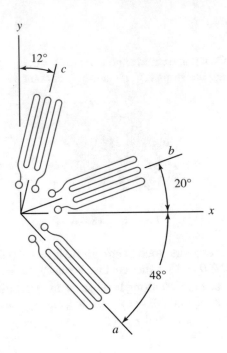

Problem 8.72

8.73 The state of plane strain at a point is $\varepsilon_x = 0.003$, $\varepsilon_y = 0.001$, and $\gamma_{xy} = -0.006$. Determine the magnitude of the maximum in-plane shear strain.

8.74 For the state of plane strain $\varepsilon_x = 0.0025$, $\varepsilon_y = 0$, and $\gamma_{xy} = -0.005$, what is the absolute maximum shear strain?

8.75 For the state of plane strain $\varepsilon_x = -0.008$, $\varepsilon_y = -0.006$, and $\gamma_{xy} = -0.012$, what is the absolute maximum shear strain?

8.76 A drill bit is made of steel with modulus of elasticity $E = 200$ GPa and Poisson's ratio $\nu = 0.28$. At a point p, the bit is subjected to a state of plane stress, and the components of strain $\varepsilon_x = -0.0024$, $\varepsilon_y = 0.0044$, and $\gamma_{xy} = -0.0030$. What is the state of stress at p?

8.77 A drill bit is made of steel with modulus of elasticity $E = 200$ GPa and Poisson's ratio $\nu = 0.28$. At a point p, the bit is subjected to a state of plane stress, and the components of strain $\varepsilon_x = 0.0038$, $\varepsilon_y = 0.0066$, and $\gamma_{xy} = -0.0048$. What is the component of strain ε_z at p?

8.78* Beginning with Eq. (8.59), show that the bulk modulus is given in terms of the Lamé constants by

$$K = \lambda + \frac{2}{3}\mu.$$

Chapter 9
Deflections of Beams

Deflections of certain types of beams, such as leaf springs in the suspensions of vehicles, diving boards, and vaulting poles, are central to their function. In contrast, most structural beams are intended simply to support loads, and deflections are not important considerations in their design. But even in such cases, we show in this chapter that calculation of deflections can be of crucial importance for another reason. It is through calculating deflections that we can determine the reactions on statically indeterminate beams.

9.1 The Second-Order Equation

Let v be the deflection of a beam's neutral axis relative to the x axis, and let θ be the angle between the neutral axis and the x axis (Fig. 9.1). Our objective is to determine v and θ as functions of x for a beam with given loads.

9.1.1 Differential Equation

By considering the deflection at x and at $x + dx$ (Fig. 9.2), we see that v and θ are related by

$$\frac{dv}{dx} = \tan\theta = \theta + \frac{1}{3}\theta^3 + \cdots,$$

where we express $\tan\theta$ in terms of its Taylor series. We will restrict our analysis to beams and loadings for which θ is small enough to neglect terms of second and higher orders, so we obtain the equation

© Springer Nature Switzerland AG 2020
A. Bedford, K. M. Liechti, *Mechanics of Materials*,
https://doi.org/10.1007/978-3-030-22082-2_9

Fig. 9.1 Deflection v and the angle θ between the neutral and x axes

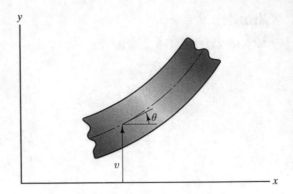

Fig. 9.2 Determining the relation between v and θ

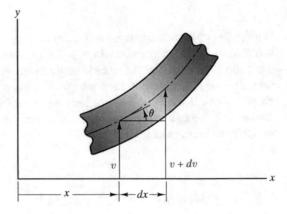

$$\frac{dv}{dx} = \theta. \tag{9.1}$$

(The magnitude of θ in the figures is greatly exaggerated.) In Fig. 9.3 we draw lines perpendicular to the neutral axis at x and at $x + dx$. The angle $d\theta$ between these lines equals the change in θ from x to $x + dx$. In terms of the radius of curvature of the neutral axis ρ and the distance ds, the angle $d\theta$ is

$$d\theta = \frac{1}{\rho} ds. \tag{9.2}$$

Notice from Fig. 9.3 that

$$dx = ds \cos\theta = ds\left(1 - \frac{1}{2}\theta^2 + \cdots\right).$$

Therefore $ds = dx$ when we neglect terms of second order and higher in θ, so we can express Eq. (9.2) as

Fig. 9.3 Relating θ to the radius of curvature of the neutral axis

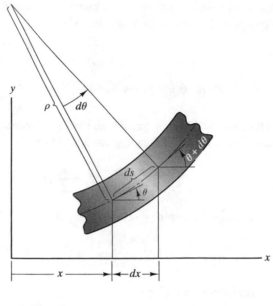

$$\frac{d\theta}{dx} = \frac{1}{\rho}.$$

Substituting Eq. (9.1) into this expression, we obtain

$$\frac{d^2v}{dx^2} = \frac{1}{\rho}. \tag{9.3}$$

In Chap. 6 we obtained an equation relating the beam's radius of curvature, the bending moment, and the flexural rigidity. From Eq. (6.10),

$$\frac{1}{\rho} = \frac{M}{EI}.$$

Substituting this result into Eq. (9.3), we obtain a relationship between the beam's deflection and the bending moment:

$$v'' = \frac{M}{EI}, \tag{9.4}$$

where the primes denote derivatives with respect to x. With this equation we can determine deflections of beams. Although there are several details we must discuss, the basic procedure is to determine the bending moment M in a beam as a function of x then integrate Eq. (9.4) twice to determine the deflection v as a function of x. (If the beam is prismatic and consists of homogeneous material, the flexural rigidity EI is a constant.) Notice that once v is known as a function of x, we can use Eq. (9.1) to determine the slope θ as a function of x.

9.1.2 Boundary Conditions

The beam in Fig. 9.4 has pin and roller supports and is subjected to a uniformly distributed load. The bending moment M as a function of x is

$$M = \frac{w_0}{2}\left(Lx - x^2\right).$$

We substitute this expression into Eq. (9.4):

$$v'' = \frac{w_0}{2EI}\left(Lx - x^2\right).$$

Integrating, we obtain

$$v' = \frac{w_0}{2EI}\left(\frac{Lx^2}{2} - \frac{x^3}{3}\right) + A,$$

where A is an integration constant. Integrating again yields

$$v = \frac{w_0}{2EI}\left(\frac{Lx^3}{6} - \frac{x^4}{12}\right) + Ax + B, \tag{9.5}$$

where B is a second integration constant.

To complete our determination of the deflection, we must evaluate the integration constants A and B by using the boundary conditions imposed by the beam's supports (Fig. 9.5). The deflection is zero at the pin support ($x = 0$). We substitute this condition into Eq. (9.5), obtaining the value of B

$$v\big|_{x=0} = B = 0.$$

The deflection is also zero at the roller support ($x = L$):

Fig. 9.4 Beam subjected to a uniformly distributed load

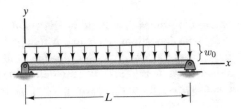

Fig. 9.5 The deflection
equals zero at each support

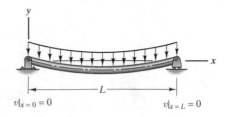

$$v|_{x=0} = 0 \qquad\qquad v|_{x=L} = 0$$

Fig. 9.6 (a) Beam
subjected to a uniformly
distributed load over part of
its length. (b) Functions
describing the distribution
of the bending moment

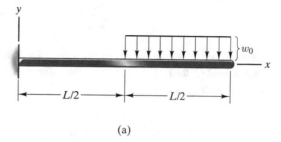

(a)

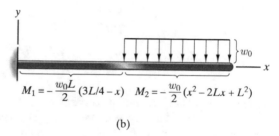

$$M_1 = -\frac{w_0 L}{2}(3L/4 - x) \qquad M_2 = -\frac{w_0}{2}(x^2 - 2Lx + L^2)$$

(b)

$$v|_{x=L} = \frac{w_0}{2EI}\left(\frac{L^4}{6} - \frac{L^4}{12}\right) + AL = 0.$$

From this equation we obtain $A = -w_0 L^3/24EI$. Substituting the values of A and B into Eq. (9.5), we obtain the solution for the beam's deflection:

$$v = -\frac{w_0 x}{24EI}\left(L^3 - 2Lx^2 + x^3\right).$$

In the previous example, we determined the beam's deflection by substituting the expression for the bending moment as a function of x into Eq. (9.4), integrating twice, and using the boundary conditions at the supports to evaluate the integration constants. A new consideration arises when we apply this procedure to the beam in Fig. 9.6(a). When we determine the bending moment as a function of x, we obtain one expression M_1 that applies for $0 \le x \le L/2$ and a different expression M_2 that applies for $L/2 \le x \le L$ (Fig. 9.6(b)). We need to determine the deflection independently for each of these regions. Let v_1 denote the beam's deflection from $x = 0$ to $x = L/2$. We substitute the expression for M_1 into Eq. (9.4):

$$v_1'' = \frac{M_1}{EI} = -\frac{w_0 L}{2EI}\left(\frac{3L}{4} - x\right).$$

Integrating twice gives

$$v_1' = -\frac{w_0 L}{2EI}\left(\frac{3Lx}{4} - \frac{x^2}{2}\right) + A,$$

$$v_1 = -\frac{w_0 L}{2EI}\left(\frac{3Lx^2}{8} - \frac{x^3}{6}\right) + Ax + B,$$

where A and B are integration constants. Letting v_2 denote the beam's deflection from $x = L/2$ to $x = L$, we substitute the expression for M_2 into Eq. (9.4) and integrate twice:

$$v_2'' = \frac{M_2}{EI} = -\frac{w_0}{2EI}\left(x^2 - 2Lx + L^2\right),$$

$$v_2' = -\frac{w_0}{2EI}\left(\frac{x^3}{3} - Lx^2 + L^2 x\right) + C,$$

$$v_2 = -\frac{w_0}{2EI}\left(\frac{x^4}{12} - \frac{Lx^3}{3} + \frac{L^2 x^2}{2}\right) + Cx + D,$$

where C and D are integration constants.

We must now identify four boundary conditions with which to evaluate the integration constants A, B, C, and D. Two conditions are imposed by the beam's fixed support. At $x = 0$, the deflection and slope equal zero (Fig. 9.7):

$$v_1\big|_{x=0} = B = 0,$$

$$v_1'\big|_{x=0} = A = 0.$$

We also know that the deflections given by the two solutions must be equal at $x = L/2$ (Fig. 9.7):

Fig. 9.7 Boundary conditions at the fixed support and where the two solutions meet

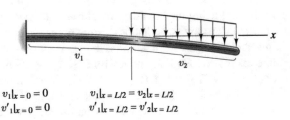

$$v_1\big|_{x=0} = 0$$
$$v_1'\big|_{x=0} = 0$$

$$v_1\big|_{x=L/2} = v_2\big|_{x=L/2}$$
$$v_1'\big|_{x=L/2} = v_2'\big|_{x=L/2}$$

$$v_1\big|_{x=L/2} = v_2\big|_{x=L/2} :$$

$$-\frac{14w_0L^4}{384EI} = -\frac{17w_0L^4}{384EI} + \frac{CL}{2} + D.$$

The fourth condition is that the slopes given by the two solutions must be equal at $x = L/2$.

$$v_1'\big|_{x=L/2} = v_2'\big|_{x=L/2} :$$

$$-\frac{6w_0L^3}{48EI} = -\frac{7w_0L^3}{48EI} + C.$$

From the four boundary conditions, we find that $A = 0$, $B = 0$, $C = w_0L^3/48EI$, and $D = -w_0L^4/384EI$. Substituting these results into our expressions for v_1 and v_2, the beam's deflection is

$$v = \begin{cases} -\dfrac{w_0L}{48EI}(9Lx^2 - 4x^3), & 0 \le x \le L/2 \\ -\dfrac{w_0}{384EI}(16x^4 - 64Lx^3 + 96L^2x^2 - 8L^3x + L^4), & L/2 \le x \le L. \end{cases}$$

$$(9.6 \text{ and } 9.7)$$

The examples we have discussed illustrate the steps required to determine a beam's deflection:

1. Determine the bending moment as a function of x. As in our second example, this may result in two or more functions M_1, M_2..., each of which applies to a different segment of the beam's length.
2. For each segment, integrate Eq. (9.4) twice to determine v_1, v_2.... If there are N segments, this step will result in $2N$ unknown integration constants.
3. Use the boundary conditions to determine the integration constants. Our two examples illustrate the most common boundary conditions encountered in determining deflections of beams. These boundary conditions are summarized in Fig. 9.8.

Study Questions

1. What is the definition of the deflection v?
2. If you know the deflection v as a function of x, how can you determine the angle θ between the neutral axis and the x axis as a function of x?
3. If you know the bending moment M in a beam as a function of x, how can you use Eq. (9.4) to determine the deflection v as a function of x? Do you need any additional information to do so?

Fig. 9.8 Common
boundary conditions

At a pin or
roller support:

At a fixed
support:

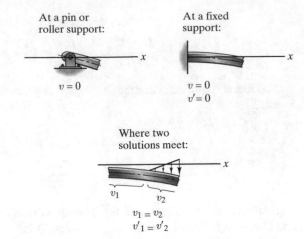

$v = 0$

$v = 0$
$v' = 0$

Where two
solutions meet:

$v_1 = v_2$
$v'_1 = v'_2$

Example 9.1 Determining a Beam Deflection
The beam in Fig. 9.9 consists of material with modulus of elasticity
$E = 72\,\text{GPa}$, the moment of inertia of its cross section is $I = 1.6\text{E} - 7\,\text{m}^4$,
and its length is $L = 4\,\text{m}$. If $w_0 = 100\,\text{N/m}$, what is the deflection at the right
end of the beam?

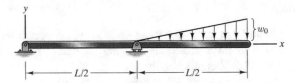

Fig. 9.9

Strategy
We must first determine the distribution of the bending moment in the beam.
In this example the bending moment must be determined for the segments
$0 \le x \le L/2$ and $L/2 \le x \le L$. We will then integrate Eq. (9.4) twice *for each
segment* to determine the deflection in each segment. Finally, we must use
boundary conditions to determine the integration constants.

Solution
Determine the bending moment as a function of x   We leave it as an
exercise to show that the distributions of the bending moment in the left and
right halves of the beam are given by the expressions in Fig. (a).

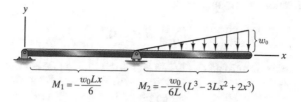

$$M_1 = -\frac{w_0 L x}{6} \qquad M_2 = -\frac{w_0}{6L}(L^3 - 3Lx^2 + 2x^3)$$

(a) Equations for the bending moment

For each segment, integrate Eq. (9.4)   For the segment $0 \leq x \leq L/2$, we obtain

$$v_1'' = \frac{M_1}{EI} = -\frac{w_0 L x}{6EI},$$

$$v_1' = -\frac{w_0 L x^2}{12EI} + A,$$

$$v_1 = -\frac{w_0 L x^3}{36EI} + Ax + B,$$

where A and B are integration constants, and for the segment $L/2 \leq x < L$, we obtain

$$v_2'' = \frac{M_2}{EI} = -\frac{w_0}{6LEI}(L^3 - 3Lx^2 + 2x^3),$$

$$v_2' = -\frac{w_0}{6LEI}\left(L^3 x - Lx^3 + \frac{x^4}{2}\right) + C,$$

$$v_2 = -\frac{w_0}{6LEI}\left(\frac{L^3 x^2}{2} - \frac{Lx^4}{4} + \frac{x^5}{10}\right) + Cx + D,$$

where C and D are integration constants.

Use the boundary conditions to determine the integration constants   We can apply the four boundary conditions shown in Fig. (b).

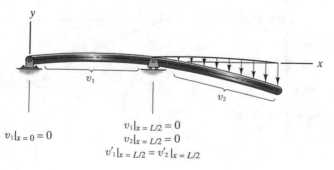

$$v_1|_{x=0} = 0$$

$$v_1|_{x=L/2} = 0$$
$$v_2|_{x=L/2} = 0$$
$$v_1'|_{x=L/2} = v_2'|_{x=L/2}$$

(b) Boundary conditions

The deflection is zero at $x = 0$:

$$v_1|_{x=0} = B = 0.$$

The deflection is zero at $x = L/2$, a condition that applies to both v_1 and v_2:

$$v_1|_{x=L/2} = -\frac{w_0L^4}{288EI} + \frac{AL}{2} = 0,$$

$$v_2|_{x=L/2} = -\frac{3w_0L^4}{160EI} + \frac{CL}{2} + D = 0.$$

The slopes of the two solutions must be equal at $x = L/2$.

$$v_1'|_{x=L/2} = v_2'|_{x=L/2} \quad :$$

$$-\frac{w_0L^3}{48EI} + A = -\frac{13w_0L^3}{192EI} + C.$$

Substituting the values of w_0, L, E, and I into these four equations and solving, we obtain $A = 0.00386$, $B = 0$, $C = 0.0299$, and $D = -0.0181$ m. Now we can use the equation for v_2 to obtain the deflection at the right end of the beam:

$$v_2 = -\frac{w_0}{6LEI}\left(\frac{L^3x^2}{2} - \frac{Lx^4}{4} + \frac{x^5}{10}\right) + Cx + D$$

$$= \frac{-100\,\text{N/m}}{(6)(4\,\text{m})(72\text{E}9\,\text{N/m}^2)(1.6\text{E-7}\,\text{m}^4)}\left[\frac{(4\,\text{m})^3(4\,\text{m})^2}{2} - \frac{(4\,\text{m})(4\,\text{m})^4}{4} + \frac{(4\,\text{m})^5}{10}\right]$$

$$+ (0.0299)(4\,\text{m}) - 0.0181\,\text{m}$$

$$= -0.0282\,\text{m}.$$

Figure (c) is a graph of the beam's deflection.

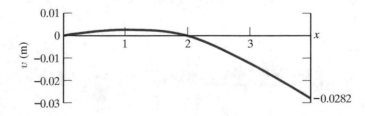

(c) Deflection as a function of x

Discussion

From Fig. 9.8 you would conclude that the beam in this example has three boundary conditions on the deflection at the roller support: $v_1|_{x\,=\,L/2} = 0$, $v_2|_{x\,=\,L/2} = 0$, and $v_1|_{x\,=\,L/2} = v_2|_{x\,=\,L/2}$. But notice that only two of these conditions are independent. Once any two of them are chosen, the third one is implied.

Problems

9.1 The prismatic beam is subjected to an upward force at the left end and has a built-in support at $x = L$. (**a**) Determine the bending moment M as a function of x. (**b**) Determine the deflection v as a function of x.

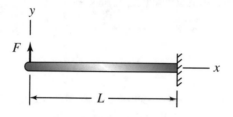

Problems 9.1–9.2

9.2 The length of the prismatic beam is $L = 2$ m, and the moment of inertia of its cross section is $I = 1.6\text{E} - 5$ m^4. It consists of steel with modulus of elasticity $E = 220$ GPa. If the force $F = 4$ kN, what is the magnitude of the resulting deflection of the beam at $x = 0$?

9.3 The prismatic beam is subjected to a uniformly distributed load and has a built-in support at $x = L$. (**a**) Determine the bending moment M as a function of x. (**b**) Determine the deflection v as a function of x.

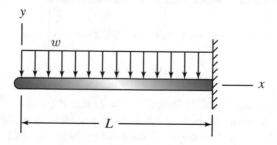

Problems 9.3–9.4

9.4 The length of the prismatic beam is $L = 48$ in, and the moment of inertia of its cross section is $I = 2.00$ in^4. The beam consists of steel with modulus of elasticity $E = 32E6$ psi. If the magnitude of the deflection of the beam at $x = 0$ is 0.2 in, what is the magnitude w of the uniformly distributed load in lb./in?

9.5 The beam has a circular cross section and is subjected to a uniformly distributed load. It consists of steel with modulus of elasticity $E = 120$ GPa. Determine the deflection and slope at $x = 1$ m.

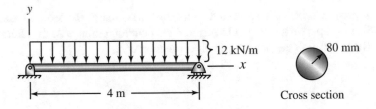

Problems 9.5–9.6

9.6 The beam has a circular cross section and is subjected to a uniformly distributed load. It consists of steel with modulus of elasticity $E = 120$ GPa. What is the beam's maximum deflection? What is the slope of the beam where the maximum deflection occurs?

9.7 The moment of inertia of the beam's cross section is $I = 2.4$ in^4, and it consists of steel with modulus of elasticity $E = 30E6$ psi. Determine the deflection at $x = 36$ in. (See the discussion of this beam and loading in Sect. 9.1.)

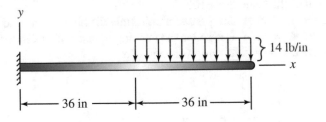

Problems 9.7–9.8

9.8 The moment of inertia of the beam's cross section is $I = 2.4$ in^4, and it consists of steel with modulus of elasticity $E = 30E6$ psi. What is the beam's maximum deflection? What is the slope of the beam where the maximum deflection occurs?

9.9 The beam consists of material with modulus of elasticity $E = 72$ GPa, and the moment of inertia of its cross section is $I = 1.6E - 7$ m^4. What is the maximum magnitude of the deflection between $x = 0$ and $x = 2$ m?

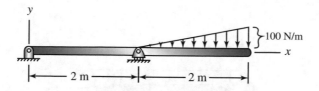

Problems 9.9–9.10

9.10* The beam consists of material with modulus of elasticity $E = 72$ GPa, and the moment of inertia of its cross section is $I = 1.6E - 7$ m^4. What is the magnitude of the deflection of the beam's right end?

For the beams in Problems 9.11–9.21, determine the deflection v as a function of x, and confirm the results in Appendix F. Each beam has length L, moment of inertia I, and modulus of elasticity E.

9.11

Problem 9.11

9.12

Problem 9.12

9.13

Problem 9.13

9.14

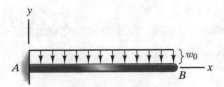

Problem 9.14

9.15

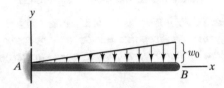

Problem 9.15

9.16

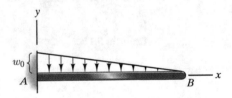

Problem 9.16

9.17

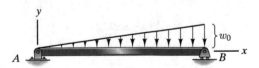

Problem 9.17

9.18∗

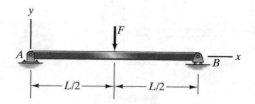

Problem 9.18

9.19∗

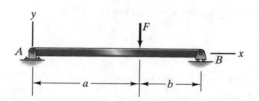

Problem 9.19

9.20∗

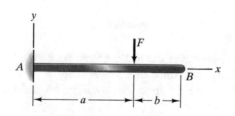

Problem 9.20

9.21∗

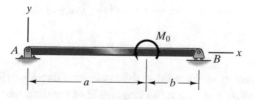

Problem 9.21

9.22 The beam is made of aluminum alloy with modulus of elasticity $E = 70$ GPa. When the force F is applied, the magnitude of the deflection at B is measured and determined to be 4.8 mm. What is the maximum resulting tensile stress in the beam?

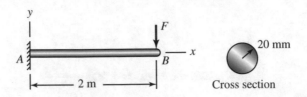

Problem 9.22

9.23 The beam is made of titanium alloy with modulus of elasticity $E = 15E6$ psi. When the uniformly distributed load of magnitude w_0 is applied, the magnitude of the deflection at B is measured and determined to be 2 in. What is the maximum resulting tensile stress in the beam?

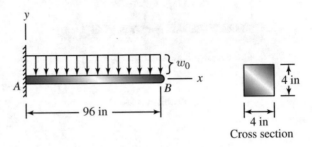

Problem 9.23

9.24* The moment of inertia of the beam's cross section is $I = 2E - 5$ m^4, and it is made of material with modulus of elasticity $E = 120$ GPa. Where is the magnitude of the beam's deflection a maximum, and what is its value?

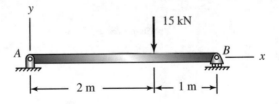

Problem 9.24

9.25 The beam consists of material with elastic modulus $E = 190$ GPa. Determine the axial position x at which the magnitude of the deflection due to the couple M_0 is a maximum.

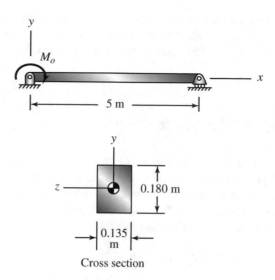

Cross section

Problems 9.25–9.27

9.26 The beam consists of material with elastic modulus $E = 80$ GPa. It is used in a structure whose design requires that the magnitude of the beam's maximum deflection be no greater than 30 mm. Determine the maximum couple M_0 that can be applied.

9.27 The beam consists of material with elastic modulus $E = 80$ GPa. The magnitude of the deflection due to the couple M_0 is measured at $x = 2.5$ m and determined to be 20 mm. What is the maximum tensile stress in the beam?

9.28 The titanium beam has elastic modulus $E = 16E6$ psi. Determine the axial position x at which the magnitude of the deflection is a maximum.

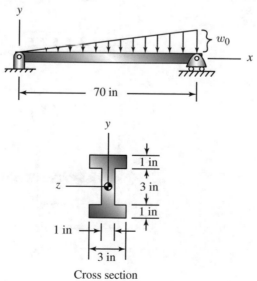

Cross section

Problems 9.28–9.29

9.29* The titanium beam has elastic modulus $E = 16E6$ psi. It is used in a structure whose design requires that the magnitude of the beam's maximum deflection be no greater than 1.0 in. Determine the maximum value of w_0 that can be applied.

9.2   Statically Indeterminate Beams

Remarkably, the procedure we used in the previous section to determine the deflection of a statically determinate beam also works for a statically indeterminate beam, yielding both the beam's deflection and the unknown reactions acting on it. For example, consider the beam in Fig. 9.10(a). From the free-body diagram of the beam (Fig. 9.10(b)), we obtain the equilibrium equations

$$\Sigma F_x = A_x = 0, \tag{9.8}$$

$$\Sigma F_y = A_y + B - w_0 L = 0, \tag{9.9}$$

$$\Sigma M_{\text{point } A} = M_A + LB - \frac{L}{2} w_0 L = 0. \tag{9.10}$$

We can't solve Eqs. (9.9) and (9.10) for the three reactions A_y, M_A, and B. This beam is statically indeterminate.

We ignore this setback and proceed to determine the beam's deflection *in terms of the unknown reactions*. From Fig. 9.11, the distribution of the bending moment is

Fig. 9.10 (a) Statically indeterminate beam. (b) Free-body diagram of the entire beam

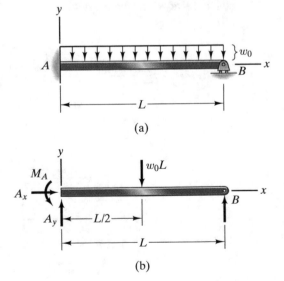

$$M = -\frac{w_0 x^2}{2} + A_y x - M_A.$$

Substituting this expression into Eq. (9.4) and integrating gives

$$EIv'' = -\frac{w_0 x^2}{2} + A_y x - M_A,$$

$$EIv' = -\frac{w_0 x^3}{6} + \frac{A_y x^2}{2} - M_A x + C,$$

$$EIv = -\frac{w_0 x^4}{24} + \frac{A_y x^3}{6} - \frac{M_A x^2}{2} + Cx + D,$$

where C and D are integration constants.

Our next step is to use the boundary conditions to evaluate the integration constants. But while there are two integration constants, we see from Fig. 9.12 that there are three boundary conditions. This is the key to the solution: *There are more boundary conditions than unknown integration constants.* (The boundary conditions are compatibility conditions imposed on the beam's deflection.) We can use the three equations obtained from the boundary conditions together with the two equilibrium

Fig. 9.11 Determining the distribution of the bending moment

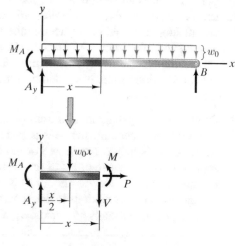

Fig. 9.12 Boundary conditions

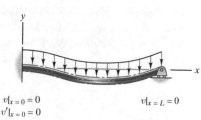

$$v|_{x=0} = 0$$
$$v'|_{x=0} = 0$$

$$v|_{x=L} = 0$$

equations (9.9) and (9.10) to determine the integration constants C and D and the unknown reactions A_y, M_A, and B.

The boundary conditions can be written as

$$EIv|_{x=0} = D = 0,$$

$$EIv'|_{x=0} = C = 0,$$

$$EIv|_{x=L} = -\frac{w_0 L^4}{24} + \frac{A_y L^3}{6} - \frac{M_A L^2}{2} + CL + D = 0.$$

We see that $C = 0$ and $D = 0$. Solving the third equation together with Eqs. (9.9) and (9.10) yields the unknown reactions: $A_y = 5w_0 L/8$, $B = 3w_0 L/8$, and $M_A = w_0 L^2/8$. We complete the solution by substituting these results into the expression for v to obtain the beam's deflection:

$$v = -\frac{w_0}{48EI}\left(2x^4 - 5Lx^3 + 3L^2x^2\right).$$

This example sheds light on the analysis of statically indeterminate structures in general. An object is statically indeterminate if it has more supports than are necessary for equilibrium. If the roller support is removed from the beam in Fig. 9.10(a), it remains in equilibrium and is statically determinate. The roller support introduces an additional reaction, which makes the beam statically indeterminate, *but also introduces an additional boundary (compatibility) condition*. Each redundant support added to a structure introduces a new compatibility condition. As a consequence, the number of combined equilibrium equations and compatibility conditions remains sufficient to determine the reactions.

In the following example, we demonstrate the steps required to analyze statically indeterminate beams:

1. Draw a free-body diagram of the entire beam and write the equilibrium equations.
2. Determine the bending moment as a function of x in terms of the unknown reactions. This may result in two or more functions M_1, $M_2 \ldots$, each of which applies to a different segment of the beam's length.
3. For each segment, integrate Eq. (9.4) twice to determine v_1, $v_2 \ldots$ ..
4. Use the boundary conditions together with the equilibrium equations to determine the integration constants and unknown reactions.

Example 9.2 Statically Indeterminate Beam
The beam in Fig. 9.13 has length $L = 4\,\text{m}$. The magnitude of the distributed load is $w_0 = 5\,\text{kN/m}$. What are the reactions at A, B, and C?

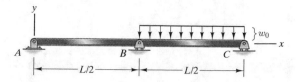

Fig. 9.13

Strategy

We will draw the free-body diagram of the beam and write the equilibrium equations. The distribution of the bending moment in the beam will then be determined *in terms of the unknown reactions on the beam.* We will then use Eq. (9.4) to determine the distribution of the displacement. By applying the boundary conditions together with the equilibrium equations, we can determine the integration constants as well as the unknown reactions.

Solution

Write the equilibrium equations   From the free-body diagram in Fig. (a), we obtain the equations

$$\Sigma F_x = A_x = 0,$$

$$\Sigma F_y - A_y + B + C - \frac{w_0 L}{2} = 0,$$ (9.11 and 9.12)

$$\Sigma M_{\text{point } A} = \frac{L}{2}B + LC - \left(\frac{3}{4}L\right)\frac{w_0 L}{2} = 0.$$

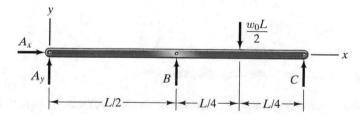

(a) Free-body diagram of the entire beam

   Determine the bending moment as a function of x   From Fig. (b), the distributions of the bending moment in the left and right halves of the beam are

$$M_1 = A_y x,$$

$$M_2 = -\frac{w_0 x^2}{2} + (w_0 L - C)x - \left(\frac{w_0 L}{2} - C\right)L.$$

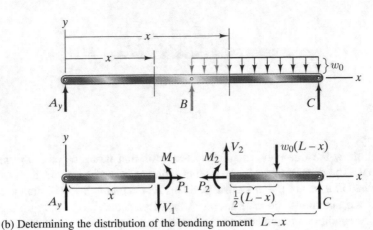

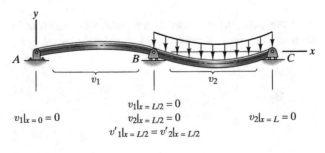

(b) Determining the distribution of the bending moment $L-x$

For each segment, integrate Eq. (9.4)   Substituting the expressions for M_1 and M_2 and integrating, we obtain

$$EIv_1'' = M_1 = A_y x,$$

$$EIv_1' = \frac{A_y x^2}{2} + G,$$

$$EIv_1 = \frac{A_y x^3}{6} + Gx + H,$$

$$EIv_2'' = M_2 = -\frac{w_0 x^2}{2} + (w_0 L - C)x - \left(\frac{w_0 L}{2} - C\right)L,$$

$$EIv_2' = -\frac{w_0 x^3}{6} + (w_0 L - C)\frac{x^2}{2} - \left(\frac{w_0 L}{2} - C\right)Lx + J,$$

$$EIv_2 = -\frac{w_0 x^4}{24} + (w_0 L - C)\frac{x^3}{6} - \left(\frac{w_0 L}{2} - C\right)\frac{Lx^2}{2} + Jx + K,$$

where G, H, J, and K are integration constants.

Use the boundary conditions together with the equilibrium equations
We can apply the five boundary conditions shown in Fig. (c).

$$v_1|_{x=L/2} = 0$$
$$v_1|_{x=0} = 0 \qquad v_2|_{x=L/2} = 0 \qquad v_2|_{x=L} = 0$$
$$v'_1|_{x=L/2} = v'_2|_{x=L/2}$$

(c) Boundary conditions

The deflection is zero at $x = 0$:

$$EIv_1|_{x=0} = H = 0. \tag{9.13}$$

Both v_1 and v_2 are zero at $x = L/2$:

$$EIv_1|_{x=L/2} = \frac{A_y L^3}{48} + \frac{GL}{2} + H = 0,$$

$$EIv_2|_{x=L/2} = -\frac{w_0 L^4}{384} + (w_0 L - C)\frac{L^3}{48} - \left(\frac{w_0 L}{2} - C\right)\frac{L^3}{8} \qquad \text{(9.14 and 9.15)}$$

$$+ \frac{JL}{2} + K = 0.$$

The slopes of the two solutions must be equal at $x = L/2$.

$$EIv_1'|_{x=L/2} = EIv_2'|_{x=L/2}:$$

$$\frac{A_y L^2}{8} + G = -\frac{w_0 L^3}{48} + (w_0 L - C)\frac{L^2}{8} - \left(\frac{w_0 L}{2} - C\right)\frac{L^2}{2} + J. \tag{9.16}$$

The deflection is zero at $x = L$:

$$EIv_2|_{x=L} = -\frac{w_0 L^4}{24}$$

$$+ (w_0 L - C)\frac{L^3}{6} - \left(\frac{w_0 L}{2} - C\right)\frac{L^3}{2} + JL + K = 0. \tag{9.17}$$

We can solve the equilibrium equations (9.11) and (9.12) together with Eqs. (9.13)–(9.17) for the integration constants G, H, J, and K and the unknown reactions A_y, B, and C. Substituting the values of L and w_0 and solving, the solutions for the reactions are $A_y = -625\,\text{N}$, $B = 6250\,\text{N}$, and $C = 4375$ N.

Discussion

If the beam in this example did not have the roller support at C, it would be statically determinate. It has one *redundant support*, or one more support than is necessary for the beam to be held in equilibrium under the type of loads to which it is subjected. (This statement does not imply that the beam would necessarily support the given loads without failing if the support at C were removed. It means that the beam will remain in equilibrium under those loads if it is sufficiently strong.) As a result, there is one more unknown reaction than can be determined from the equilibrium equations. But the support at C also introduces a new boundary condition, which allowed us to determine both the unknown reactions and the beam's deflection.

Problems

9.30 The prismatic beam has elastic modulus E and moment of inertia I. Determine the reactions at A and B in terms of the applied couple M_0.

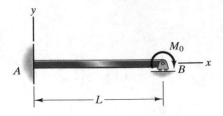

Problems 9.30–9.31

9.31 The prismatic beam has elastic modulus E and moment of inertia I. Determine the deflection as a function of x in terms of the applied couple M_0.

9.32 The prismatic beam has elastic modulus E and moment of inertia I. Determine the reactions at A and B.

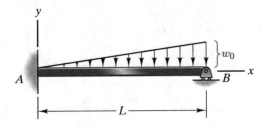

Problems 9.32–9.33

9.33 The prismatic beam has elastic modulus E and moment of inertia I. Determine the deflection as a function of x.

9.34 The length of the structural beam is $L = 144$ in. It is built-in at both ends and is subjected to a uniform distributed load $w_0 = 170$ lb/in. The free-body diagram of the beam is shown. Determine the reactions F_0 and M_0.

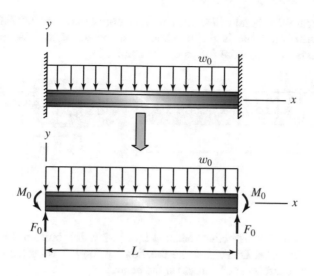

Problems 9.34–9.36

9.35 The length of the structural beam is $L = 144$ in. The moment of inertia of its cross section is $I = 28$ in^4, and the modulus of elasticity of its material is $E = 32E6$ psi. It is built-in at both ends and is subjected to a uniform distributed load $w_0 = 170$ lb/in. The free-body diagram of the beam is shown. What is the maximum magnitude of its deflection?

9.36 The structural beam is built-in at both ends and is subjected to a uniform distributed load w_0. The free-body diagram of the beam is shown. The beam's length is $L = 144$ in. Determine the value of x at which the magnitude of the beam's slope is a maximum.

9.37 The beam is made of material that will safely support a normal stress of 110 ksi. Based on this criterion, what is the maximum safe value of w_0? (Assume that the supports exert no axial forces on the beam.)

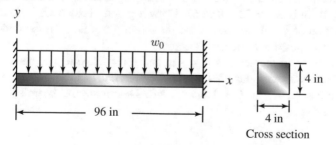

Problem 9.37

9.38 The beam's material will safely support a normal stress of 220 MPa. Based on this criterion, what is the maximum safe value of w_0? (Assume that the supports exert no axial forces on the beam.)

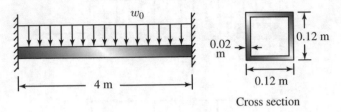

Cross section

Problem 9.38

9.39 The length of the structural beam is $L = 144$ in. It is built-in at both ends and $w_0 = 170$ lb/in. Determine the reactions at the left wall. (Assume that the supports exert no axial forces on the beam.)

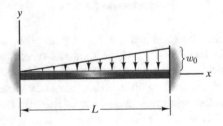

Problems 9.39–9.41

9.40 The prismatic beam has elastic modulus E and moment of inertia I. Determine the deflection at $x = L/2$.

9.41* The prismatic beam has length $L = 10$ m, elastic modulus $E = 210$ GPa, and moment of inertia $I = 8E\text{-}6$ m^4. The magnitude of the distributed load at $x = 10$ m is $w_0 = 12$ kN/m. **(a)** Draw a graph of the beam's deflection as a function of x. **(b)** Determine the maximum magnitude of the deflection and the axial position at which it occurs.

9.42* The length of the beam is $L = 4$ m and $w_0 = 5$ kN/m. The beam consists of material with modulus of elasticity $E = 72$ GPa, and the moment of inertia of its cross section is $I = 1.6E\text{-}7$ m^4. Determine the beam's deflection at $x = 1$ m.

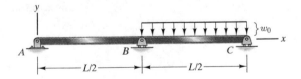

Problems 9.42–9.43

9.43* The length of the beam is $L = 4\,\text{m}$ and $w_0 = 5\,\text{kN/m}$. The beam consists of material with modulus of elasticity $E = 72\,\text{GPa}$, and the moment of inertia of its cross section is $I = 1.6\text{E} \text{-} 7\,\text{m}^4$. What is the maximum magnitude of the beam's deflection between $x = 0$ and $x = 2\,\text{m}$? At what value of x does it occur?

9.44* The length of the structural beam is $L = 144$ in. The moment of inertia of its cross section is $I = 28\,\text{in}^4$, and the modulus of elasticity of its material is $E = 32\text{E}6$ psi. It is built-in at both ends and $w_0 = 170\,\text{lb/in}$. Determine the reactions at A. Assume that the supports exert no axial forces on the beam. (See the figure for Problems 9.44–9.46.)

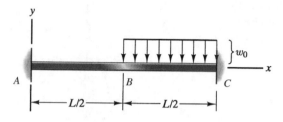

Problems 9.44–9.46

9.45* The length of the structural beam is $L = 144$ in. The moment of inertia of its cross section is $I = 6\,\text{in}^4$, and the modulus of elasticity of its material is $E = 32\text{E}6$ psi. It is built-in at both ends and $w_0 = 400\,\text{lb/in}$. Determine the deflection of the beam at $x = 36$ in. (See the figure for Problems 9.44–9.46.)

9.46* The length of the structural beam is $L = 144$ in. The moment of inertia of its cross section is $I = 6\,\text{in}^4$, and the modulus of elasticity of its material is $E = 32\text{E}6$ psi. It is built-in at both ends and $w_0 = 400\,\text{lb/in}$. Determine the deflection of the beam at $x = 108$ in. (See the figure for Problems 9.44–9.46.)

9.3 Singularity Functions

We introduced singularity functions in Sect. 5.4 and used them to determine the shear force and bending moment distributions in beams. (A review of that section is recommended.) Unlike traditional approaches, which can result in different sets of equations for different portions of a beam's axis, singularity functions allow us to obtain equations for V and M that apply to an entire beam. This technique is even more advantageous for determining deflections of beams with complex loadings. It yields a single equation for the deflection as a function of x, which reduces the number of integration constants and boundary conditions we must contend with.

The singularity functions discussed in Sect. 5.4, and their integrals, are summarized in Table 9.1. With these functions we can express the reactions and loads on a beam as a distributed load. Consider the beam in Fig. 9.14. A delta function

Table 9.1 Singularity functions

Type of function	w	$\int_0^x w\,dx$
Step function	$\langle x-a\rangle^0$	$\langle x-a\rangle$
Delta function	$\langle x-a\rangle^{-1}$	$\langle x-a\rangle^0$
Dipole	$\langle x-a\rangle^{-2}$	$\langle x-a\rangle^{-1}$
Macaulay functions ($n \geq 0$)	$\langle x-a\rangle^n = \begin{cases} 0 & \text{if } x < a, \\ (x-a)^n & \text{if } x \geq a. \end{cases}$	$\frac{1}{n+1}\langle x-a\rangle^{n+1}$

Fig. 9.14 A cantilever beam with a distributed load and its free-body diagram

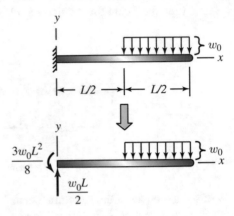

$w = \langle x - a\rangle^{-1}$ exerts a downward load of unit magnitude at $x = a$, so the contribution of the force exerted by the fixed support to the distributed load is

$$-\frac{w_0 L}{2}\langle x\rangle^{-1}.$$

A dipole $w = \langle x - a\rangle^{-2}$ exerts a counterclockwise couple of unit magnitude at $x = a$, so the contribution of the couple to the distributed load is

$$\frac{3w_0 L^2}{8}\langle x\rangle^{-2}.$$

The distributed load on the beam can be expressed in terms of the Macaulay function with $n = 0$, which is the step function:

$$w_0\left\langle x - \frac{L}{2}\right\rangle^0.$$

The loads on the beam are therefore represented by the distributed load

$$w = \frac{3w_0 L^2}{8}\langle x\rangle^{-2} - \frac{w_0 L}{2}\langle x\rangle^{-1} + w_0\left\langle x - \frac{L}{2}\right\rangle^0.$$

Using this result, we can integrate the equation $dV/dx = -w$ to determine the shear force distribution in the beam

$$V = \int_0^x -w\,dx = -\frac{3w_0L^2}{8}\langle x\rangle^{-1} + \frac{w_0L}{2}\langle x\rangle^0 - w_0\left\langle x - \frac{L}{2}\right\rangle,$$

and then integrate the equation $dM/dx = V$ to determine the bending moment distribution:

$$M = \int_0^x V\,dx = -\frac{3w_0L^2}{8}\langle x\rangle^0 + \frac{w_0L}{2}\langle x\rangle - \frac{w_0}{2}\left\langle x - \frac{L}{2}\right\rangle^2.$$

We have obtained a single equation that describes the bending moment throughout the beam in Fig. 9.14. We substitute it into Eq. (9.4) and integrate the resulting equation twice to obtain the deflection:

$$v'' = \frac{M}{EI} = \frac{w_0}{EI}\left[-\frac{3L^2}{8}\langle x\rangle^0 + \frac{L}{2}\langle x\rangle - \frac{1}{2}\left\langle x - \frac{L}{2}\right\rangle^2\right],$$

$$v' = \frac{w_0}{EI}\left[-\frac{3L^2}{8}\langle x\rangle + \frac{L}{4}\langle x\rangle^2 - \frac{1}{6}\left\langle x - \frac{L}{2}\right\rangle^3\right] + A\langle x\rangle^0,$$

$$v = \frac{w_0}{EI}\left[-\frac{3L^2}{16}\langle x\rangle^2 + \frac{L}{12}\langle x\rangle^3 - \frac{1}{24}\left\langle x - \frac{L}{2}\right\rangle^4\right] + A\langle x\rangle + B\langle x\rangle^0,$$

where A and B are integration constants. From the boundary conditions $v = 0$ at $x = 0$ and $v' = 0$ at $x = 0$, we determine that A and B are zero, so the deflection is

$$v = \frac{w_0}{EI}\left[-\frac{3L^2}{16}\langle x\rangle^2 + \frac{L}{12}\langle x\rangle^3 - \frac{1}{24}\left\langle x - \frac{L}{2}\right\rangle^4\right].$$

Thus we obtain a single equation for the deflection throughout the beam, shown in Fig. 9.15.

Compare this analysis using singularity functions with our treatment of the same beam in Sect. 9.1. Because we had different equations for the bending moment for $0 \le x \le L/2$ and for $L/2 \le x \le L$, we obtained different equations for the deflection in the left and right halves of the beam, each containing two integration constants. To evaluate the integration constants, we needed to use the two boundary conditions at $x = 0$, and also the conditions that the deflections and slopes of the solutions in the left and right halves must match at $x = L/2$. Singularity functions substantially simplified this example, and they become even more advantageous for beams with loadings of greater complexity (see Example 9.3).

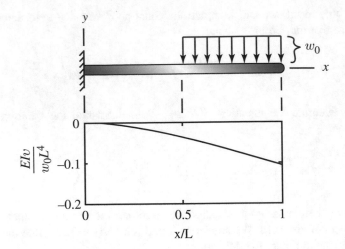

Fig. 9.15 Deflection of the beam

**Example 9.3 Determining a Beam Deflection Using Singularity
Functions**
The beam in Fig. 9.16 consists of material with modulus of elasticity
$E = 28\text{E}6$ psi. The moment of inertia of its cross section is $I = 0.06$ in^4, and
its length is $L = 120$ in. The force $F = 100$ lb, and the magnitude of the
distributed load is $w_0 = 10$ lb/in. What is the deflection at the right end of the
beam?

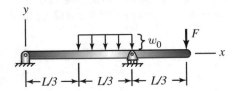

Fig. 9.16

Strategy
We will first use singularity functions to determine the bending moment in the
beam as a function of x. Then we can integrate Eq. (9.4) to determine the
deflection as a function of x.

Solution
Bending Moment   The free-body diagram of the beam is shown in Fig. (a).
From the equilibrium equations

$$\Sigma F_x = A_x = 0,$$

$$\Sigma F_y = A_y + B - F - \frac{w_0 L}{3} = 0,$$

$$\Sigma M_{\text{point } A} = \frac{2L}{3}B - LF - \frac{L}{2}\left(\frac{w_0 L}{3}\right) = 0,$$

we obtain

$$A_x = 0,$$

$$A_y = \frac{w_0 L}{12} - \frac{F}{2} = 50\,\text{lb},$$

$$B = \frac{w_0 L}{4} + \frac{3F}{2} = 450 \text{ lb.}$$

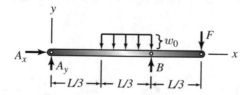

(a) Free-body diagram of the beam

We can express the forces acting on the beam in terms of delta functions. A delta function $\langle x - a \rangle^{-1}$ exerts a *downward* load of unit magnitude at $x = a$, so the contributions to the distributed load due to the forces on the beam are

$$-A_y \langle x \rangle^{-1} - B\left\langle x - \frac{2L}{3}\right\rangle^{-1} + F\langle x - L \rangle^{-1}.$$

For the distributed load on the beam, we use a step function representing a downward distributed load of magnitude w_0 that begins at $x = L/3$ and superimpose a step function representing an *upward* distributed load that begins at $x = 2L/3$:

$$w_0 \left\langle x - \frac{L}{3}\right\rangle^0 - w_0\left\langle x - \frac{2L}{3}\right\rangle^0.$$

The total distributed load is therefore

$$w = -A_y \langle x \rangle^{-1} - B \left\langle x - \frac{2L}{3} \right\rangle^{-1} + F \langle x - L \rangle^{-1} + w_0 \left\langle x - \frac{L}{3} \right\rangle^{0}$$
$$- w_0 \left\langle x - \frac{2L}{3} \right\rangle^{0}.$$

We integrate the equation $dV/dx = -w$ to determine the shear force distribution

$$V = \int_0^x -w \, dx$$
$$= A_y \langle x \rangle^{0} + B \left\langle x - \frac{2L}{3} \right\rangle^{0} - F \langle x - L \rangle^{0} - w_0 \left\langle x - \frac{L}{3} \right\rangle + w_0 \left\langle x - \frac{2L}{3} \right\rangle,$$

and then integrate the equation $dM/dx = V$ to determine the bending moment distribution:

$$M = \int_0^x V \, dx$$
$$= A_y \langle x \rangle + B \left\langle x - \frac{2L}{3} \right\rangle - F \langle x - L \rangle - \frac{w_0}{2} \left\langle x - \frac{L}{3} \right\rangle^{2} + \frac{w_0}{2} \left\langle x - \frac{2L}{3} \right\rangle^{2}.$$

Deflection   We substitute the bending moment into Eq. (9.4), writing it as

$$EIv'' = M$$
$$= A_y \langle x \rangle + B \left\langle x - \frac{2L}{3} \right\rangle - F \langle x - L \rangle - \frac{w_0}{2} \left\langle x - \frac{L}{3} \right\rangle^{2} + \frac{w_0}{2} \left\langle x - \frac{2L}{3} \right\rangle^{2}.$$

We integrate this equation twice to obtain the deflection

$$EIv' = \frac{A_y}{2} \langle x \rangle^{2} + \frac{B}{2} \left\langle x - \frac{2L}{3} \right\rangle^{2} - \frac{F}{2} \langle x - L \rangle^{2}$$
$$- \frac{w_0}{6} \left\langle x - \frac{L}{3} \right\rangle^{3} + \frac{w_0}{6} \left\langle x - \frac{2L}{3} \right\rangle^{3} + C \langle x \rangle^{0},$$

$$EIv = \frac{A_y}{6} \langle x \rangle^{3} + \frac{B}{6} \left\langle x - \frac{2L}{3} \right\rangle^{3} - \frac{F}{6} \langle x - L \rangle^{3}$$
$$- \frac{w_0}{24} \left\langle x - \frac{L}{3} \right\rangle^{4} + \frac{w_0}{24} \left\langle x - \frac{2L}{3} \right\rangle^{4} + C \langle x \rangle + D \langle x \rangle^{0},$$

where C and D are integration constants. From the boundary conditions that $v = 0$ at $x = 0$ and $v = 0$ at $x = 2L/3$, we obtain

$$C = -\frac{2A_yL^2}{27} + \frac{w_0L^3}{1296},$$
$$D = 0.$$

The deflection of the beam as a function of x is therefore

$$v = \frac{1}{EI}\left[\frac{A_y}{6}\langle x\rangle^3 + \frac{B}{6}\left\langle x - \frac{2L}{3}\right\rangle^3 - \frac{F}{6}\langle x - L\rangle^3 - \frac{w_0}{24}\left\langle x - \frac{L}{3}\right\rangle^4 + \frac{w_0}{24}\left\langle x - \frac{2L}{3}\right\rangle^4 \right.$$
$$\left. - \left(\frac{2A_yL^2}{27} - \frac{w_0L^3}{1296}\right)\langle x\rangle\right].$$

The deflection is shown in Fig. (b). Setting $x = 120$ in, the deflection at the right end is $v = -0.952$ in.

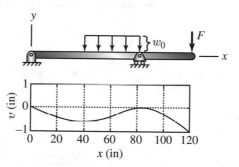

(b) Graph of the deflection

Discussion

Solving this example without the use of singularity functions would require three different equations to describe the bending moment throughout the beam, which would result in three equations for the deflection containing six integration constants to be determined. Singularity functions become increasingly advantageous as the complexity of loading of the beam increases.

Problems

Problems 9.47–9.54 are to be solved using singularity functions.

9.47 The beam has pin and roller supports and is subjected to a force F. Determine the deflection v as a function of x.

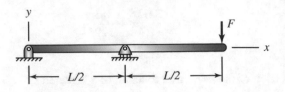

Problem 9.47

9.48 The beam has pin and roller supports and is subjected to a counterclockwise couple M_0. Determine the deflection v as a function of x.

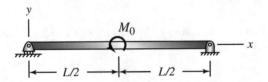

Problem 9.48

9.49 The beam has pin and roller supports and is subjected to two forces. Determine the deflection v as a function of x.

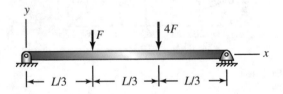

Problem 9.49

9.50 The beam consists of aluminum alloy with modulus of elasticity $E = 72$ GPa, and its moment of inertia is $I = 1.4\text{E} - 5$ m^4. Determine the deflection of the right end of the beam.

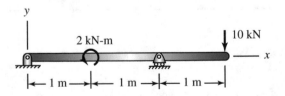

Problem 9.50

9.51 The beam consists of material with modulus of elasticity $E = 1.6\text{E}6$ psi, and its moment of inertia is $I = 0.125$ in^4. Determine the deflection of the beam at $x = 10$ in.

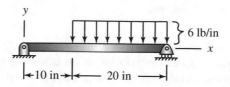

Problem 9.51

9.52 The beam consists of aluminum alloy with modulus of elasticity $E = 200$ GPa, and its moment of inertia is $I = 8.2E - 6$ m^4. Determine the deflection of the right end of the beam.

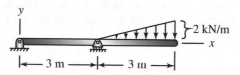

Problem 9.52

9.53∗ The modulus of elasticity of the beam's material is $E = 30E6$ psi, and the moment of inertia of its cross section is $I = 0.35$ in^4. Plot a graph of the beam's deflection as a function of x.

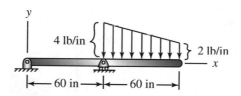

Problem 9.53

9.54 The beam has a fixed support at the left end. Determine the value of the force F for which the deflection of the right end of the beam is zero.

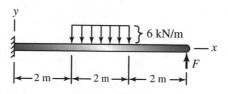

Problem 9.54

9.4   Moment-Area Method

When the deflection or slope of a beam must be determined only at specific locations
along its axis, a technique called the moment-area method is sometimes quicker than
the other approaches we have described. This method is based on two theorems that
relate the slopes and deflections of a beam at two axial positions.

9.4.1   First Theorem

Because we assume the deflection v and slope dv/dx to be small, the angle θ between
a beam's neutral axis and the x axis (Fig. 9.17) is equal to the slope of the deflection:
$dv/dx = \theta$. Therefore we can write Eq. (9.4) as

$$\frac{d}{dx}\left(\frac{dv}{dx}\right) = \frac{d\theta}{dx} = \frac{M}{EI}. \tag{9.18}$$

Let A and B be two positions on the beam's axis with axial coordinate x_A and x_B,
and let θ_A and θ_B be the values of the slope at those positions (Fig. 9.18). We
integrate Eq. (9.18) with respect to x from A to B

Fig. 9.17 The angle θ
between the beam's neutral
axis and the x axis is equal to
the slope of the deflection

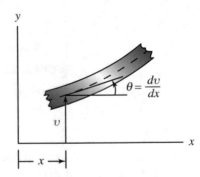

Fig. 9.18 The deflections
and slopes at A and B

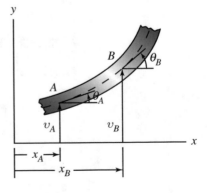

Fig. 9.19 Area defined by
the graph of *M/EI*

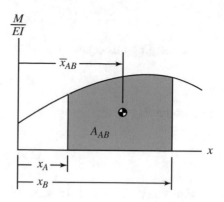

$$\int_{\theta_A}^{\theta_B} d\theta = \int_{x_A}^{x_B} \frac{M}{EI}\, dx,$$

which yields the result

$$\theta_B - \theta_A = A_{AB}, \tag{9.19}$$

where

$$A_{AB} = \int_{x_A}^{x_B} \frac{M}{EI}\, dx \tag{9.20}$$

is the area defined by the graph of *M/EI* from x_A to x_B (Fig. 9.19). Equation (9.19) is
the first moment-area theorem. It allows us to determine the change in the slope of a
beam's neutral axis between two axial positions by calculating the area defined by
the graph of *M/EI* between those positions.

9.4.2   Second Theorem

To obtain the second theorem, we multiply Eq. (9.4) by x and integrate with respect
to x from A to B:

$$\int_{x_A}^{x_B} x\frac{d^2v}{dx^2}\, dx = \int_{x_A}^{x_B} x\frac{M}{EI}\, dx. \tag{9.21}$$

Consider a function f of x. The differential of the product of x and f is

$$d(xf) = xdf + f\,dx.$$

Rearranging this equation as

$$xdf = d(xf) - f\,dx$$

and integrating yields

$$\int xdf = xf - \int f\,dx. \tag{9.22}$$

Equation (9.22) is called *integration by parts*. We apply integration by parts to the integral on the left side of Eq. (9.21) with $f = dv/dx$, obtaining

$$\int_{x_A}^{x_B} x\underbrace{\frac{d^2v}{dx^2}\,dx}_{df} = \left[x\underbrace{\frac{dv}{dx}}_{f} \right]_{x_A}^{x_B} - \int_{x_A}^{x_B} \underbrace{\frac{dv}{dx}}_{f}\,dx \tag{9.23}$$

$$= [x\theta]_{x_A}^{x_B} - \int_{v_A}^{v_B} dv$$

$$= x_B\theta_B - x_A\theta_A - v_B + v_A.$$

Let $\bar{x}_{AB}$ denote the axial position of the centroid of the area defined by the graph of M/EI from x_A to x_B (Fig. 9.19):

$$\bar{x}_{AB} = \frac{\int_{x_A}^{x_B} x\frac{M}{EI}dx}{\int_{x_A}^{x_B} \frac{M}{EI}dx} = \frac{\int_{x_A}^{x_B} x\frac{M}{EI}dx}{A_{AB}}.$$

Using this expression and Eq. (9.23), we can write Eq. (9.21) as

$$v_B - v_A = x_B\theta_B - x_A\theta_A - \bar{x}_{AB}A_{AB}. \tag{9.24}$$

The term $\bar{x}_{AB}A_{AB}$ can be regarded as the "moment" about the origin of the area A_{AB}. Equation (9.24) is one form of the second moment-area theorem. If the slope of a beam's neutral axis is known at two positions, we can use Eq. (9.24) to determine the change in the deflection between those positions by calculating the moment about the origin of the area A_{AB} defined by the graph of M/EI between those positions.

In traditional presentations of the moment-area method, the second theorem has been expressed in terms of the quantity $t_{B/A}$ illustrated in Fig. 9.20(a). It is defined to be the deflection of the neutral axis at B relative to the point at which the line tangent to the neutral axis at A intersects a vertical line through B. Figure 9.20(b) illustrates that $t_{B/A}$ is given in terms of the deflections at A and B and the slope at A by

Fig. 9.20 (a) Definition of
the variable $t_{B/A}$. (b)
Relating $t_{B/A}$ to the
deflections and slopes at
A and B

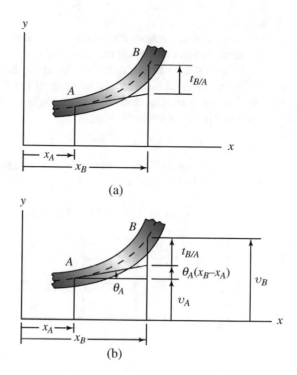

(a)

(b)

$$t_{B/A} = v_B - v_A - \theta_A(x_B - x_A).$$

Substituting Eq. (9.24) into this expression yields the equation

$$t_{B/A} = (x_B - \bar{x}_{AB})A_{AB}.$$

Although this form of the second moment-area theorem is simpler in form, we find the form given by Eq. (9.31) to be easier to explain and apply.

Example 9.4 Application of the Moment-Area Method
For the beam in Fig. 9.21, use the moment-area method to determine the slope and deflection at B.

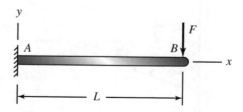

Fig. 9.21

Strategy
The slope and deflection of the beam at the built-in support A are both zero. Therefore, we can determine the slope at B from Eq. (9.19) and then determine the deflection at B from Eq. (9.24). To apply those equations, we must first determine the distribution of the bending moment M in the beam and use it to calculate the area A_{AB}, defined by the graph of M/EI from x_A to x_B, and the position $\bar{x}_{AB}$ of the centroid of A_{AB}.

Solution
By applying equilibrium to the free-body diagram in Fig. (a), the bending moment as a function of x is

$$M = F(x - L).$$

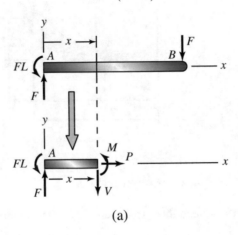

(a)

(a) Free-body diagram obtained by cutting the beam at an arbitrary position x

In Fig. (b) we draw the graph of the function

$$\frac{M}{EI} = \frac{F}{EI}(x - L).$$

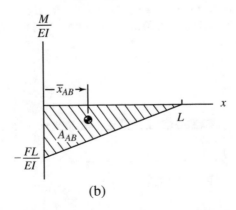

(b)

(b) Graph of M/EI

From the graph we see that

$$A_{AB} = \frac{1}{2}(L)\left(-\frac{FL}{EI}\right) = -\frac{FL^2}{2EI}$$

and

$$\bar{x}_{AB} = \frac{1}{3}L.$$

From Eq. (9.19), the slope at B is

$$\theta_B = \theta_A + A_{AB}$$
$$= -\frac{FL^2}{2EI}.$$

Now that θ_B is known, we can determine the deflection at B from Eq. (9.24).

$$v_B = v_A + x_B\theta_B - x_A\theta_A - \bar{x}_{AB}A_{AB}$$
$$= (0) + (L)\left(-\frac{FL^2}{2EI}\right) - (0) - \left(\frac{1}{3}L\right)\left(-\frac{FL^2}{2EI}\right)$$
$$= -\frac{FL^3}{3EI}.$$

Discussion
Observe that the slope and deflection at B given by the moment-area method agree with the results in Appendix F.

Example 9.5 Application of the Moment-Area Method
For the beam in Fig. 9.22, use the moment-area method to determine the deflection at B.

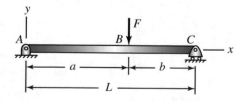

Fig. 9.22

Strategy

The deflection is zero at A and C. The deflection at B and the slopes at A, B, and C are unknown. By writing Eqs. (9.19) and (9.24) for the positions A and B and also for positions B and C, we can obtain four equations in four unknowns.

Solution

By applying equilibrium to the free-body diagrams in Fig. (a), the equations for the bending moment as a function of x are

$$M = \frac{Fb}{L}x \qquad 0 \le x \le a,$$

$$M = \frac{Fa}{L}(L - x) \qquad a \le x \le L.$$

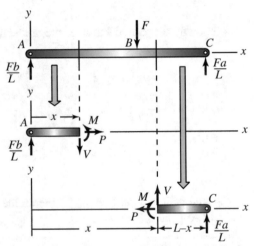

(a) Free-body diagrams for determining the bending moment distribution

In Fig. (b) we draw the graph of M/EI as a function of x. From the graph we see that

$$A_{AB} = \frac{1}{2}(a)\left(\frac{Fab}{LEI}\right) = \frac{Fa^2 b}{2LEI},$$

$$\bar{x}_{AB} = \frac{2}{3}a,$$

$$A_{BC} = \frac{1}{2}(b)\left(\frac{Fab}{LEI}\right) = \frac{Fab^2}{2LEI},$$

$$\bar{x}_{BC} = a + \frac{1}{3}b.$$

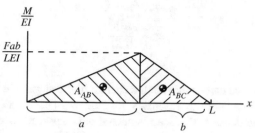

(b) Graph of M/EI

We write Eq. (9.26) for positions A and B and for positions B and C:

$$\theta_B - \theta_A = A_{AB}, \tag{1}$$

$$\theta_C - \theta_B = A_{BC}. \tag{2}$$

We also write Eq. (9.24) for positions A and B and for B and C:

$$v_B - v_A = x_B\theta_B - x_A\theta_A - \bar{x}_{AB}A_{AB}, \tag{3}$$

$$v_C - v_B = x_C\theta_C - x_B\theta_B - \bar{x}_{BC}A_{BC}. \tag{4}$$

Summing Eqs. (3) and (4) yields

$$v_C - v_A = x_C\theta_C - x_A\theta_A - \bar{x}_{AB}A_{AB} - \bar{x}_{BC}A_{BC}.$$

Because $v_A = 0$, $v_C = 0$ and $x_A = 0$, we can solve this equation for θ_C:

$$\theta_C = \frac{1}{x_C}\left(\bar{x}_{AB}A_{AB} + \bar{x}_{BC}A_{BC}\right)$$

$$= \frac{1}{L}\left[\left(\frac{2}{3}a\right)\left(\frac{Fa^2b}{2LEI}\right) + \left(a + \frac{1}{3}b\right)\left(\frac{Fab^2}{2LEI}\right)\right]$$

$$= \frac{Fab(L + a)}{6LEI}.$$

From Eq. (2), the slope at B is

$$\theta_B = \theta_C - A_{BC}$$

$$= \frac{Fab}{3LEI}(a - b).$$

Using this result, we can solve Eq. (3) for the deflection at B:

$$v_B = x_B \theta_B - \bar{x}_{AB} A_{AB}$$

$$= -\frac{Fa^2 b^2}{3LEI}.$$

Discussion

With Eq. (9.19) we can determine the change in the slope of a beam's neutral axis between two axial positions. With Eq. (9.24) we can determine the change in the deflection of a beam's neutral axis between two axial positions when the slopes at those positions are known. In this example the slope of the beam was not known at any point. But by writing Eqs. (9.19) and (9.24) for points A and B *and* for points B and C, we were able to determine the deflection at B.

Problems

Problems 9.55–9.61 are to be solved by the moment-area method.

9.55 The beam is subjected to a clockwise couple M_0 at B. Determine the slope at B and confirm the result in Appendix F.

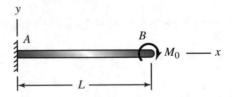

Problems 9.55–9.56

Strategy: Determine the bending moment M in the beam as a function of x, and then use Eq. (9.19) to determine the slope.

9.56 The beam is subjected to a clockwise couple M_0 at B. Determine the deflection at B and confirm the result in Appendix F.

9.57 The beam is subjected to a uniformly distributed load. Determine the slope at B and confirm the result in Appendix F.

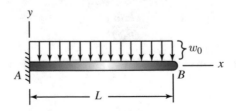

Problems 9.57–9.58

9.58 The beam is subjected to a uniformly distributed load. Determine the deflection at B and confirm the result in Appendix F.

9.59 The beam has a circular cross section and is subjected to a uniformly distributed load. It consists of steel with modulus of elasticity $E = 200$ GPa. Determine the slope at A.

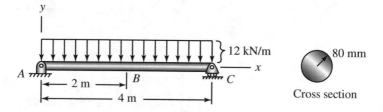

Problems 9.59–9.60

9.60 The beam has a circular cross section and is subjected to a uniformly distributed load. It consists of steel with modulus of elasticity $E = 200$ GPa. Determine the deflection at B.

9.61∗ The beam is subjected to a clockwise couple M_0. Determine the slope and deflection at B.

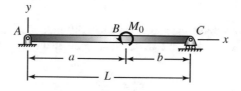

Problem 9.61

9.5 Superposition

Determining deflections of beams is a time-consuming process even when the loads are relatively simple. For the convenience of structural engineers, deflections of prismatic beams with typical supports and simple loads are available in tables such as Appendix F. Furthermore, we will show that the solutions in such tables can be superimposed to obtain deflections of beams with more complicated loadings.

Consider the beam and loading in Fig. 9.23(a). Let the bending moment in the beam be denoted by M_a, and let v_a be its deflection. The bending moment and deflection satisfy Eq. (9.4):

$$v_a'' = \frac{M_a}{EI}. \tag{9.25}$$

Fig. 9.23 Superimposing
the loads (**a**) and (**b**) results
in (**c**)

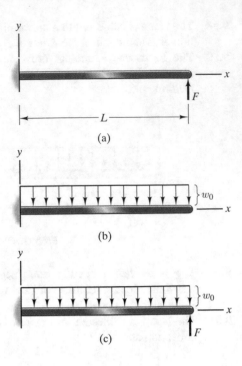

(a)

(b)

(c)

In Fig. 9.23(b) we subject the same beam to a different load. Let M_b and v_b denote
the resulting bending moment and deflection. They also satisfy Eq. (9.4):

$$v_b'' = \frac{M_b}{EI}. \tag{9.26}$$

In Fig. 9.23(c) we superimpose the loads in Fig. 9.23(a) and (b). Summing
Eqs. (9.25) and (9.26), we conclude that the superimposed deflections and bending
moments also satisfy Eq. (9.4):

$$(v_a + v_b)'' = \frac{M_a + M_b}{EI}.$$

This result, a consequence of the linearity of the differential equation for the
displacement, confirms that we can obtain the deflection resulting from the loads in
Fig. 9.23(c) by summing the deflections resulting from the loads in Fig. 9.23(a) and (b).
Using the deflections given in Appendix F, we obtain

$$v_a + v_b = \frac{Fx^2}{6EI}(3L - x) - \frac{w_0 x^2}{24EI}\left(6L^2 - 4Lx + x^2\right).$$

The reduction in effort achieved by this approach is obvious. Notice that to determine the deflection of a given beam by superposition, the loading must be matched by the superimposed loads, and the boundary conditions must be satisfied by the superimposed displacements.

Example 9.6 Beam Deflection by Superposition
Use superposition to determine the deflection of the beam in Fig. 9.24.

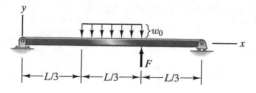

Fig. 9.24

Strategy
We can obtain this loading by superimposing the loads in Figs. (a), (b), and (c), so we can determine the deflection by superimposing the deflections for these three loads.

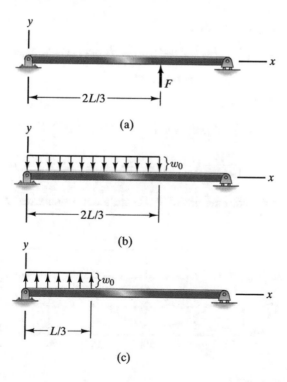

(a), (b), (c) Three loads that can be superimposed to obtain the desired load

Solution

Using the results in Appendix F, the deflections due to the loads in Figs. (a), (b), and (c) are

$$
v_a = \begin{cases} \dfrac{Fx}{162EI}\left(8L^2 - 9x^2\right), & 0 \le x \le 2L/3 \\[2ex] \dfrac{F}{162EI}\left(18x^3 - 54Lx^2 + 44L^2x - 8L^3\right), & 2L/3 \le x \le L \end{cases}
$$

$$
v_b = \begin{cases} -\dfrac{w_0x}{1944EI}\left(64L^3 - 144Lx^2 + 81x^3\right), & 0 \le x \le 2L/3 \\[2ex] -\dfrac{w_0L}{1944EI}\left(72x^3 - 216Lx^2 + 160L^2x - 16L^3\right), & 2L/3 \le x \le L \end{cases}
$$

$$
v_c = \begin{cases} \dfrac{w_0x}{1944EI}\left(25L^3 - 90Lx^2 + 81x^3\right), & 0 \le x \le L/3 \\[2ex] \dfrac{w_0L}{1944EI}\left(18x^3 - 54Lx^2 + 37L^2x - L^3\right). & L/3 \le x \le L \end{cases}
$$

Superimposing these results, we obtain the beam's deflection.

$0 \le x \le L/3$:
$$
v = \frac{Fx}{162EI}\left(8L^2 - 9x^2\right) - \frac{w_0x}{1944EI}\left(39L^3 - 54Lx^2\right).
$$

$L/3 \le x \le 2L/3$:
$$
v = \frac{Fx}{162EI}\left(8L^2 - 9x^2\right) - \frac{w_0}{1944EI}\left(81x^4 - 162Lx^3 \right.
$$
$$
\left. + 54L^2x^2 + 27L^3x + L^4\right).
$$

$2L/3 \le x \le L$:
$$
v = \frac{F}{162EI}\left(18x^3 - 54Lx^2 + 44L^2x - 8L^3\right)
$$
$$
- \frac{w_0L}{1944EI}\left(54x^3 - 162Lx^2 + 123L^2x - 15L^3\right).
$$

A nondimensional measure of the beam's deflection when $F = w_0L/3$ is shown in Fig. (d).

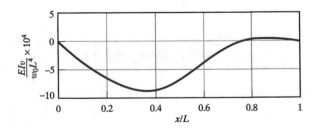

(d) Graph of the deflection

Discussion

Notice how we obtained a uniformly distributed load acting over the middle third of the beam by superimposing the loadings in Figs. (b) and (c). The upward load in Fig. (c) canceled part of the downward load in Fig. (b). With the results in Appendix F and stratagems such as the one we used in this example, you can match or approximate a large variety of loadings and boundary conditions.

Example 9.7 Statically Indeterminate Beam by Superposition
Use superposition to determine the deflection of the beam in Fig. 9.25.

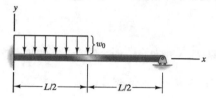

Fig. 9.25

Strategy

The beam is statically indeterminate. If we superimpose the loads in Figs. (a) and (b), we can express the beam's deflection in terms of the unknown reaction B exerted by the roller support. Then we can determine B from the condition that the deflection equals zero at $x = L$.

Solution

Using the results in Appendix F, the deflection due to the load in Fig. (a) is

$$
v_a =
\begin{cases}
-\dfrac{w_0 x^2}{48EI}\left(3L^2 - 4Lx + 2x^2\right), & 0 \le x \le L/2 \\[3mm]
-\dfrac{w_0 L^3}{384EI}\left(8x - L\right), & L/2 \le x \le L
\end{cases}
$$

and the deflection due to the unknown reaction in Fig. (b) is

$$
v_b = \frac{Bx^2}{6EI}\left(3L - x\right).
$$

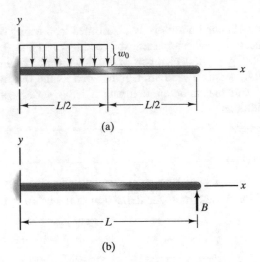

(a), (b) Superimposing two loads to determine the deflection in terms of B

The beam's deflection in terms of B is

$$
v = v_a + v_b =
\begin{cases}
-\dfrac{w_0 x^2}{48EI}\left(3L^2 - 4Lx + 2x^2\right) + \dfrac{Bx^2}{6EI}(3L - x), & 0 \le x \le L/2 \\[2ex]
-\dfrac{w_0 L^3}{384EI}(8x - L) + \dfrac{Bx^2}{6EI}(3L - x). & L/2 \le x \le L
\end{cases}
$$

To determine B, we apply the boundary condition at $x = L$:

$$
v\big|_{x=L} = -\frac{w_0 L^3}{384EI}(8L - L) + \frac{BL^2}{6EI}(3L - L) = 0.
$$

Solving, we obtain $B = 7w_0 L/128$. Substituting this result, the beam's deflection is

$$
v =
\begin{cases}
-\dfrac{w_0 x^2}{768EI}\left(27L^2 - 57Lx + 32x^2\right), & 0 \le x \le L/2 \\[2ex]
-\dfrac{w_0 L}{768EI}\left(7x^3 - 21Lx^2 + 16L^2 x - 2L^3\right). & L/2 \le x \le L
\end{cases}
$$

Discussion
You can apply the approach in this example to most statically indeterminate beam problems. Replace supports by their (unknown) reactions until a statically determinate problem is obtained. Use superposition to determine the deflection of the statically determinate beam in terms of the unknown reactions. Then use the boundary conditions imposed by the supports that were replaced to determine the unknown reactions.

Problems

Solve Problems 9.62–9.73 by superposition using the results in Appendix F. Each beam has moment of inertia I and modulus of elasticity E.

9.62 Determine the deflection of the beam as a function of x.

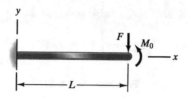

Problem 9.62

9.63 Determine the deflection of the beam as a function of x.

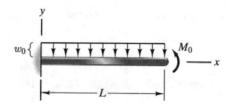

Problem 9.63

9.64 The beam's length is $L = 60$ in and the moment of inertia of its cross section is $I = 0.2$ in^4. It consists of material with modulus of elasticity $E = 10E6$ psi. If $F = 2000$ lb and $w_0 = 52$ lb/in, what is the beam's displacement at $x = 30$ in?

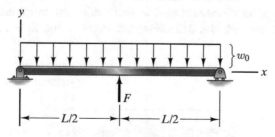

Problems 9.64–9.65

9.65∗ The beam's length is $L = 60$ in and the moment of inertia of its cross section is $I = 0.2$ in^4. It consists of material with modulus of elasticity $E = 10E6$ psi. The force $F = 2000$ lb and $w_0 = 52$ lb/in.

Plot a graph of the beam's displacement for $0 \le x \le 30$ in.

9.66 The beam's length is $L = 4$ m and the moment of inertia of its cross section is $I = 6.75\mathrm{E}\text{-}8 \text{ m}^4$. It consists of material with modulus of elasticity $E = 70\mathrm{E}9$ Pa. The distributed load is $w_0 = 2$ kN/m. Determine the reactions at A, B, and C.

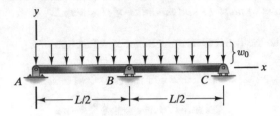

Problems 9.66–9.67

9.67 The beam's length is $L = 4$ m and the moment of inertia of its cross section is $I = 6.75\mathrm{E}\text{-}8 \text{ m}^4$. It consists of material with modulus of elasticity $E = 70\mathrm{E}9$ Pa. The distributed load is $w_0 = 2$ kN/m. Plot a graph of the beam's displacement for $0 \le x \le 2$ m.

9.68 Determine the deflection of the beam as a function of x.

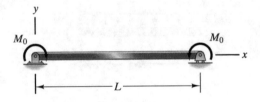

Problem 9.68

9.69 The beam's length is $L = 80$ in and the moment of inertia of its cross section is $I = 0.2 \text{ in}^4$. It consists of material with modulus of elasticity $E = 10\mathrm{E}6$ psi. The magnitude of the distributed load $w_0 = 52$ lb/in. Determine the values of F and M_0 for which both the deflection and slope of the right end of the beam are zero.

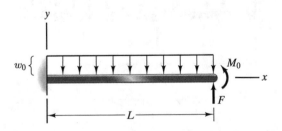

Problem 9.69

9.70 Determine the deflection of the beam at $x = L/2$. (Assume that the supports do not exert axial loads on the beam.)

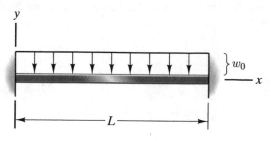

Problem 9.70

9.71 The beam's length is $L = 36$ in and the moment of inertia of its cross section is $I = 0.015$ in^4. It consists of material with modulus of elasticity $E = 26E6$ psi. The magnitude of the distributed load $w_0 = 20$ lb/in. The constant of the linear spring is $k = 18$ lb/in. The linear spring is unstretched when the beam is unloaded. What is the magnitude of the force exerted on the beam by the spring when the distributed load is applied?

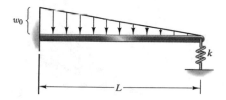

Problem 9.71

9.72 The beam's length is $L = 4$ m and the moment of inertia of its cross section is $I = 6.75E-8$ m^4. It consists of material with modulus of elasticity $E = 70E9$ Pa. The distributed load is $w_0 = 2$ kN/m. Determine the reactions at B and C.

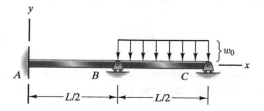

Problems 9.72–9.73

9.73 The beam's length is $L = 4$ m and the moment of inertia of its cross section is $I = 6.75\text{E-}8$ m^4. It consists of material with modulus of elasticity $E = 70\text{E}9$ Pa. The distributed load is $w_0 = 2$ kN/m. Plot a graph of the beam's displacement for $0 \leq x \leq 2$ m.

Chapter Summary

The Second-Order Equation

Let v be the deflection of a beam's neutral axis relative to the x axis, and let θ be the angle between the neutral axis and the x axis (Fig. (a)). For small deflections, v and θ are related by

$$\frac{dv}{dx} = \theta. \tag{9.1}$$

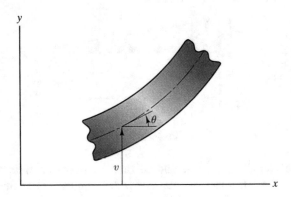

The deflection is related to the bending moment by

$$v'' = \frac{M}{EI}, \tag{9.4}$$

where the primes denote derivatives with respect to x. Determining a beam's deflection using Eq. (9.4) requires three steps:

1. Determine the bending moment as a function of x in terms of the loads and reactions acting on the beam. This may result in two or more functions M_1, $M_2 \ldots$, each of which applies to a different segment of the beam's length.
2. For each segment, integrate Eq. (9.4) twice to determine v_1, v_2, $\ldots$. If there are N segments, this step will result in $2N$ unknown integration constants.
3. Use the boundary conditions to determine the integration constants and, if the beam is statically indeterminate, the unknown reactions. Common boundary conditions are summarized in Fig. 9.8.

Singularity Functions

If the distribution of the bending moment in a beam is determined in terms of singularity functions as described in Sect. 5.4, Eq. (9.4) can be integrated to determine the beam's deflection in terms of singularity functions. For beams with more complex loads, this approach reduces the number of integration constants that must be determined and results in a single equation for the deflection throughout the beam.

Moment-Area Method

Let A and B be two positions on a beam's axis with axial coordinates x_A and x_B, and let θ_A and θ_B be the values of the slope at those positions (Fig. (b)). The first moment-area theorem states that the change in the slope from A to B is

$$\theta_B - \theta_A = A_{AB}, \tag{9.19}$$

where A_{AB} is the area defined by the graph of M/EI from x_A to x_B (Fig. (c)). The second moment-area theorem states that the change in the deflection from A to B is

$$v_B - v_A = x_B\theta_B - x_A\theta_A - \bar{x}_{AB}A_{AB}, \tag{9.24}$$

where $\bar{x}_{AB}$ is the axial position of the centroid of the area defined by the graph of M/EI from x_A to x_B (Fig. (c)). The term $\bar{x}_{AB}A_{AB}$ can be regarded as the "moment" about the origin of the area A_{AB}

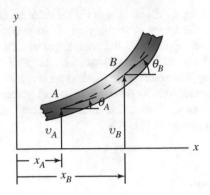

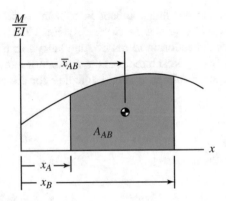

Superposition

Deflections of prismatic beams with typical supports and simple loads are available in tables such as the one in Appendix F. The solutions in such tables can be superimposed to obtain deflections of beams with more complicated loads. To determine the deflection of a given beam by superposition, the loading must be matched by the superimposed loads, and the boundary conditions must be satisfied by the superimposed displacements.

Review Problems

9.74 The beam has length L, moment of inertia I, and modulus of elasticity E. Use Eq. (9.4) to determine the deflection v as a function of x and confirm the results in Appendix F.

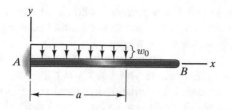

Problem 9.74

9.75* The beam has length L, moment of inertia I, and modulus of elasticity E. Use Eq. (9.4) to determine the deflection v as a function of x and confirm the results in Appendix F.

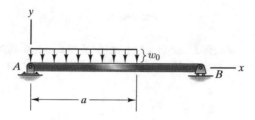

Problem 9.75

9.76 The prismatic beam has elastic modulus E and moment of inertia I. What are the reactions at A and B? In solving the problem, use superposition to determine the beam's deflection. (See the figure for Problems 9.76–9.79.)

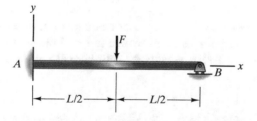

Problems 9.76–9.79

9.77 The prismatic beam has elastic modulus E and moment of inertia I. Use Eq. (9.4) to determine the deflection as a function of x in the region $0 \leq x \leq L/2$. (See the figure for Problems 9.76–9.79.)

9.78 The prismatic beam has elastic modulus E and moment of inertia I. Use singularity functions to determine the beam's deflection at $x = 3L/4$. (See the figure for Problems 9.76–9.79.)

9.79∗ The beam has length $L = 4$ m and a solid circular cross section with 80-mm radius. If $F = 5$ kN, what is the maximum resulting tensile stress in the beam? (See the figure for Problems 9.76–9.79.)

9.80 The beam has length $L = 60$ in, moment of inertia $I = 1.4$ in^4, and modulus of elasticity $E = 12E6$ psi. If you don't want the magnitude of the beam's deflection to exceed 1 in, what is the maximum allowable magnitude of M_0? (Use Appendix F to determine the beam's deflection and slope.)

Problem 9.80

9.81 The beam has length L, moment of inertia I, and modulus of elasticity E. Use singularity functions to determine the deflection v at the beam's right end.

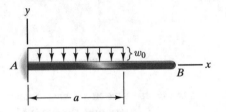

Problem 9.81

9.82 The beam consists of material with modulus of elasticity $E = 72\,\mathrm{GPa}$, the moment of inertia of its cross section is $I = 1.6\mathrm{E}\text{-}7$ m^4, and its length is $L = 6$ m. If $F = 100\,\mathrm{N}$, what is the deflection at $x = 3$ m?

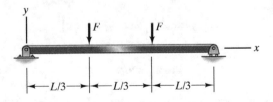

Problems 9.82–9.83

9.83 The beam has length L, moment of inertia I, and modulus of elasticity E. Use singularity functions to determine the deflection v as a function of x.

9.84 Use equilibrium and superposition to solve Problem 9.30.

9.85 Use equilibrium and superposition to solve Problem 9.32.

Chapter 10
Buckling of Columns

Bars subjected to compressive loads must be designed so that they do not fail by buckling laterally. We analyze the buckling of axially loaded bars in this chapter. Because load-bearing columns of buildings must support large compressive axial loads, this subject is traditionally called *buckling of columns*.

10.1 Euler Buckling Load

Suppose that a bar with the dimensions shown in Fig. 10.1 is subjected to a compressive axial load P. If the material is steel with yield stress $\sigma_Y = 520\,\text{MPa}$, the value of P that can be applied without exceeding the yield stress is

$$
\begin{aligned}
P &= \sigma_Y A \\
&= (520\text{E}6 \text{ N/m}^2)(0.012 \text{ m})(0.0005 \text{ m}) \\
&= 3120 \text{ N} \quad (701 \text{ lb}).
\end{aligned}
$$

The bar we have described is a common hacksaw blade. If one is held between the palms (Fig. 10.2(a)) and an increasing compressive force is exerted, it quickly collapses into a bowed shape (Fig. 10.2(b)). Although the axial force exerted is tiny compared to the force necessary to exceed the yield stress of the material, the blade fails as a structural element. It will not support the exerted compressive load.

Clearly, the criterion for preventing failure of structural elements that we have used until now—making sure that loads do not cause the yield stress of the material (or a defined allowable stress) to be exceeded—does not apply in this situation. The hacksaw blade fails by geometric instability or *buckling*. Buckling can occur whenever a slender structural member—a thin bar or a thin-walled plate—is subjected to compression.

© Springer Nature Switzerland AG 2020
A. Bedford, K. M. Liechti, *Mechanics of Materials*,
https://doi.org/10.1007/978-3-030-22082-2_10

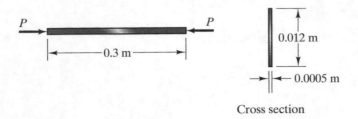

Cross section

Fig. 10.1 Applying compressive axial load to a bar

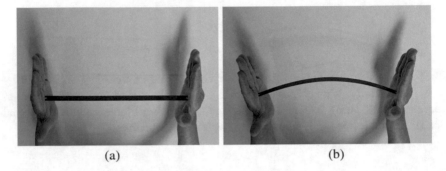

(a)   (b)

Fig. 10.2 (a) Applying compressive axial load to a hacksaw blade. (b) A small compressive load causes the blade to collapse into a bowed shape

In this section we derive the buckling load for a prismatic column subjected to axial forces at the ends. We begin by assuming that the column has already buckled (Fig. 10.3(a)) and seek to determine the value of P necessary to hold it in equilibrium. We can accomplish this by proceeding to determine the distribution of the column's deflection in terms of P. In Fig. 10.3(b) we introduce a coordinate system and obtain a free-body diagram by passing a plane through the column at an arbitrary position x. Solving for the bending moment yields $M = -Pv$, where v is the column's deflection at x. We substitute this expression into Eq. (9.4) $v'' = M/EI$, and write the resulting equation as

$$v'' + \lambda^2 v = 0, \tag{10.1}$$

where

$$\lambda^2 = \frac{P}{EI}. \tag{10.2}$$

The general solution of the second-order differential Eq. (10.1) is

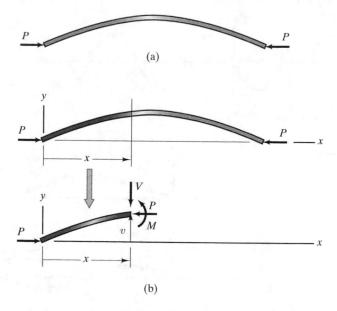

Fig. 10.3 (a) Buckled column in equilibrium. (b) Determining the bending moment as a function of x

$$v = B \sin \lambda x + C \cos \lambda x, \tag{10.3}$$

where B and C are constants we must determine from the boundary conditions. [It is easy to confirm that this expression satisfies Eq. (10.1).] From the boundary condition that the deflection equals zero at $x = 0$, $v|_{x \,=\, 0} = 0$, we see that $C = 0$. The deflection is also zero at $x = L$.

$$v|_{x=L} = 0:$$
$$B \sin \lambda L = 0.$$

This boundary condition is satisfied if $B = 0$, but in that case, the solution for the deflection reduces to $v = 0$. [Notice that this solution does indeed satisfy Eq. (10.1) and the boundary conditions, but we are seeking the buckled solution.] If $B \neq 0$, the boundary condition at $x = L$ requires that

$$\sin \lambda L = 0. \tag{10.4}$$

The parameter λ depends on P, so it is from this equation that we can determine the axial load. Here something interesting happens. As Fig. 10.4 indicates, Eq. (10.4) has not one but an infinite number of roots for λ. It is satisfied if

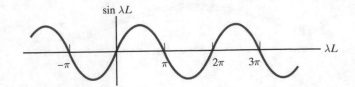

sin λL

$-\pi$ π 2π 3π λL

Fig. 10.4 Roots of $\sin\lambda L = 0$

Fig. 10.5 Deflection
distributions for increasing
values of n

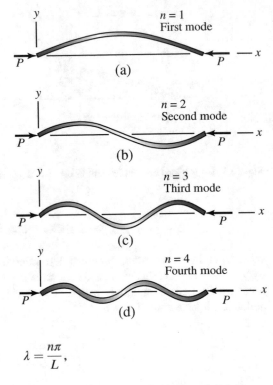

$n = 1$
First mode

(a)

$n = 2$
Second mode

(b)

$n = 3$
Third mode

(c)

$n = 4$
Fourth mode

(d)

$$\lambda = \frac{n\pi}{L},$$

where n is *any integer*. Substituting this expression into Eqs. (10.2) and (10.3), we obtain the axial load and deflection:

$$P = \frac{n^2\pi^2 EI}{L^2}, \tag{10.5}$$

$$v = B \sin \frac{n\pi x}{L}. \tag{10.6}$$

What are all these solutions? Consider the solution for $n = 1$. As x increases from 0 to L, the argument of the sine in Eq. (10.6) increases from 0 to π. This is the buckled solution we have been seeking (Fig. 10.5(a)), and the corresponding value of P is

$$P = \frac{\pi^2 EI}{L^2}. \tag{10.7}$$

Notice that we have not determined the value of B, which is the column's maximum deflection. The load P given by Eq. (10.7) is the force necessary to hold the column in the buckled state for an arbitrarily small value of B, so it is interpreted as the buckling load. This result was obtained by Leonhard Euler in 1744 and is called the *Euler buckling load*.

When $n = 2$, the argument of the sine in Eq. (10.6) increases from 0 to 2π as x increases from 0 to L, resulting in the deflection in Fig. 10.5(b). The deflections for $n = 3$ and $n = 4$ are shown in Figs. 10.5(c and d). (Our analysis requires that the column's slope remains small, so the deflections in Fig. 10.5 are exaggerated.) The solutions for the various values of n are all valid in the sense that they satisfy Eq. (10.1) and the boundary conditions. They are called the *buckling modes* and are referred to as the first mode ($n = 1$), second mode ($n = 2$), and so on. Although the buckled shape of the column could theoretically correspond to any of these modes, we know from experience that it will buckle in the first mode. The higher modes are unstable. But they are not merely of academic interest. The column can be provided with supports so that the lowest mode in which it can buckle is the second or a higher mode (Fig. 10.6). As the figure emphasizes, this greatly increases the column's buckling load.

We can now return to the example with which we began this discussion, the hacksaw blade in Fig. 10.1. The value of the axial load P necessary to exceed the yield stress of the material was 3120 N or 701 lb. Let us calculate the buckling load. When the blade buckles, the axis of its cross section about which it bends is obvious (Fig. 10.7), so the moment of inertia about the z axis is

Fig. 10.6 Preventing a column from buckling in a lower mode

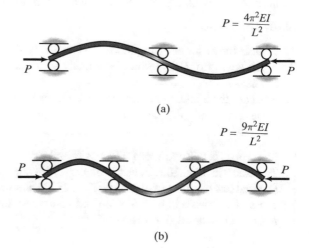

$$P = \frac{4\pi^2 EI}{L^2}$$

(a)

$$P = \frac{9\pi^2 EI}{L^2}$$

(b)

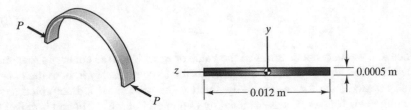

Fig. 10.7 Orientation of the cross section of the buckled blade

$$I = \frac{1}{12}bh^3$$

$$= \frac{1}{12}(0.012 \text{ m})(0.0005 \text{ m})^3$$

$$= 1.25\text{E-}13 \text{ m}^4.$$

(Generally, if it is free to do so, *a buckling column will bend about the principal axis of its cross section which has the smaller moment of inertia.*) If the modulus of elasticity of the steel blade is $E = 200$ GPa, its Euler buckling load is

$$P = \frac{\pi^2 EI}{L^2}$$

$$= \frac{\pi^2 (200\text{E9 N/m}^2)(1.25\text{E-}13 \text{ m}^4)}{(0.3 \text{ m})^2}$$

$$= 2.74\text{N} \quad (0.616\text{lb}).$$

The buckling load of the hacksaw blade is one-tenth of 1% of the compressive load necessary to exceed the yield stress of the material. This illustrates the relative vulnerability to buckling of bars subjected to compression.

Study Questions

1. What is the Euler buckling load? How can it be determined for a given column?
2. If you solve Eq. (10.4) for λ, how many solutions do you obtain? What are the solutions?
3. What are the buckling modes for a column that is free at the ends? Are the higher modes physically meaningful?

Example 10.1 Buckling of a Truss Member
The truss in Fig. 10.8 consists of bars with the cross section shown. The material has modulus of elasticity $E = 72$ GPa and will safely support a normal stress of 270 MPa in both tension and compression. What is the largest force F that can be applied to the truss?

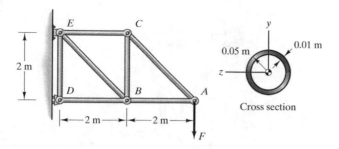

Fig. 10.8

Strategy

The truss must first be analyzed to determine the axial loads in the members in terms of F. We can then identify the member in which the magnitude of the axial load is greatest and determine the value of F that will subject that member to the allowable normal. We must then identify the members that are subject to compression and determine the smallest value of F that will cause a member to buckle.

Solution

We leave it as an exercise to show that the axial loads in the members are

$$AB : F \quad (C)$$
$$AC : \sqrt{2}F \quad (T)$$
$$BC : F \quad (C)$$
$$BD : 2F \quad (C)$$
$$BE : \sqrt{2}F \quad (T)$$
$$CE : F \quad (T)$$
$$DE : 0$$

where (T) and (C) denote tension and compression. The magnitude of the axial load is greatest in member BD. Setting the magnitude of the normal stress in member BD equal to 270 MPa

$$\frac{2F}{A} = \frac{2F}{\pi(0.05 \text{ m})^2 - \pi(0.04 \text{ m})^2} = 270\text{E6 N/m}^2,$$

we obtain $F = 382$ kN. This is the largest value of F that will not exceed the normal stress the material will safely support.

We must now consider buckling. Three members, AB, BC, and BD, are subjected to compression. These three members are of equal length, and the compressive load is greatest in member BD, so it is member BD with which we must be concerned with regard to buckling. The moment of inertia of the cross section is

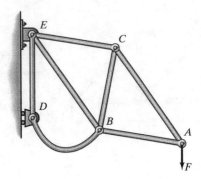

Fig. 10.9 Collapse of the truss by buckling of member *BD*

$$I = \frac{1}{4}\pi(0.05 \text{ m})^4 - \frac{1}{4}\pi(0.04 \text{ m})^4$$
$$= 2.90\text{E-6 m}^4,$$

so the Euler buckling load of member *BD* is

$$P = \frac{\pi^2 EI}{L^2}$$
$$= \frac{\pi^2(72\text{E9 N/m}^2)(2.90\text{E-6 m}^4)}{(2 \text{ m})^2}$$
$$= 515 \text{ kN}.$$

Equating this value to $2F$ (the compressive load in member *BD*), we determine that member *BD* buckles when $F = 257$ kN.

We have found that a force $F = 382$ kN causes the maximum normal stress in the truss to equal the allowable value, while a force $F = 257$ kN will cause member *BD* to buckle. If an increasing force F is applied, the truss fails, not by failure of the material but by geometric instability (Fig. 10.9). Thus it is the value $F = 257$ kN, with a suitable factor of safety imposed, that determines the largest force that can be applied to the truss.

Discussion
The largest force the truss in this example could support was determined by the buckling load of the members subjected to compression. To achieve light structures while preventing buckling, trusses are sometimes designed with tension members that are relatively thin in comparison to the compression members. This approach is strikingly evident in R. Buckminster Fuller's *tensegrity* structures (Fig. 10.10). The compression members, supported by tension members consisting of thin cords, can appear suspended in the air.

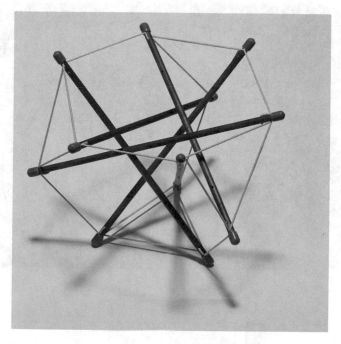

Fig. 10.10 Tensegrity structure with thick compression members and slender tension members. (Model by Design Science Toys, Ltd.)

Problems

10.1 The column is bronze with modulus of elasticity $E = 16E6$ psi. Determine its Euler buckling load.

P ⟶ |←———— 120 in ————→| ⟵ P 2 in

Cross section

Problems 10.1–10.2

10.2 The column is steel with modulus of elasticity $E = 28E6$ psi. What is its second-mode buckling load?

10.3 The architectural column is steel with modulus of elasticity $E = 200$ GPa. Its height is 4 m. Its cross section is shown. Determine its Euler buckling load.

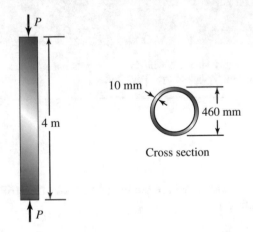

Problem 10.3

10.4 The column is aluminum alloy with modulus of elasticity $E = 70\,\text{GPa}$. Determine its Euler buckling load.

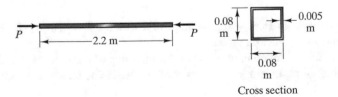

Problems 10.4–10.5

10.5 The column is aluminum alloy with modulus of elasticity $E = 70\,\text{GPa}$. Suppose that you want the outer dimensions of the column to remain $0.08\,\text{m} \times 0.08\,\text{m}$, but you want to increase the wall thickness so that the Euler buckling load is 300 kN. What wall thickness is required?

10.6 The column is aluminum alloy with modulus of elasticity $E = 70\,\text{GPa}$. In order to increase its buckling load, it is provided with lateral supports as shown. What is the buckling load?

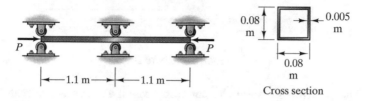

Cross section

Problem 10.6

10.7 Confirm that Eq. (10.3) satisfies Eq. (10.1) for any values of the constants B and C.

10.8 The column has a solid circular cross section with radius R and consists of material with modulus of elasticity E. The material will safely support an allowable normal stress σ_{allow}. Suppose that you want to achieve an optimal design in the sense that the compressive axial load P that subjects the material to a normal stress of magnitude σ_{allow} is equal to the column's Euler buckling load. Show that this is achieved by choosing the dimensions R and L so that

$$\frac{R}{L} = \frac{2}{\pi}\sqrt{\frac{\sigma_{\text{allow}}}{E}}.$$

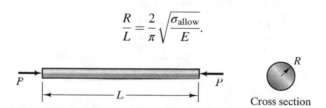

Cross section

Problem 10.8

10.9 The bar AB has a solid circular cross section with 15-mm radius. It consists of material with modulus of elasticity $E = 14\,\text{GPa}$. If the force F is gradually increased, at what value of F will the bar AB buckle?

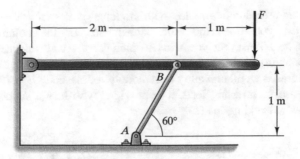

Problems 10.9–10.10

10.10 The bar AB has a solid circular cross section and consists of material with modulus of elasticity $E = 14$ GPa. If you want to design bar AB so that it doesn't buckle until the force $F = 10$ kN, what should its radius be?

10.11∗ The bars of the truss have the cross section shown. The material has modulus of elasticity $E = 190$ GPa and will safely support a normal stress of 250 MPa in both tension and compression. What is the largest force F that can be applied to the truss? (See Example 10.1.)

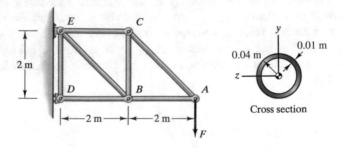

Problem 10.11

10.12 The bars of the truss have the cross section shown and consist of material with modulus of elasticity $E = 2$E6 psi. If the force F is gradually increased, at what value will the structure fail due to buckling?

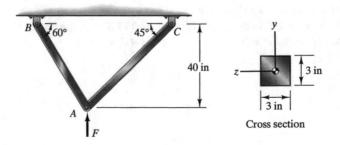

Problems 10.12–10.13

10.13 The bars of the truss have the cross section shown and consist of 2014-T6 aluminum alloy. If the force F is gradually increased, at what value will the structure fail due to buckling?

10.14 The bars of the truss have the cross section shown. The outer radius of the hollow circular cross section is $R = 15$ mm. The bars consist of material with modulus of elasticity $E = 70$ GPa. If the force F is gradually increased, at what value will the structure fail due to buckling?

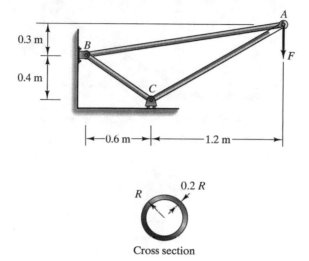

Cross section

Problems 10.14–10.15

10.15* The bars of the truss have the cross section shown and consist of material with modulus of elasticity $E = 70$ GPa. If you don't want the structure to fail due to buckling until $F = 7.5$ kN, what is the required minimum value of the outer radius R of the hollow circular cross section?

10.16* The bars of the truss have a solid circular cross section with 30-mm radius and consist of material with modulus of elasticity $E = 70$ GPa. What is the smallest value of the mass m that will cause the structure to fail due to buckling?

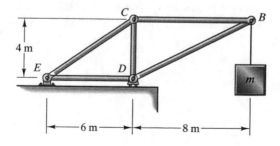

Problem 10.16

10.17* The bars of the truss have the cross section shown and consist of material with modulus of elasticity $E = 2E6$ psi. If $b = 4$ in and the force F is gradually increased, at what value of F will the structure fail due to buckling?

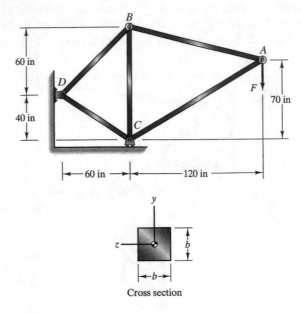

Cross section

Problems 10.17–10.18

10.18* The bars of the truss have the cross section shown and consist of material with modulus of elasticity $E = 2E6$ psi. The bars consist of material that will safely support an allowable normal stress $\sigma_{\text{allow}} = 1000$ psi in tension or compression. Suppose that you want to achieve an optimal design in the sense that the force F which causes the maximum normal stress in the truss to equal σ_{allow} is equal to the smallest force F that causes buckling of a member. Determine the necessary value of the dimension b and the value of F at which failure occurs.

10.19 Bars AB and CD have a solid circular cross section with 20-mm radius. They consist of material with modulus of elasticity $E = 14$ GPa. If the force F is gradually increased, at what value does the structure fail due to buckling?

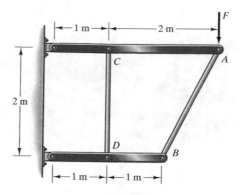

Problems 10.19–10.20

10.20 Bars AB and CD have solid circular cross sections and consist of material with modulus of elasticity $E = 14$ GPa. They consist of material with an allowable normal stress (in tension or compression) of 8 GPa. Suppose that you don't want the structure to fail until the force F reaches 4 kN but you want the radius of each of the bars AB and CD to be as small as possible. What are the radii of the two bars?

10.21 The system supports half of the weight of the 680-kg excavator. Member AC has the cross section shown and consists of material with modulus of elasticity $E = 73$ GPa. What value of the dimension b would cause member AC to buckle with the stationary system in the position shown?

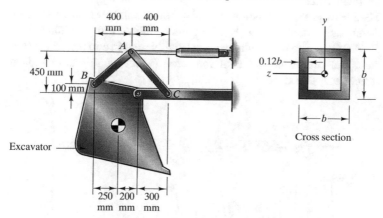

Problem 10.21

10.22 The link AB of the pliers has the cross section shown and is made of steel with modulus of elasticity $E = 190$ GPa. Determine the value of the force F that would cause failure of the link by buckling.

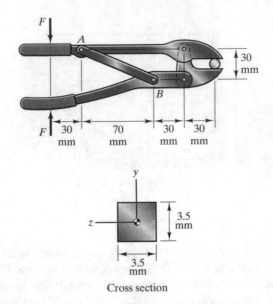

Cross section

Problems 10.22–10.23

10.23 Suppose that the link AB of the pliers has the cross section shown and is made of 7075-T6 aluminum alloy. Determine the value of the force F that would cause failure of the link by buckling and compare your answer to that of Problem 10.22.

10.24 The identical vertical bars A, B, and C have a solid circular cross section with 5-mm radius and are made of aluminum alloy with modulus of elasticity $E = 70$ GPa. The horizontal bar is relatively rigid in comparison to the vertical bars. Determine the value of the force F that would cause failure of the structure by buckling.

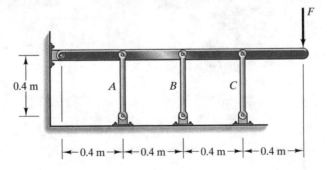

Problems 10.24–10.25

10.25∗ The identical vertical bars A, B, and C have a solid circular cross section with 5-mm radius and are made of aluminum alloy with modulus of elasticity $E = 70\,\text{GPa}$. The horizontal bar is relatively rigid in comparison to the vertical bars. Determine the magnitudes of the axial forces in the three vertical bars
(a) if $F = 2200\,\text{N}$; (b) if $F = 2800\,\text{N}$.

10.26 The column is steel with modulus of elasticity $E = 200$ GPa. It has a C200 × 20.5 American Standard Channel cross section (see Appendix E). Determine its Euler buckling load.

Problem 10.26

10.2 Other End Conditions

The Euler buckling load is derived under the assumption that the column is free to rotate at the ends where the compressive axial loads are applied. That is, there are no reactions at the ends other than the applied axial loads. Various types of end supports are used in applications of columns subjected to compressive axial loads, and the choice of supports can greatly affect the resulting buckling load. In some of these cases, the buckling load can be determined by analyzing the column's deflection as we did in Sect. 10.1.

10.2.1 Analysis of the Deflection

The column in Fig. 10.11(a) is free at the left end and has a fixed support at the right end. We begin by assuming that it is buckled (Fig. 10.11(b)) and seek to determine the value of axial force P necessary to hold it in equilibrium. In Fig. 10.11(c), we introduce a coordinate system with its origin at the column's left end and obtain a free-body diagram by passing a plane through the column at an arbitrary position x. This free-body diagram is identical to the one we obtained in deriving the Euler buckling load (Fig. 10.3(b)), so the steps leading to Eq. (10.3) are unchanged. The deflection is governed by

$$v = B \sin \lambda x + C \cos \lambda x, \tag{10.8}$$

where B and C are constants we must determine from the boundary conditions and

Fig. 10.11 (a) Column
with fixed support. (b)
Buckled column in
equilibrium. (c)
Determining the bending
moment as a function of x

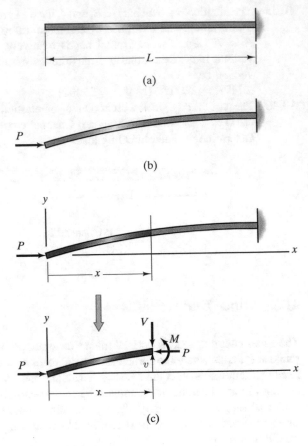

$$\lambda^2 = \frac{P}{EI}. \tag{10.9}$$

Substituting the boundary condition $v|_{x = 0} = 0$ into Eq. (10.8), we find that the constant $C = 0$. Because of the fixed support at the right end, the slope is zero at $x = L$.

$$v'|_{x=L} = 0:$$
$$\lambda B \cos \lambda L = 0.$$

This boundary condition requires that

$$\cos \lambda L = 0. \tag{10.10}$$

From Fig. 10.12, we see that this equation is satisfied if

$$\lambda = \frac{\pi}{2L}, \frac{3\pi}{2L}, \frac{5\pi}{2L}, \dots.$$

Fig. 10.12 Roots of $\cos\lambda L = 0$

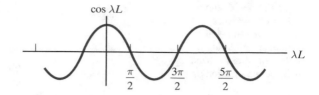

Substituting $\lambda = \pi/2L$ into Eqs. (10.9) and (10.8), we obtain

$$P = \frac{\pi^2 EI}{4L^2},$$

(10.11)

$$v = B\sin\frac{\pi x}{2L}.$$

(10.12)

As x increases from O to L, the argument of the sine in Eq. (10.12) increases from 0 to $\pi/2$. This is the buckled solution shown in Fig. 10.11(b), and Eq. (10.11) is the buckling load. (Notice that the buckling load is one-fourth of the Euler buckling load for a column of equal length.) This is the first buckling mode for a prismatic column supported in this way. We leave it as an exercise to determine the buckling loads and distributions of the deflection for higher modes (see Problem 10.36).

As another example, the column in Fig. 10.13(a) has a roller support that prevents lateral deflection at the left end and a fixed support at the right end. We assume that it is buckled (Fig. 10.13(b)) and seek to determine the value of axial force P necessary to hold it in equilibrium. In Fig. 10.13(c), we introduce a coordinate system and obtain a free-body diagram by passing a plane through the column at an arbitrary position x. The force V_0 is the unknown reaction at the roller support, which equals the shear force at x. Solving for the bending moment and substituting the resulting expression into the equation $v'' = M/EI$, we write the resulting equation as

$$v'' + \lambda^2 v = \frac{V_0}{EI}x,$$

(10.13)

where

$$\lambda^2 = \frac{P}{EI}.$$

(10.14)

Equation (10.13) is nonhomogeneous, because the term on the right side does not contain v or one of its derivatives. Its general solution consists of the sum of the homogeneous and particular solutions:

$$v = v_h + v_p.$$

Fig. 10.13 (a) Column
with roller and fixed
supports. (b) Buckled
column in equilibrium.
(c) Determining the bending
moment as a function of x

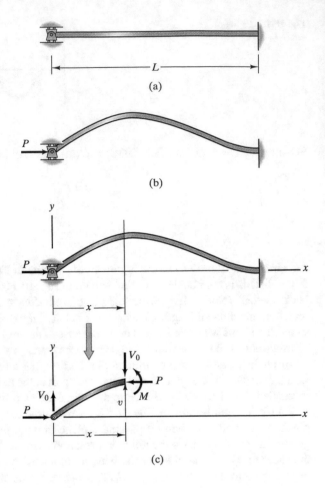

The homogeneous solution is the general solution of Eq. (10.13) with the right side
set equal to zero, which we introduced in Sect. 10.1:

$$v_h = B \sin \lambda x + C \cos \lambda x,$$

where B and C are constants. The particular solution is one that satisfies Eq. (10.13).
The nonhomogeneous term is a polynomial in x, so we seek a particular solution in
the form of a polynomial of the same order: $v_p = a_0 + a_1 x$, where a_0 and a_1 are
constants we must determine. Substituting this expression into Eq. (10.13), we write
the resulting equation as

$$\lambda^2 a_0 + \left(\lambda^2 a_1 - \frac{V_0}{EI} \right) x = 0.$$

This equation is satisfied over an interval of x only if $a_0 = 0$ and $a_1 = V_0/\lambda^2 EI$,
yielding the particular solution

$$v_{\mathrm{p}} = \frac{V_0}{\lambda^2 EI} x.$$

[Confirm that this is a particular solution by substituting it into Eq. (10.13).] The general solution of Eq. (10.13) is

$$v = v_{\mathrm{h}} + v_{\mathrm{p}} = B \sin \lambda x + C \cos \lambda x + \frac{V_0}{\lambda^2 EI} x. \tag{10.15}$$

From the boundary condition $v|_{x=0} = 0$, we obtain $C = 0$. The boundary conditions at $x = L$ are

$$v|_{x=L} = B \sin \lambda L + \frac{V_0 L}{\lambda^2 EI} = 0, \tag{10.16}$$

$$v'|_{x=L} = \lambda B \cos \lambda L + \frac{V_0}{\lambda^2 EI} = 0. \tag{10.17}$$

We solve the first of these equations for the reaction V_0:

$$V_0 = -\frac{\lambda^2 EI \sin \lambda L}{L} B. \tag{10.18}$$

Substituting this result into Eq. (10.17), we find that the boundary conditions at $x = L$ are satisfied only if

$$\sin \lambda L - \lambda L \cos \lambda L = 0. \tag{10.19}$$

We also substitute Eq. (10.18) into Eq. (10.15), obtaining the distribution of the deflection in the form

$$v = B \left(\sin \lambda x - \frac{x}{L} \sin \lambda L \right). \tag{10.20}$$

Equation (10.19) is called the *characteristic equation* for this problem. For each value of λ that satisfies the characteristic equation, Eq. (10.14) determines the buckling load, and Eq. (10.20) determines the shape of the buckled column. [Eq. (10.10) is the characteristic equation for a column with free and fixed ends.] In this case we must determine the roots of the characteristic equation numerically. Figure 10.14 is a graph of the characteristic function $f(\lambda L) = \sin \lambda L - \lambda L \cos \lambda L$. The first four roots, determined numerically, are $\lambda L = 4.493$, 7.725, 10.904, and 14.066. The shapes of the resulting buckling modes and the associated buckling loads are shown in Fig. 10.15.

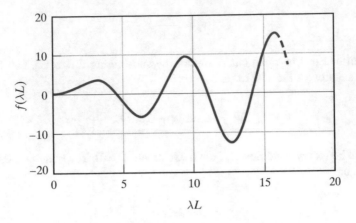

Fig. 10.14   Graph of the characteristic function

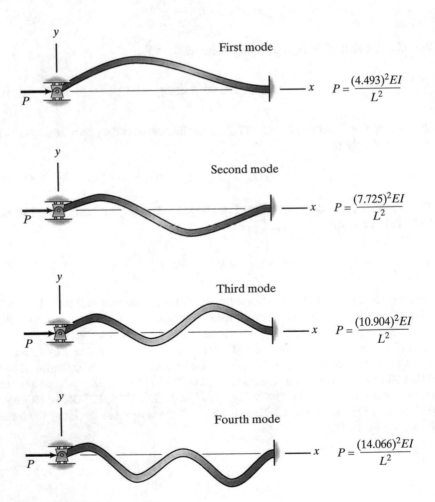

Fig. 10.15   First four buckling modes

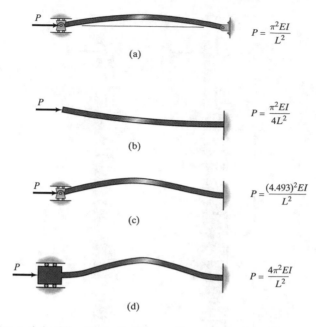

$$P = \frac{\pi^2 EI}{L^2}$$

(a)

$$P = \frac{\pi^2 EI}{4L^2}$$

(b)

$$P = \frac{(4.493)^2 EI}{L^2}$$

(c)

$$P = \frac{4\pi^2 EI}{L^2}$$

(d)

Fig. 10.16   First-mode buckling loads of columns with common end conditions

The common end conditions that can be analyzed in this way are shown in Fig. 10.16 together with their buckling loads.

Example 10.2 Buckling with Various End Conditions
The column in Fig. 10.17 supports an axial load P. The base of the column is built in. The support at the top prevents lateral deflection. The support at the top allows rotation of the column in the x–y plane but prevents rotation in the x–z plane. The moments of inertia of the cross section about the y and z axes are 5E - 6 m^4 and 15E - 6 m^4, respectively. The modulus of elasticity is $E = 70$ GPa. If P is gradually increased, at what value will the column buckle?

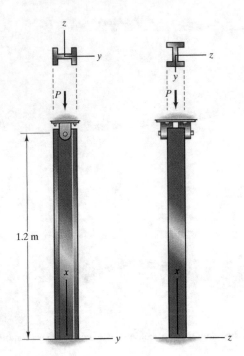

Fig. 10.17

Strategy
If it is free to do so, a column subjected to axial compression buckles by bending about the principal axis of the cross section that has the smaller moment of inertia. In this example, the top support of the column will cause the column to buckle as shown in Fig. 10.16(c) if it buckles by bending in the *x–y* plane (Fig. (a)) but will cause it to buckle as shown in Fig. 10.16(d) if it buckles by bending in the *x–z* plane (Fig. (b)). We must determine which of these possibilities yields the lowest buckling load.

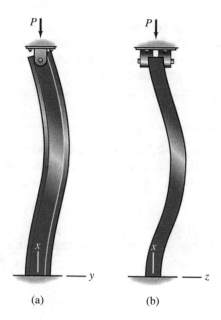

(a)   (b)

(a) Geometry of buckling in the x–y plane. **(b)** Geometry of buckling in the x–z plane

Solution

If the column buckles by bending in the x–y plane (Fig. (a)), it fails by bending about the z axis. From Fig. 10.16(c), the buckling load is

$$P = \frac{(4.493)^2 EI}{L^2}$$

$$= \frac{(4.493)^2 (70E9 \text{ N/m}^2)(15E\text{-}6 \text{ m}^4)}{(1.2 \text{ m})^2}$$

$$= 14.72 \text{ MN}.$$

If the column buckles by bending in the x–z plane (Fig. (b)), it fails by bending about the y axis. From Fig. 10.16(d), the buckling load is

$$P = \frac{4\pi^2 EI}{L^2}$$

$$= \frac{4\pi^2 (70E9 \text{ N/m}^2)(5E\text{-}6 \text{ m}^4)}{(1.2 \text{ m})^2}$$

$$= 9.60 \text{ MN}.$$

We see that the column will buckle as shown in Fig. (b), and the buckling load is 9.60 MN. That is, the column does buckle by bending about the principal axis of the cross section that has the smaller moment of inertia.

Discussion
This example demonstrates that the supports of a column can be designed to influence the manner in which it buckles. Generally, constraining the lateral or rotational motions of the ends of a column causes it to buckle in a higher mode and therefore increases its buckling load.

Example 10.3 Buckling with Various End Conditions
The column in Fig. 10.18 has a support that prevents lateral deflection and rotation at the left end and a fixed support at the right end. Determine its buckling load.

Fig. 10.18

Solution
In Fig. (a) we introduce a coordinate system and obtain a free-body diagram by passing a plane through the buckled column at an arbitrary position x.

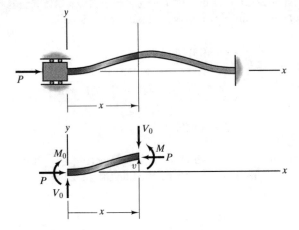

(**a**) Determining the bending moment

Solving for the bending moment and substituting it into the equation $v'' = M/EI$, we write the resulting equation as

$$v'' + \lambda^2 v = \frac{V_0}{EI}x + \frac{M_0}{EI},\tag{1}$$

where

$$\lambda^2 = \frac{P}{EI}. \tag{2}$$

The homogeneous solution of Eq. (1) is

$$v_h = B \sin \lambda x + C \cos \lambda x,$$

where B and C are constants. The nonhomogeneous term in Eq. (1) is a polynomial in x, so we seek a particular solution in the form of a polynomial of the same order: $v_p = a_0 + a_1 x$. Substituting this expression into Eq. (1) and writing the resulting equation as

$$\left(\lambda^2 a_0 - \frac{M_0}{EI}\right) + \left(\lambda^2 a_1 - \frac{V_0}{EI}\right) x = 0,$$

we find that $a_0 = M_0/\lambda^2 EI$ and $a_1 = V_0/\lambda^2 EI$, yielding the particular solution

$$v_p = \frac{M_0}{\lambda^2 EI} + \frac{V_0}{\lambda^2 EI} x.$$

The general solution of Eq. (1) is therefore

$$v = v_h + v_p = B \sin \lambda x + C \cos \lambda x + \frac{M_0}{\lambda^2 EI} + \frac{V_0}{\lambda^2 EI} x, \tag{3}$$

and the slope is

$$v' = \lambda B \cos \lambda x - \lambda C \sin \lambda x + \frac{V_0}{\lambda^2 EI}. \tag{4}$$

The boundary conditions at $x = 0$ are

$$v\big|_{x=0} = C + \frac{M_0}{\lambda^2 EI} = 0, \tag{5}$$

$$v'\big|_{x=0} = \lambda B + \frac{V_0}{\lambda^2 EI} = 0. \tag{6}$$

We solve these equations for M_0 and V_0 and substitute the results into Eqs. (3) and (4), obtaining

$$v = (\sin \lambda x - \lambda x)B + (\cos \lambda x - 1)C, \tag{7}$$

$$v' = (\lambda \cos \lambda x - \lambda)B - \lambda C \sin \lambda x. \tag{8}$$

The boundary conditions at $x = L$ are

$$v|_{x=L} = (\sin \lambda L - \lambda L)B + (\cos \lambda L - 1)C = 0, \qquad (9)$$

$$v'|_{x=L} = (\lambda \cos \lambda L - \lambda)B - \lambda C \sin \lambda L = 0. \qquad (10)$$

Solving the first of these equations for B gives

$$B = \frac{1 - \cos \lambda L}{\sin \lambda L - \lambda L} C. \qquad (11)$$

Substituting this result into Eq. (10) yields the characteristic equation:

$$\lambda L \sin \lambda L + 2 \cos \lambda L - 2 = 0. \qquad (12)$$

We also substitute Eq. (11) into Eq. (7), obtaining the distribution of the deflection in the form

$$v = C \left[\frac{1 - \cos \lambda L}{\sin \lambda L - \lambda L} (\sin \lambda x - \lambda x) + \cos \lambda x - 1 \right]. \qquad (13)$$

Figure 10.19 is a graph of the characteristic function $f(\lambda L) = \lambda L \sin \lambda L + 2 \cos \lambda L - 2$. The first four roots, determined numerically, are $\lambda L = 2\pi$, 8.987, 12.566, and 15.451. The shapes of the first three buckling modes (obtained from Eq. (10.13)) and the associated buckling loads (obtained from Eq. (2)) are shown in Fig. 10.20.

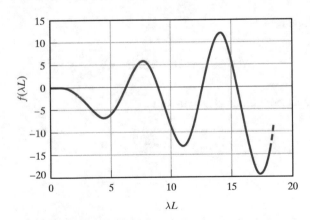

Fig. 10.19 Graph of the characteristic function

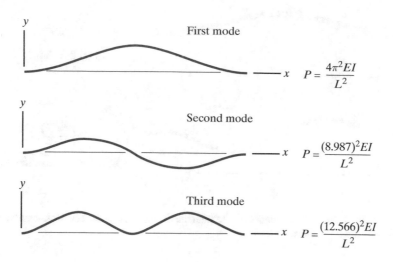

Fig. 10.20 Graph of the characteristic function

Discussion
In this example we showed how the buckling loads and buckled shapes of a column with particular end conditions can be determined. The analysis was not trivial. Fortunately, for practicing engineers results such as these have been tabulated for columns with common end conditions. Some of them are shown in Fig. 10.16.

10.2.2   Effective Length

In Sect. 10.1 we analyzed a prismatic column that is free at the ends and subjected to compressive axial load, obtaining the Euler buckling load (Fig. 10.21(a)). We also obtained the buckling load of the second mode from the same analysis (Fig. 10.21(b)), but there is a simpler way to determine the buckling load of the second mode.

Let us assume that the column is buckled in the second mode. In Fig. 10.22, we cut it by a plane at the midpoint and draw the resulting free-body diagrams. Each of these free-body diagrams is buckled in the first mode. We can therefore determine the second-mode buckling load by calculating the Euler buckling load of a column of length $L_e = L/2$:

$$
\begin{aligned}
P &= \frac{\pi^2 EI}{L_e^2} \\
&= \frac{\pi^2 EI}{(L/2)^2} \\
&= \frac{4\pi^2 EI}{L^2}.
\end{aligned}
$$

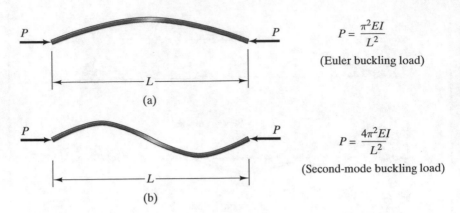

$$P = \frac{\pi^2 EI}{L^2}$$

(Euler buckling load)

(a)

$$P = \frac{4\pi^2 EI}{L^2}$$

(Second-mode buckling load)

(b)

Fig. 10.21 First two buckling modes of a column that is free at the ends. The buckling load of the first mode is the Euler buckling load

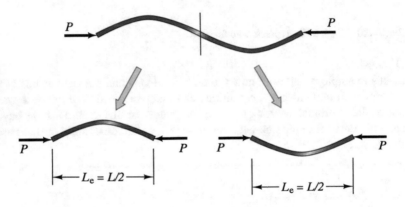

Fig. 10.22 Obtaining free-body diagrams by dividing the column at its midpoint

We see that the second-mode buckling load of the column of length L is equal to the Euler buckling load of a column of length $L_e = L/2$.

This example motivates a concept called the effective length. Suppose that a given column buckles in a particular mode, and the buckling load is P. The *effective length* L_e is defined to be the length of a column of the same flexural rigidity whose Euler buckling load equals P:

$$P = \frac{\pi^2 EI}{L_e^2}. \tag{10.21}$$

If P is known for a given column and buckling mode, this equation can be used to determine the effective length. Alternatively, in some cases, the effective length can be determined or approximated by observation, and Eq. (10.21) can be used to determine P. The ratio of the effective length to the column's actual length is denoted by K and called the *effective length factor*:

$$K = \frac{L_e}{L}.$$ (10.22)

How can we determine the effective length by observation? The distribution of the deflection of a column that is free at the ends and buckled in the first mode, from which the Euler buckling load is derived (Fig. 10.21(a)), is a half-cycle of a sine function. When such a column buckles in the second mode (Fig. 10.21(b)), its deflection forms two half-cycles, each of length L/2. The buckling load of the column in the second mode equals the Euler buckling load of these half-cycles (Fig. 10.22), so the effective length is L/2. Thus we obtain the effective length by dividing L by the number of half-cycles. This approach can be applied whenever the deflection of a buckled column consists of an identifiable number of half-cycles (see Example 10.4).

Example 10.4 Application of the Effective Length
The column in Fig. 10.23 has a support that prevents lateral deflection and rotation at the left end and a fixed support at the right end. Figure 10.20 shows the deflection distributions of the first three buckling modes of this column. For the first and third modes, determine the value of the effective length factor K and verify the buckling loads in Fig. 10.20.

Fig. 10.23

Strategy
By counting the number of half-cycles of sine functions in the deflection distribution for each mode, we can determine its effective length. Then we can determine the effective length factor from Eq. (10.22) and obtain the buckling load from Eq. (10.21).

Solution
First mode The distribution of the deflection consists of four quarter-cycles of a sine function (Fig. (a)) or two half-cycles. The effective length is therefore $L_e = L/2$. The effective length factor is

$$K = \frac{L_e}{L} = \frac{\left(\frac{L}{2}\right)}{L} = 0.5,$$

and the buckling load is

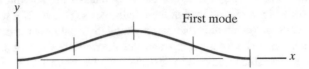

$$P = \frac{\pi^2 EI}{L_e^2}$$
$$= \frac{\pi^2 EI}{(L/2)^2}$$
$$= \frac{4\pi^2 EI}{L^2}.$$

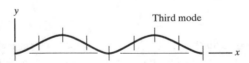

(a) First mode divided into quarter-cycles

Third mode The distribution of the deflection consists of eight quarter-cycles of a sine function (Fig. (b)) or four half-cycles. The effective length is $L_e = L/4$. The effective length factor is

$$K = \frac{L_e}{L} = \frac{\left(\frac{L}{4}\right)}{L} = 0.25,$$

and the buckling load is

$$P = \frac{\pi^2 EI}{L_e^2}$$
$$= \frac{\pi^2 EI}{(L/4)^2}$$
$$= \frac{(12.566)^2 EI}{L^2}.$$

(b) Third mode divided into quarter-cycles

Discussion

If you compare this example to Example 10.3, in which we analytically determined the buckling loads for a column with the same end conditions, you will gain appreciation for the concept of the effective length. By very brief calculations, we have obtained the same buckling loads that we obtained in Example 10.3 by solving the differential equation governing the displacement. However, using the effective length in this way can require intuition and judgment.

Problems

10.27 The architectural column has modulus of elasticity $E = 1.8E6$ psi. It has a fixed support at the base. The support at the top prevents both lateral deflection and rotation. Determine the buckling load P. (See the figure for Problems 10.27–10.29.)

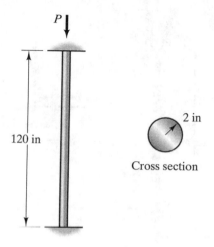

Problems 10.27–10.29

10.28 The architectural column has modulus of elasticity $E = 1.8E6$ psi. It has a fixed support at the base. The architect's specifications for the column would prevent both lateral deflection and rotation at the top, but the contractor installs the column in such a way that only lateral deflection is prevented, and the top of the column is free to rotate. What is the resulting buckling load? Compare your answer to the answer to Problem 10.27. (See the figure for Problems 10.27–10.29.)

10.29 The architectural column has modulus of elasticity $E = 1.8E6$ psi. It has a fixed support at the base. The architect's specifications for the column would prevent both lateral deflection and rotation at the top, but the contractor installs the column in such a way that the column is free to rotate and to deflect laterally at the top. What is the resulting buckling load? Compare your answer to the answer to Problem 10.27. (See the figure for Problems 10.27–10.29.)

10.30 The column is aluminum alloy with modulus of elasticity $E = 70$ GPa. The width of the hollow square cross section is $b = 0.08$ m. The column has a fixed support at the base. The support at the top prevents lateral deflection but allows rotation in the x–z plane. Determine the buckling load P.

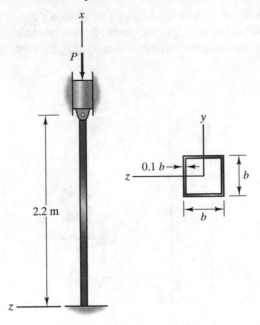

Problems 10.30–10.31

10.31 The column is aluminum alloy with modulus of elasticity $E = 70$ GPa. It has a fixed support at the base. The support at the top prevents lateral deflection but allows rotation in the x–z plane. Determine the required width b of the hollow square cross section so that the column's buckling load is 700 kN.

10.32 The column supports an axial load P. The supports at the top and bottom prevent lateral deflection and prevent rotation in the x–z plane but allow rotation in the x–y plane. The moments of inertia of the cross section about the y and z axes are 12 in^4 and 36 in^4, respectively. The elastic modulus is $E = 10$E6 psi. What is the column's buckling load? Does it buckle by bending in the x–y plane or the x–z plane?

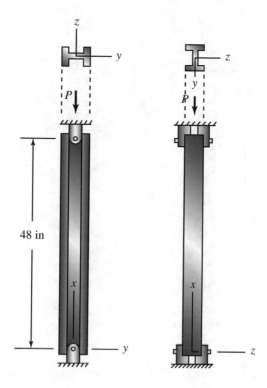

Problem 10.32

10.33 The column supports an axial load P. The base of the column is built in. The support at the top prevents lateral deflection. The support at the top allows rotation of the column in the x–y plane but prevents rotation in the x–z plane. The dimensions $b = 0.08$ m and $h = 0.14$ m. The modulus of elasticity is $E = 12$ GPa. What is the column's buckling load? Does it buckle by bending in the x–y plane or the x z plane? (See the figure for Problems 10.33–10.35.)

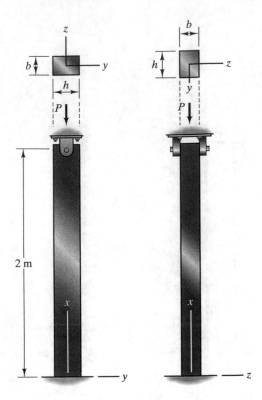

Problems 10.33–10.35

10.34 The column supports an axial load P. The base of the column is built in. The support at the top prevents lateral deflection. The support at the top allows rotation of the column in the x–y plane but prevents rotation in the x–z plane. The dimensions $b = 0.12$ m and $h = 0.14$ m. The modulus of elasticity is $E = 12$ GPa. What is the column's buckling load? Does it buckle by bending in the x–y plane or the x–z plane? (See the figure for Problems 10.33–10.35.)

10.35 For the column shown, determine the ratio b/h if the value of P that causes the column to buckle in the x–y plane is equal to the value of P that causes it to buckle in the x–z plane. (See the figure for Problems 10.33–10.35.)

10.36∗ The column is free at the left end and has a fixed support at the right end. Determine the buckling loads for modes two, three, and four and draw graphs of the distributions of the deflection. (See the discussion of this problem in Sect. 10.2.)

Problem 10.36

10.37* The column has a support that prevents lateral deflection and rotation at the left end and a fixed support at the right end. Using the results in Example 10.3, determine the buckling load for the fourth buckling mode and draw a graph of the distribution of the deflection.

Problem 10.37

10.38* The rectangular platform in Fig. (a) is supported by four identical columns of length L. The platform is loaded with weights in such a way that the axial force on each column is the same until the columns buckle (Fig. (b)). The connections of the columns to the floor and the platform behave like fixed supports. By modifying the analysis in Example 10.3, determine the buckling load of each column for the first buckling mode.

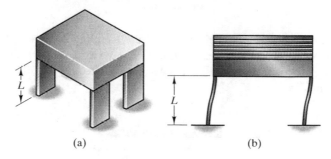

(a)   (b)

Problem 10.38

10.39 Suppose that the rectangular platform described in Problem 10.38 is provided with horizontal supports as shown. The four columns are made of steel with modulus of elasticity 28E6 psi. Their length is $L = 14$ in and the dimensions of their cross section are 1/16 in × 3 in. How much weight will the platform support without buckling?

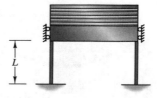

Problems 10.39

10.40 The first four buckling modes of a column of length L that is free to bend at the ends are shown. What are the effective length and effective length factor of each mode?

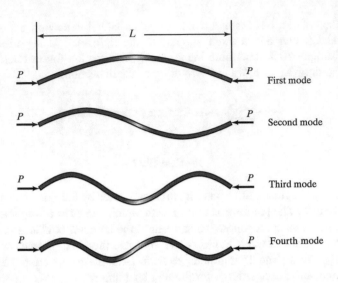

Problem 10.40

10.41 The distributions of the deflection and the buckling loads are given for the first three modes of the column shown. Use the expressions for the buckling loads to calculate the effective lengths and the effective length factors for the three modes.

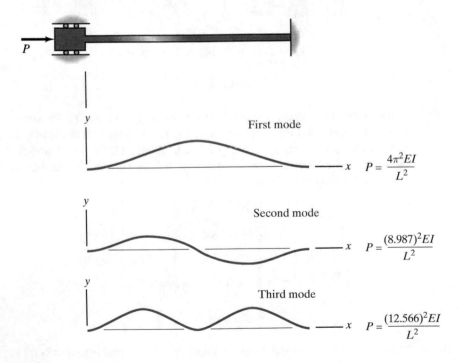

Problem 10.41

10.42 The prismatic column of length L has a fixed support at the right end and is free at the left end. It is shown buckled in the first mode. What is its effective length? Use the effective length to determine the buckling load.

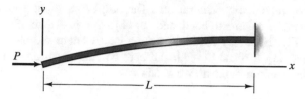

Problem 10.42

10.43 The fourth buckling mode of the column is shown. (a) Use the figure to determine the approximate effective length. (b) Use the approximate effective length to determine the buckling load.

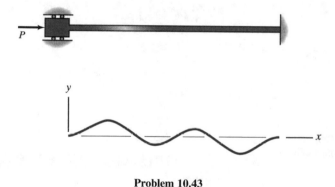

Problem 10.43

10.3   Eccentric Loads

In determining buckling loads of columns in Sect. 10.2, we assumed that compressive loads were applied precisely at the centroids of the cross sections of perfectly prismatic columns. In practice, the lines of action of the loads will often be offset from the centroid to some extent, either unintentionally or to satisfy some design requirement, and the column's neutral axis will not be perfectly straight. These variations in the geometry of the classical buckling problem have important quantitative and qualitative effects on a column's response to compressive loads. In this section we illustrate these effects by analyzing a prismatic column that is free to rotate at the ends and is subjected to *eccentric loads*, loads that are offset relative to the centroid of the cross section.

10.3.1   Analysis of the Deflection

The prismatic column in Fig. 10.24 is subjected to axial loads P that are displaced a distance e from the column's neutral axis. The couples resulting from the displaced loads will cause the column to bend, and our objective is to determine the resulting deflection. In Fig. 10.25, we introduce a coordinate system and obtain a free-body diagram by passing a plane through the bent column at an arbitrary position x. Solving for the bending moment, we obtain

$$M = -P(v + e). \tag{10.23}$$

[The exact bending moment is $M = -P[v + e\,\cos(\theta|_{x=0})]$, but the column's slope is assumed small, so we can make the approximation $\cos(\theta|_{x=0}) = 1$.] We substitute Eq. (10.23) into the equation $v'' = M/EI$ and write the resulting equation as

$$v'' + \lambda^2 v = -\lambda^2 e, \tag{10.24}$$

where

Fig. 10.24 Column subjected to eccentric axial loads

Fig. 10.25 Determining the bending moment

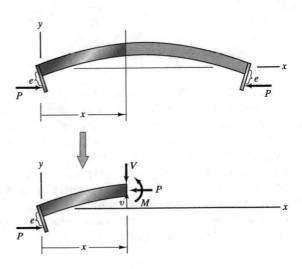

$$\lambda^2 = \frac{P}{EI}. \tag{10.25}$$

The general solution of Eq. (10.24) is

$$v = B \sin \lambda x + C \cos \lambda x - e. \tag{10.26}$$

Using the boundary conditions $v|_{x=0} = 0$ and $v|_{x=L} = 0$ to evaluate the constants B and C, we obtain the deflection:

$$v = e\left(\frac{1 - \cos \lambda L}{\sin \lambda L} \sin \lambda x + \cos \lambda x - 1\right). \tag{10.27}$$

By substituting the identities

$$\begin{aligned}
\sin \lambda L &= 2 \sin \frac{\lambda L}{2} \cos \frac{\lambda L}{2}, \\
\cos \lambda L &= \cos^2 \frac{\lambda L}{2} - \sin^2 \frac{\lambda L}{2}
\end{aligned} \tag{10.28}$$

into Eq. (10.27), we can write it as

$$v = e\left(\tan \frac{\lambda L}{2} \sin \lambda x + \cos \lambda x - 1\right). \tag{10.29}$$

For a given value of P, Eqs. (10.25) and (10.29) determine the distribution of the column's deflection. The maximum deflection occurs at the column's midpoint:

$$v_{\max} = v|_{x=L/2} = e\left(\sec \frac{\lambda L}{2} - 1\right). \tag{10.30}$$

But what is the relationship between these results and the column's buckling load? If increasing compressive loads are applied to the column at the centroid of its cross section (Fig. 10.26(a)), there is no deflection until the value of P reaches the buckling load. When P equals the buckling load, the deflection is indeterminate because the column is in equilibrium for any value of its amplitude (Fig. 10.26(b)). (Of course, the magnitude is limited by the underlying assumption that the column's slope is small.) In contrast, when increasing eccentric loads are applied, the column begins bending immediately, and the amplitude of the deflection increases as P increases (Fig. 10.26(c)). These phenomena are illustrated by Fig. 10.27, in which the maximum deflection is plotted as a function of P for different constant values of the eccentricity e. When $e = 0$, $v_{\max} = 0$ until P equals the buckling load, after which the deflection is indeterminate. For any finite value of e, $v_{\max}$ increases as P increases and increases without bound as P approaches the buckling load. As $e \to 0$, the column's behavior predictably approaches the behavior observed when the loads are applied at the centroid.

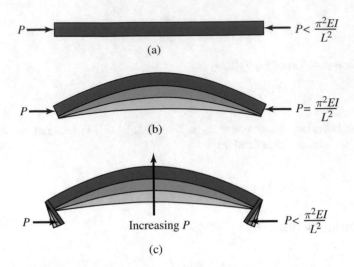

(a)

(b)

Increasing P

(c)

Fig. 10.26 (a) Column subjected to axial loads applied at the centroid of the cross section. (b) At the buckling load, the deflection is indeterminate. (c) With eccentric loads, the amplitude of the deflection increases as P increases

Fig. 10.27 Graph of the maximum deflection as a function of the axial load

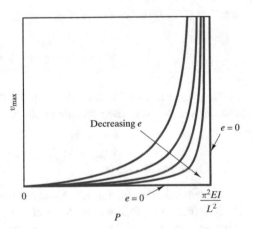

10.3.2   Secant Formula

We have seen that a column subjected to increasing eccentric compressive loads deflects as soon as load is applied and the amplitude of the deflection grows without bound as the buckling load is approached. Because the normal stress due to bending depends on the column's radius of curvature, the normal stress also grows without bound as the buckling load is approached. As a result, the focus in design shifts from the buckling load to the maximum deflection and normal stress that can be allowed.

The maximum deflection of the column resulting from a given value of P can be determined from Eqs. (10.25) and (10.30), so we only need to determine the

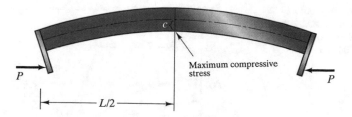

Fig. 10.28 Location of the maximum compressive stress

maximum stress. From Eq. (10.23), we see that the magnitude of the bending moment in the column is a maximum at the midpoint where the deflection is a maximum:

$$M_{\max} = P(v_{\max} + e).$$

By using Eqs. (10.25) and (10.30), we can write this expression as

$$\begin{aligned} M_{\max} &= Pe \sec \frac{\lambda L}{2} \\ &= Pe \sec \sqrt{\frac{PL^2}{4EI}}. \end{aligned} \tag{10.31}$$

Let c be the distance from the column's neutral axis to the location on the cross section where the maximum compressive stress occurs (Fig. 10.28). The maximum normal stress in the column is the sum of the compressive stress resulting from the axial load and the maximum compressive stress due to bending. Its magnitude is

$$\sigma_{\max} = \frac{P}{A} + \frac{M_{\max}c}{I}.$$

Substituting Eq. (10.31) and introducing the *radius of gyration* $k = \sqrt{I/A}$, we can write the maximum stress as

$$\sigma_{\max} = \frac{P}{A}\left[1 + \frac{ec}{k^2}\sec\left(\frac{L}{2k}\sqrt{\frac{P}{EA}}\right)\right]. \tag{10.32}$$

This equation is called the *secant formula*. It expresses the magnitude of the maximum stress as a function of P/A, E, and two dimensionless parameters, the *eccentricity ratio* ec/k^2 and the *slenderness ratio* L/k.

When the column's dimensions, the modulus of elasticity, and the eccentricity are known and the maximum allowable compressive stress is specified, Eq. (10.32) can be solved numerically for the maximum allowable axial load. As an example, we consider steel columns with modulus of elasticity $E = 200\,\text{GPa}$ and assume that the

Fig. 10.29 Graph of P/A as a function of the slenderness ratio for $E = 200$ GPa and $\sigma_{max} = 250$ MPa

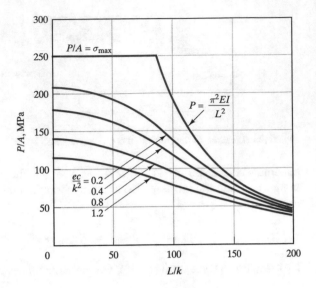

maximum allowable compressive stress is $\sigma_{max} = 250$ MPa. In Fig. 10.29, the solution for P/A obtained from Eq. (10.32) is plotted as a function of the slenderness ratio for different constant values of the eccentricity ratio. When the eccentricity is zero, the allowable values of P/A are bounded by the buckling load and σ_{max}.

Example 10.5 Column with Eccentric Loads
The column in Fig. 10.30 has the cross section shown, length $L = 0.76$ m, and is subjected to compressive loads with eccentricity $e = 0.0046$ m. The material is steel with modulus of elasticity $E = 200$ GPa. **(a)** If the allowable compressive stress is 250 MPa, what is the largest allowable value of P? **(b)** If the compressive load determined in part (a) is applied, what is the column's maximum deflection?

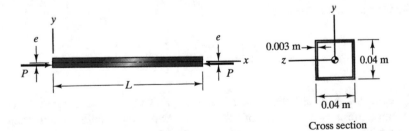

Cross section

Fig. 10.30

Strategy
(a) The modulus of elasticity and allowable compressive stress are the values on which Fig. 10.29 is based. By calculating the column's eccentricity and

slenderness ratios, we can estimate the value of P/A from Fig. 10.29. Alternatively, we can determine P/A by numerical solution of the secant formula, Eq. (10.32). **(b)** Once P is known, the maximum deflection is given by Eq. (10.30).

Solution
(a) The column's cross-sectional area is

$$A = (0.04 \text{ m})^2 - [0.04 \text{ m} - 2 (0.003 \text{ m})]^2$$
$$= 0.000444 \text{ m}^2,$$

and the moment of inertia about the z axis is

$$I = \frac{1}{12}(0.04 \text{ m})^4 - \frac{1}{12}[0.04 \text{ m} - 2 (0.003 \text{ m})]^4$$
$$= 1.020 \times 10^{-7} \text{ m}^4.$$

The radius of gyration is

$$k = \sqrt{\frac{I}{A}} = 0.0152 \text{ m}.$$

Using this result, we calculate the eccentricity ratio

$$\frac{ec}{k^2} = \frac{(0.0046 \text{ m}) (0.02 \text{ m})}{(0.0152 \text{ m})^2}$$
$$= 0.401$$

and the slenderness ratio

$$\frac{L}{k} = \frac{0.76 \text{ m}}{0.0152 \text{ m}}$$
$$= 50.1.$$

With these values, we estimate from Fig. 10.29 that $P/A = 165$ MPa. By solving Eq. (10.32) numerically, we obtain $P/A = 163.3$ MPa. Using the latter value, the largest allowable value of P is

$$P = (163\text{E}6 \text{ N}/\text{m}^2) (0.000444 \text{ m}^2) = 72.5 \text{ kN}.$$

(b) The value of the parameter λ is

$$\lambda = \sqrt{\frac{P}{EI}}$$

$$= \sqrt{\frac{72.5\text{E3 N}}{(200\text{E9 N/m}^2)(1.020\text{E-7 m}^4)}}$$

$$= 1.89\,\text{m}^{-1}.$$

From Eq. (10.30), the maximum deflection is

$$v_{max} = e\left(\sec\frac{\lambda L}{2} - 1\right)$$

$$= (0.0046\,\text{m})\left[\sec\frac{(1.89\,\text{m}^{-1})(0.76\,\text{m})}{2} - 1\right]$$

$$= 0.00150\,\text{m}.$$

Discussion
We should emphasize that the behavior of a column with an eccentric load is *qualitatively* different from that of a column that is loaded at the neutral axis. In the case of an eccentric load, there is no abrupt buckling load. The column begins deforming as soon as any load is applied and gradually becomes more deformed as the load is increased. Because virtually all columns will have some degree of eccentricity, what is the motivation for calculating the buckling load? As indicated by Fig. 10.27, even when eccentricity is present, the buckling load provides an approximation for the value of the load at which the deformation of the column would increase catastrophically. Furthermore, as in this example, the buckling load can be supplemented by a constraint on the axial load that is based on stress.

Problems

10.44 The column is subjected to compressive loads $P = 400\,\text{kN}$ with eccentricity $e = 0.5\,\text{m}$. The column's length is $L = 6\,\text{m}$, its cross-sectional area is $A = 0.122\,\text{m}^2$, and its moment of inertia is $I = 0.00125\,\text{m}^4$. The modulus of elasticity of the material is 72 GPa. What is the column's maximum deflection?

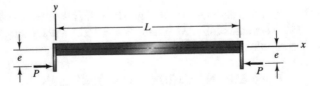

Problems 10.44–10.45

10.45 The column is subjected to compressive loads $P = 400 \, \text{kN}$ with eccentricity $e = 0.5 \, \text{m}$. The column's length is $L = 6 \, \text{m}$, its cross-sectional area is $A = 0.122 \, \text{m}^2$, and its moment of inertia is $I = 0.00125 \, \text{m}^4$. The modulus of elasticity of the material is 72 GPa. The distance from the column's neutral axis to the location on the cross section where the maximum compressive stress occurs is $c = 0.175 \, \text{m}$. What is the maximum normal stress in the column?

10.46 The area of the column's cross section is $A = 140 \, \text{in}^2$ and its moment of inertia is $I = 1730 \, \text{in}^4$. The modulus of elasticity of the material is $E = 11\text{E}6 \, \text{psi}$. What is the column's maximum deflection?

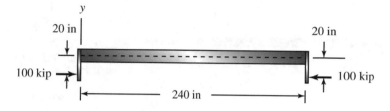

Problems 10.46–10.47

10.47 The area of the column's cross section is $A = 140 \, \text{in}^2$ and its moment of inertia is $I = 1730 \, \text{in}^4$. The modulus of elasticity of the material is $E = 11\text{E}6 \, \text{psi}$. The distance from the column's neutral axis to the location on the cross section where the maximum compressive stress occurs is $c = 6 \, \text{in}$. What is the maximum normal stress in the column?

10.48 The column is subjected to compressive loads $P = 40 \, \text{kN}$ with eccentricity $e = 4.6 \, \text{mm}$. The column's length is $L = 760 \, \text{mm}$ and the modulus of elasticity of the material is $E = 200 \, \text{GPa}$. What is the column's maximum deflection?

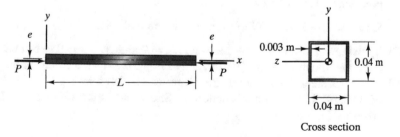

Cross section

Problems 10.48–10.49

10.49 The column has eccentricity $e = 9.2 \, \text{mm}$ and length $L = 760 \, \text{mm}$. The modulus of elasticity of the material is $E = 200 \, \text{GPa}$. If the allowable compressive stress of the material is 250 MPa, what is the largest allowable value of the axial load P?

Strategy: Use Fig. 10.29 to determine the largest allowable value of P.

10.50 The column has the cross section shown and its length is $L = 2.5$ m. It is subjected to compressive loads with eccentricity $e = 0.00625$ m. The material is steel with modulus of elasticity $E = 200$ GPa. If the compressive load is $P = 200$ kN, what is the column's maximum deflection? (See the figure for Problems 10.50–10.52.)

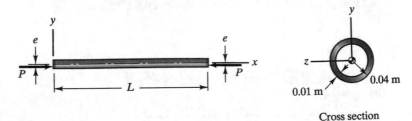

Cross section

Problems 10.50–10.52

10.51 The column has the cross section shown and its length is $L = 2.5$ m. It is subjected to compressive loads with eccentricity $e = 0.00625$ m. The material is steel with modulus of elasticity $E = 200$ GPa. If the compressive load is $P = 200$ kN, what is the maximum normal stress in the column? (See the figure for Problems 10.50–10.52.)

10.52 The column has the cross section shown and its length is $L = 2.5$ m. It is subjected to compressive loads with eccentricity $e = 0.00625$ m. The material is steel with modulus of elasticity $E = 200$ GPa. If the allowable compressive stress is 250 MPa, what is the largest allowable value of P? (See the figure for Problems 10.50–10.52.)

Strategy: Use Fig. 10.29 to estimate the value of P/A.

10.53 By using the boundary conditions $v|_{x = 0} = 0$ and $v|_{x = L} = 0$ to evaluate the constants B and C in Eq. (10.26), derive Eq. (10.27) for the deflection.

10.54 By substituting the trigonometric identities (10.28) into Eq. (10.27), derive Eq. (10.29) for the column's deflection. Show that the deflection at $x = L/2$ is given by Eq. (10.30).

Chapter Summary

Buckling Loads

The buckling load for a prismatic column of length L that is free at the ends and buckles as shown in Fig. (a) is called the *Euler buckling load*:

$$P = \frac{\pi^2 EI}{L^2}. \tag{10.7}$$

Figure (a) is the first buckling mode for a column that is free at the ends. The distributions of the deflection for the second, third, and fourth modes are shown in Fig. 10.5. The buckling load of the nth mode is

$$P = \frac{n^2 \pi^2 EI}{L^2}. \tag{10.5}$$

First-mode buckling loads of columns with other common end conditions are shown in Fig. 10.16.

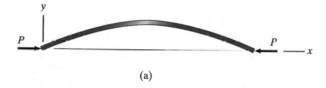

(a)

Effective Length

Suppose that a given column buckles in a particular mode and the buckling load is P. The *effective length* L_e is the length of a column of the same flexural rigidity whose Euler buckling load equals P:

$$P = \frac{\pi^2 EI}{L_e^2}. \tag{10.21}$$

The ratio of the effective length to the column's actual length is denoted by K and called the *effective length factor*:

$$K = \frac{L_e}{L} \qquad (10.22)$$

For example, if a column of length L that is free at the ends buckles in the second mode, the effective length is $L_e = L/2$ (Fig. (b)). The effective length factor is

$$K = \frac{L_e}{L} = \frac{\left(\frac{L}{2}\right)}{L} = 0.5$$

and the buckling load is

$$P = \frac{\pi^2 EI}{L_e^2} = \frac{4\pi^2 EI}{L^2}.$$

In some cases, the effective length of a buckled column can be determined or approximated by observation, and Eq. (10.21) can be used to determine the buckling load.

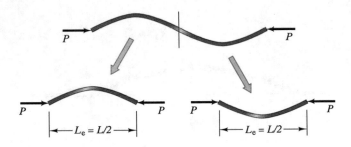

Eccentric Loads and the Secant Formula

Consider a prismatic column that is free to rotate at the ends and is subjected to eccentric loads (Fig. (c)). The distribution of the deflection is

$$v = e\left(\tan \frac{\lambda L}{2} \sin \lambda x + \cos \lambda x - 1\right) \qquad (10.29)$$

and the maximum deflection is

$$v_{max} = v|_{x=L/2} = e\left(\sec \frac{\lambda L}{2} - 1\right), \qquad (10.30)$$

where $\lambda^2 = P/EI$.

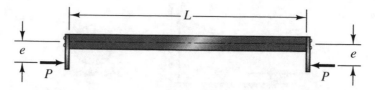

The maximum compressive stress in the column in Fig. (c) is given by the *secant formula*

$$\sigma_{max} = \frac{P}{A}\left[1 + \frac{ec}{k^2}\sec\left(\frac{L}{2k}\sqrt{\frac{P}{EA}}\right)\right],\qquad (10.32)$$

where $k = \sqrt{I/A}$ is the radius of gyration of the cross section. The parameter ec/k^2 is the *eccentricity ratio* and L/k is the *slenderness ratio*. When the column's dimensions, the modulus of elasticity, and the eccentricity are known and the maximum allowable compressive stress is specified, Eq. (10.32) can be solved numerically for the maximum allowable axial load.

Review Problems

10.55 The column is steel with elastic modulus $E = 220$ GPa. Determine its Euler buckling load.

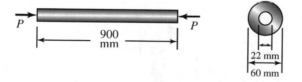

Cross section

Problems 10.55–10.56

10.56 The column is steel with elastic modulus $E = 220$ GPa. Determine its second-mode buckling load.

10.57 The bar AB consists of material with elastic modulus $E = 70$ GPa. It has a hollow circular cross section with radii $R_o = 20$ mm and $R_i = 8$ mm. Can a horizontal force $F = 120$ kN be applied at A without causing bar AB to fail by buckling?

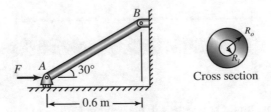

Cross section

Problems 10.57–10.58

10.58 Suppose that the circular cross section of the bar AB has an outer radius $R_o = 20$ mm and you want to choose the inner radius R_i so that bar AB will buckle when $F = 100$ kN. What is the radius R_i?

10.59 The column supports an axial load P. The dimensions of its rectangular cross section are $b = 180$ mm and $h = 360$ mm. The column has a fixed support at its base. The support at the top allows rotation, prevents lateral deflection in the x–y plane, and allows lateral deflection in the x–z plane. The elastic modulus is $E = 200$ GPa. What is the column's buckling load? Does it buckle by bending in the x–y plane or the x–z plane?

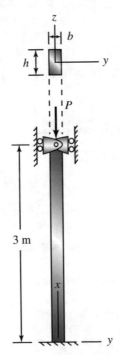

Problems 10.59–10.60

10.60 Suppose that you want to choose the dimensions b and h of the cross section of the column so that the buckling load of the column if the buckles by bending in the x–y plane is equal to the buckling load if it buckles by bending in the x–z plane. Determine the necessary value of the ratio h/b.

10.61 The cylindrical tube is subjected to eccentric axial loads $P = 120$ kN. It consists of aluminum alloy with modulus of elasticity $E = 72$ MPa. What is the maximum resulting deflection?

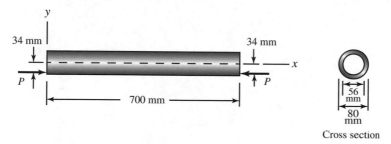

Problems 10.61–10.62

10.62 The cylindrical tube is subjected to eccentric axial loads $P = 120$ kN. It consists of aluminum alloy with modulus of elasticity $E = 72$ MPa. What is the maximum normal stress in the tube?

Chapter 11
Energy Methods

When an archer draws a bow and shoots an arrow, it demonstrates that energy can be stored in an object when work is done to deform it relative to a reference state. The energy in the flexed bow is transferred into kinetic energy of the arrow. Many advanced techniques used in mechanics of materials, including finite elements, are based on energy methods. We now introduce some of the fundamental concepts underlying these methods and apply them to simple structural elements.

11.1 Work and Energy

Our objective is to describe a class of problems in solid mechanics that can be analyzed by considering the energy stored in deformed elastic materials. To do so, we must first define the work done by a force or couple.

11.1.1 Work

Suppose that an object is subjected to a force $\mathbf{F}$ (Fig. 11.1(a)). If the point of the object to which $\mathbf{F}$ is applied undergoes an infinitesimal displacement represented by a vector $d\mathbf{r}$ (Fig. 11.1(b)), the *work* done on the object by the force is defined to be the dot product

$$dW = \mathbf{F} \cdot d\mathbf{r}.$$

The dimensions of work are (force) × (length). In US customary units, work is usually expressed in in-lb or ft-lb. In SI units, work is expressed in N-m or joules (J). Let us introduce a coordinate system oriented as shown in Fig. 11.1(c), with the

© Springer Nature Switzerland AG 2020
A. Bedford, K. M. Liechti, *Mechanics of Materials*,
https://doi.org/10.1007/978-3-030-22082-2_11

Fig. 11.1 (a) Object
subjected to a force.
(b) Displacement of the
point of application.
(c) Introducing a
coordinate system

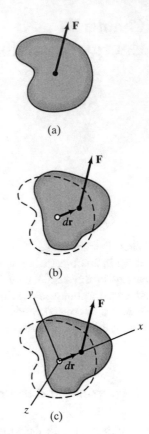

(a)

(b)

(c)

x axis parallel to $d\mathbf{r}$. Expressing the force and displacement vectors in terms of their
components as $\mathbf{F} = F_x\mathbf{i} + F_y\mathbf{j} + F_z\mathbf{k}$ and $d\mathbf{r} = dx\,\mathbf{i}$, the work is

$$dW = \left(F_x\mathbf{i} + F_y\mathbf{j} + F_z\mathbf{k}\right) \cdot (dx\,\mathbf{i})$$
$$= F_x\,dx.$$

We see that *only the component of force parallel to the displacement does work.*
Because we will only be concerned with the component of $\mathbf{F}$ parallel to the
displacement, let $F_x = F$. If the point to which F is applied moves a finite distance
along the x axis from $x = 0$ to $x = x_0$ (Fig. 11.2), the work done is

$$W = \int_0^{x_0} F\,dx.$$

In a graph of F as a function of x, the work done equals the "area" between the
force and the x axis (Fig. 11.3(a)). Two particular cases are of interest:

Fig. 11.2 Path of the point
of application along the
x axis

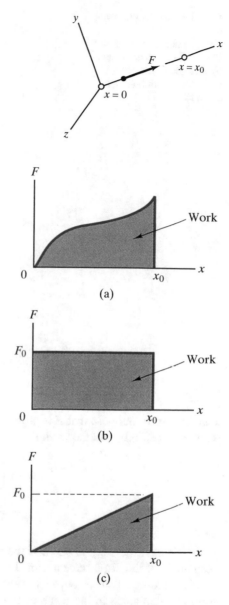

Fig. 11.3 (**a**) The work
equals the area defined by
the graph of F as a function
of x. (**b**) Work done by a
constant force. (**c**) Work
done by a linear force

Constant Force If the force is a constant $F = F_0$, the work is simply the product of
F_0 and the magnitude of the displacement (Fig. 11.3(b)):

$$W = F_0 x_0.$$

Fig. 11.4 (**a**) Couple acting
on an object. (**b**)
Infinitesimal rotation of the
object. (**c**) Object acted upon
by a couple of moment
M that rotates through an
angle $d\theta$

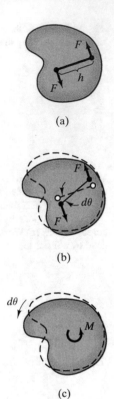

(a)

(b)

(c)

Linear Force Suppose that F is proportional to x and its value is F_0 when $x = x_0$.
Then $F = (F_0/x_0)x$, and the work is

$$W = \int_0^{x_0} \frac{F_0}{x_0} x \, dx$$
$$= \frac{1}{2} F_0 x_0.$$

In this case the work is one-half the product of the value of F at $x = x_0$ and the
magnitude of the displacement (Fig. 11.3(c)).

Now suppose that an object is subjected to a couple (Fig. 11.4(a)). Let both
forces be contained in the plane of the page, so that the couple exerts a counterclockwise moment $M = Fh$. If the object undergoes a motion that causes the
couple to rotate through a counterclockwise angle $d\theta$ (Fig. 11.4(b)), the work done
is $dW = F\left(\frac{1}{2}h \, d\theta\right) + F\left(\frac{1}{2}h \, d\theta\right) = M \, d\theta$. Thus when a couple of moment M rotates
through an angle $d\theta$ in the same direction as the couple (Fig. 11.4(c)), the work
done is

$$dW = M \, d\theta. \tag{11.1}$$

If the angle varies from $\theta = 0$ to $\theta = \theta_0$, the work done is

$$W = \int_0^{\theta_0} M \, d\theta.$$

As in the case of the work done by a force, two cases are of interest:

Constant Couple If the moment is a constant $M = M_0$, the work is the product of M_0 and the magnitude of the angular displacement:

$$W = M_0 \theta_0.$$

Linear Couple Suppose that M is proportional to θ and its value is M_0 when $\theta = \theta_0$. Then $M = (M_0/\theta_0)\theta$, and the work is

$$W = \int_0^{\theta_0} \frac{M_0}{\theta_0} \theta \, d\theta$$
$$= \frac{1}{2} M_0 \theta_0.$$

The work is one-half the product of the value of M at $\theta = \theta_0$ and the magnitude of the angular displacement.

11.1.2 Strain Energy

Suppose that we apply a force F to a linear spring as shown in Fig. 11.5. By definition, the relation between F and the resulting stretch of the spring is

$$F = kx, \tag{11.2}$$

where k is the spring constant. Let the force required to stretch the spring an amount x_0 be $F_0 = kx_0$. Because the force required to stretch the spring is a linear function of x, we have seen that the work W done in stretching the spring an amount x_0 is one-half the product of the force necessary to cause the displacement and the magnitude of the displacement (Fig. 11.6):

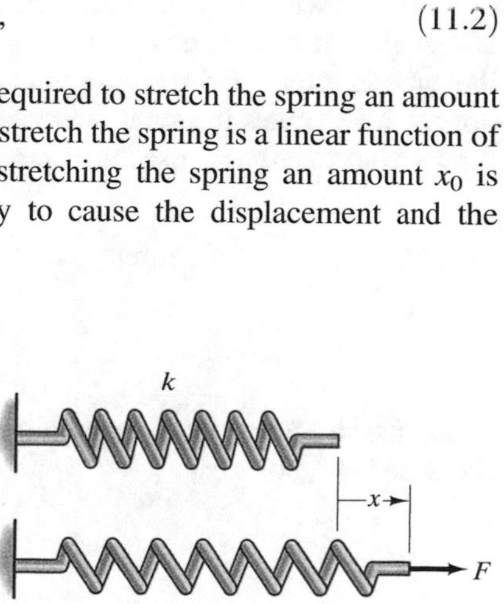

Fig. 11.5 Stretching a linear spring

Fig. 11.6 Work done in stretching a linear spring an amount x_0

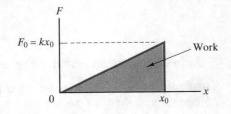

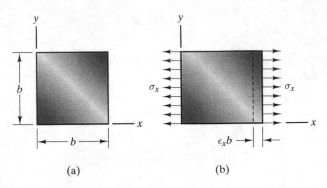

(a)   (b)

Fig. 11.7 Subjecting a cube of material to normal stress

$$W = \frac{1}{2} F_0 x_0.$$

The work done is "stored" in the spring, in the sense that the spring is capable of doing an equal amount of work as it contracts to its original length. We say that the work is stored in the form of potential energy. We will show that potential energy can be stored within any object consisting of elastic material when the object is deformed. This kind of potential energy is called *strain energy*.

We begin with a cube of elastic material with dimensions $b \times b \times b$ (Fig. 11.7(a)) and subject it to a normal stress σ_x (Fig. 11.7(b)). The tensile force exerted on the cube is $\sigma_x(b)^2$, and the change in its length in the x direction is $\varepsilon_x b$. The work done, which equals the strain energy stored in the cube, is one-half the product of the force necessary to cause the displacement and the magnitude of the displacement:

$$\text{Strain energy} = U = \frac{1}{2} \left[\sigma_x(b)^2 \right] \varepsilon_x b$$

$$= \frac{1}{2} \sigma_x \varepsilon_x b^3.$$

Let the strain energy per unit volume of material be denoted by u. The volume of the cube is b^3, so

$$u = \frac{1}{2} \sigma_x \varepsilon_x.$$

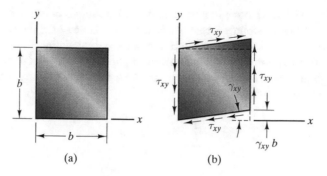

Fig. 11.8 Subjecting the cube to shear stress

Now let us subject the cube to a shear stress τ_{xy} (Fig. 11.8) and determine the strain energy per unit volume. How much work is done on the cube? Comparing the deformed shape of the cube to its original shape as shown in Fig. 11.8(b), we see that work is done only by the shear stress acting on the cube's right face. (The force exerted on each infinitesimal element of the top and bottom faces is perpendicular to the motion of the element and so does no work.) The force exerted on the right face is $\tau_{xy}(b)^2$, and its displacement in the y direction is $\gamma_{xy}b$. The resulting strain energy is

$$U = \frac{1}{2}\left[\tau_{xy}(b)^2\right]\gamma_{xy}b$$
$$= \frac{1}{2}\tau_{xy}\gamma_{xy}b^3,$$

so the strain energy per unit volume of material is

$$u = \frac{1}{2}\tau_{xy}\gamma_{xy}.$$

We can obtain the strain energy per unit volume of an isotropic linearly elastic material subjected to a general state of stress by superimposing the strain energies due to the individual components of stress. The result is

$$u = \frac{1}{2}\left(\sigma_x\varepsilon_x + \sigma_y\varepsilon_y + \sigma_z\varepsilon_z + \tau_{xy}\gamma_{xy} + \tau_{yz}\gamma_{yz} + \tau_{xz}\gamma_{xz}\right). \tag{11.3}$$

We have calculated the strain energy by equating it to the work done in deforming a linearly elastic material. Although many materials closely approximate elastic behavior under appropriate conditions, all real materials exhibit some internal energy dissipation, and as a result only part of the work done in deforming them is stored in the form of strain energy. For example, a metal wire becomes warm when it is flexed back and forth. Some of the work done in bending is converted into heat instead of strain energy. The work done in permanently deforming an object is obviously not stored as recoverable strain energy. In fact, the existence of a recoverable strain energy is sometimes used as the definition of an elastic material.

Suppose that the state of stress is known at a point, and we want to determine the strain energy per unit volume u at that point. By using the stress-strain relations for an isotropic elastic material,

$$\varepsilon_x = \frac{1}{E}\sigma_x - \frac{v}{E}(\sigma_y + \sigma_z), \tag{11.4}$$

$$\varepsilon_y = \frac{1}{E}\sigma_y - \frac{v}{E}(\sigma_x + \sigma_z), \tag{11.5}$$

$$\varepsilon_z = \frac{1}{E}\sigma_z - \frac{v}{E}(\sigma_x + \sigma_y), \tag{11.6}$$

$$\gamma_{xy} = \frac{1}{G}\tau_{xy}, \tag{11.7}$$

$$\gamma_{yz} = \frac{1}{G}\tau_{yz}, \tag{11.8}$$

$$\gamma_{xz} = \frac{1}{G}\tau_{xz}, \tag{11.9}$$

we can determine the state of strain and then use Eq. (11.3) to determine u. Or, if we know the state of strain at a point, we can solve Eqs. (11.4)–(11.9) for the state of stress and then use Eq. (11.3) to determine u.

11.1.3   Applications

Deflections of trusses and beams can be determined by equating the work done by external forces and couples during their application to the resulting strain energy. Here we consider applications in which only a single external force or couple does work. Although this restriction allows us to introduce energy methods in a simple context, it greatly limits the situations we can consider. We remove this restriction in Sect. 11.2.

11.1.3.1   Axially Loaded Bars

In Fig. 11.9 an axial load P acts on a prismatic bar of length L, causing its length to increase an amount δ. Suppose that we know P and want to determine δ. Because the change in length of the bar is a linear function of the axial load, the work done by the external load P is one-half the product of the force and the change in the bar's length:

Fig. 11.9 Applying an
axial load to a bar

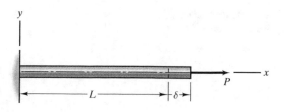

$$W = \frac{1}{2} P\delta. \qquad (11.10)$$

If we can determine the strain energy in the bar, we can equate it to the external work and solve for δ. In terms of the coordinate system shown, we know that the only nonzero stress component in the bar is $\sigma_x = P/A$, where A is the bar's cross-sectional area. From Eq. (11.4), the strain component $\varepsilon_x = (1/E)\sigma_x$. Applying Eq. (11.3), the strain energy per unit volume of the bar is

$$u = \frac{1}{2}\sigma_x \varepsilon_x$$

$$= \frac{1}{2E}\sigma_x^2$$

$$= \frac{P^2}{2EA^2}.$$

Multiplying this expression by the bar's volume AL, the strain energy of the axially loaded bar is

$$U = \frac{P^2 L}{2EA}. \qquad (11.11)$$

Equating this result to Eq. (11.10) and solving for δ, we obtain

$$\delta = \frac{PL}{EA}.$$

Of course we already obtained this result, in a simpler way, in Chap. 3. But we are now in a position to consider an example that dramatically demonstrates the power of energy methods.

In Fig. 11.10 a vertical load F acts at point B of the truss. Let the vertical component of the resulting displacement of point B be denoted by v. Suppose that we know F and want to determine v. We can analyze the truss in the usual way to determine the axial loads in the individual members (assuming that the change in the geometry of the truss due to the application of F is small) and thereby determine the

Fig. 11.10 Subjecting a
truss to a force F

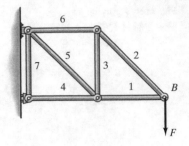

change in length of each member. But using that information to determine the
displacement v would be very difficult.

Instead, we can determine v by equating the work done by the force F to the
resulting strain energy in the truss. Assuming that the vertical displacement at B is a
linear function of the vertical load applied at B, the work done by F is

$$W = \frac{1}{2}Fv. \tag{11.12}$$

Let the axial loads in the members be $P_1, P_2, \ldots, P_7$. Once we have determined
the axial loads, we can calculate the strain energy in each member from Eq. (11.11).
The strain energy in the ith member is

$$U_i = \frac{P_i^2 L_i}{2E_i A_i}.$$

We equate the work done by the external force to the total strain energy,

$$\frac{1}{2}Fv = \sum_{i=1}^{7} \frac{P_i^2 L_i}{2E_i A_i},$$

and solve for v:

$$v = \frac{1}{F} \sum_{i=1}^{7} \frac{P_i^2 L_i}{E_i A_i}.$$

Example 11.1 Determining Displacement of a Truss Joint
The truss in Fig. 11.11 is subjected to a vertical force F at B. The members
each have elastic modulus E and cross-sectional area A. What is the vertical
component of the resulting displacement of point B?

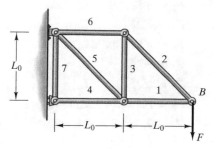

Fig. 11.11

Strategy
Let the vertical component of the displacement of B be denoted by v. We can determine v by equating the work done by the external load F to the sum of the strain energies in the members. The strain energy in an axially loaded bar is given by Eq. (11.11).

Solution
The work done by F is

$$W = \frac{1}{2}Fv.$$

Let P_i be the axial load in the ith member. From Eq. (11.11), the strain energy in the ith member is

$$U_i = \frac{P_i^2 L_i}{2EA}.$$

We equate the work to the sum of the strain energies,

$$\frac{1}{2}Fv = \sum_{i=1}^{7} \frac{P_i^2 L_i}{2EA},$$

and solve for v:

$$v = \frac{1}{FEA} \sum_{i=1}^{7} P_i^2 L_i. \tag{11.13}$$

Determining the axial loads in the members and their lengths, we obtain

$$P_1 = -F, \qquad L_1 = L_0$$
$$P_2 = \sqrt{2}F, \qquad L_2 = \sqrt{2}L_0$$
$$P_3 = -F, \qquad L_3 = L_0$$
$$P_4 = -2F, \qquad L_4 = L_0$$
$$P_5 = \sqrt{2}F, \qquad L_5 = \sqrt{2}L_0$$
$$P_6 = F, \qquad L_6 = L_0$$
$$P_7 = 0, \qquad L_7 = L_0.$$

Substituting these values into Eq. (11.13), the vertical displacement is

$$v = \frac{\left(7 + 4\sqrt{2}\right)FL_0}{EA}.$$

Discussion
Although this result was easy to obtain and the example is an important type of
problem, notice that we have determined the displacement of the point where
F is applied, and F is the only external force or couple that does work. The
procedure we use in this section cannot be used to determine the displacements of other points of the truss or be applied to more complicated loadings.
We consider such problems in Sect. 11.2.

11.1.3.2   Beams

Consider a prismatic beam whose cross section is symmetric about the vertical (y)
axis. Our objective is to determine the strain energy in the beam resulting from the
normal stress due to bending. Figure 11.12 shows an element of a beam with length
dx and cross-sectional area dA. This element is subjected to normal stress

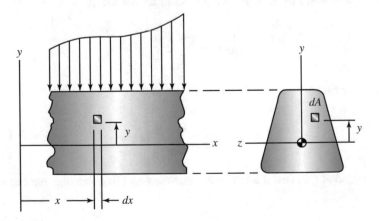

Fig. 11.12   Element of a beam

$$\sigma_x = -\frac{My}{I}, \tag{11.14}$$

where y is the element's position relative to the neutral axis and M is the bending moment at the axial position x. From Eqs. (11.3) and (11.4), the strain energy per unit volume associated with the normal stress is

$$u = \frac{1}{2}\sigma_x\varepsilon_x$$
$$= \frac{1}{2E}\sigma_x^2.$$

Multiplying this expression by the volume of the element and substituting Eq. (11.14) give the strain energy of the element:

$$u\,dx\,dA = \frac{M^2y^2}{2EI^2}\,dx\,dA.$$

Integrating this result with respect to A yields the strain energy of a "slice" of the beam of width dx, which we denote by dU:

$$dU = \int_A u\,dx\,dA = \frac{M^2}{2EI^2}\,dx\int_A y^2\,dA.$$

Recognizing that $\int_A y^2\,dA = I$, we obtain the strain energy of an element of the beam of width dx (Fig. 11.13) in the form

$$dU = \frac{M^2}{2EI}\,dx. \tag{11.15}$$

In general, the strain energy in a beam also includes contributions due to the stresses caused by the axial load P and the shear force V. But we will consider only beams that are not subjected to axial loads, and for slender beams the strain energy due to the shear force is normally negligible in comparison to that resulting from the bending moment.

Fig. 11.13 Element of a beam of width dx

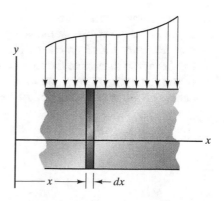

Fig. 11.14 Cantilever beam
subjected to a force

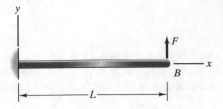

To determine the strain energy of an entire beam, we must integrate Eq. (11.15) with respect to x over the beam's length. For example, the bending moment in the prismatic cantilever beam in Fig. 11.14 is

$$M = F(L - x).$$

By substituting this expression into Eq. (11.15) and integrating from $x = 0$ to $x = L$, we obtain the strain energy:

$$U = \int_0^L \frac{[F(L - x)]^2}{2EI} dx$$
$$= \frac{F^2 L^3}{6EI}.$$

Let v be the beam's deflection at B. The work done by the external force F is one-half the product of the force and the deflection at B:

$$W = \frac{1}{2} Fv.$$

Equating the work and the strain energy and solving for v, we obtain

$$v = \frac{FL^3}{3EI}.$$

Study Questions

1. What is strain energy?
2. If you know the state of stress at a point of an isotropic elastic material, how can you determine the strain energy per unit volume at that point?
3. How can you determine the strain energy in an axially loaded bar?
4. If you know the distribution of the bending moment in a beam, how can you determine the strain energy of the beam?

Example 11.2 Determining a Beam Displacement
The simply supported beam in Fig. 11.15 is subjected to a vertical force F.
What is the beam's deflection at the point where F is applied?

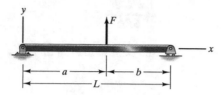

Fig. 11.15

Strategy
We can determine the deflection by equating the work done by the external
load F to the strain energy in the beam. Because the bending moment in the
beam is described by different equations in the regions $0 \leq x \leq a$ and $a \leq x \leq L$,
we must determine the strain energy by integrating Eq. (11.15) over these two
regions separately.

Solution
From Fig. (a), the bending moment in the region $0 \leq x \leq a$ is

$$M = -\frac{bF}{L}x.$$

(a) Determining the bending moment in the region $0 \leq x \leq a$

Integrating Eq. (11.15), the strain energy contained in this region of the
beam is

$$U_1 = \frac{1}{2EI} \int_0^a M^2 \, dx$$

$$= \frac{1}{2EI} \int_0^a \frac{b^2 F^2}{L^2} x^2 \, dx$$

$$= \frac{a^3 b^2 F^2}{6EIL^2}.$$

From Fig. (b), the bending moment in the region $a \leq x \leq L$ is

$$M = -\frac{aF}{L}(L - x).$$

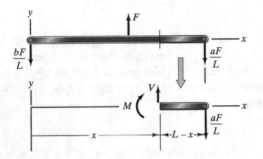

(b) Determining the bending moment in the region $a \leq x \leq L$

The strain energy contained in this region of the beam is

$$U_2 = \frac{1}{2EI} \int_a^L M^2 \, dx$$

$$= \frac{1}{2EI} \int_a^L \frac{a^2 F^2}{L^2} (L - x)^2 \, dx$$

$$= \frac{a^2 F^2}{2EIL^2} \left(\frac{L^3}{3} - L^2 a + La^2 - \frac{a^3}{3} \right).$$

Let v be the beam's deflection at the point where F is applied. The work done by F equals the total strain energy:

$$\frac{1}{2}Fv = U_1 + U_2$$

$$= \frac{a^2 F^2}{2EIL^2} \left(\frac{ab^2}{3} + \frac{L^3}{3} - L^2 a + La^2 - \frac{a^3}{3} \right).$$

Solving for v, the beam's deflection at the point where F is applied is

$$v = \frac{a^2 F}{EIL^2}\left(\frac{ab^2}{3} + \frac{L^3}{3} - L^2 a + La^2 - \frac{a^3}{3}\right).$$

Discussion

You should verify this result by using Eq. (9.4) and compare the amount of effort required with this example.

Problems

11.1 When the elastic bar is subjected to an axial load $P = 20\,\text{kN}$, the resulting stretch of the bar is $\delta = 3\,\text{mm}$. How much strain energy is stored in the bar?

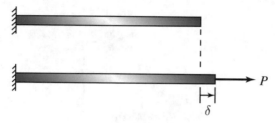

Problem 11.1

11.2 The bar has cross-sectional area $A = 0.0003\,\text{m}^2$ and is made of aluminum alloy with modulus of elasticity $E = 80\,\text{GPa}$. It is subjected to 200-N axial loads as shown. (a) How much strain energy is stored in the bar? (b) How much work is done on the bar by the 200-N loads?

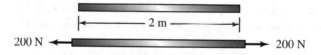

Problems 11.2–11.3

11.3 The bar has cross-sectional area $A = 0.0003\,\text{m}^2$ and is made of rubber with modulus of elasticity $E = 2\,\text{MPa}$. How much work is done on the bar by the 200-N loads? What is the ratio of the work done on the rubber bar to the work done on the aluminum bar in Problem 11.2?

11.4 The bar has cross-sectional area $A = 2\,\text{in}^2$ and is made of steel with modulus of elasticity $E = 28\text{E}6\,\text{psi}$. It is subjected to a 30-kip axial load as shown. (a) How much strain energy is stored in the bar? (b) How much work is done on the bar by the 30-kip load?

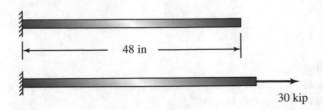

48 in

30 kip

Problem 11.4

11.5 The wooden bar AB has cross-sectional area $A = 0.8\,\text{in}^2$ and elastic modulus $E = 1.4\text{E6}\,\text{psi}$. A downward force $F = 3500\,\text{lb}$ is applied at B. Determine the resulting deflection v of point B (**a**) without using strain energy and (**b**) using strain energy.

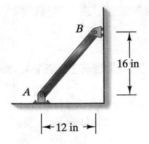

B

16 in

A

$\leftarrow$ 12 in $\rightarrow$

Problem 11.5

11.6 The aluminum bar AB has cross-sectional area $A = 0.002\,\text{m}^2$ and elastic modulus $E = 72\,\text{GPa}$. The horizontal bar is rigid in comparison with bar AB. A counterclockwise couple $M = 1.2\,\text{MN}\cdot\text{m}$ is applied at C. Use strain energy to determine the angle through which the horizontal bar rotates.

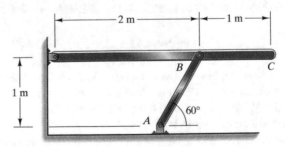

$\leftarrow$ 2 m $\rightarrow$ $\leftarrow$ 1 m $\rightarrow$

B C

1 m

$60°$

A

Problem 11.6

11.7 Each bar has length L, cross-sectional area A, and elastic modulus E. If a downward force F is applied at B, determine the resulting displacement v of point B (**a**) without using strain energy and (**b**) using strain energy.

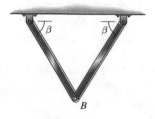

Problems 11.7–11.8

11.8 The angle $\beta = 50°$. Each bar has length $L = 2\,\text{m}$, cross-sectional area $A = 0.0006\,\text{m}^2$, and modulus of elasticity $E = 70\,\text{GPa}$. What downward force applied at B will cause B to move 4 mm?

11.9 Each bar has cross-sectional area $A = 1.2\,\text{in}^2$ and elastic modulus $E = 1.8\text{E}6\,\text{psi}$. If a downward force $F = 5\,\text{kip}$ is applied at B, use strain energy to determine the vertical component v of the displacement of point B.

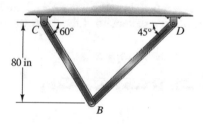

Problem 11.9

11.10 The truss is subjected to a vertical force F at B. The members each have modulus of elasticity E and cross-sectional area A. What is the vertical component of the resulting displacement of point B? (See Example 11.1.)

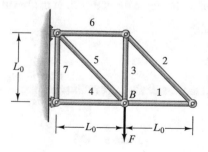

Problem 11.10

11.11* The truss is made of steel bars with cross-sectional area $A = 0.003\,\text{m}^2$ and elastic modulus $E = 220\,\text{GPa}$. Determine the vertical component v of the displacement of point B if the suspended mass $m = 5000\,\text{kg}$.

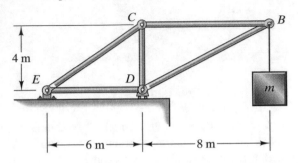

Problem 11.11

11.12 The beam has length $L = 60\,\text{in}$, modulus of elasticity $E = 16\text{E6}\,\text{psi}$, and moment of inertia $I = 6.75\,\text{in}^4$. The force $F = 20,000\,\text{lb}$. Use strain energy to determine the beam's deflection at the point where F is applied.

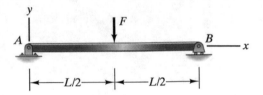

Problem 11.12

11.13 The beam has length $L = 50\,\text{in}$, modulus of elasticity $E = 12\text{E6}\,\text{psi}$, and moment of inertia $I = 8\,\text{in}^4$. The couple $M_0 = 40,000\,\text{in - lb}$. Use strain energy to determine the angle through which the beam rotates at B.

Problem 11.13

11.14 The beam has length L, modulus of elasticity E, and moment of inertia I. Use strain energy to determine the beam's deflection at the point where F is applied.

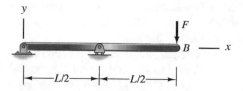

Problem 11.14

11.15 The horizontal beam has moment of inertia I and elastic modulus E. The vertical bar has cross-sectional area A and elastic modulus E. A downward force F is applied at B. Use strain energy to determine the deflection at B.

Strategy: Draw separate free-body diagrams of the beam and bar, and use strain energy to determine the deflection of the beam and the change in length of the bar.

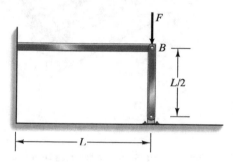

Problem 11.15

11.16 A material is subjected to a uniform state of plane stress $\sigma_x = 400\,\text{MPa}$, $\sigma_y = -200\,\text{MPa}$, $\tau_{xy} = 300\,\text{MPa}$. Using a strain gauge rosette, it is determined experimentally that $\varepsilon_x = 0.00239$, $\varepsilon_y = -0.00162$, and $\gamma_{xy} = 0.00401$. What is the strain energy per unit volume of the material?

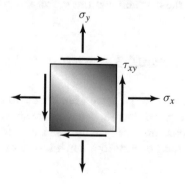

Problems 11.16–11.17

11.17 A sample of 2014-T6 aluminum alloy is subjected to a uniform state of plane stress $\sigma_x = 40\,\text{MPa}$, $\sigma_y = -30\,\text{MPa}$, and $\tau_{xy} = 30\,\text{MPa}$. What is the strain energy per unit volume of the material?

11.18 The bar has a solid circular cross section and consists of material with shear modulus G. It is subjected to an axial torque T at the end. Show that the strain energy stored in the bar is $T^2 L/2GJ$, where J is the polar moment of inertia of the cross section.

Strategy: Calculate the work done by the torque T in twisting the end of the bar and use Eq. (4.7).

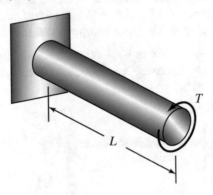

Problems 11.18–11.19

11.19∗ From Eq. (4.9), the shear stress in the bar as a function of distance r from the bar's axis is $\tau = Tr/J$. From Eq. (11.3), the strain energy per unit volume in the bar is therefore

$$u = \frac{1}{2}\tau\gamma = \frac{1}{2G}\tau^2 = \frac{T^2 r^2}{2GJ^2}.$$

Integrate this expression over the bar's volume to determine the strain energy stored in the bar, and confirm the answer to Problem 11.18.

11.2 Castigliano's Second Theorem

In applying strain energy in Sect. 11.1, we were limited to situations in which only a single external force or couple did work, and we were able to determine the resulting deflection or angular displacement only at the point at which the force or couple was applied. We can apply strain energy to a much broader class of problems using a result presented by Italian engineer Alberto Castigliano in 1879. We will sketch a proof of his result and then present applications.

11.2.1 Derivation

Consider an object consisting of elastic material that is subjected to N external forces $F_1, F_2, \ldots, F_N$ (Fig. 11.16). Let u_i be the component of the deflection of the point of application of the ith force F_i in the direction of F_i, and assume that u_i is a linear function of F_i. Our objective is to determine u_i. We assume that the strain energy U resulting from the application of the N forces can be expressed as a function of the forces:

$$U = U(F_1, F_2, \ldots, F_N).\tag{11.16}$$

If we increase the magnitude of the ith force F_i by an amount ΔF_i and assume that Eq. (11.16) can be expressed as a Taylor series in terms of each of its arguments, we can write the increased strain energy as

$$U(F_1, F_2, \ldots, F_i + \Delta F_i, \ldots, F_N)$$
$$= U(F_1, F_2, \ldots, F_N) + \frac{\partial U}{\partial F_i}\Delta F_i + O(\Delta F_i^2),\tag{11.17}$$

where the notation $O(\Delta F_i^2)$ means "terms of order two or greater in ΔF_i."

Now suppose that the force ΔF_i is applied before the forces $F_1, F_2, \ldots, F_N$. If Δu_i is the component of the displacement of the point of application of ΔF_i in the direction of ΔF_i, the work done is $\frac{1}{2}\Delta F_i \Delta u_i$. (Notice that $\Delta u_i \rightarrow 0$ as $\Delta F_i \rightarrow 0$.) If we then apply the forces $F_1, F_2, \ldots, F_N$, the force ΔF_i undergoes the additional displacement u_i, doing work $\Delta F_i u_i$. The resulting strain energy must equal the work done by the force ΔF_i plus the strain energy associated with the application of the forces $F_1, F_2, \ldots, F_N$:

$$\text{Strain energy} = \frac{1}{2}\Delta F_i \Delta u_i + \Delta F_i u_i + U(F_1, F_2, \ldots, F_N).\tag{11.18}$$

Fig. 11.16 The N forces $F_1, F_2, \ldots, F_N$

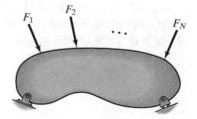

But the strain energy does not depend on the order in which the forces are applied, so Eqs. (11.17) and (11.18) must be equal:

$$
U(F_1, F_2, \ldots, F_N) + \frac{\partial U}{\partial F_i} \Delta F_i + O\left(\Delta F_i^2\right)
$$
$$
= \frac{1}{2} \Delta F_i \Delta u_i + \Delta F_i u_i + U(F_1, F_2, \ldots, F_N).
$$

Dividing this equation by ΔF_i and taking the limit as $\Delta F_i \to 0$, we obtain

$$
u_i = \frac{\partial U}{\partial F_i}. \tag{11.19}
$$

This result is called *Castigliano's second theorem*. It can be used to determine the component of the deflection of the point of application of a given force F_i in the direction of F_i. In the following section, we show how this result can also be used to determine the deflection of a point at which no force acts.

We have simplified the derivation of this theorem by assuming that the object was subjected only to forces. But as we will demonstrate in the following section, it can also be used when both forces and couples act on an object. In that case, the angle of rotation of the point of application of a given couple M_i in the direction of M_i is given by

$$
\theta_i = \frac{\partial U}{\partial M_i}. \tag{11.20}
$$

Castigliano's second theorem is based on the assumption that the work done by external forces and couples is equal to the resulting strain energy. In the case of a structure, this requires that no net work be done at connections between members when forces and couples are applied. For example, the theorem does not hold if work is done by friction in pinned joints.

11.2.2   Applications

We can use Castigliano's second theorem to determine deflections of trusses and beams, and we are not limited to applications in which only a single external force or couple does work. Furthermore, we can determine the deflection or rotation at a point at which no force or couple acts.

11.2.2.1 Axially Loaded Bars

In Fig. 11.17 an elastic bar of length L is subjected to an axial load P. From Eq. (11.11), the strain energy stored in the bar is

$$U = \frac{P^2 L}{2EA}.$$

Applying Castigliano's second theorem, we obtain the familiar expression for the displacement of the right end of the bar:

$$\delta = \frac{\partial U}{\partial P} = \frac{PL}{EA}.$$

Suppose that the truss in Fig. 11.18(a) consists of bars with the same elastic modulus and cross-sectional area and that a downward force F is applied at joint A, resulting in horizontal and vertical deflections u and v (Fig. 11.18(b)). We can use

Fig. 11.17 Subjecting a bar to an axial load

Fig. 11.18 Subjecting a truss to a downward force F

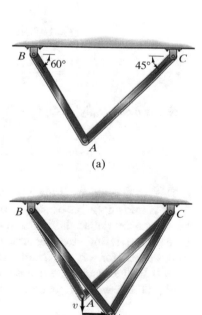

Fig. 11.19 Joint A

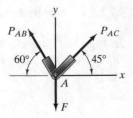

Castigliano's second theorem to determine the vertical deflection v. From the free-body diagram of joint A (Fig. 11.19), we obtain the equilibrium equations

$$\Sigma F_x = -P_{AB} \cos 60^\circ + P_{AC} \cos 45^\circ = 0,$$
$$\Sigma F_y = P_{AB} \sin 60^\circ + P_{AC} \sin 45^\circ - F = 0.$$

The solutions of these equations for the axial loads in the bars are

$$P_{AB} = \frac{F \cos 45^\circ}{D}, \quad P_{AC} = \frac{F \cos 60^\circ}{D},$$

where $D = \sin 45^\circ \cos 60^\circ + \cos 45^\circ \sin 60^\circ$. Using these expressions for the axial loads, the strain energy in the truss is

$$U = \frac{P_{AB}^2 L_{AB}}{2EA} + \frac{P_{AC}^2 L_{AC}}{2EA}$$
$$= \frac{F^2 \left(L_{AB} \cos^2 45^\circ + L_{AC} \cos^2 60^\circ \right)}{2EAD^2}.$$

Now we can apply Castigliano's second theorem to determine v:

$$v = \frac{\partial U}{\partial F} = \frac{F \left(L_{AB} \cos^2 45^\circ + L_{AC} \cos^2 60^\circ \right)}{EAD^2}.$$

This is the same result obtained by a different method in Example 3.6.

By employing a bit of sleight of hand, we can also use Castigliano's second theorem to determine the horizontal deflection u. There is no horizontal external force at joint A. But we can assume that a horizontal force F' acts at A in addition to the downward force F, use Castigliano's second theorem to determine u, and then set $F' = 0$ in the resulting expression for u. From the free-body diagram of joint A (Fig. 11.20), we now obtain the axial loads

$$P_{AB} = \frac{F \cos 45^\circ + F' \sin 45^\circ}{D}, \quad P_{AC} = \frac{F \cos 60^\circ - F' \sin 60^\circ}{D}.$$

Fig. 11.20 Joint A with an
assumed horizontal force F'

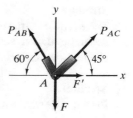

The strain energy in the truss is

$$U = \frac{P_{AB}^2 L_{AB}}{2EA} + \frac{P_{AC}^2 L_{AC}}{2EA}$$
$$= \frac{(F \cos 45° + F' \sin 45°)^2 L_{AB} + (F \cos 60° - F' \sin 60°)^2 L_{AC}}{2EAD^2}.$$

Applying Castigliano's second theorem, the deflection u is

$$u = \frac{\partial U}{\partial F'} = \frac{(F \cos 45° + F' \sin 45°)L_{AB} \sin 45°}{EAD^2}$$
$$- \frac{(F \cos 60° - F' \sin 60°)L_{AC} \sin 60°}{EAD^2}.$$

Setting $F' = 0$ in this expression, the horizontal displacement due to the vertical force F alone is

$$u = \frac{F(L_{AB} \sin 45° \cos 45° - L_{AC} \sin 60° \cos 60°)}{EAD^2},$$

which is the same result obtained in Example 3.6.

Example 11.3 Applying Castigliano's Theorem to a Truss
The truss in Fig. 11.21 is subjected to a vertical force F at B. The members each have elastic modulus E and cross-sectional area A. Use Castigliano's second theorem to determine the vertical component of the displacement of point C.

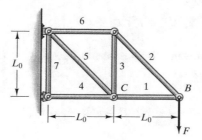

Fig. 11.21

Strategy

There is no external force at C, but we can determine the vertical displacement by assuming that a vertical force F' acts at C, applying Castigliano's second theorem, and then setting $F' = 0$ in the resulting expression for the displacement.

Solution

Placing a downward force F' at C (Fig. (a)), the axial loads in the members and their lengths are

$$
\begin{aligned}
P_1 &= -F, & L_1 &= L_0 \\
P_2 &= \sqrt{2}F, & L_2 &= \sqrt{2}L_0 \\
P_3 &= -F, & L_3 &= L_0 \\
P_4 &= -(2F + F'), & L_4 &= L_0 \\
P_5 &= \sqrt{2}(F + F'), & L_5 &= \sqrt{2}L_0 \\
P_6 &= F, & L_6 &= L_0 \\
P_7 &= 0, & L_7 &= L_0.
\end{aligned}
$$

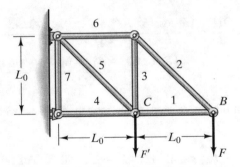

(a) Assuming that a vertical force F' acts at C

Using these values, the strain energy in the truss is

$$
U = \sum_{i=1}^{7} \frac{P_i^2 L_i}{2EA}
$$

$$
= \frac{1}{2EA}\Big[F^2 L_0 + 2\sqrt{2}F^2 L_0 + F^2 L_0 + (2F + F')^2 L_0
$$

$$
+ 2\sqrt{2}(F + F')^2 L_0 + F^2 L_0\Big].
$$

Let v be the downward displacement at C. Applying Castigliano's second theorem yields

$$
v = \frac{\partial U}{\partial F'} = \frac{1}{2EA}\Big[2(2F + F')L_0 + 4\sqrt{2}(F + F')L_0\Big].
$$

Setting $F' = 0$ in this expression, we obtain

$$v = \frac{2(1 + \sqrt{2})FL_0}{EA}.$$

Discussion

The "phantom load" we used in this example is a powerful technique. Notice that it can be used to determine the deflection of any joint of the truss in any direction.

11.2.2.2 Beams

The cantilever beam in Fig. 11.22 is loaded by a force at the right end. The deflection and the counterclockwise slope at the right end of the beam are

$$v_B = \frac{FL^3}{3EI}, \quad \theta_B = \frac{FL^2}{2EI}.$$

We can use Castigliano's second theorem to obtain these results. The bending moment in the beam is

$$M = F(L - x).$$

Substituting this expression into Eq. (11.15) and integrating from $x = 0$ to $x = L$, the strain energy is

$$U = \int_0^L \frac{[F(L - x)]^2}{2EI} dx$$
$$= \frac{F^2 L^3}{6EI}.$$

Applying Castigliano's second theorem, the deflection at the right end of the beam is

$$v_B = \frac{\partial U}{\partial F} = \frac{FL^3}{3EI}.$$

Fig. 11.22 Cantilever beam subjected to a force

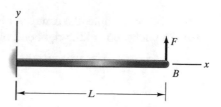

Fig. 11.23 Applying a
couple M' at the right end of
the beam

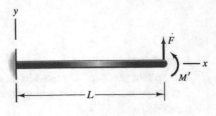

Fig. 11.24 Cantilever beam
subjected to a
distributed load

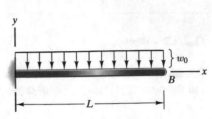

To determine the slope at the right end of the beam, we can assume that a couple
M' acts at the right (Fig. 11.23), calculate the slope, and then set $M' = 0$. In this case
the bending moment is

$$M = F(L - x) + M'.$$

Substituting this expression into Eq. (11.15), the strain energy is

$$U = \int_0^L \frac{[F(L - x) + M']^2}{2EI} dx$$
$$= \frac{1}{2EI} \left[\frac{F^2 L^3}{3} + FL^2 M' + L(M')^2 \right].$$

Applying Eq. (11.20), the slope at the right end of the beam is

$$\theta_B = \frac{\partial U}{\partial M'} = \frac{1}{2EI} \left(FL^2 + 2LM' \right).$$

Setting $M' = 0$ in this expression, we obtain

$$\theta_B = \frac{FL^2}{2EI}.$$

As another example, the cantilever beam in Fig. 11.24 is loaded by a distributed
force. The deflection at the right end of this beam given in Appendix F is

Fig. 11.25 Assuming that a force F' acts at the right end

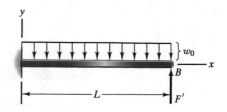

$$v_B = -\frac{w_0 L^4}{8EI}.$$

We can obtain this result by assuming that a force F' acts at the end of the beam (Fig. 11.25), applying Castigliano's second theorem to determine the deflection, and then setting $F' = 0$. The bending moment in the beam is

$$M = -\frac{1}{2} w_0 (L - x)^2 + F'(L - x).$$

From Eq. (11.15), the strain energy is

$$U = \int_0^L \frac{\left[-\frac{1}{2} w_0 (L - x)^2 + F'(L - x) \right]^2}{2EI} dx$$

$$= \frac{1}{2EI} \left[\frac{w_0^2 L^5}{20} - \frac{w_0 L^4 F'}{4} + \frac{L^3 (F')^2}{3} \right].$$

The deflection at the right end of the beam is

$$v_B = \frac{\partial U}{\partial F'} = \frac{1}{2EI} \left(-\frac{w_0 L^4}{4} + \frac{2L^3 F'}{3} \right). \tag{11.21}$$

Setting $F' = 0$ in this expression, we obtain

$$v_B = -\frac{w_0 L^4}{8EI}.$$

Example 11.4 Applying Castigliano's Theorem to a Beam
The beam in Fig. 11.26 is statically indeterminate. Use Castigliano's second theorem to determine the reaction at B.

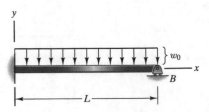

Fig. 11.26

Strategy
In Fig. (a) we show the unknown reaction B exerted on the beam by the roller support. We can use Castigliano's second theorem to determine the deflection at the right end of the beam due to the distributed load and the force B and then determine B by applying the boundary (compatibility) condition that the deflection of the right end of the beam is zero.

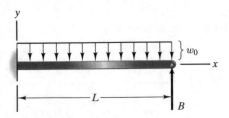

(a) Showing the reaction exerted by the roller support

Solution
We have already applied Castigliano's second theorem to a cantilever beam subjected to a uniformly distributed load and a point force at the right end (Fig. 11.25). The resulting deflection of the right end of the beam is given by Eq. (11.21). By setting $F' = B$ in that expression, we obtain the deflection of the right end of the beam in Fig. (a):

$$v_B = \frac{1}{2EI}\left(-\frac{w_0 L^4}{4} + \frac{2L^3 B}{3}\right).$$

Then from the boundary condition $v_B = 0$, we obtain the unknown reaction at the roller support:

$$B = \frac{3w_0 L}{8}.$$

Discussion
Compare the solution given in this example with the solution in Sect. 9.2. Energy methods often provide quicker solutions even for relatively simple problems.

Problems

Solve Problems 11.20–11.29 by using Castigliano's second theorem. Unless otherwise stated, beams have length L, modulus of elasticity E, and moment of inertia I.

11.20 The bar AB has cross-sectional area $A = 0.8\,\text{in}^2$ and modulus of elasticity $E = 1.4\text{E}6\,\text{psi}$. A downward force $F = 3500\,\text{lb}$ is applied at B. Determine the resulting deflection of point B.

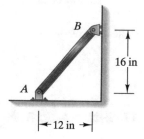

Problem 11.20

11.21 Each bar has length $L = 24\,\text{in}$, cross-sectional area $A = 1.2\,\text{in}^2$, and elastic modulus $E = 1.8\text{E}6\,\text{psi}$. The angle $\beta = 50°$. If a downward force $F = 12\,\text{kip}$ is applied at B, determine the resulting displacement of point B.

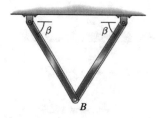

Problem 11.21

11.22 The beam is subjected to a uniformly distributed load. Determine the slope at B.

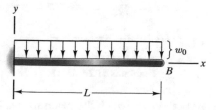

Problem 11.22

11.23 Determine the beam's deflection at the point where the force F is applied.

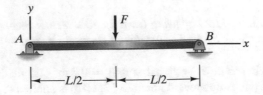

Problem 11.23

11.24 Determine the beam's deflection at B.

Problem 11.24

11.25 Determine the angle through which the beam rotates at the point where the couple M_0 is applied.

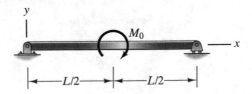

Problem 11.25

11.26 Determine the beam's deflection at the point where the force F is applied.

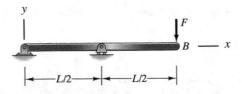

Problems 11.26–11.27

11.27 Determine the angle through which the beam rotates at the point where the force F is applied.

11.28 Determine the reaction at the roller support of the statically indeterminate beam.

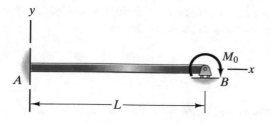

Problem 11.28

11.29 Determine the reaction at the roller support of the statically indeterminate beam.

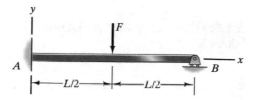

Problem 11.29

Chapter Summary

Work

If the point to which the force F in Fig. (a) is applied moves along the x axis from $x = 0$ to $x = x_0$, the work done is

$$W = \int_0^{x_0} F \, dx.$$

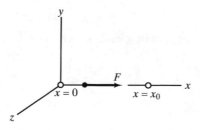

If the force is a constant $F = F_0$, the work is $W = F_0 x_0$. If F is proportional to x and its value is F_0 when $x = x_0$, the work is $W = \frac{1}{2} F_0 x_0$.

When a couple of moment M rotates through an angle from $\theta = 0$ to $\theta = \theta_0$ in the same direction as the couple, the work done is

$$W = \int_0^{\theta_0} M \, d\theta.$$

If the moment is a constant $M = M_0$, the work is $W = M_0 \theta_0$. If M is proportional to θ and its value is M_0 when $\theta = \theta_0$, the work is $W = \frac{1}{2} M_0 \theta_0$.

Strain Energy

The strain energy per unit volume of an isotropic linearly elastic material subjected to a general state of stress is

$$u = \frac{1}{2} \left(\sigma_x \varepsilon_x + \sigma_y \varepsilon_y + \sigma_z \varepsilon_z + \tau_{xy} \gamma_{xy} + \tau_{yz} \gamma_{yz} + \tau_{xz} \gamma_{xz} \right). \qquad (11.3)$$

Axially Loaded Bars

The strain energy of a prismatic bar of length L and cross-sectional area A that is subjected to an axial load P is

$$U = \frac{P^2 L}{2EA}. \qquad (11.11)$$

Beams

The strain energy of an element of a beam of width dx (Fig. (b)) is

$$dU = \frac{M^2}{2EI} \, dx. \qquad (11.15)$$

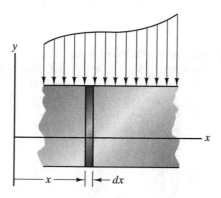

Castigliano's Second Theorem

Consider an object of elastic material that is subjected to N external forces $F_1, F_2, \ldots,$ F_N (Fig. (c)). Let u_i be the component of the deflection of the point of application of the ith force F_i in the direction of F_i. Then

$$u_i = \frac{\partial U}{\partial F_i},\qquad (11.19)$$

where U is the strain energy resulting from the application of the N forces.

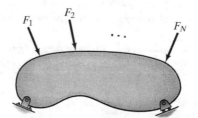

If couples also act on an object, the angle of rotation of the point of application of the ith couple M_i in the direction of M_i is given by

$$\theta_i = \frac{\partial U}{\partial M_i}.\qquad (11.20)$$

Review Problems

11.30 The two bars have cross-sectional area $A = 1.4\,\text{in}^2$ and modulus of elasticity $E = 12\text{E}6\,\text{psi}$. If an upward force $F = 20,000\,\text{lb}$ is applied at B, what is the vertical component v of the deflection of point B?

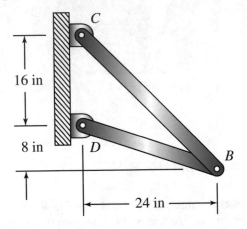

Problems 11.30–11.31

11.31 The two bars have cross-sectional area $A = 1.4\,\text{in}^2$ and modulus of elasticity $E = 12\text{E}6\,\text{psi}$. If an upward force $F = 20,000\,\text{lb}$ is applied at B, what is the horizontal component u of the deflection of point B?

11.32 The truss is made of steel bars with cross-sectional area $A = 0.003\,\text{m}^2$ and modulus of elasticity $E = 220\,\text{GPa}$. The suspended mass $m = 5000\,\text{kg}$. Determine the vertical component v of the deflection of point B.

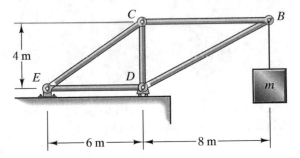

Problems 11.32–11.33

11.33* The truss is made of steel bars with cross-sectional area $A = 0.003\,\text{m}^2$ and modulus of elasticity $E = 220\,\text{GPa}$. The suspended mass $m = 5000\,\text{kg}$. Determine the horizontal component u of the deflection of point B.

11.34 Use strain energy to determine the angle through which the beam rotates at
A, and confirm the value given in Appendix F.

Problem 11.34

11.35∗ Use strain energy to determine the angle through which the beam rotates
at A.

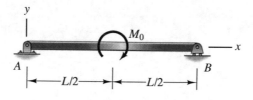

Problem 11.35

Chapter 12
Criteria for Failure and Fracture

The failures we will consider are those due to a single overload or to smaller repeated or cyclic loads. In both cases, we distinguish between situations in which a structural component does or does not contain preexisting flaws or cracks. When it does not, it is sufficient to rely on stress analyses of the kind developed in previous chapters and apply failure criteria based on them. When a structural component does contain a crack, we show how to use fracture mechanics to determine the state of stress and predict when failure will occur.

12.1 Stress Concentrations

Either intentionally to satisfy specific design requirements or as a result of processing or manufacturing flaws, structural members may have geometric features (including holes, notches, and corners) that result in localized regions of high stress called *stress concentrations*. They are of great concern in structural design, because the stresses in these regions can be of much larger magnitude than the stresses to which the member would be subjected in the absence of the geometric features causing them. For example, in the stepped bar subjected to tension in Fig. 12.1, the magnitude of the maximum stress near the step substantially exceeds the value predicted by the equation $\sigma = P/A$, although the stress approaches this value at an axial distance of the order of one diameter away from the step. In this section we give examples of geometric features that give rise to such stress concentrations and present empirical information of the types available to structural designers for analyzing them.

© Springer Nature Switzerland AG 2020
A. Bedford, K. M. Liechti, *Mechanics of Materials*,
https://doi.org/10.1007/978-3-030-22082-2_12

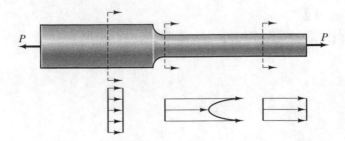

Fig. 12.1 Stepped rod in tension showing the normal stress distributions before, after, and at the step

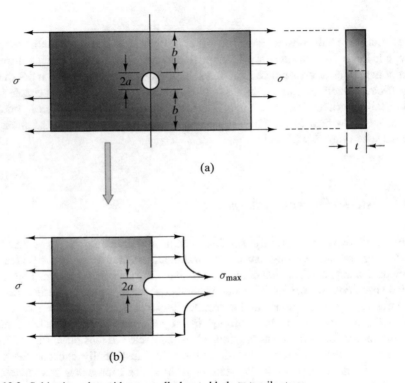

Fig. 12.2 Subjecting a bar with a centrally located hole to tensile stress

12.1.1   Axially Loaded Bars

Structural members may be designed with holes, for example, to decrease weight or to allow a bolt to be inserted, or holes may be present inadvertently due to manufacturing flaws. In Fig. 12.2(a), a bar of rectangular cross section containing

Fig. 12.3 Stress concentration factor C for tensile loading of a thin bar with a centrally located hole

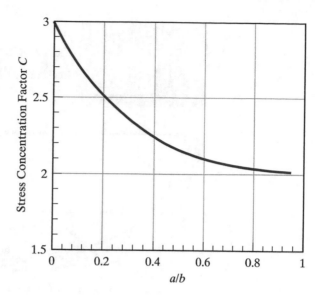

a hole of radius a is subjected to a tensile stress σ. The hole creates a stress concentration (Fig. 12.2(b)). The ratio of the maximum stress σ_{max} to the nominal stress σ is denoted by C and is called the *stress concentration factor*:

$$C = \frac{\sigma_{max}}{\sigma}. \tag{12.1}$$

Stress concentration factors must be determined by advanced analytical or numerical methods. Under the assumption that the width $2(a + b)$ of the bar is much larger than its thickness t, values of C are shown as a function of a/b in Fig. 12.3. (Because the width of the bar is assumed to be much greater than its thickness, we could have anticipated that the stress concentration factor should depend only on the dimensionless quantity a/b.) Notice that even a microscopic hole resulting from a manufacturing flaw results in a maximum stress three times the nominal stress. This is one example showing why designers choose conservative factors of safety.

We have seen in Fig. 12.1 that a stress concentration occurs in a bar with a stepped cross section. Figure 12.4 shows stress concentration factors for stepped circular bars subjected to tensile load as a function of the ratio of the radius a of the *fillet* to the smaller diameter d_2. Figure 12.5 shows equivalent stress concentration factors for thin rectangular bars. As the radius a approaches zero, the stress concentration factor approaches infinity, which emphasizes the importance of incorporating fillets (literally "rounding inside corners") in design.

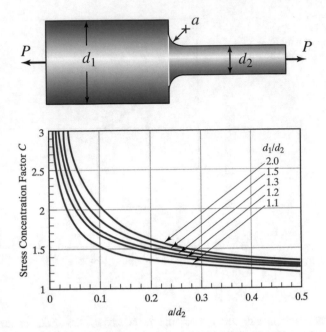

Fig. 12.4 Stress concentration factor for a circular bar with a shoulder fillet under axial loading. The nominal stress is the normal stress in the part of the bar with the smaller diameter

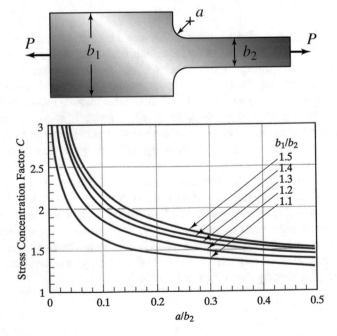

Fig. 12.5 The stress concentration factor for a flat bar with shoulder fillets. The nominal stress is the normal stress in the part of the bar with the smaller height

Fig. 12.6 Stress distributions before, after, and at a step in diameter in a rod subjected to torsion

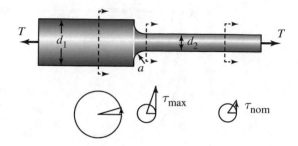

12.1.2   Torsion

In Chap. 4 we used Eq. (4.9),

$$\tau = \frac{Tr}{J},$$

(12.2)

to determine the shear stress in a circular bar subjected to a torque T, where J is the polar moment of inertia of the cross section and r is the radial distance from the bar's axis. In a stepped bar (Fig. 12.6), the distribution of stress is approximated by Eq. (12.2) at cross sections that are not near the step, but a stress concentration exists at the step. The stress concentration factor is defined in terms of the shear stress,

$$C = \frac{\tau_{max}}{\tau_{nom}},$$

(12.3)

where τ_{nom} is the maximum shear stress in the part of the bar with the smaller diameter. As can be seen in Fig. 12.7, the stress concentration factor depends on the ratios of the dimensions in the problem: the ratio of the diameters and the ratio of the fillet radius to the smaller diameter.

12.1.3   Bending

In Chap. 6 we used Eq. (6.12),

$$\sigma_x = -\frac{My}{I},$$

(12.4)

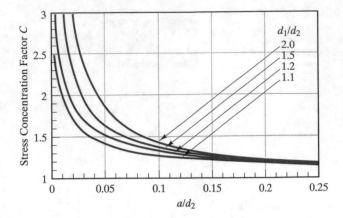

Fig. 12.7 Stress concentration factor for a stepped bar subjected to torsion. The nominal stress is the maximum shear stress in the part of the bar with the smaller diameter

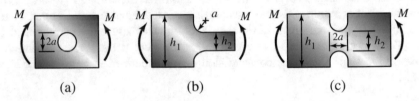

Fig. 12.8 Examples of beam geometries that result in stress concentrations

to determine the normal stress in a prismatic beam subjected to bending moment M, where I is the moment of inertia of the cross section and y is the distance from the neutral axis. As examples of stress concentrations in beams, we consider beams having rectangular cross sections that are subjected to a bending moment M for the cases of beams containing holes, having steps in height, or containing notches (Fig. 12.8). The height of the rectangular cross section is assumed to be large compared to its width.

For the case of a beam containing a hole (Fig. 12.8(a)), let the hole be centered at the neutral axis. When the hole diameter $2a$ is small compared to the beam's height, the stress is approximated by the nominal linear distribution except near the hole (Fig. 12.9(a)). Although amplification of the stress occurs near the hole, the increased value does not exceed the maximum stress resulting from the nominal distribution. For larger holes $\left(2a/h_1 > \frac{1}{2}\right)$, the stress concentration dominates the linear distribution, and the resulting maximum stress is approximately twice the maximum nominal stress.

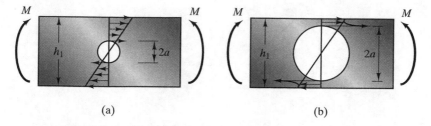

Fig. 12.9 Stress distributions in beams with small and large holes

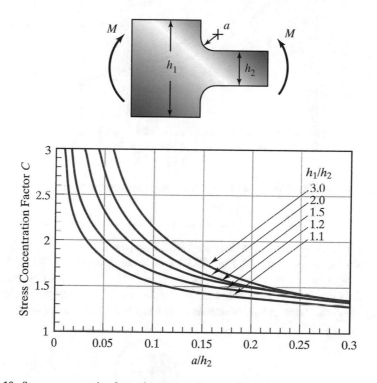

Fig. 12.10 Stress concentration factor for a stepped beam subjected to bending. The nominal stress is the maximum stress in the part of the bar with the smaller height

Figure 12.10 shows stress concentration factors $C = \sigma_{max}/\sigma_{nom}$ for a stepped beam subjected to bending moment as a function of the ratio of the fillet radius a to the smaller height h_2. The stress σ_{nom} is the maximum stress magnitude resulting from Eq. (12.4) in the part of the beam with the smaller height. Stress concentration factors for a notched beam are shown in Fig. 12.11.

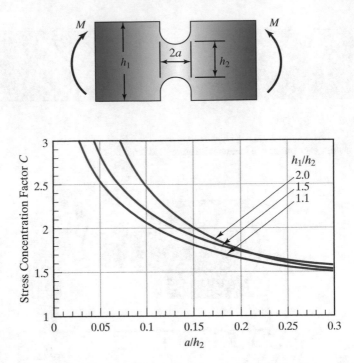

Fig. 12.11 Stress concentration factor for a notched beam subjected to bending

Example 12.1 Maximum Axial Load in a Stepped Bar
The thin rectangular bar in Fig. 12.12 has a step change in height from
$b_1 = 3$ in to $b_2 = 2$ in with a fillet radius $a = 0.2$ in. The bar's thickness is
$t = 0.01$ in. If the material can be subjected to an allowable stress
$\sigma_{\text{allow}} = 40$ ksi, what is the maximum axial load P that can be applied?

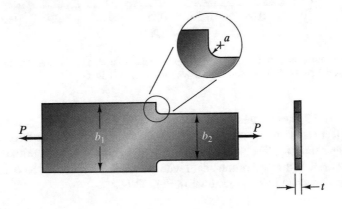

Fig. 12.12

Strategy

We must determine the stress concentration factor for an axially loaded bar with this geometry and fillet radius. Then by expressing the nominal stress (the normal stress in the part of the bar with the smaller height) in terms of P, we can determine the value of P for which the maximum stress due to the stress concentration is equal to the allowable stress.

Solution

From Fig. 12.5, the stress concentration factor for $b_1/b_2 = 3\,\text{in}/2\,\text{in} = 1.5$ and $a/b_2 = 0.2\,\text{in}/2\,\text{in} = 0.1$ is $C = 2.24$. The nominal stress in terms of P is

$$\sigma_{nom} = \frac{P}{b_2 t},$$

so the maximum stress resulting from the stress concentration in terms of P is

$$\sigma_{max} = C\sigma_{nom} = \frac{2.24P}{b_2 t}.$$

We equate this stress to the allowable stress.

$$\frac{2.24P}{b_2 t} = \sigma_{allow} :$$

$$\frac{2.24P}{(2\,\text{in})(0.01\,\text{in})} = 40{,}000\,\text{lb/in}^2.$$

Solving for P, the maximum permissible axial load is 357 lb.

Discussion

As we demonstrate in this example, the first step in problems involving stress concentrations is to identify the graph for the stress concentration factor that corresponds to the component, loading, and notch geometry being considered. The stress concentration factor C for this example is presented in Fig. 12.5 as a function of the parameter a/b_2, and a series of curves are plotted for different values of the parameter b_1/b_2. If the particular value of b_1/b_2 you need is not plotted, you will need to interpolate between curves. Linear interpolation is usually adequate in cases such as this.

Problems

12.1 The plate is 0.005 m thick and has a central hole with 0.1-m diameter. If the axial load $P = 45\,\text{kN}$, what is the maximum stress in the plate?

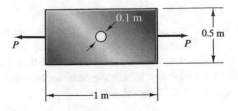

Problem 12.1

12.2 The thin sheet of width W has a central hole with 0.5-inch radius and is subjected to a nominal tensile stress $\sigma = 10\,\text{ksi}$. If the maximum stress in the sheet is to be limited to 25 ksi, what is the minimum required value of W?

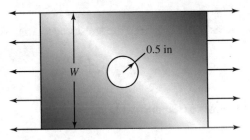

Problem 12.2

12.3 A thin sheet is to have a width of 8.25 in and a central hole of 0.75-in diameter. The design calls for an allowable nominal stress $\sigma = 18\,\text{ksi}$ with a safety factor of 1.2 on the allowable nominal stress. There is a choice of using aluminum alloys 2014-T6, 6061-T6, or 7075-T6. If the cost of the material is proportional to yield stress (see Appendix B), determine the most cost-effective alloy that could be used.

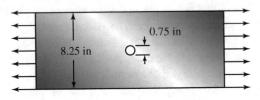

Problem 12.3

12.4 The sheet shown is 5 mm in thickness. The length of the sheet (not shown to scale) is large in comparison to its 15-cm width. It contains four equally spaced holes each having a 1-cm diameter. The overall load on the sheet is $P = 64\,\text{kN}$, and the allowable normal stress of the material is 200 MPa. Determine whether this hole spacing scheme is satisfactory. Assume that the holes do not interact with one another, i.e., the sheet can be divided into four separate sheets.

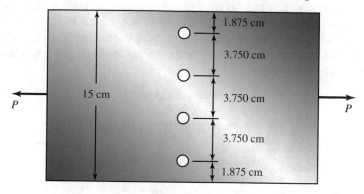

Problem 12.4

12.5 The diameters of the stepped rod are $d_1 = 2.625$ in and $d_2 = 1.5$ in. A fillet radius of 0.225 in is used to transition from one cross section to the other. If the rod is subjected to a 40,000-lb axial load, what is the resulting maximum stress? (It may be necessary to interpolate between d_1/d_2 values in Fig. 12.4.)

Problem 12.5

12.6 The radius of the fillet in the stepped circular bar is equal to the radius of the smaller section: $a = d_2/2$. Show that making the fillet a full quarter circle results in the largest stress concentration. What is the value of the stress concentration factor in that case?

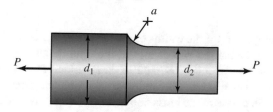

Problem 12.6

12.7   The tensile specimen has a gage section with 1/2-in diameter. The 1-in fillet radius transitions to a shoulder with 5/8-in diameter. The specimen fails at the transition region where the maximum stress concentration occurs. Determine the ratio of the failure load to the value the failure load would be if the specimen failed in the middle of the gage section. For values of a/d_2 greater than 0.5, assume that the stress concentration factor can be approximated by its value at $a/d_2 = 0.5$.

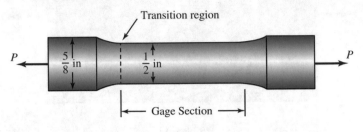

Problem 12.7

12.8   The stepped rod must carry a load $P = 37$ kips. It is made of a material that yields at 70 ksi. If the rod has a fillet with radius $a = 0.24$ in that makes a quarter circle as shown, determine the largest value that the diameter d_2 can have without yielding occurring in the rod.

Strategy: Begin by guessing a solution. Consider the d_1/d_2 ratios shown in Fig. 12.4.

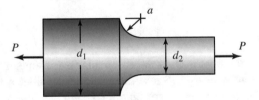

Problem 12.8

12.9   Solve Example 12.1 if the widths of the bar are $b_1 = 1.2$ in and $b_2 = 1.0$ in, the thickness of the bar is $t = 0.05$ in, and the fillet radius is $a = 0.3$ in.

12.10   The stepped bar is to be designed to carry a load $P = 4$ kN. It has widths $b_1 = 3$ cm and $b_2 = 2$ cm, fillet radius $a = 0.4$ cm, and thickness $t = 1$ mm. It is to be made of one of the brass alloys (80% Cu or Naval) listed in Appendix B. The cost of each alloy is proportional to its yield stress. Which alloy is optimum in terms of cost if yielding is to be avoided?

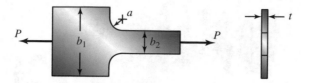

Problems 12.10–12.11

12.11 The stepped bar carries a load $P = 20,000$ lb. It has a yield stress of 80 ksi, dimensional ratios $b_1/b_2 = 1.4$ and $a/b_2 = 0.131$, and thickness $t = \frac{1}{4}$ in. Determine the value of the fillet radius a that would lead to the onset of yielding under these conditions.

12.12 The diameters of the stepped rod are $d_1 = 1.8$ in and $d_2 = 1.5$ in. If the rod is to be subjected to a torque of 28.7 in-kip and the allowable shear stress is 65 ksi, what fillet radius should be used?

Problems 12.12–12.13

12.13 The stepped rod is made of a material that has an allowable shear stress of 11.9 MPa. The diameters are $d_1 = 40$ mm and $d_2 = 20$ mm and the fillet radius is 1 mm. Determine the allowable torque that can be applied to the rod.

12.14 The 1.2-m diameter hub of a turbine steps down to a shaft that has a radius of 0.4 m through a fillet radius of 12 cm. If the allowable shear stress of the material is 25.3 MPa, what is the maximum power rating of the turbine at 50 rpm? (The power transmitted by the turbine equals $T\omega$, where T is the torque and ω is the angular velocity of the turbine in rad/s.)

12.15 The drive shaft of a race car has diameter $d_2 = 1.2$ in. It steps up to a larger diameter d_1 with a fillet radius of 0.06 inches. The allowable shear stress of the material is 18 ksi. Determine what the diameter d_1 should be if the engine develops its maximum power of 320 horsepower at 5600 rpm. (See Problem 12.14. A horsepower is 550 ft-lb/s.)

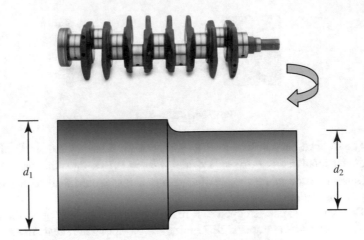

Problem 12.15 (*Photograph from iStock by schlol*)

12.16 The simply supported beam supports a container whose weight W is much greater than the weight of the beam. The beam has a rectangular cross section with a height of 400 mm and a width of 150 mm. The hole in the middle of the beam has a radius of 50 mm. The beam's material will support a normal stress of 187.5 MPa in tension and compression. Determine the maximum allowable value of W, assuming a factor of safety of 3 in the design.

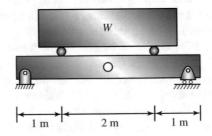

Problems 12.16–12.17

12.17 The simply supported beam supports a container whose weight W is much greater than the weight of the beam. The beam has a rectangular cross section with a height of 400 mm and a width of 150 mm. The hole in the middle of the beam has a radius of 110 mm. The beam's material will support a normal stress of 187.5 MPa in tension and compression. Determine the maximum allowable value of W, assuming a factor of safety of 3 in the design. Compare your answer to that of Problem 12.16.

12.18 The 10-m beam has a rectangular cross section and supports a uniform load. The height of the rectangular cross section is 500 mm and its width is 50 mm.

In case (a), the beam has a single hole with 120-mm radius at its midpoint. In case (b), the beam has two holes with 130-mm radius located at 1.5 m from the ends. Which case is safer?

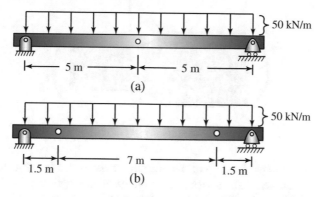

Problem 12.18

12.19 The stepped beam has heights $h_1 = 0.5$ m and $h_2 = 0.167$ m. It has width $t = 50$ mm and a fillet radius of 16.7 mm. The material being considered for the beam yields at 270 MPa. Would this choice result in yielding of the material if the bending moment $M = 30$ kN-m?

Problems 12.19–12.20

12.20 The stepped beam has heights $h_1 = 400$ mm and $h_2 = 200$ mm, width $t = 80$ mm, and a fillet radius of 30 mm. The material being used for the beam has a tensile strength of 235 MPa. What bending moment M can the beam support if the safety factor of the design is 1.5?

12.21∗ The design of the stepped beam calls for a quarter circular fillet as shown, steel with a yield stress of 90 ksi and a factor of safety of 4. Determine the necessary height h if the beam will be subjected to a moment $M = 150$ ft-kip.

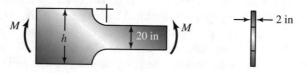

Problem 12.21

12.22 The stepped beam has a fillet that subtends a $45°$ arc. If $h_1 = 0.88$ m and $h_2 = 0.80$ m, what are the fillet radius and the stress concentration factor?

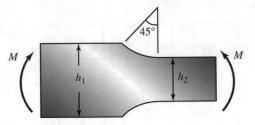

Problem 12.22

12.23 A double-notched beam with height ratio $h_1/h_2 = 1.5$ is subjected to a pure bending moment. As a designer, you have the option of using either a semicircular notch (Fig. (a)) or a deep notch with $a/h_2 = 0.05$ (Fig. (b)). Show that the allowable bending moment is increased by approximately 80% when the semicircular notch is used.

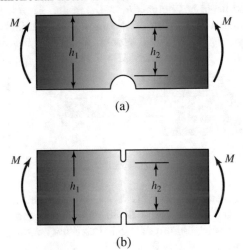

(a)

(b)

Problem 12.23

12.24 The wooden beam has a yield stress of 60 MPa and a width (the dimension perpendicular to the page) of 20 mm. The design calls for two semicircular notches in the top and bottom surfaces of the beam. The dimension $h_1 = 0.55$ m and the distance between the bottoms of the notches is $h_2 = 0.5$ m. What is the smallest bending moment M that will cause yielding of the material?

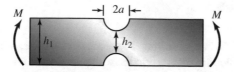

Problem 12.24

12.25 The height $h_1 = 40$ in and the beam has a width (the dimension perpendicular to the page) of 2 in. A test indicates that if the notch radius $a = 2$ in, the beam fails when the bending moment $M = 311$ ft-kip. Determine the bending moment at which the beam would fail if the notch radius is (a) $a = 3$ in and (b) $a = 4$ in.

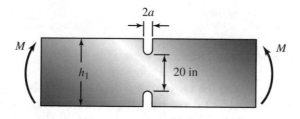

Problem 12.25

12.2   Failure

The structural designer may make compromises in satisfying many requirements and constraints, including material availability, machinability, cost, weight, and aesthetic concerns, but avoiding failure is essential. In previous chapters we have presented many examples of the design of structural members to prevent failure under specified loads. The members we considered—bars subjected to axial loads and torques, pressure vessels, and beams subjected to lateral loads—had simple geometries and simple states of stress. When there was only one component of stress, we were able to apply simple failure criteria. When two or more components of stress are present at a point, we need failure criteria that can account for multiple stresses. In addition, many structures have complicated geometries that require advanced analytical or numerical methods of analysis and result in general states of stress. In this section we discuss criteria that have been introduced to prevent failure in materials subjected to general states of stress. We first consider *overloads*, or monotonic loading, and then repeated loads or *fatigue*.

12.2.1   Overloads

The mechanism of failure resulting from an overload depends on the material. Broadly dividing structural metals into those that are brittle and those that are ductile, failure in brittle materials occurs when a component breaks or fractures. In contrast, failure in ductile materials is usually defined to occur with the onset of yielding.

In previous chapters we have introduced failure criteria based on simple states of stress. As we discussed in Chap. 3, yielding in a bar subjected to axial load takes place when the yield stress σ_Y has been exceeded, and fracture occurs when the ultimate stress σ_U occurs. We will now describe criteria that can be used to predict when failure will occur in objects subjected to more complex states of stress, such as those in bars subjected to combined axial load and torsion or in the walls of pressure vessels.

12.2.1.1   Maximum Normal Stress Criterion

The simplest way to extend the concept of failure used in bars subjected to axial loads to more complex states of stress is to compare the maximum principal stress at a point to the ultimate stress determined under axial loading. This is called the maximum normal stress criterion. (See our discussion of principal stresses in Sects. 7.2, 7.3, and 7.4.) Expressed mathematically, this criterion states that failure will occur when

$$\text{Max}(|\sigma_1| \ |\sigma_2| \ |\sigma_3|) = \sigma_U. \tag{12.5}$$

The use of absolute values means the material is assumed to have the same strength in tension and compression. Because the implied mechanism of failure in the material is one of separation rather than sliding (shear), this criterion applies to brittle materials, which do not yield or undergo much plastic deformation.

In the case of plane stress ($\sigma_3 = 0$), the maximum normal stress criterion can be described in a simple graphical manner. The values of σ_1 and σ_2 for which failure will not occur are bounded by the square region defined by the lines (Fig. 12.13)

$$\sigma_1 = \pm\sigma_U, \quad \sigma_2 = \pm\sigma_U. \tag{12.6}$$

Notice that there is no interaction between the principal stresses in this criterion—its boundaries are horizontal and vertical. In a three-dimensional stress state, the safe stress regime is enclosed in the cube bounded by the planes

$$\sigma_1 = \pm\sigma_U, \quad \sigma_2 = \pm\sigma_U, \quad \sigma_3 = \pm\sigma_U. \tag{12.7}$$

Fig. 12.13 Graphical
representation of the
maximum normal stress
criterion under plane stress

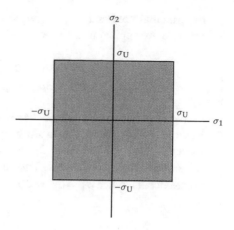

Example 12.2 Application of the Maximum Normal Stress Criterion
In Fig. 12.14 a bar of ultrahigh-strength steel is subjected to an axial stress σ
and a lateral pressure p. It is a brittle material with an ultimate normal stress of
300 ksi in tension or compression. Determine the pressure level p at which the
bar will fail if the pressure and axial stress are applied proportionally with the
ratio $\sigma - 2p$.

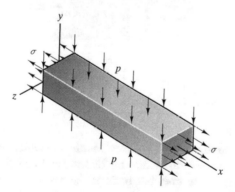

Fig. 12.14

Strategy
Because the material is brittle and fails at the same stress in tension and
compression, the maximum normal stress criterion applies. We can determine
the principal stresses and use Eq. (12.5) to determine the level of p at which
failure occurs. We will also determine the pressure at which failure occurs
graphically.

Solution
In terms of the coordinate system in Fig. 12.14, the only nonzero components
of stress in the bar are $\sigma_x = \sigma$, $\sigma_y = -p$. The bar is in a state of plane stress, and

the principal stresses are $\sigma_1 = \sigma$, $\sigma_2 = -p$, $\sigma_3 = 0$. Because the axial stress and pressure are applied in the ratio $\sigma = 2p$, we can write Eq. (12.5) as

$$\text{Max}(|\sigma_1| \ |\sigma_2| \ |\sigma_3|) = \sigma_U :$$
$$\text{Max}(|2p|, \ |-p|, \ 0) = 300 \, \text{ksi}.$$

We see that failure occurs when $p = 150 \, \text{ksi}$. This is demonstrated in Fig. (a), which shows the loading path and failure in σ_1 versus σ_2 space. Because $\sigma_2 = -p$ and $\sigma_1 = 2p$ as the loading is applied, the loading path follows the line $\sigma_2 = -\frac{1}{2}\sigma_1$. The loading path reaches the boundary, and failure occurs, when $\sigma_1 = 2p = 300 \, \text{ksi}$.

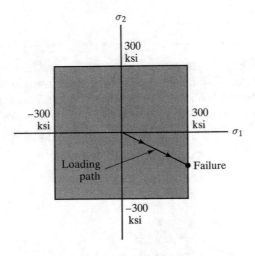

(a) Graphical representation of the loading path and failure

Discussion

This example demonstrates *proportional loading*, in which the components of stress at a point increase with increasing load and maintain the same ratio at all load levels. This is often true, but perhaps the best example is a cylindrical pressure vessel, in which the hoop stress is twice the axial stress at all pressure levels (see Example 12.4).

12.2.1.2   Mohr's Failure Criterion

There are many materials whose fracture strengths in tension and compression differ, which require a modified version of Fig. 12.13. The failure boundaries in the first and third quadrants are qualitatively the same as before, but the ultimate tensile stress σ_U^t and ultimate compressive stress σ_U^c are different (Fig. 12.15(a)). The question that then arises is the nature of the failure boundaries in the second and fourth quadrants. Mohr suggested that they should be the straight lines joining the boundaries in the

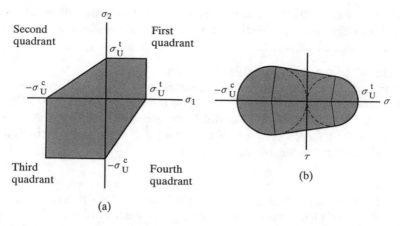

Fig. 12.15 (a) Graphical representation of Mohr's failure criterion under plane stress conditions. (b) Rationale for the failure boundaries in the second and fourth quadrants

first and third quadrants in Fig. 12.15(a). The resulting failure boundary is *Mohr's failure criterion*. The rationale for his choice is shown in Fig. 12.15(b). The right circle is the Mohr's circle mapping the states of stress when $\sigma_1 = \sigma_U^t, \sigma_2 = 0$. The left circle is the Mohr's circle mapping the states of stress when $\sigma_1 = 0, \sigma_2 = -\sigma_U^c$. Mohr postulated that the failure boundary in $\sigma - \tau$ space is defined by these two circles and the straight lines tangent to them in Fig. 12.15(b). When σ_1 and σ_2 are of opposite sign, values that lie on the straight lines in the second and fourth quadrants of Fig. 12.15(a) yield Mohr's circles that are tangent to the straight lines in Fig. 12.15(b).

Figure 12.15(a) indicates that while there is no interaction between the values of σ_1 and σ_2 at which fracture occurs in the first and third quadrants, interaction does exist in the second and fourth quadrants. In the first quadrant, failure occurs when

$$\sigma_1 = \sigma_U^t \quad \text{or} \quad \sigma_2 = \sigma_U^t, \tag{12.8}$$

and in the third quadrant, it occurs when

$$\sigma_1 = -\sigma_U^c \quad \text{or} \quad \sigma_2 = -\sigma_U^c. \tag{12.9}$$

In the second quadrant, failure occurs when σ_1 and σ_2 lie on the straight line

$$\sigma_2 = \sigma_U^t + \frac{\sigma_U^t}{\sigma_U^c}\sigma_1, \tag{12.10}$$

and in the fourth quadrant, it occurs when σ_1 and σ_2 lie on the straight line

$$\sigma_2 = -\sigma_U^c + \frac{\sigma_U^c}{\sigma_U^t}\sigma_1. \tag{12.11}$$

As in the case of the maximum normal stress criterion, the failure mechanism being modeled is fracture, so Mohr's failure criterion also applies to brittle materials.

Example 12.3 Application of Mohr's Failure Criterion
Preliminary design of a mechanism indicates that a point on the surface of a component will be in the state of plane stress $\sigma_x = \sigma_y = -75$ ksi, $\tau_{xy} = 175$ ksi. The material fails according to Mohr's criterion with ultimate tensile and compressive strengths of $\sigma_U^t = 150$ ksi and $\sigma_U^c = 300$ ksi, respectively. Is the design safe with respect to this state of stress?

Strategy
We can use Eq. (7.15) to determine the principal stresses σ_1 and σ_2 for the given state of plane stress. Then, knowing the quadrant of Fig. 12.15(a) in which the stress state lies, we can determine whether it is within the failure boundary from Eqs. (12.8) to (12.11).

Solution
The principal stresses are

$$\sigma_1, \sigma_2 = \frac{\sigma_x + \sigma_y}{2} \pm \sqrt{\left(\frac{\sigma_x - \sigma_y}{2}\right)^2 + \tau_{xy}^2}$$

$$= \frac{-75\,\text{ksi} - 75\,\text{ksi}}{2} \pm \sqrt{\left(\frac{-75\,\text{ksi} + 75\,\text{ksi}}{2}\right)^2 + (175\,\text{ksi})^2}$$

$$= 100\,\text{ksi}, -250\,\text{ksi}.$$

The state of stress lies in the fourth quadrant. Substituting $\sigma_1 = 100$ ksi into Eq. (12.11), the value of σ_2 corresponding to the failure boundary is

$$\sigma_2|_{\text{boundary}} = -\sigma_U^c + \frac{\sigma_U^c}{\sigma_U^t}\sigma_1 = -300\,\text{ksi} + \left(\frac{300\,\text{ksi}}{150\,\text{ksi}}\right)100\,\text{ksi} = -100\,\text{ksi}.$$

Because the design value $\sigma_2 = -250$ ksi is below this value, the design state of stress lies outside the failure boundary, and the material would fail if the design was implemented.

Discussion
Determination of the principal stresses is the key to applying Mohr's failure criterion. A point on a free surface of an object (i.e., a surface that is not subjected to external loads) is always in a state of plane stress, because the traction vector on the free surface is zero. That simplifies the determination of the principal stresses.

12.2.1.3 Tresca Criterion

Here we consider a failure criterion that is valid for ductile materials, which yield before reaching their ultimate tensile strengths. Even if a material does not fracture upon reaching its yield stress, permanent deformation can occur, and in many structural components, proper function can be lost at that point. At the same time, there are structures, such as oil pipelines in deep water, that are intentionally allowed to undergo significant plastic deformation during installation or fabrication.

The plastic deformation that is initiated when the yield stress is reached takes place through a process of slip or shear deformation. It is therefore natural to expect criteria for failure by yielding to be expressed in terms of shear stress. Conceptually, then, the simplest experiment to determine yielding in ductile structural materials would be a shear test, such as subjecting a tube to torsion. If a thin-walled tube of median radius R and wall thickness t is subjected to a torque T, the resulting shear stress is

$$\tau = \frac{T}{2\pi R^2 t}. \tag{12.12}$$

In this way the yield stress in shear τ_Y can be determined. If we define the onset of yield as failure, τ_Y is also the maximum shear stress of the material.

However, not all materials are easily available in the form of thin-walled cylinders, and subjecting bars to axial loads has become the common benchmark for determining yield strength. As we have seen, when a bar is subjected to a normal stress σ by axial load, the maximum resulting shear stress is $\sigma/2$, so the shear stress at the onset of yielding is related to the normal stress by

$$\tau_Y = \frac{\sigma_Y}{2}. \tag{12.13}$$

This criterion can be extended to an arbitrary state of stress by assuming that yielding occurs when the absolute maximum shear stress is equal to τ_Y:

$$\text{Max}\left(\left|\frac{\sigma_1 - \sigma_2}{2}\right|, \left|\frac{\sigma_2 - \sigma_3}{2}\right|, \left|\frac{\sigma_1 - \sigma_3}{2}\right|\right) = \tau_Y. \tag{12.14}$$

This is the *Tresca criterion* for failure, which is also called the *maximum shear stress criterion*. By using the relationship (12.13), the Tresca criterion can be expressed in terms of the tensile yield stress:

$$\text{Max}(|\sigma_1 - \sigma_2|, |\sigma_2 - \sigma_3|, |\sigma_1 - \sigma_3|) = \sigma_Y. \tag{12.15}$$

This criterion can be visualized easily in the case of plane stress. Then Eq. (12.15) reduces to

Fig. 12.16 Failure
boundaries for the Tresca
(hexagon) and von Mises
(ellipse) failure criteria
under plane stress

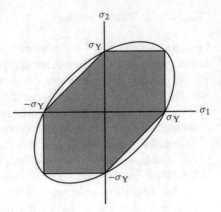

$$\text{Max}(|\sigma_1 - \sigma_2|, |\sigma_2|, |\sigma_1|) = \sigma_Y. \tag{12.16}$$

Consequently, in a plot of σ_2 versus σ_1, the safe region is bounded by the lines

$$\sigma_1 - \sigma_2 = \pm\sigma_Y, \quad \sigma_1 = \pm\sigma_Y, \quad \sigma_2 = \pm\sigma_Y. \tag{12.17}$$

These lines form the hexagon shown in Fig. 12.16. (We discuss the ellipse
circumscribing the hexagon below.) Stress states that give rise to values of σ_1 and
σ_2 lying on or outside the hexagon cause yielding. This boundary therefore defines
the maximum shear stress criterion in plane stress.

Notice from Eq. (12.15) that for a given state of stress, the normal stress

$$\sigma_T = \text{Max}(|\sigma_1 - \sigma_2|, |\sigma_2 - \sigma_3|, |\sigma_1 - \sigma_3|) \tag{12.18}$$

can be compared to the yield stress σ_Y to obtain a measure of how close the material
is to failure. Called the *Tresca equivalent stress*, it motivates the definition of the
Tresca factor of safety:

$$\text{FS}_T = \frac{\sigma_Y}{\sigma_T}. \tag{12.19}$$

As we have seen in previous chapters, a given safety factor can be prescribed as
a design objective. Calculating the safety factor for a given state of stress is also
useful because it provides a quick measure of how near a component is to failure.
(Failure occurs when FS $= 1$.)

12.2.1.4   von Mises Criterion

The von Mises criterion for failure of ductile materials states that yielding occurs
when

$$\frac{1}{2}\left[(\sigma_1 - \sigma_2)^2 + (\sigma_2 - \sigma_3)^2 + (\sigma_1 - \sigma_3)^2\right] = \sigma_Y^2. \quad (12.20)$$

This relation can be derived from strain energy considerations and is also known as the *distortional energy criterion*. In plane stress the failure boundary simplifies to

$$\sigma_1^2 - \sigma_1\sigma_2 + \sigma_2^2 = \sigma_Y^2, \quad (12.21)$$

which describes the ellipse in Fig. 12.16. The Tresca and von Mises criteria coincide under conditions of uniaxial tension and compression and in equibiaxial tension and compression. Otherwise, the Tresca criterion is slightly more conservative (i.e., the stresses required to cause failure are generally smaller).

We see from Eq. (12.20) that the *von Mises equivalent stress*, defined by

$$\sigma_M = \frac{1}{\sqrt{2}}\sqrt{(\sigma_1 - \sigma_2)^2 + (\sigma_2 - \sigma_3)^2 + (\sigma_1 - \sigma_3)^2}, \quad (12.22)$$

can be compared to σ_Y as a measure of how close the material is to failure. It can be expressed in terms of the components of stress as

$$\sigma_M = \frac{1}{\sqrt{2}}\sqrt{(\sigma_x - \sigma_y)^2 + (\sigma_y - \sigma_z)^2 + (\sigma_x - \sigma_z)^2 + 6\left(\tau_{xy}^2 + \tau_{yz}^2 + \tau_{xz}^2\right)}. \quad (12.23)$$

Thus, if we have a solution to a particular problem, we can form the von Mises equivalent stress directly from the components of stress and map regions of yielding within a component without first having to determine the principal stresses. The *von Mises factor of safety* is defined by

$$FS_M = \frac{\sigma_Y}{\sigma_M}. \quad (12.24)$$

Example 12.4 Tresca and von Mises Safety Factors
A cylindrical pressure vessel with hemispherical ends has radius $R = 2$ m and wall thickness $t = 10$ mm and is made of steel with yield stress $\sigma_Y = 1800$ MPa. It is internally pressurized at $p = 2$ MPa. Compare the Tresca and von Mises safety factors.

Strategy
We must calculate the principal stresses in the vessel (see Sect. 7.6) and then determine the factors of safety from Eqs. (12.19) and (12.24).

Solution

Neglecting the stress normal to the vessel wall, the principal stresses are the hoop and axial stresses

$$\sigma_1 = \frac{pR}{t}, \quad \sigma_2 = \frac{pR}{2t}.$$

From Eq. (12.18), the Tresca equivalent stress is

$$\begin{aligned}
\sigma_T &= \text{Max}(|\sigma_1 - \sigma_2|, |\sigma_2 - \sigma_3|, |\sigma_1 - \sigma_3|) \\
&= \text{Max}\left(\left|\frac{pR}{t} - \frac{pR}{2t}\right|, \left|\frac{pR}{2t}\right|, \left|\frac{pR}{t}\right|\right) \\
&= \frac{pR}{t} \\
&= \frac{(2\,\text{MPa})(2\,\text{m})}{0.01\,\text{m}} \\
&= 400\,\text{MPa},
\end{aligned}$$

so the Tresca safety factor is

$$\text{FS}_T = \frac{\sigma_Y}{\sigma_T} = \frac{1800\,\text{MPa}}{400\,\text{MPa}} = 4.50.$$

From Eq. (12.22), the von Mises equivalent stress is

$$\begin{aligned}
\sigma_M &= \frac{1}{\sqrt{2}}\sqrt{(\sigma_1 - \sigma_2)^2 + (\sigma_2 - \sigma_3)^2 + (\sigma_1 - \sigma_3)^2} \\
&= \frac{1}{\sqrt{2}}\sqrt{\left(\frac{pR}{t} - \frac{pR}{2t}\right)^2 + \left(\frac{pR}{2t}\right)^2 + \left(\frac{pR}{t}\right)^2} \\
&= \frac{\sqrt{3}\,pR}{2t} \\
&= \frac{\sqrt{3}(2\,\text{MPa})\,(2\,\text{m})}{2(0.01\,\text{m})} \\
&= 346\,\text{MPa}.
\end{aligned}$$

Therefore, the von Mises safety factor is

$$\text{FS}_M = \frac{\sigma_Y}{\sigma_M} = \frac{1800\,\text{MPa}}{346\,\text{MPa}} = 5.20.$$

The von Mises safety factor is 15.6% larger than that based on the Tresca equivalent stress. This reflects the fact that the von Mises failure envelope generally lies outside the Tresca envelope (Fig. 12.16).

Discussion

We neglected the normal stress perpendicular to the wall of the cylindrical pressure vessel because its magnitude is smaller than that of the hoop stress by the factor R/t, which should be at least 10 for this analysis to be applicable and is usually greater. This meant that the state of stress was one of plane stress with its attendant simplifications. Notice that while the Tresca criterion is the more conservative of the two failure criteria, other constraints, such as the need to minimize the weight of an aircraft component, might lead you to choose the von Mises criterion instead.

12.2.2 Repeated Loads

On the basis of our discussion in the preceding section, it would appear that safe design of a component requires only that stresses remain below a suitable failure criterion. But it has been found that components subjected to repeated loads can fail even though the associated stress levels are well below the yield stress. The failure mechanism is that a small amount of damage to the material is produced each time a repeated load is applied. Although the amount of damage done in each repetition, or *cycle*, is insufficient to cause failure, the damage accumulates and eventually results in failure. The term *fatigue* was first applied in the mid-nineteenth century by engineers dealing with this type of failure in stagecoach and railroad car axles.

Fatigue is generally divided into three categories: low cycle, high cycle, and fatigue crack growth. In *low-cycle fatigue*, stress levels exceed the yield stress, with the result that the number of cycles to failure is relatively low ($<10^3$). Breaking a piece of wire by bending it back and forth demonstrates failure by low-cycle fatigue. (This experiment can be conducted with a paper clip.) *High-cycle fatigue* can occur when stress levels are lower than the yield stress, and failure may require 10^3 to 10^6 cycles. As we discuss in Sect. 12.3, *fatigue crack growth* refers to the growth of discrete cracks under repeated loading. In this section we consider high-cycle fatigue.

12.2.2.1 S-N Curves

The resistance of materials to high-cycle fatigue is determined by subjecting them to stress levels of constant frequency (cycles per second) and amplitude in axial load or bending experiments. In applications loads may be random in frequency and amplitude, but the use of constant values in testing is desirable both for simplicity and from the point of view of making comparisons.

Figure 12.17 shows two examples of the stress resulting from cyclical loading with constant frequency and amplitude. In case (a) the mean value of the stress is zero and in case (b) it is not. Denoting the maximum and minimum values of the stress by σ_{max} and σ_{min}, the *stress amplitude* σ_a and *mean stress* σ_m are

$$\sigma_a = \frac{\sigma_{max} - \sigma_{min}}{2}, \qquad \sigma_m = \frac{\sigma_{max} + \sigma_{min}}{2}. \qquad (12.25)$$

To determine the fatigue strength of a specimen subjected to cyclic loading with zero mean value, the number of cycles N required to fail the specimen is recorded for a given stress amplitude or *fatigue strength*. The fatigue strength is plotted as a function of N (Fig. 12.18), yielding what is called an *S-N* or *endurance curve*. Notice that the fatigue strength decreases as the number of cycles is increased. For some materials, notably steels, there is a distinct stress limit σ_{fat} below which the specimen or component will have essentially infinite fatigue life. This is known as the *fatigue limit* or *endurance limit* and becomes another stress level to be considered in design. Keeping stress levels below the fatigue limit is known as *safe life design*. In some

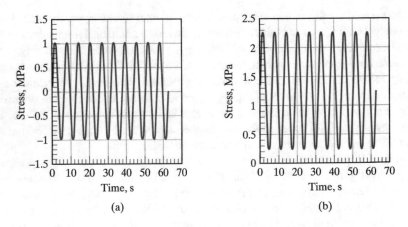

Fig. 12.17 Constant stress amplitude loading: (**a**) with zero mean stress; (**b**) with nonzero mean stress

Fig. 12.18 *S-N* (endurance) curve

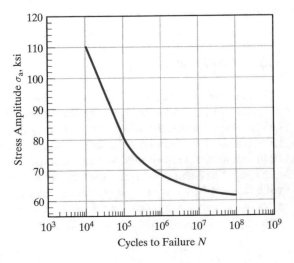

situations, safe life design is incompatible with other design constraints (such as weight), and a finite fatigue life must be accepted.

An empirical expression used to fit *S-N* curves is

$$\sigma_a = \sigma_{fat} + \frac{b}{N^c},\tag{12.26}$$

where the fatigue limit σ_{fat} and the parameters b and c are material properties that measure a material's resistance to high-cycle fatigue. (For the example shown in Fig. 12.18, $\sigma_{fat} = 60$ ksi, $b = 2000$ ksi, and $c = 0.4$.) Once these values have been determined by fatigue testing, the fatigue strength or allowable stress level for a specified lifetime can be determined from Eq. (12.26). Alternatively, if the stress amplitude is the specified quantity, Eq. (12.23) can be solved for the corresponding lifetime:

$$N = \left(\frac{b}{\sigma_a - \sigma_{fat}}\right)^{1/c}.\tag{12.27}$$

Fatigue life is also affected by the mean stress level. When there is a mean stress σ_m, the maximum stress in each cycle is brought closer to the yield stress, which leads to more damage per cycle and decreases fatigue life. The main effect of mean stress is to depress the fatigue strength, so the *S-N* curves shift downward with increasing mean stress level (Fig. 12.19).

The manner in which the fatigue strength is affected by mean stress depends on the material, and as a consequence three different empirical relationships have been developed. Denoting the fatigue limit in the presence of mean stress by σ'_{fat}, the *Goodman relation* is

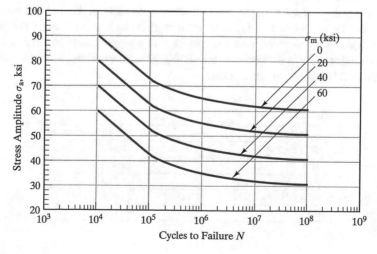

Fig. 12.19 Effect of mean stress on an *S-N* curve

$$\sigma'_{\text{fat}} = \sigma_{\text{fat}} \left(1 - \frac{\sigma_m}{\sigma_U} \right).$$ (12.28)

According to this expression, the fatigue limit reduces linearly, based on the ultimate tensile stress of the material. Another equation used is the *Gerber parabola*, which is also based on the ultimate tensile stress of the material but uses a quadratic expression:

$$\sigma'_{\text{fat}} = \sigma_{\text{fat}} \left[1 - \left(\frac{\sigma_m}{\sigma_U} \right)^2 \right].$$ (12.29)

The third relation that is used is called the *Soderberg line*:

$$\sigma'_{\text{fat}} = \sigma_{\text{fat}} \left(1 - \frac{\sigma_m}{\sigma_Y} \right).$$ (12.30)

It uses the yield stress of the material as the basis for adjusting the endurance limit in the presence of mean stress. These relationships have been derived from what are called *constant life diagrams*, in which the normalized stress amplitude is plotted as a function of mean stress with lifetime as the third parameter. Depending on the material, such plots are usually either approximately linear or approximately quadratic, which explains the origin of these empirical expressions. Once σ'_{fat} has been determined for a given material and value of mean stress, it can be substituted for σ_{fat} in Eq. (12.26) to determine the stress amplitude as a function of the number of cycles to failure.

Example 12.5 Application of an *S-N* Curve

Fatigue testing of a steel alloy at zero mean stress results in the *S-N* data shown in Table 12.1. This material is to be used in a rod with 1-in diameter that will be subjected to an axial load varying between equal levels in tension and compression. The rod must withstand 500,000 cycles. Determine the maximum load level that can be applied.

Table 12.1 *S-N* data

Cycles to failure (1000 cycles)	Stress amplitude (ksi)
1.0	190.7
10.0	135.1
100.0	100.0
1000.0	77.9

Strategy

By drawing a graph of the measured stress amplitude as a function of the number of cycles to failure, we can estimate the stress amplitude corresponding to 500,000 cycles and thereby determine the maximum load amplitude.

Solution

We plot the *S-N* curve in semilog form in Fig. (a). The fatigue strength (stress amplitude) corresponding to 500,000 cycles is approximately $\sigma_a = 84\,\text{ksi}$. Because this is a case of zero mean stress, the maximum and minimum stresses and therefore the maximum and minimum loads have the same amplitude. Therefore, the maximum load amplitude is

$$\begin{aligned} P_{\max} = -P_{\min} &= \pi R^2 \sigma_a \\ &= \pi (0.5\,\text{in})^2 \left(84\text{E3}\,\text{lb/in}^2\right) \\ &= 66{,}000\,\text{lb.} \end{aligned}$$

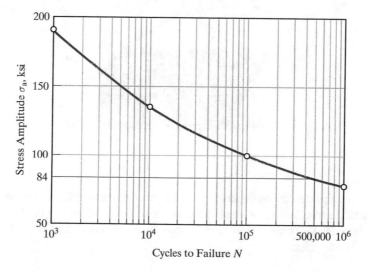

(a) *S-N* curve

Discussion

Plotting the *S-N* curve in semilog form made it easier for us to determine the stress amplitude corresponding to 500,000 cycles. Draw a normal graph of the S-N data in Table 12.1, and try to determine the stress amplitude corresponding to 500,000 cycles.

Example 12.6 Application of the Goodman Relation
The I-beam of a gantry crane (Fig. 12.20) is used for loading and unloading 80,000-lb containers in a factory. The beam has fixed supports. Its dimensions and moment of inertia are $L = 20$ ft, $h = 12.5$ in and $I = 285$ in^4. The beam's material has ultimate stress $\sigma_U = 80$ ksi, and its S-N curve is described by
Equation (12.26) with $\sigma_{fat} = 10$ ksi, $b = 120$ ksi, and $c = 0.15$. Mean stress effects can be accounted for through the Goodman relation, Eq. (12.28). Determine how many years the crane can be used if the beam is the critical structural element and 20 containers per day are lifted in a 200-working day year.

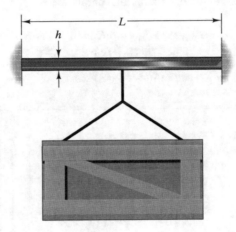

Fig. 12.20

Strategy
We must first determine the maximum range of stresses to which the beam is subjected and calculate its amplitude and mean value. Then we can determine σ'_{fat} from the Goodman relation and substitute it into Eq. (12.27) to obtain the number of cycles.

Solution
We leave it as an exercise to show that the maximum bending moment in the beam due to the weight W of the container supported at the beam's midpoint is $M = WL/8$. The maximum magnitude of the resulting tensile stress occurs at $y = h/2$:

$$
\begin{aligned}
\sigma_{max} &= \frac{My}{I} \\
&= \frac{WL(h/2)}{8I} \\
&= \frac{(80,000\,\text{lb})\,(240\,\text{in})\,(6.25\,\text{in})}{8\,(285\,\text{in}^4)} \\
&= 52,600\,\text{psi}.
\end{aligned}
$$

The stress cycles between this value and zero (when the container is released), so the stress amplitude and mean stress are

$$\sigma_a = \sigma_m = \frac{\sigma_{max}}{2} = 26{,}320\,\text{psi}.$$

Applying the Goodman relation yields

$$\sigma'_{fat} = \sigma_{fat}\left(1 - \frac{\sigma_m}{\sigma_U}\right)$$

$$= (10{,}000\,\text{psi})\left(1 - \frac{26{,}320\,\text{psi}}{80{,}000\,\text{psi}}\right)$$

$$= 6710\,\text{psi}.$$

We determine the number of cycles from Eq. (12.27) with σ'_{fat} substituted for σ_{fat}:

$$N = \left(\frac{b}{\sigma_a - \sigma'_{fat}}\right)^{1/c}$$

$$= \left(\frac{120{,}000\,\text{psi}}{26{,}320\,\text{psi} - 6710\,\text{psi}}\right)^{1/0.15}$$

$$= 176{,}000\,\text{cycles}.$$

Dividing N by 4000 cycles/year gives a useful lifetime of 44 years.

Discussion
Notice how important it was to determine the state of stress in the component. In each case, the nature of the loading, whether axial, torsional, bending, or combinations thereof, must be properly assessed and the appropriate formulas for the stress state applied. As this example demonstrates, the manner in which the component is utilized determines the stress amplitude and mean stress. The number of cycles applied in a particular situation is often given in terms of rate (e.g., cycles per day, month, or year).

12.2.2.2 Cumulative Damage

Cyclic loading on a structural component may not have a single amplitude. Components are generally subjected to a spectrum of amplitudes. But it is often possible to divide the loading into segments, or blocks, each of which is of approximately constant amplitude. A certain amount of damage will occur during each loading block, and the question that arises is how to quantify the cumulative damage.

Suppose that N is the number of cycles to failure of a particular component at a given stress amplitude and mean stress. Let the *damage D* done in subjecting the component to n cycles at that stress amplitude and mean stress be defined by

$$D = \frac{n}{N}.$$

Notice that failure occurs at $D = 1$ if the loading continues until $n = N$. This is called *Miner's law*, which also states that the cumulative damage resulting from blocks of loading of different amplitudes and mean stresses can be obtained by summing the damages of the individual blocks, irrespective of the order in which they are applied, and that failure occurs when $D = 1$. In practice, the sequence in which blocks of loads are applied does have an effect in certain materials. Nevertheless, this simple approach works surprisingly well. Suppose that a component is subjected to a sequence of blocks of loading. Let n_i be the number of cycles applied in the ith block, and let N_i be the number of cycles required to fail the undamaged component at that stress amplitude and mean stress. The resulting damage is $D_i = n_i/N_i$. The accumulated damage after k blocks of loading is

$$D = \sum_{i=1}^{k} D_i = \sum_{i=1}^{k} \frac{n_i}{N_i}. \qquad (12.31)$$

We demonstrate the application of Miner's law in the following example.

Example 12.7 Cumulative Damage of a Component
The numbers of cycles applied to a structural component at several levels of stress amplitude are listed in Table 12.2. The loading blocks all consisted of stress cycles with zero mean stress, and the *S-N* curve is described by Eq. (12.26) with $\sigma_{\text{fat}} = 60\,\text{ksi}$, $b = 1200\,\text{ksi}$, and $c = 0.4$. How many additional cycles of loading could be applied to the component at a stress amplitude of 85 ksi?

Table 12.2 Cycles applied to a structural component

Loading block	Stress amplitude	Cycles applied
i	(ksi)	(1000 cycles)
1	80	2.79
2	75	12.59
3	70	58.37
4	65	124.93

Strategy
The first step is to determine how many cycles would be required to fail the component at each of the stress levels in the table. We can then use Eq. (12.31) to determine the accumulated damage. Knowing how much damage remains to be accumulated before failure occurs, we can calculate the necessary number of cycles at a stress amplitude of 85 ksi.

Solution

From Eq. (12.27), the number of cycles to failure for the first block of loading is

$$N_1 = \left(\frac{b}{\sigma_a - \sigma_{\text{fat}}}\right)^{1/c} = \left(\frac{1,200,000\,\text{psi}}{80,000\,\text{psi} - 60,000\,\text{psi}}\right)^{1/0.4} = 27,900\,\text{cycles}.$$

The damage done in this loading block is therefore

$$D_1 = \frac{n_1}{N_1} = \frac{2790}{27,900} = 0.1.$$

Repeating these steps for each loading block, the results are summarized in Table 12.3. After the four load blocks have been applied, the accumulated damage is

$$D = \sum_{i=1}^{4} D_i = 0.83.$$

Table 12.3 Loading block damages

Loading block i	Stress amplitude (ksi)	Cycles to failure N_i (1000 cycles)	Cycles applied n_i (1000 cycles)	Damage D_i
1	80	27.9	2.79	0.1
2	75	57.2	12.59	0.22
3	70	157.7	58.37	0.37
4	65	892.3	124.93	0.14

The damage remaining before failure is $1 - D = 0.17$. The number of cycles to failure at 85 ksi is

$$N = \left(\frac{1,200,000\,\text{psi}}{85,000\,\text{psi} - 60,000\,\text{psi}}\right)^{1/0.4} = 16,000\,\text{cycles}.$$

To determine the number of cycles n that can be applied until failure, we set

$$\frac{n}{N} = 0.17,$$

obtaining $n = 2710\,\text{cycles}$.

Discussion

Here you can see that cumulative damage problems require repeated application of the steps followed in Example 12.6. The damage in each loading block can then be determined from the ratio n_i/N_i. If the damage becomes greater than one during any loading block, you can conclude that failure occurred. However, a check of your calculations would also be in order.

Problems

12.26 A thin sheet of brittle material is subjected to the state of plane stress $\sigma_x = 125$ ksi, $\sigma_y = 75$ ksi, $\tau_{xy} = 100$ ksi. The ultimate stress of the material (in tension and compression) is $\sigma_U = 175$ ksi. Determine whether the sheet will fail according to the maximum normal stress criterion.

12.27 A load applied to a machine component results in the state of plane stress $\sigma_x = 80$ MPa, $\sigma_y = 100$ MPa, $\tau_{xy} = 60$ MPa. The component is made of a brittle high-strength steel that follows the maximum normal stress criterion with $\sigma_U = 200$ MPa. If increasing the load increases each stress component proportionally, determine the percentage increase that can be applied before the component fails.

12.28 A closed, thin-walled cylinder is to be pressurized internally at 6000 psi. (Atmospheric pressure is 14.7 psi.) The mean radius and thickness of the cylinder are 2 and 0.1 in., respectively. The material being considered for the design has an ultimate stress (in tension and compression) $\sigma_U = 180$ ksi. Will the material suffice?

12.29 A point on the surface of a material is subjected to the state of plane stress $\sigma_x = -50$ ksi, $\sigma_y = -35$ ksi, $\tau_{xy} = 40$ ksi. The material is brittle and has tensile and compressive ultimate stresses $\sigma_U^t = 80$ ksi, $\sigma_U^c = 140$ ksi. Is this a safe state of stress according to Mohr's failure criterion?

12.30 A structural component is being designed using a brittle material that follows Mohr's failure criterion with $\sigma_U^t = 220$ MPa and $\sigma_U^c = 300$ MPa. The component is expected to be in the state of plane stress $\sigma_x = 120$ MPa, $\sigma_y = -160$ MPa, $\tau_{xy} = -75$ MPa. Is this design concept satisfactory?

12.31 A material that follows Mohr's failure criterion has a compressive ultimate stress of 2 GPa, which is four times the tensile ultimate stress. A load is applied that causes the principal stresses to be in the ratio $\sigma_2/\sigma_1 = -4$, with $\sigma_1 \geq 0$ at all times. Determine the value of σ_2 at failure.

12.32 A thin sheet of material is loaded in such a way that $\sigma_x = -95$ ksi, $\sigma_y = 75$ ksi, $\tau_{xy} = 29.5$ ksi. It is made of a brittle material with tensile and compressive ultimate stresses of 150 and 300 ksi, respectively. If increasing the load multiplies each plane stress component proportionally, determine the percentage increase that can be applied before failure occurs according to Mohr's failure criterion.

12.33 A thin sheet of material is made of a brittle material that follows Mohr's failure criterion with tensile and compressive strengths σ_U^t and σ_U^c, respectively. It is subjected to the proportional state of plane stress $\sigma_x = \sigma$,

$\sigma_y = -\alpha\sigma$, $\tau_{xy} = \alpha\sigma$, where σ is positive. Determine the value of α for which the second principal stress $\sigma_2 = 0$. Under this condition, for what value of σ will the sheet fail?

12.34 Determine the factor of safety when a ductile material that follows the Tresca criterion is in the state of plane stress $\sigma_x = 50\,\text{ksi}$, $\sigma_y = 30\,\text{ksi}$, $\tau_{xy} = 25\,\text{ksi}$. The yield stress of the material is $\sigma_Y = 80\,\text{ksi}$.

12.35 A thin sheet of material is loaded in plane stress, and yielding occurs when $\sigma_x = 80\,\text{MPa}$, $\sigma_y = 35\,\text{MPa}$, $\tau_{xy} = 45\,\text{MPa}$. According to the Tresca criterion, what is the tensile yield stress σ_Y of the material?

12.36* A rod with a solid circular cross section is subjected to a bending moment M and a torque T. Obtain an expression for the radius R of the rod according to the Tresca criterion if the yield stress of the material is σ_Y and a safety factor of 2.5 is required.

12.37* The L-shaped bar has a solid circular cross section with 5-mm radius. The yield stress of the material is $\sigma_Y = 300\,\text{MPa}$, and it follows the von Mises failure criterion. If a safety factor of 2 is desired, determine the maximum value of the vertical load P that can be applied.

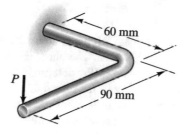

60 mm

P

90 mm

Problem 12.37

12.38* The open cylindrical tank has height $h_S = 25\,\text{ft}$, inner radius $R = 20\,\text{ft}$, and wall thickness $t = 1/4\,\text{in}$. It is made of steel that has weight density $490\,\text{lb/ft}^3$ and yield stress $\sigma_Y = 110\,\text{ksi}$ and follows the von Mises yield criterion. Based on the state of stress in the cylindrical wall due to pressure and the tank's weight, determine the factor of safety when the tank is filled with water (weight density $62.4\,\text{lb/ft}^3$) to a depth $h_W = 20\,\text{ft}$. (See the discussion of pressure vessels in Sect. 7.6.)

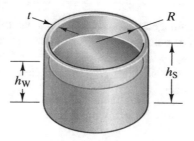

t

R

h_W

h_S

Problem 12.38

12.39   The thin cylindrical shell is subjected to a compressive axial load P and an internal pressure p. (External pressure is negligible.) It has internal radius R and thickness t ($t \ll R$) and is made of material with yield stress σ_Y. If the axial load is proportional to the pressure, $P = \alpha p$, where α is a constant, determine the allowable pressure p so that the von Mises factor of safety is no less than 2. (See the discussion of pressure vessels in Sect. 7.6.)

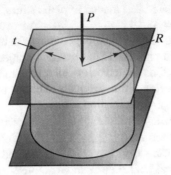

Problem 12.39

12.40   A thin sheet consists of material that follows the von Mises yield criterion with yield stress $\sigma_Y = 270$ MPa. The sheet is subjected to plane stress. If $\sigma_x = 100$ MPa and $\sigma_y = -100$ MPa, determine the value of the shear stress τ_{xy} required to cause yielding of the material.

12.41∗   A series of fatigue tests have been conducted on a material. The tests were conducted at zero mean stress level and resulted in the data tabulated below. If the endurance limit σ_{fat} was 15 ksi, find an empirical expression of the form of Eq. (12.26) that fits the data. How many cycles could be applied to the material at a stress amplitude of 35 ksi?

Stress amplitude (ksi)	Cycles to failure
52.7	1.000E3
38.8	1.000E4
30.0	1.000E5
24.5	1.000E6
21.0	1.000E7

12.42   A sample of undamaged material has been subjected to the blocks of cyclic loading tabulated below, all at zero mean stress. The parameters in Eq. (12.26) are $\sigma_{fat} = 15.0$ ksi, $b = 150$ ksi, $c = 0.2$. Determine how many more zero mean stress cycles could be applied to the material at a stress amplitude of 35 ksi.

Stress amplitude (ksi)	Cycles applied
42	7.938E2
28	5.318E4
25	8.353E4
22	1.536E6

12.43 A sample of undamaged material has been subjected to the blocks of cyclic loading tabulated below, all at a level of mean stress equal to 20% of the ultimate tensile stress. The parameters $\sigma_{fat} = 15.0$ ksi, $b = 150$ ksi, $c = 0.2$. If the material follows the Goodman relation, determine how many more cycles could be applied to the material at a stress amplitude of 35 ksi and the same level of mean stress.

Stress amplitude (ksi)	Cycles applied
42	7.188E2
28	1.231E4
25	7.363E4
22	1.595E5

12.44 The titanium alloy rod has length $L = 15$ in and a circular cross section with radius $R = 0.5$ in. The axial load $P_a = 98$ kip is constant. The load P_b is applied cyclically with zero mean value and the amplitudes tabulated below. The material has an ultimate stress of 250 ksi and an endurance limit $\sigma_{fat} = 100$ ksi and follows the Gerber parabola for mean stress effects. Its S-N curve is described by Eq. (12.26) with constants $b = 200$ ksi, $c = 0.15$. Determine how much damage is caused by each block of loading. How many additional cycles of the load P_b could be applied at an amplitude of 548 lb before fatigue failure would occur?

Amplitude of P_b (lb)	Cycles applied
610	1.000E3
560	3.850E4
518	2.000E6

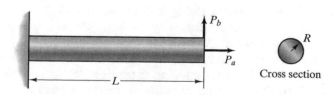

Problem 12.44

12.45∗ The spar of an airplane's wing is cantilevered from the fuselage and has a length of 20 ft. The height and width of the spar are 20 and 1 in, respectively. The wing's lift during flight is represented by a uniformly distributed load of 52 lb/in. Fatigue due to gust loading is modeled by an additional uniformly distributed load whose amplitude varies cyclically with zero mean value. The amplitude of the resulting cyclic displacement of the free end of the spar is 8.64 in. The spar consists of aluminum alloy with elastic modulus $E = 10E6$ psi and yield stress 100 ksi. The S-N curve of the material is shown, and mean stress effects are accounted for through the Soderberg line (Eq. 12.30). If there are 355 cycles per flight, how much fatigue damage results from 5000 flights?

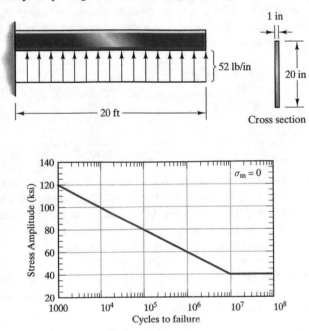

Problem 12.45

12.46∗ A bridge carries cars, trucks, and trains. It has been in service for 3 years and has carried the numbers of vehicles tabulated below on a daily basis. The main structural element of the bridge is a simply supported rectangular beam. The height, width, and span of the beam are 0.25, 0.1, and 250 m, respectively. The weight carried by the beam for each type of vehicle is tabulated. The S-N curve of the bridge material is described by Eq. (12.26) with $b = 11$ GPa, $c = 0.2$, and mean stress effects are accounted for by the Soderberg line. Laboratory testing determines that $\sigma_{fat} = 500$ MPa and $\sigma_Y = 1200$ MPa. The level of traffic in the first 3 years was greater than originally anticipated, so the rail traffic was diverted. How much longer can the bridge be used by cars and trucks?

	Weight	
Vehicle	W (kN)	Vehicles/day
Auto	18.3	5000
Truck	26.0	100
Train	37.6	30

12.47 A turbine blade material has ultimate tensile stress 820 MPa and an endurance limit of 414 MPa. The constants b and c in the equation $\sigma_a = \sigma_{fat} + b/N^c$ are 4.67 GPa and 0.28, respectively. Mean stress effects are accounted for via the Goodman relation. The mass density of the material is 5000 kg/m^3 and the length of the blades is $L = 1.5$ m. The turbine has been operated at the angular frequencies noted below. The normal stress in the blades is given as a function of the axial coordinate x by $\sigma = \rho\omega^2(L^2 - x^2)/2$. Determine how many times the turbine can be operated at angular frequency $\omega = 400$ rad/s.

Cycles	Frequency (rad/s)
868	430
1002	460
451	490
75	520

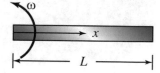

Problem 12.47

12.48* The beam is of length $L = 10$ ft. It has a rectangular cross section that has height $h = 10$ in and a width of 2 in. The beam is subjected to repeated uniform loadings with magnitude q_0 that then drop to zero on unloading. The different values of q_0 in the table correspond to different load histories. The material that the beam is made of has a yield stress of 80 ksi. Its fatigue behavior follows the form $\sigma_a = \sigma'_{fat} + b/N^c$, where mean stress effects are accounted for through the Soderberg line. The endurance limit is 12 ksi, and the constants $b = 50$ ksi and $c = 0.2$. It is subjected to the loading blocks given below. How many more cycles could it withstand with load magnitude $q_0 = 134$ lb/in?

q_0 (lb/in)	Cycles
100	2.39E9
180	714
155	6.16E3

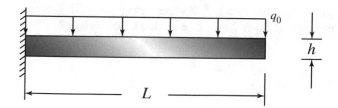

Problem 12.48

12.49* The pressure vessel shown has radius $R = 72$ in and wall thickness
$t = 0.075$ in. The internal pressure fluctuates between p_i and zero. (Assume
that the external pressure is zero.) The ultimate tensile stress and endurance
limit of the vessel's material are 140 and 70 ksi, respectively, and the
constants in Eq. (12.26) are $b = 800$ ksi and $c = 0.15$. Mean stress effects
are accounted for by the Goodman line. How many more cycles can be
applied at $p_i = 280$ psi if the pressure history shown has already been
applied? (Base your analysis on the damage caused by the largest component of principal normal stress.)

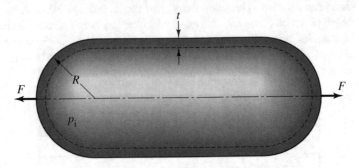

Pressure p_i (psi)	Cycles
250	5.47E4
257	3.52E4
261	6.35E4
280	2.31E4

Problem 12.49

12.50   A model of a truck's axle has a solid circular cross section and carries a load
W as shown. One-half of the load is exerted at $L = 10$ in from each end of the
axle. Fatigue tests carried out on the axle material found that the S-N curve
followed the empirical law $\sigma_a = \sigma_{fat} + b/N^c$, where the endurance limit of the
material was 50 ksi and the constants $b = 165$ ksi and $c = 0.18$. Mean stress
effects were accounted for by the Goodman relation, where the ultimate
tensile stress was 90 ksi. The axle carries the three types of loads listed.
(Assume that the axle is subjected to no load when W is removed.) Load type
1 is carried 220 times a year, type 2 is carried 80 times a year, and type 3 is
carried 40 times a year. After 2 years of service, there was concern about
fatigue failure and load type 3 was no longer carried. How many additional
years can load types 1 and 2 be carried before failure?

Load Type	Weight, W (lbs)
1	147,340
2	176,393
3	188,417

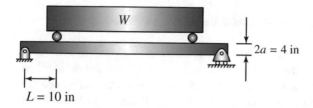

Problem 12.50

12.3   Fracture

Although the subdiscipline of mechanics of materials known as fracture mechanics essentially originated with a paper by A. A. Griffith in 1920, substantial growth of the field did not begin until the 1960s. This growth was motivated by a series of catastrophic structural failures, including the literal splitting in two of Navy T2 tankers while at anchor in calm but cold conditions and crashes of two Comet jet airliners, in which it became clear that the cause was growth of cracks from preexisting flaws. Such flaws had always been present in structural materials as a consequence of the manufacturing processes used, stress concentrations, and repeated loading. But their effects were accentuated in these more recent failures because the structures consisted of new high-strength materials that were much more sensitive to the presence of small cracks. As a result, the structures failed at much lower stress levels than had been anticipated by engineers following the procedures described in our discussion of failure in Sect. 12.2. In the materials science community, crack sensitivity has led to a classic balancing act in the design of new materials in which high strength does not necessarily mean that a material is *tough* or resistant to crack growth. The organization of this section mirrors our discussion of failure in that overloads and repeated loads are considered separately.

12.3.1   Overloads and Fast Crack Growth

We consider an overload to be a one-time event in which load levels have exceeded a safe allowable value. This is in contrast to repeated loading, which can cause failure at much lower load levels. In first thinking about cracks in materials, it might appear that as soon as there is a crack in a structure, it will fail. In fact, some structures can tolerate quite sizable cracks before failing, which makes it clear that severity of cracks can be ranked. We therefore need to develop a "crack meter," a quantitative measure of crack severity. To do so, we must return to the concept of stress concentrations.

12.3.1.1   Stress Concentration due to an Elliptical Hole

We have already seen that circular holes in components give rise to stress concentrations. The ellipse is a useful hole geometry for us to consider here, because at one extreme it can be made a circle while at the other extreme it becomes a crack. We can relate the former extreme to what we already know, and with the latter extreme, we can explore new ground.

In Fig. 12.21, a uniform tensile stress is applied to the ends of a sheet containing an elliptical hole. The major axis of the ellipse ($2a$) is perpendicular to the direction of the applied stress. We assume that the width and height of the sheet are large compared to the dimensions of the hole. The maximum normal stress occurs at the root of the ellipse ($x = \pm a$) and is given by

$$\sigma_{max} = \sigma \left(1 + \frac{2a}{b} \right). \tag{12.32}$$

Setting $a = b$ in Eq. (12.32), we recover the stress concentration factor for a circular hole: $C = \sigma_{max}/\sigma = 3$. In terms of the *root radius* $\rho = b^2/a$ of the ellipse, the maximum stress is

Fig. 12.21 Applying tension to a sheet containing an elliptical hole

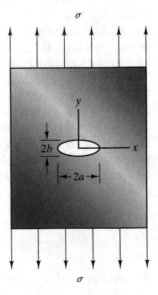

$$\sigma_{\max} = \sigma \left(1 + 2\sqrt{\frac{a}{\rho}} \right). \tag{12.33}$$

When the crack length (the major axis of the ellipse) is large compared to the root radius, the stress concentration factor is approximately

$$C = 2\sqrt{\frac{a}{\rho}}. \tag{12.34}$$

For an infinitely sharp crack ($\rho \to 0$), the stress concentration factor is infinite, a result we will soon encounter again. Infinite stresses are predicted because our solution assumes elastic material behavior. In reality the material yields and the stresses remain finite.

Equation (12.34) provides justification for the common practice of drilling holes at the tips of sharp cracks (e.g., in the metal skin of an airplane's wing) in order to arrest their growth. The radius of the hole drilled is much larger than the notch radius of the crack. As a result, the stress concentration is significantly reduced, "arresting" the crack.

12.3.1.2 Stress Distribution near a Crack Tip

Figure 12.22 is a schematic of the region very near a crack tip. With reference to the coordinate system shown, the components of plane stress are

$$\sigma_x = \frac{K}{\sqrt{2\pi r}} \cos \frac{\theta}{2} \left(1 - \sin \frac{\theta}{2} \cos \frac{3\theta}{2} \right),$$

$$\sigma_y = \frac{K}{\sqrt{2\pi r}} \cos \frac{\theta}{2} \left(1 + \sin \frac{\theta}{2} \cos \frac{3\theta}{2} \right), \tag{12.35}$$

$$\tau_{xy} = \frac{K}{\sqrt{2\pi r}} \cos \frac{\theta}{2} \sin \frac{\theta}{2} \cos \frac{3\theta}{2},$$

where r and θ are the polar coordinates in Fig. 12.22 and K is a constant called the *stress intensity factor* whose value depends on the geometry and loading of the cracked component.

Equations (12.35) imply that the stresses near all cracks have the same dependence on r and θ. These solutions predict that the stresses are infinite at the crack tip,

Fig. 12.22 Region near the crack tip

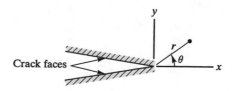

Crack faces

just as we found for an elliptical hole when it reduces to an infinitely sharp crack ($\rho \to 0$), but in fact yielding of the material will occur very near the crack tip. Let σ_Y be the yield stress of the material, and let us assume that the *crack opening stress*, the component σ_y, governs yielding. As the crack tip is approached along the x axis ($\theta = 0$), σ_y increases as the radial distance r decreases and is equal to the yield stress when the radial distance is

$$r_p = \frac{1}{2\pi} \left(\frac{K}{\sigma_Y} \right)^2. \tag{12.36}$$

If r_p is much smaller than the smallest characteristic dimension of the cracked component (e.g., the thickness of a cracked plate), the state of stress within a neighborhood of the crack tip is described by Eq. (12.35) except very near the crack tip (Fig. 12.23). To avoid boundary effects, the radius D of this neighborhood must be significantly smaller than any in-plane dimensions of the cracked component. For example, in the cracked plate in Fig. 12.24, D must be small in comparison with the length L and width W of the plate.

Fig. 12.23 Yielded zone and region of dominance of Eq. (12.35)

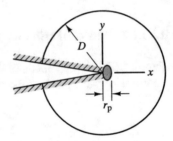

Fig. 12.24 Plate with a central crack under tension

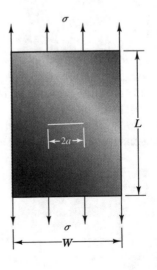

Because the r and θ dependence is the same near all cracks, the only parameter distinguishing one crack from another is the stress intensity factor K. It is, in essence, the crack meter that we suggested must exist. We will show that it allows us to make quantitative comparisons between the severities of cracks in different structural components under different loads. Basically, the larger the value of K, the more severe is the crack.

12.3.1.3   Stress Intensity Factor Solutions

The solution leading to Eq. (12.35) focuses on the region of the crack tip without concern for where the crack is located within a component, the overall shape or geometry of the component, or the applied loads. This explains why the stress intensity factor K appears as an undetermined constant. To determine K requires the development of solutions that are specific to each cracked component and its loading. There was a period in the development of fracture mechanics when a great deal of emphasis was placed on obtaining such solutions. Analytical (rather than numerical) solutions were preferred because varying parameters was simpler. However, it was found that analytical solutions could be obtained only for relatively simple geometries.

Numerical techniques and dimensional analysis have now extended the range of problems that can be solved, and handbooks exist in which many solutions are cataloged. We give examples of stress intensity factors in this section, and handbooks contain many others. When an existing solution cannot be found, finite element codes with special "crack elements" and fracture mechanics post-processing packages are available that can be used to solve very specific and complex crack problems.

12.3.1.4   Centrally Cracked Plate

The simplest solution for a stress intensity factor is for a plate loaded in tension by a uniform normal stress σ and containing a centrally placed crack whose plane is perpendicular to the loading direction (Fig. 12.24). The crack length is assumed to be small in comparison to the plate's dimensions ($W, L \gg 2a$). This assumption means that the outer boundary is effectively at infinity, so the crack length $2a$ is the only geometrical parameter in the problem. The stress intensity factor is

$$K = \sigma\sqrt{\pi a}. \tag{12.37}$$

Thus the stress intensity factor is proportional to the applied stress and to the square root of the crack length. For a given level of stress, the longer the crack becomes, the more severe it becomes.

Fig. 12.25 Central load
P on a crack face

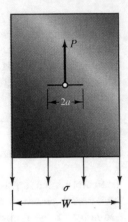

12.3.1.5 Central Load on a Crack Face

This problem can occur when cracks emanate from a bolt hole. In Fig. 12.25, a concentrated force P is exerted by a bolt or rivet at the top of the bolt hole, which is at the center of a crack of length $2a$. The diameter of the hole is assumed to be small in comparison to the crack length $2a$. Under these conditions the stress intensity factor is

$$K = \frac{P}{b\sqrt{\pi a}},\tag{12.38}$$

where b is the thickness of the plate. We see once again that the stress intensity factor is proportional to the applied load. But in this case a appears in the denominator, and although K can be very large for short cracks, it decreases as the crack grows. This suggests the possibility of spontaneous crack arrest.

In many cases stress intensity factors can be expressed in the form

$$K = \sigma\sqrt{\pi a}\, Q\!\left(\frac{a}{W}\right),\tag{12.39}$$

where σ represents the stress applied to the component, a is the crack length, and $Q(a/W)$ is a nondimensional function of the crack length a normalized by an in-plane dimension W (such as the width of a plate) that is called the *configuration factor*. Comparing Eqs. (12.37) and (12.39), we see that $Q = 1$ for a centrally cracked plate, so this problem is the prototype for the expression (12.39). Many stress intensity factors given in handbooks are expressed in the form of Eq. (12.39).

12.3.1.6 Fracture Criterion

Up to this point, we have dealt with the stress analysis of cracks. Doing so has allowed us to compare the severities of different crack configurations but says nothing as to whether or not a given crack will grow. Because this section deals with overloads, the type of crack growth we now consider is known as *fast crack growth*. We will show subsequently that slow crack growth can occur as a result of repeated loads and requires a different criterion.

The philosophy underlying the establishment of a fast crack growth criterion is very much akin to the use of the yield stress to establish the onset of yielding. Because we use the stress intensity factor to measure crack severity, an experiment can be conducted to measure the stress level and crack length at which a crack begins to grow (becomes a fast crack). These values can be substituted into the equation for the stress intensity factor, establishing the value of the stress intensity factor at which the crack begins to grow. This critical value of stress intensity factor is called the *fracture toughness* K_c. Once determined for a particular material, the fracture toughness can be used to predict the onset of fast crack growth in other structural components made of the same material. Because Eq. (12.39) is a general expression for the stress intensity factor that applies for any crack configuration, the criterion for fast crack growth is

$$\sigma\sqrt{\pi a}\,Q\!\left(\frac{a}{W}\right) = K_c. \tag{12.40}$$

From the left-hand side of this equation, we see that there are two possible conditions for fast crack growth. The first one arises when the crack length a is fixed at a given value and the stress level for fast crack growth is to be established. Then Eq. (12.40) can be solved for the stress level σ at which a crack will become a fast crack before it can be seen by nondestructive inspection techniques or before it emerges from behind an overlapping component such as a washer covering the region around a hole. The second criterion for fast crack growth arises when the stress σ is fixed at some level. In this case it is often necessary to establish how long a crack can become before it becomes a fast crack by solving Eq. (12.40) for a. We will show in the following section that Eq. (12.40) can also be used to determine when a slow crack becomes a fast crack under cyclic loading.

The two criteria we have discussed bring out an important difference between yielding and fast crack growth. In the former we need only be concerned about stress levels. In the latter, both stress level and crack length must be considered. Another complication is that for many materials the fracture toughness is dependent on the thickness for thin sections, although it becomes constant as the thickness increases. We will assume that components are sufficiently thick that K_c does not depend on the thickness, in which case it is called the *plane strain fracture toughness*.

Example 12.8 Fast Crack Growth

The steel plate in Fig. 12.26 has an edge crack of length $a = 2\,\text{cm}$ midway along its length. The plate is 5 mm thick and 20 cm wide and has a length L sufficiently large that the loads P can be assumed to result in a uniform stress distribution across the width of the plate in the neighborhood of the crack. The stress intensity factor obtained from a handbook for this configuration is

$$K = 1.12\sigma\sqrt{\pi a}. \quad (a/W \le 0.13, L/W \ge 2) \tag{1}$$

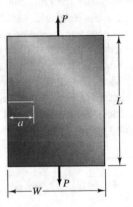

Fig. 12.26

An alternative expression that applies for all values of $\alpha = a/W$ is

$$K = \sigma\sqrt{\pi a}\left[0.265(1-\alpha)^4 + \frac{0.857 + 0.265\alpha}{(1-\alpha)^{3/2}}\right]. \quad (L/W \ge 2) \tag{2}$$

The fracture toughness of the plate is $K_c = 66\,\text{MPa-m}^{1/2}$. Determine the load level that will cause the crack to grow.

Strategy

We can use the simpler expression for the stress intensity factor because $a/W = 0.1$. The stress in the expression for the stress intensity factor is $\sigma = P/Wb$, where b is the thickness of the plate. Because the fracture toughness is known, we can solve for the value of P that will cause crack growth.

Solution

We substitute the fracture toughness and the expression $\sigma = P/Wb$ into Eq. (1),

$$K_c = 1.12 \left(\frac{P}{Wb}\right)\sqrt{\pi a}:$$

$$66\text{E}6\,\text{Pa-m}^{1/2} = 1.12 \left[\frac{P}{(0.2\,\text{m})\,(0.005\,\text{m})}\right]\sqrt{\pi(0.02\,\text{m})}.$$

Solving, we obtain the critical load level $P = 0.235\,\text{MN}$.

Discussion
Use Eq. (2) instead of Eq. (1) to determine the load level that will cause the crack to grow in this example, and compare your result to the one we obtained by using Eq. (1).

Example 12.9 Fast Crack Growth
The block of weight W in Fig. 12.27 is supported by two parallel beams, each with semispan $L = 20\,\text{ft}$ and rectangular cross section of height $h = 10\,\text{in}$ and width $b = 1\,\text{in}$. The weight of the block is $W = 15,000\,\text{lb}$. A crack of length a exists at the bottom of one of the beams at its midpoint as shown. The stress intensity factor obtained from a handbook for this configuration is

$$K = 1.12 \left(\frac{6M}{h^2b}\right)\sqrt{\pi a}, \quad (a/h \le 0.4) \tag{1}$$

where M is the bending moment at the cross section containing the crack. An alternative expression that applies for all values of $\alpha = a/h$ and large values of h/L is

$$K = \frac{6M}{h^2b}\sqrt{\pi a}\,\sqrt{\frac{2}{\pi\alpha}\tan\frac{\pi\alpha}{2}\left\{\frac{0.923 + 0.199[1 - \sin(\pi\alpha/2)]^4}{\cos(\pi\alpha/2)}\right\}}. \tag{2}$$

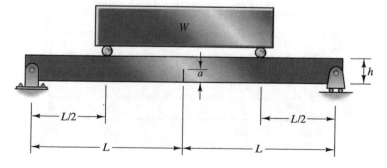

Fig. 12.27

The fracture toughness of the beam material is $K_c = 120\,\text{ksi-in}^{1/2}$. Determine the largest crack length a that could be tolerated before fast crack growth would occur.

Strategy

We will first obtain a solution for the crack length using the simpler Eq. (1) for the stress intensity factor. If the resulting crack length is not within the range of validity of that expression, we must determine the crack length from Eq. (2).

Solution

Each beam supports half the weight of the container and its contents, so the bending moment at the cross section containing the crack is

$$M = \frac{WL}{8}$$
$$= \frac{(15,000\,\text{lb})\,(240\,\text{in})}{8}$$
$$= 450,000\,\text{in-lb}.$$

Substituting this value and the fracture toughness into Eq. (1),

$$120,000 \text{ psi-in}^{1/2} = 1.12 \left[\frac{6(450,000 \text{ in-lb})}{(10 \text{ in})^2 (1 \text{ in})}\right] \sqrt{\pi a},$$

and solving, we obtain a critical crack length $a = 5.01$ in. This value is beyond the range of validity of Eq. (1), so we must obtain the solution from Eq. (2).

From a graph of the values of K given by Eq. (2) as a function of α (Fig. (a)), we estimate that $K = 120\,\text{ksi-in}^{1/2}$ at $\alpha = 0.40$. Using software designed to solve nonlinear algebraic equations, we obtain $\alpha = 0.406$. The critical crack length is $a = 4.06$ in.

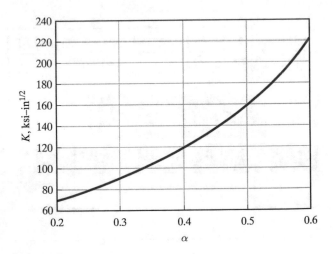

(a) Graph of K as a function of α

Discussion

Notice that we needed an expression for the stress intensity factor for the particular cracked component and loading in this example. For the problems you must deal with in this section, these expressions will be given. However, in engineering practice you will need to consult a handbook of stress intensity factors or develop an expression by a finite element analysis of the cracked component. If a handbook expression is not available, it is sometimes possible to use the expression for a cracked component that is a close relative of the problem you are considering. In some cases, handbook expressions can be superimposed to obtain the stress intensity factor for a desired component and loading.

12.3.2   Repeated Loads and Slow Crack Growth

We have seen in Sect. 12.2 that repeated loads can cause failure at stress levels that are lower than the yield stress of the material. When cracks are present, an analogous situation arises in the sense that crack growth may occur at stress intensity factor levels that are lower than the fracture toughness K_c. Repeated loading under these conditions can result in rates of crack growth that are much lower than those we discussed in the preceding section and are referred to as *slow crack growth*. Slow crack growth can be influenced by elevated temperatures and diffusive environments, but for this introduction we consider only the effects of repeated loading.

The photograph of the airliner in Fig. 12.28 shows how multiple damage sites, combined with repeated loads, caused slow growth of fatigue cracks that eventually

Fig. 12.28 An Aloha Airlines Boeing 737 lands safely despite the loss of part of its fuselage (*Photograph from Associated Press by Robert Nichols*)

resulted in fast crack growth and the loss of part of the fuselage. In this case the damage sites or initial flaws were induced by corrosion. The aircraft was used for short but frequent fights among the Hawaiian Islands, so the number of pressurization/depressurization cycles accumulated more quickly than usual.

12.3.2.1 Paris Law

The repeated application of loads is also called *cyclic loading*, and the ensuing crack growth is known as *fatigue crack growth*. In the simplest case, the repeated loads have equal maximum and minimum values over time (Fig. 12.29), which is called *constant amplitude loading*. For example, constant amplitude loading can occur in the structure of an airliner whose cabin is pressurized to the same degree from flight to flight. In contrast, turbulent airflow and other loadings can subject the airliner to an irregular pattern of repeated loads. We first consider constant amplitude loading, which provides a basis for our subsequent analysis of more complex load histories.

Let us describe a fatigue crack growth experiment on the plate with a central crack shown in Fig. 12.24, whose stress intensity factor is given by Eq. (12.37). Instead of the monotonically increasing level of stress that would be used to measure K_c, the applied stress is varied sinusoidally as in Fig. 12.29. The change in applied stress over each cycle is

$$\Delta\sigma = \sigma_{max} - \sigma_{min}.$$

In such experiments the crack length is measured as a function of the number of loading cycles N, yielding data that can be presented as shown in Fig. 12.30. Observe that increasing the stress amplitude causes the crack to grow faster.

However, Fig. 12.30 does not give a universal picture of a material's resistance to fatigue crack growth. A better measure is the rate at which a crack grows in a given material. We will see that rate information allows crack growth predictions to be made. From our discussion of fast cracks, it can be anticipated that the crack growth rate da/dN depends not only on stress amplitude but also crack length, which brings

Fig. 12.29 Example of the stress resulting from constant amplitude loading

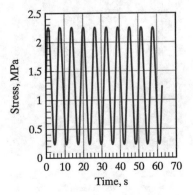

Fig. 12.30 Crack growth history for various stress amplitudes

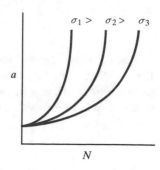

Fig. 12.31 Crack growth per cycle as a function of ΔK

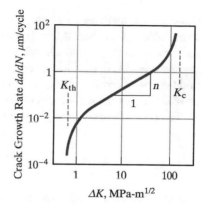

the stress intensity factor to mind. The crack length does not increase very much over one loading cycle (usually $<10^{-4}$ in), so from Eq. (12.37) the change in the stress intensity factor over one cycle is given approximately by

$$\Delta K = K_{max} - K_{min} = (\sigma_{max} - \sigma_{min})\sqrt{\pi a} = \Delta\sigma\sqrt{\pi a}.$$

When the results of a fatigue crack growth experiment are presented as crack growth per cycle da/dN as a function of the change in stress intensity factor per cycle ΔK, the curves corresponding to different levels of stress lie on a single universal curve that measures the material's resistance to fatigue crack growth (Fig. 12.31). This curve, which is sigmoidal in shape, has two limits. The limit at low ΔK values is the threshold toughness K_{th}. The upper limit is the fracture toughness K_c, at which fast crack growth begins.

A crack subjected to stress intensity factors lower than K_{th} will never grow, which makes it possible to design structural components so that the possibility of crack growth is very small. In less critical circumstances, it may be permissible to tolerate slow crack growth in order to reduce structural weight.

Because Fig. 12.31 is a log-log graph, the linear portion of the curve indicates that da/dN and ΔK are related through a power law:

$$\frac{da}{dN} = A(\Delta K)^n. \tag{12.41}$$

This equation is named the *Paris law* after an early pioneer in fracture research. The constants A and n are material properties reflecting resistance to fatigue crack growth. For steels, $n = 3$, and for polymers, $n = 6$ to 10. We will show that once A and n are determined for a given material, Eq. (12.41) provides a basis for predicting crack growth.

12.3.2.2   Predicting Slow Crack Growth

Slow crack growth may take years to become fast, so if slowly growing cracks can be tolerated despite other factors (appearance, leaks, etc.), a structure may be able to continue operating quite safely. Nevertheless, inspections must be carried out at regular intervals to monitor the progress of crack growth. The ability to predict paths and lengths of cracks helps establish the necessary frequency of inspections and also provides an estimate of the lifetimes of cracked components.

To predict the crack length, we write Eq. (12.41) as

$$dN = \frac{da}{A(\Delta K)^n}$$

and integrate, obtaining a relation between the number of cycles N and the crack length a:

$$N = \frac{1}{A} \int_{a_0}^{a} \frac{da}{(\Delta K)^n}. \tag{12.42}$$

The limit a_0 is the initial crack length from which the cycle count is made. We say it is where the "clock" starts running. Its value often results from the resolution of the equipment used to detect the presence of cracks. The time required for cracks to reach this length, which is often substantial, is not accounted for. In effect, this introduces a degree of conservatism into the design process.

Equation (12.42) determines the number of cycles N required to reach a given crack length a. The number of cycles at which failure occurs is obtained by setting a equal to the critical crack length for fast crack growth. The critical crack length is determined from Eq. (12.40) with $\sigma = \sigma_{max}$, the maximum magnitude of the constant amplitude loading. Examples 12.10 and 12.11 demonstrate these procedures.

Example 12.10 Slow Crack Growth
A bolt with diameter $d = 0.5$ in is subjected to a cyclic tensile service load that varies from zero to 10,000 lb. Due to a manufacturing flaw, the bolt has a crack initiating from a thread with 0.01-in depth. The stress intensity factor

obtained from a handbook for such a circumferential crack in an axially loaded bar is

$$K = 1.12\sigma\sqrt{\pi a}, \quad (2a/d < 0.21) \tag{1}$$

where a is the crack depth and σ is the normal stress in the bar. The fracture toughness of the bolt material is $K_c = 23$ ksi-in$^{1/2}$. Its resistance to fatigue crack growth can be expressed by the Paris law with $A = 1.0E\text{-}18$ in/cycle and $n = 3$, where crack growth rate is in inches/cycle and the stress intensity factor is in psi-in$^{1/2}$. How many load cycles can the bolt withstand?

Strategy
By substituting the fracture toughness and the maximum stress due to the constant amplitude loading into Eq. (1), we can determine the critical crack length for fast crack growth. Then we can determine the number of load cycles to failure from Eq. (12.42).

Solution
The maximum normal stress due to the constant amplitude loading is

$$\sigma_{max} = \frac{10,000\,lb}{\pi(0.5\,in)^2/4} = 50,900\,psi.$$

Substituting this value and the fracture toughness into Eq. (1),

$$K_c = 1.12\sigma_{max}\sqrt{\pi a}:$$
$$23,000\,psi\text{-}in^{1/2} = 1.12\,(50,900\,psi)\sqrt{\pi a},$$

we find that the critical crack length is $a = 0.052$ in. We also use Eq. (1) to obtain the change in the stress intensity factor over one cycle:

$$\Delta K = K_{max} - K_{min} = 1.12(\sigma_{max} - \sigma_{min})\sqrt{\pi a}$$
$$= 1.12\,(50,900\,psi - 0)\sqrt{\pi a}.$$

The initial depth of the crack is the depth of the thread: $a_0 = 0.01$ in. From Eq. (12.42), the number of cycles to failure is therefore

$$N = \frac{1}{A}\int_{a_0}^{a} \frac{da}{(\Delta K)^n}$$
$$= \frac{1}{1.0E - 18}\int_{0.010}^{0.052} \frac{da}{[(1.12)\,(50,900)\sqrt{\pi a}]^3}$$
$$= 10,800\,cycles.$$

Discussion
Notice how we used the fracture toughness to determine the critical crack length, the length of the crack at which fast crack growth begins and failure occurs.

Example 12.11 Slow Crack Growth
The plate in Fig. 12.25 is 1 cm thick, its width is $W = 50$ cm, and the hole is 2 cm in diameter. The load P is cyclic and varies from zero to 30 kN. The stress intensity factor for this configuration is given by Eq. (12.38). The resistance to fatigue crack growth satisfies the Paris law with $A = 5E{-}9$ m/cycle, $n = 4$, and $K_{th} = 5$ MPa-m$^{1/2}$, where crack growth rate is in m/cycle. The crack emanating from the hole is initially obscured by a washer with a 4-cm outer diameter. How many cycles are required for a visible crack to stop growing?

Strategy
Equation (12.38) indicates that the stress intensity factor decreases with increasing crack length, which raises the possibility of crack arrest as the stress intensity factor drops below K_{th}. The length of the arrested crack is

$$a_{th} = \frac{1}{\pi}\left(\frac{P_{max}}{bK_{th}}\right)^2 = \frac{1}{\pi}\left[\frac{30E3}{(0.01)(5E6)}\right]^2 = 0.115\,\text{m},$$

so the crack stops growing before reaching the edges of the plate. We can use Eq. (12.42) to determine how many loading cycles occur from the time the crack becomes visible until it stops growing.

Solution
From Eq. (12.38), the change in the stress intensity factor over one cycle is

$$\Delta K = K_{max} - K_{min} = \frac{P_{max} - P_{min}}{b\sqrt{\pi a}}$$
$$= \frac{30E3}{(0.01)\sqrt{\pi a}}\ \text{Pa-m}^{1/2}$$
$$= \frac{30E{-}3}{(0.01)\sqrt{\pi a}}\ \text{MPa-m}^{1/2}.$$

The crack first becomes visible when it grows beyond the outer diameter of the washer, so $a_0 = 0.02$ m. From Eq. (12.42), the number of cycles to crack arrest is

$$N = \frac{1}{A} \int_{a_0}^{a_{th}} \frac{da}{(\Delta K)^n}$$

$$= \frac{1}{5 \times 10^{-9}} \int_{0.020}^{0.115} \left[\frac{(0.01)\sqrt{\pi a}}{30\text{E-3}} \right]^4 da$$

$$= 1.23\text{E4 cycles.}$$

Discussion

In problems of this kind, it is useful for strategy purposes to examine how the stress intensity factor varies with the crack length. If it increases with increasing crack length, fast crack growth is likely at some stage, i.e., $K \rightarrow K_c$. On the other hand, if the stress intensity factor decreases with increasing crack length, then $K \rightarrow K_{th}$, and the crack may stop at some stage. These considerations help you to establish the limits of the integral in Eq. (12.42). Also, you should pay particular attention to the units in these problems. For example, the fracture toughness may be given in ksi-in$^{1/2}$ but the Paris law expressed in terms of psi-in$^{1/2}$.

Review Problems

12.51 The steel plate is 1.5 m wide and 20 cm thick. It is subjected to loads $P = 100$ MN. The length L is sufficiently large that the loads can be assumed to result in a uniform stress distribution across the width of the plate in the neighborhood of the central crack. If the fracture toughness of the material is 170 MPa-m$^{1/2}$, determine how large a central crack can be tolerated.

Problem 12.51

12.52 The 7075-T651 aluminum plate has a 6-in central crack. The fracture toughness of this alloy is 22 ksi-in$^{1/2}$. Determine the maximum stress σ that can be applied to the plate. Notice that the stress intensity factor given in Eq. (12.37) only applies to small cracks (up to 20% of the width). A more accurate expression (0.3% error for $2a/W \leq 0.7$) is

$$K = \sigma\sqrt{\pi a}\sqrt{\sec\left(\frac{\pi a}{2W}\right)}.$$

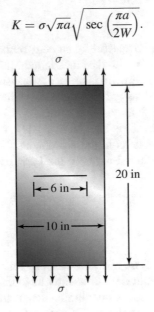

Problem 12.52

12.53 Three thin sheets with the same dimensions (5-in width, 0.1-in thickness) are subjected to tensile stress. Sheet (a) has a central hole with 0.5-in diameter, sheet (b) has a 0.5-in central crack, and sheet (c) has no defects. They are made of a material (300-M, 300 °C temper) steel) whose yield stress and fracture toughness are 252 ksi and 59 ksi-in$^{1/2}$, respectively. Compare the values of σ that result in failure in the three cases.

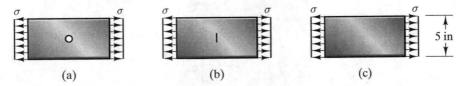

Problems 12.53–12.54

12.54 Compare the failure stresses of the plates if they are made of 2024-T351 aluminum, which has a yield stress 47 ksi and fracture toughness 31 ksi-in$^{1/2}$.

(You should find that the sheet with the hole has the lowest failure load in this case. This is because the aluminum is a relatively tough or crack growth-resistant material. To understand why, compare the plastic zone sizes in the steel and the aluminum using Eq. (12.36). The plastic zone in the aluminum is almost eight times as large as in the steel, which helps resist crack growth.)

12.55 The edge-cracked plate has a thickness of 1.5 inches and a fracture toughness of 90 ksi-in$^{1/2}$. (See Example 12.8.) What crack length a would result in fast crack growth if loads $P = 175$ kip were applied?

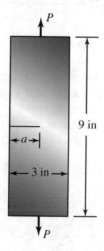

Problem 12.55

12.56 The edge-cracked beam is to be subjected to a bending moment $M = 0.21$ MN-m. The material that has been chosen for the design has a fracture toughness of 35 MPa-m$^{1/2}$. If the nondestructive technique that is used to detect cracks has a resolution of 5 mm, what minimum height h should the beam have?

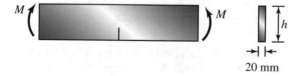

Problem 12.56

12.57 The beam has a rectangular cross section with 2-in width and consists of material with a fracture toughness of 26 ksi-in$^{1/2}$. It is subjected to a 100 ft-kip counterclockwise couple at the left end and an 8.33-kip downward load at the right end. Due to corrosion, a crack that is midway between the two supports is growing very slowly from the top surface of the beam. If the load and moment remain the same, what will be the length of the crack when it becomes a fast crack?

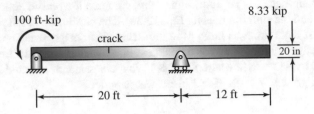

Problem 12.57

12.58 The 1-in thick plate is loaded by a weight $W = 125$ kip. The weight is supported by two links and a pin that passes through a 0.25-in diameter hole in the plate. The fracture toughness of the material is 40 ksi-in$^{1/2}$. The plate has a symmetrical crack that emanates from the hole. If the crack is 3 in long when the weight is suspended, will the plate fail?

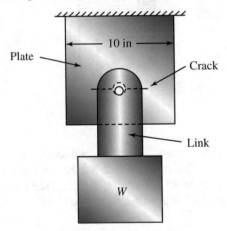

Problem 12.58

12.59 The double lap joint carries a load $P = 16$ kN. The bolt hole has a diameter of 20 mm. The center member is 50 mm in width and 2 mm in thickness. The outer straps are 40 mm wide. (a) If a symmetrical crack forms at the bolt hole in the center member, whose fracture toughness is 33 MPa-m$^{1/2}$, will the crack growth arrests before failure occurs? (b) If the answer to part (a) is yes, will the arrested crack be visible?

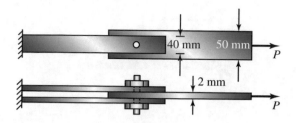

Problem 12.59

12.60 The titanium strut has radius $b = 10\,\text{mm}$ and a circumferential crack with depth a and is subjected to tensile loads $P = 200\,\text{kN}$. The fracture toughness of the titanium is $66\,\text{MPa-m}^{1/2}$, and the stress intensity factor for this configuration is given by

$$K = \frac{1.12P}{\pi b^2}\sqrt{\pi a}.$$

Nondestructive inspection techniques are used that can resolve a crack with depth $a = 1\,\text{mm}$. Will inspection reveal any cracks before fast crack growth occurs? Explain.

Problem 12.60

12.61 In the fracture toughness test shown, a beam with a crack of length $a = 5\,\text{in}$ (called a *double cantilever beam* specimen) is subjected to the load P. The height $2h = 1\,\text{in}$ and the thickness (the dimension into the page) is $b = 0.5\,\text{in}$. The stress intensity factor for this configuration is

$$K = \frac{2\sqrt{3}\,Pa}{bh^{3/2}}.$$

If fast crack growth initiates when $P = 460\,\text{lb}$, what is the material's fracture toughness?

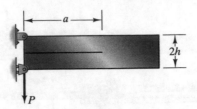

Problem 12.61

12.62 A strip of the material tested in Problem 12.61 is bonded to rigid bars at its top and bottom edges and has an edge crack of length a. The length $L = 5$ in, height $2h = 1$ in, and thickness $b = 0.2$ in. A weight $W = 100,000$ lb is suspended from the bottom bar. The stress intensity factor of this configuration is

$$K = \frac{W}{bL}\left(1 - e^{-5a/h}\right)\sqrt{h(1 - \nu^2)},$$

where $\nu = 0.3$ is the Poisson's ratio of the aluminum. Determine the critical crack length for fast crack growth.

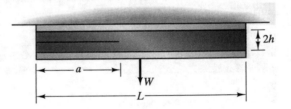

Problem 12.62

12.63 The wedge-shaped blade of an axe is being driven into a piece of wood of thickness $2h = 1$ in. The blade imparts a displacement $\Delta = 0.25$ in when the crack length $a = 6$ in. The stress intensity factor for this situation is

$$K = \frac{\sqrt{3}\,E\,\Delta\,h^{3/2}}{4a^2},$$

where $E = 20E6$ psi is the elastic modulus of the wood. Determine how much more the crack will grow under this displacement if the fracture toughness of the wood is 8 ksi-in$^{1/2}$. (Notice that the stress intensity decreases with increasing crack length, which explains why further insertions of the blade would be required to split the wood completely.)

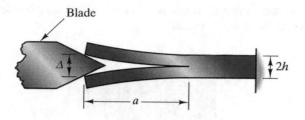

Problem 12.63

12.64 The wall of a cylindrical pressure vessel contains an elliptical surface flaw with width $2c$ which may grow through the wall to produce a leak. Alternatively, it may become a fast crack before reaching the other surface of the wall, in which case it will grow axially in an explosive manner and destroy the vessel with other severe consequences. Thus it is preferable that the vessel leaks before it breaks. To design for this, we check whether the critical half-crack length c_0 under the action of the hoop stress (based on Eq. (12.37) for the stress intensity factor of a centrally cracked plate),

$$ c_0 = \frac{1}{\pi} \left(\frac{K_c}{\sigma_h} \right)^2, $$

is greater than the wall thickness. The material being considered for a vessel with 2-m inner diameter and 15-mm wall thickness has a fracture toughness of $200\,\text{MPa-m}^{1/2}$. Determine whether the vessel satisfies this "leak before break" design philosophy if the internal pressure will be 5 MPa greater than the external pressure.

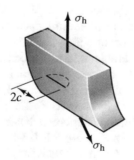

Problem 12.64

12.65 The large sheet under constant-amplitude cyclic loading contains a central crack with length $2a = 2$ in. The change in applied stress over each cycle is 20 ksi. The material follows the Paris law for fatigue crack growth rate with constants $A = 1E - 20$ and $n = 3$, where crack growth rate is in inches per cycle and stress intensity factor is in psi-in$^{1/2}$. How many cycles are required for the crack to double in size?

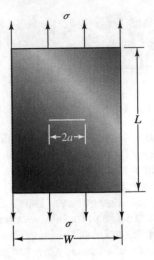

Problem 12.65

12.66 In Problem 12.65, suppose that the sheet contains an edge crack with length $a = 1$ in instead of a central crack. (See Example 12.8. Assume that $a/W \leq 0.13$, $L/W \geq 2$.) How many cycles are required for the crack to double in size?

12.67 The titanium strut described in Problem 12.60 is subjected to tensile loads P that cycle between 0 and 200 kN. The material follows the Paris law with constants $A = 1E-13$ and $n = 3.5$, where crack growth rate is in meters per cycle and stress intensity factor is in MPa-m$^{1/2}$. If a 1-mm deep circumferential crack is detected by nondestructive evaluation techniques, determine how many more cycles occur before the strut fails.

12.68 The strip described in Problem 12.61 is loaded cyclically by suspending a weight W from the roller support and removing it. The material follows the Paris law with constants $A = 1E-18$ and $n = 3$, where crack growth rate is in inches per cycle and stress intensity factor is in psi-in$^{1/2}$. If microscopic measurements of the fracture surface indicate that the crack growth rate is 8E-6 inches per cycle when the crack is 0.5 in long, what is the weight W?

12.69 The sheet of width $W = 0.5$ m and thickness $b = 15$ mm contains a quarter-circular crack of radius a. The sheet is loaded in tension, and the stress intensity factor is

$$K = \frac{0.722P\sqrt{\pi a}}{bW}$$

as long as $a/b < 0.25$ and $a/W < 0.35$. The load P cycles between 0 and 3 MN. The fracture toughness of the material is $30\,\text{MPa-m}^{1/2}$, and its Paris law constants are $A = 3\text{E-11}$ and $n = 3.7$, where crack growth rate is in meters per cycle and stress intensity factor is in $\text{MPa-m}^{1/2}$. Determine how many cycles it will take the sheet to fail if the initial crack radius is 0.5 mm and the crack maintains its quarter-circular shape until failure.

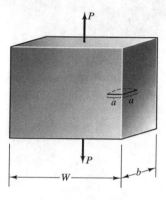

Problem 12.69

12.70 The figure shows cracks at opposite sides of a 0.5-in diameter hole in a plate. A bolt inserted into the hole is larger than the hole and exerts the loads P. The stress intensity factor for this configuration is $K = \frac{P}{b\sqrt{\pi a}}$, where $b = 0.25$ in is the thickness of the sheet.

(a) If the fracture toughness of the material is $35\,\text{ksi-in}^{1/2}$ and the bolt imparts loads $P = 6000\,\text{lb}$, will fast crack growth occur?

(b) An external load applied to the plate causes the loads on the face of the hole to drop to $P = 3000\,\text{lb}$. The external load is applied and released in a repeated manner, and when it released the loads, $P = 6000\,\text{lb}$ irrespective of crack length. If the bolt head obscures a 2-in diameter portion of the sheet, determine whether a fatigue crack will arrest before it becomes visible, given a threshold stress intensity factor of $12\,\text{ksi-in}^{1/2}$.

(c) Suppose that the diametrically opposed cracks are initially 0.005 in long. Determine the number of cycles at which they would stop growing if the material follows the Paris law with constants $A = 0.5\text{E-21}$ and $n = 4$. Crack growth rates are in inches per cycle and stress intensity factor is in $\text{psi-in}^{1/2}$.

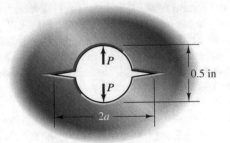

Problem 12.70

Chapter Summary

Stress Concentrations

Structural members may have geometric features (including holes, notches, and corners) that result in localized regions of high stress called *stress concentrations*. The ratio C of the maximum stress resulting from a stress concentration to a defined nominal stress is called the *stress concentration factor*. Stress concentration factors are given for an axially loaded bar in Figs. 12.3, 12.4, and 12.5, for a stepped circular bar subjected to torsion in Fig. 12.7, and for a rectangular beam subjected to bending in Figs. 12.10 and 12.11.

Failure

Failure in brittle materials occurs when a component breaks or fractures, whereas in ductile materials failure is usually defined to occur with the onset of yielding.

The *maximum normal stress criterion* states that fracture of a brittle material occurs when

$$\text{Max}(|\sigma_1| \ |\sigma_2| \ |\sigma_3|) = \sigma_U, \tag{12.5}$$

where $\sigma_1, \sigma_2, \sigma_3$ are the principal stresses and σ_U is the ultimate stress. In plane stress, the values of σ_1 and σ_2 for which fracture will not occur are bounded by the square region shown in Fig. 12.13. It is bounded by the lines

$$\sigma_1 = \pm\sigma_U, \quad \sigma_2 = \pm\sigma_U. \tag{12.6}$$

When the ultimate stress of a brittle material subjected to plane stress differs in tension and compression, *Mohr's failure criterion* states that fracture occurs when σ_1 and σ_2 lie on the boundaries shown in Fig. 12.15(a) and described by Eqs. (12.8)–(12.11).

The *Tresca criterion* states that yielding of a ductile material occurs when the absolute maximum shear stress equals the yield stress in shear τ_Y. It can also be expressed in terms of the tensile yield stress σ_Y:

$$\text{Max}(|\sigma_1 - \sigma_2| \; |\sigma_2 - \sigma_3| \; |\sigma_1 - \sigma_3|) = \sigma_Y. \tag{12.15}$$

The Tresca factor of safety is

$$\text{FS}_T = \frac{\sigma_Y}{\sigma_T}, \tag{12.19}$$

where

$$\sigma_T = \text{Max}(|\sigma_1 - \sigma_2|, |\sigma_2 - \sigma_3|, |\sigma_1 - \sigma_3|). \tag{12.18}$$

In plane stress, the safe region according to the Tresca criterion is bounded by the hexagon in Fig. (a).

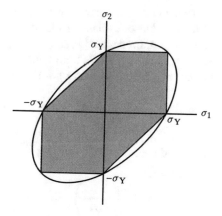

The *von Mises criterion* states that yielding of a ductile material occurs when

$$\frac{1}{2}\left[(\sigma_1 - \sigma_2)^2 + (\sigma_2 - \sigma_3)^2 + (\sigma_1 - \sigma_3)^2\right] = \sigma_Y^2. \tag{12.20}$$

The von Mises factor of safety is

$$FS_M = \frac{\sigma_Y}{\sigma_M},$$ (12.24)

where σ_M is the *von Mises equivalent stress*

$$\sigma_M = \frac{1}{\sqrt{2}} \sqrt{(\sigma_1 - \sigma_2)^2 + (\sigma_2 - \sigma_3)^2 + (\sigma_1 - \sigma_3)^2}.$$ (12.22)

In plane stress, the safe region according to the von Mises criterion is bounded by the ellipse in Fig. (a).

Fatigue is subdivided into low-cycle, high-cycle, and fatigue crack growth. In low-cycle fatigue, stress levels exceed the yield stress, and the number of cycles to failure is relatively low ($<10^3$). High-cycle fatigue can occur when stress levels are lower than the yield stress, and failure may require 10^3 to 10^6 cycles.

Denoting the maximum and minimum values of the stress in constant frequency and amplitude loading by σ_{max} and σ_{min}, the *stress amplitude* σ_a and *mean stress* σ_m are

$$\sigma_a = \frac{\sigma_{max} - \sigma_{min}}{2}, \quad \sigma_m = \frac{\sigma_{max} + \sigma_{min}}{2}.$$ (12.25)

An *S-N curve*, or *endurance curve*, is a graph of the stress amplitude or *fatigue strength* as a function of the number of cycles N to failure at zero mean stress. For some materials, there is a stress amplitude, the *fatigue limit* σ_{fat}, below which fatigue life is essentially infinite. Keeping stress levels below the fatigue limit is known as *safe life design*. An empirical expression used to fit *S-N* curves is

$$\sigma_a = \sigma_{fat} + \frac{b}{N^c}.$$ (12.26)

Solving this equation for N yields

$$N = \left(\frac{b}{\sigma_a - \sigma_{fat}} \right)^{1/c}.$$ (12.27)

Empirical equations that can be used to determine the effect of a constant level of mean stress on the *S-N* curve include the *Goodman relation* (12.28), the *Gerber parabola* (12.29), and the *Soderberg line* (12.30).

Suppose that a component is subjected to a sequence of blocks of loading. Let n_i be the number of cycles applied in the ith block, and let N_i be the number of cycles required to fail the undamaged component at that stress amplitude and mean stress. *Miner's law* states that the accumulated damage after k blocks of loading is

$$D = \sum_{i=1}^{k} D_i = \sum_{i=1}^{k} \frac{n_i}{N_i}, \qquad (12.31)$$

and that failure occurs when $D = 1$.

Fracture

The *stress intensity factor K* is a measure of the magnitude of the stress in the neighborhood of a crack tip (see Eq. 12.35). Stress intensity factors for a centrally cracked plate subjected to tension and a plate subjected to a central load on a crack face are given by Eqs. (12.37) and (12.38), respectively. In many cases stress intensity factors can be expressed in the form

$$K = \sigma\sqrt{\pi a}\, Q \left(\frac{a}{W}\right), \qquad (12.39)$$

where σ represents the stress, a is the crack length, and $Q(a/W)$ is a function called the *configuration factor*.

The value of the stress intensity factor at which *fast crack growth* begins is called the *fracture toughness* K_c. From Eq. (12.39), the criterion for fast crack growth is

$$\sigma\sqrt{\pi a}\, Q \left(\frac{a}{W}\right) = K_c. \qquad (12.40)$$

Let ΔK be the change in the stress intensity factor over one cycle in a cracked component subjected to cyclic loading. The derivative of the crack length a with respect to the number N of loading cycles is given by the *Paris law*

$$\frac{da}{dN} = A(\Delta K)^n. \qquad (12.41)$$

The constants A and n are material properties reflecting resistance to fatigue crack growth. Rearranging Eq. (12.41) and integrating give a relation between the number of cycles N and the crack length a:

$$N = \frac{1}{A} \int_{a_0}^{a} \frac{da}{(\Delta K)^n}. \qquad (12.42)$$

The number of cycles at which failure occurs is obtained by setting a equal to the critical crack length for fast crack growth. The critical crack length is determined from Eq. (12.40) with $\sigma = \sigma_{max}$, the maximum magnitude of the constant amplitude loading.

Appendix A: Results from Mathematics

Algebra

Quadratic Equations

The solutions of the quadratic equation

$$ax^2 + bx + c = 0$$

are

$$x = \frac{-b \pm \sqrt{b^2 - 4ac}}{2a}.$$

Natural Logarithms

The natural logarithm of a positive real number x is denoted by $\ln x$. It is defined to be the number such that

$$e^{\ln x} = x,$$

where $e = 2.7182\ldots$ is the base of natural logarithms.

© Springer Nature Switzerland AG 2020
A. Bedford, K. M. Liechti, *Mechanics of Materials*,
https://doi.org/10.1007/978-3-030-22082-2

Logarithms have the following properties:

$$\ln(xy) = \ln x + \ln y,$$

$$\ln\left(\frac{x}{y}\right) = \ln x - \ln y,$$

$$\ln y^x = x \ln y.$$

Trigonometry

The trigonometric functions for a right triangle (Fig. A.1) are

$$\sin \alpha = \frac{1}{\csc \alpha} = \frac{a}{c}, \quad \cos \alpha = \frac{1}{\sec \alpha} = \frac{b}{c}, \quad \tan \alpha = \frac{1}{\cot \alpha} = \frac{a}{b}.$$

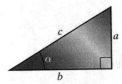

Fig. A.1

The sine and cosine satisfy the relation

$$\sin^2\alpha + \cos^2\alpha = 1,$$

and the sine and cosine of the sum and difference of two angles satisfy

$$\sin(\alpha + \beta) = \sin \alpha \cos \beta + \cos \alpha \sin \beta,$$
$$\sin(\alpha - \beta) = \sin \alpha \cos \beta - \cos \alpha \sin \beta,$$
$$\cos(\alpha + \beta) = \cos \alpha \cos \beta - \sin \alpha \sin \beta,$$
$$\cos(\alpha - \beta) = \cos \alpha \cos \beta + \sin \alpha \sin \beta.$$

The *law of cosines* for an arbitrary triangle (Fig. A.2) is

$$c^2 = a^2 + b^2 - 2ab \cos \alpha_c,$$

and the *law of sines* is

$$\frac{\sin \alpha_a}{a} = \frac{\sin \alpha_b}{b} = \frac{\sin \alpha_c}{c}.$$

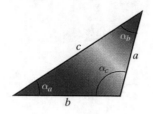

Fig. A.2

Derivatives

$$\frac{d}{dx} x^n - n x^{n-1}$$

$$\frac{d}{dx} e^x = e^x$$

$$\frac{d}{dx} \ln x = \frac{1}{x}$$

$$\frac{d}{dx} \sin x = \cos x$$

$$\frac{d}{dx} \cos x = -\sin x$$

$$\frac{d}{dx} \tan x = \frac{1}{\cos^2 x}$$

$$\frac{d}{dx} \sinh x = \cosh x$$

$$\frac{d}{dx} \cosh x = \sinh x$$

$$\frac{d}{dx} \tanh x = \frac{1}{\cosh^2 x}$$

Integrals

$$\int x^n \, dx = \frac{x^{n+1}}{n+1} \quad (n \neq -1)$$

$$\int \frac{dx}{x} = \ln x$$

$$\int \frac{dx}{a + bx^2} = \frac{1}{(ab)^{1/2}} \arctan \frac{(ab)^{1/2}x}{a}$$

$$\int \frac{dx}{a - bx^2} = \frac{1}{2(ab)^{1/2}} \ln \frac{a + (ab)^{1/2}x}{a - (ab)^{1/2}x}$$

$$\int \frac{x \, dx}{a - bx^2} = -\frac{1}{2b} \ln \left(a - bx^2\right)$$

$$\int (a + bx)^{1/2} \, dx = \frac{2}{3b} (a + bx)^{3/2}$$

$$\int x(a + bx)^{1/2} \, dx = -\frac{2(2a - 3bx)(a + bx)^{3/2}}{15b^2}$$

$$\int (1 + a^2x^2)^{1/2} \, dx = \frac{1}{2} \left\{ x(1 + a^2x^2)^{1/2} + \frac{1}{a} \ln \left[x + \left(\frac{1}{a^2} + x^2 \right)^{1/2} \right] \right\}$$

$$\int x(1 + a^2x^2)^{1/2} \, dx = \frac{a}{3} \left(\frac{1}{a^2} + x^2 \right)^{3/2}$$

$$\int x^2(1 + a^2x^2)^{1/2} \, dx = \frac{1}{4} ax \left(\frac{1}{a^2} + x^2 \right)^{3/2} - \frac{1}{8a^2} x(1 + a^2x^2)^{1/2}$$

$$- \frac{1}{8a^3} \ln \left[x + \left(\frac{1}{a^2} + x^2 \right)^{1/2} \right]$$

$$\int (1 - a^2x^2)^{1/2} \, dx = \frac{1}{2} \left[x(1 - a^2x^2)^{1/2} + \frac{1}{a} \arcsin(ax) \right]$$

$$\int (a^2 - x^2)^{1/2} \, dx = \frac{1}{2} \left[x(a^2 - x^2)^{1/2} + a^2 \arcsin \left(\frac{x}{a} \right) \right]$$

$$\int x(1 - a^2x^2)^{1/2} \, dx = -\frac{a}{3} \left(\frac{1}{a^2} - x^2 \right)^{3/2}$$

$$\int x(a^2 - x^2)^{1/2} \, dx = -\frac{1}{3} (a^2 - x^2)^{3/2}$$

$$\int x^2 (a^2 - x^2)^{1/2} \, dx = -\frac{1}{4} x (a^2 - x^2)^{3/2}$$
$$+ \frac{1}{8} a^2 \left[x (a^2 - x^2)^{1/2} + a^2 \arcsin \frac{x}{a} \right]$$

$$\int \frac{dx}{(1 + a^2 x^2)^{1/2}} = \frac{1}{a} \ln \left[x + \left(\frac{1}{a^2} + x^2 \right)^{1/2} \right]$$

$$\int \frac{dx}{(1 - a^2 x^2)^{1/2}} = \frac{1}{a} \arcsin ax \quad \text{or} \quad -\frac{1}{a} \arccos ax$$

$$\int \sin x \, dx = -\cos x$$

$$\int \cos x \, dx = \sin x$$

$$\int \sin^2 x \, dx = -\frac{1}{2} \sin x \, \cos x + \frac{1}{2} x$$

$$\int \cos^2 x \, dx = \frac{1}{2} \sin x \, \cos x + \frac{1}{2} x$$

$$\int \sin^3 x \, dx = -\frac{1}{3} \cos x \left(\sin^2 x + 2 \right)$$

$$\int \cos^3 x \, dx = \frac{1}{3} \sin x \left(\cos^2 x + 2 \right)$$

$$\int \cos^4 x \, dx = \frac{3}{8} x + \frac{1}{4} \sin 2x + \frac{1}{32} \sin 4x$$

$$\int \sin^n x \, \cos x \, dx = \frac{(\sin x)^{n+1}}{n + 1} \quad (n \neq -1)$$

$$\int \sinh x \, dx = \cosh x$$

$$\int \cosh x \, dx = \sinh x$$

$$\int \tanh x \, dx = \ln \cosh x$$

$$\int e^{ax} dx = \frac{e^{ax}}{a}$$

$$\int x e^{ax} dx = \frac{e^{ax}}{a^2} (ax - 1)$$

Taylor Series

The Taylor series of a function $f(x)$ is

$$f(a+x) = f(a) + f'(a)x + \frac{1}{2!}f''(a)x^2 + \frac{1}{3!}f'''(a)x^3 + \cdots,$$

where the primes indicate derivatives.

Some useful Taylor series are

$$e^x = 1 + x + \frac{x^2}{2!} + \frac{x^3}{3!} + \cdots,$$

$$\sin(a+x) = \sin a + (\cos a)x - \frac{1}{2}(\sin a)x^2 - \frac{1}{6}(\cos a)x^3 + \cdots,$$

$$\cos(a+x) = \cos a - (\sin a)x - \frac{1}{2}(\cos a)x^2 + \frac{1}{6}(\sin a)x^3 + \cdots,$$

$$\tan(a+x) = \tan a + \left(\frac{1}{\cos^2 a}\right)x + \left(\frac{\sin a}{\cos^3 a}\right)x^2$$

$$+ \left(\frac{\sin^2 a}{\cos^4 a} + \frac{1}{3\cos^2 a}\right)x^3 + \cdots.$$

Appendix B: Material Properties

This appendix summarizes the properties of selected materials. These values may be used as approximations for preliminary design. However, in final design calculations, you should try to use values obtained from the actual materials specified in your design. Material properties can sometimes be obtained from the sources supplying the materials, or it may be necessary to have measurements made from samples of the specified materials.

© Springer Nature Switzerland AG 2020
A. Bedford, K. M. Liechti, *Mechanics of Materials*,
https://doi.org/10.1007/978-3-030-22082-2

Table B.1 Elastic moduli of selected materials

Material	Modulus of elasticity E		Shear modulus G		Poisson's ratio ν
	10^6 psi	GPa	10^6 psi	GPa	
Aluminum	10	70	3.8	26	0.33
Aluminum alloys	10–12	70–80	3.8–4.4	26–30	0.33
2014-T6	10.6	73	4	28	0.33
6061-T6	10	70	3.8	26	0.33
7075-T6	10.4	72	3.9	27	0.33
Brick (compression)	1.5–3.5	10–24			
Cast iron	12–25	80–170	4.5–10	31–69	0.2–0.3
Gray cast iron	14	97	5.6	39	0.25
Concrete (compression)	2.6–4.4	18–30			0.1–0.2
Copper	17	115	6.2	43	0.35
Copper alloys	14–18	96–120	5.2–6.8	36–47	0.33–0.35
Brass	14–16	96–110	5.2–6	36–41	0.34
80% Cu, 20% Zn	15	100	5.5	38	0.33
Naval brass	15	100	5.5	38	0.33
Bronze	14–17	96–120	5.2–6.3	36–44	0.34
Manganese bronze	15	100	5.6	39	0.35
Glass	7–12	50–80	2.9–5	b20–33	0.20–0.27
Magnesium	5.8	40	2.2	15	0.34
Nickel	30	210	11.4	80	0.31
Nylon	0.3–0.4	2–3		0.4	
Rubber	0.0001–0.0006	0.001–0.004	0.00004–0.0002	0.0003–0.0014	0.44–0.50
Steel	28–32	190–220	10.8–12.3	75–85	0.28–0.30
Stone (compression)					
Granite	6–10	40–70			0.2–0.3
Marble	7–14	50–100			0.2–0.3
Titanium	16	110	6.0	41	0.33
Titanium alloys	15–18	100–124	5.6–6.8	39–47	0.33
Tungsten	52	360	22	150	0.2
Wood (bending)					
Ash	1.5–1.6	10–11			
Oak	1.6–1.8	11–12			
Southern pine	1.6–2	11–14			
Wrought iron	28	190	10.9	75	0.3

Table B.2 Yield and ultimate stresses of selected materials

Material	Yield stress σ_Y		Ultimate stress σ_U		Percent elongation (2-in. gauge length)
	10^3 psi	MPa	10^3 psi	GPa	
Aluminum	3	20	10	70	60
Aluminum alloys	6–75	40–520	15–80	100–560	2–45
2014-T6	60	410	70	480	13
6061-T6	40	270	45	310	17
7075-T6	70	480	80	550	11
Brick (compression)			1–10	7–70	
Cast iron (compression)			50–200	340–1400	
Cast iron (tension)	17–41	120–280	10–70	70–480	1
Gray cast iron	17	120	20–58	140–400	1
Concrete (compression)			1.5–10	10–70	
Copper					
Hard-drawn	49	340	55	380	10
Soft-annealed	8	55	33	230	50
Copper alloys					
Beryllium copper, hard	109	750	120	830	4
Brass	12–80	80–540	36–90	240–600	4–60
80% Cu, 20% Zn, hard	67	450	84	580	4
80% Cu, 20% Zn, soft	13	90	43	300	50
Naval brass, hard	58	400	84	580	15
Naval brass, soft	25	170	59	410	50
Bronze	12–100	82–700	30–120	200–830	5–60
Manganese bronze, hard	65	450	90	620	10
Manganese bronze, soft	25	170	65	450	35
Glass (plate)			9	65	
Magnesium	3–10	20–68	15–25	100–170	5–15

(continued)

Table B.2 (continued)

Material	Yield stress σ_Y		Ultimate stress σ_U			Percent elongation (2-in. gauge length)
	10^3 psi	MPa	10^3 psi	GPa		
Nickel	20–90	140–620	45–110	310–760		2–50
Nylon			6–10	40–72		50
Rubber	0.3–1	2–7	1–3	7–20		100–800
Steel, structural	30–104	200–720	50–118	340–820		10–40
ASTM-A36	36	250	60	400		30
ASTM-A572	50	340	70	500		20
ASTM-A514	100	700	120	830		15
Stone (compression)						
Granite			10–40	70–280		
Marble			8–25	50–180		25
Titanium	60	400	70	500		
Titanium alloys	110–125	760–860	130–140	900–960		10
Tungsten			200–600	1400–4000		0–4
Wood (bending)						
Ash	6–10	40–70	8–13	50–90		
Oak	6–8	40–55	8–13	50–90		
Southern pine	6–8	40–55	8–13	50–90		
Wood (compression parallel to grain)						
Ash	4–6	30–40	5–8	30–50		
Oak	4–6	30–40	5–8	30–50		
Southern pine	4–8	30–50	6–10	40–70		
Wrought iron	30	210	48	330		30

Table B.3 Densities and coefficients of thermal expansion of selected materials

Material	Density ρ		Coefficient of thermal expansion α	
	slug/ft^3	kg/m^3	$10^{-6}\,{}^\circ\mathrm{F}^{-1}$	$10^{-6}\,{}^\circ\mathrm{C}^{-1}$
Aluminum	5.2	2700	13.3	23.9
Aluminum alloys	4.9–5.4	2500–2800	13–13.4	23–24
Copper	17.4	9000	9.2	16.6
Copper alloys				
Brass	16.3–16.9	8400–8700	10.6–11.8	19–21
Bronze	14.3–17.2	7400–8900	9.9–11.6	18–21
Cast iron	13.9	7200	5.6–6.7	10–12
Gray cast iron	13.6–13.8	7000–7100	5.6	10
Concrete	2.9–4.7	1500–2400	4–8	7–14
Glass	4.6–5.8	2400–3000	3–6	5–11
Magnesium	3.4	1740	14	25
Magnesium alloys	3.4–3.5	1760–1830	14–16	26–29
Nickel	16.7	8600	7.2	13
Rubber	1.7–3.9	900–2000	70–110	130–200
Steel	14.9–15.2	7700–7830	6–10	10–18
Stone	3.9–5.1	2000–2900	3–5	5–9
Titanium	8.8	4540	4.7	8.5
Tungsten	37.4	19,300	2.4	4.3
Wrought iron	14–15	7400–7800	7	12

Appendix C: Centroids and Moments of Inertia

Centroids and moments of inertia of areas arise repeatedly in analyses of problems in solid mechanics. These quantities are defined and discussed in this appendix. Centroids and moments of inertia of specific areas are tabulated in Appendix D.

Centroids of Areas

Consider an area A in the x-y plane [Fig. C.3(a)]. The coordinates of the centroid of A are defined by

$$\bar{x} = \frac{\int_A x\,dA}{\int_A dA},$$

$$\bar{y} = \frac{\int_A y\,dA}{\int_A dA}, \qquad\qquad (\text{C.1})$$

where x and y are the coordinates of the differential element of area dA [Fig. C.1 (b)]. The subscript A on the integral signs means that the integration is carried out over the entire area.

If an area is symmetric about an axis, its centroid lies on the axis [Fig. C.2(a)], and if an area is symmetric about two axes, its centroid lies at their intersection [Fig. C.2(b)].

© Springer Nature Switzerland AG 2020
A. Bedford, K. M. Liechti, *Mechanics of Materials*,
https://doi.org/10.1007/978-3-030-22082-2

Fig. C.1 (**a**) Centroid of an
area *A*. (**b**) Coordinates of *dA*

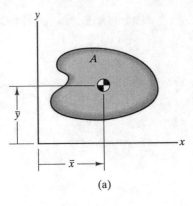

(a)

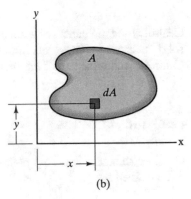

(b)

Fig. C.2 (**a**) Area that is
symmetric about an axis.
(**b**) Area with two axes of
symmetry

(a)

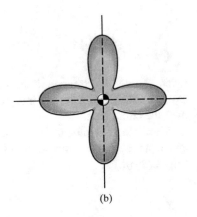

(b)

Example C.1 Centroid of an Area by Integration

Determine the coordinates of the centroid of the triangular area in Fig. C.3.

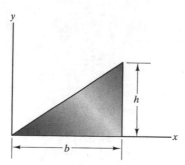

Fig. C.3

Solution

Use of a Strip Element

The x coordinate of the centroid can be obtained by using an element of area dA in the form of a vertical strip of width dx [Fig. (a)]. The area $dA = (h/b)\, x\, dx$, so

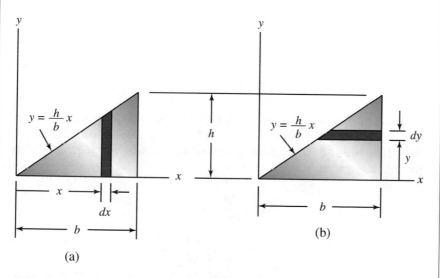

(a) Vertical strip element dA for determining $\bar{x}$. (b) Horizontal strip element dA for determining $\bar{y}$

$$\bar{x} = \frac{\int_A x\, dA}{\int_A dA}$$

$$= \frac{\int_0^b \frac{h}{b} x^2\, dx}{\int_0^b \frac{h}{b} x\, dx}$$

$$= \frac{\frac{h}{b}\left[\frac{x^3}{3}\right]_0^b}{\frac{h}{b}\left[\frac{x^2}{2}\right]_0^b}$$

$$= \frac{2}{3} b.$$

In the same way, the y coordinate of the centroid can be obtained by using an element of area dA in the form of a horizontal strip of height dy [Fig. (b)]. The area $dA = [b - (b/h)y]\, dy$, giving the result

$$\bar{y} = \frac{\int_A y\, dA}{\int_A dA}$$

$$= \frac{\int_0^h \left[by - \frac{b}{h} y^2\right] dy}{\int_0^h \left[b - \frac{b}{h} y\right] dy}$$

$$= \frac{1}{3} h.$$

An alternative way to determine $\bar{y}$ that is often useful is to use the vertical strip in Fig. (a), so that $dA = (h/b)x\, dx$, and set the y coordinate of dA equal to the y coordinate of the midpoint (centroid) of the strip, $y = [(h/b)x]/2$. This procedure gives

$$\bar{y} = \frac{\int_A y\, dA}{\int_A dA}$$

$$= \frac{\int_0^b \left(\frac{h}{2b} x\right)\left(\frac{h}{b} x\, dx\right)}{\int_0^b \frac{h}{b} x\, dx}$$

$$= \frac{1}{3} h.$$

Use of Double Integration

Using the element of area $dA = dx\,dy$ in Fig. (c), the x coordinate of the centroid is

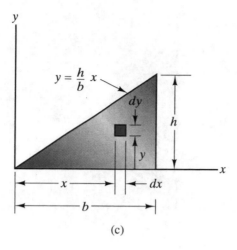

(c)

(c) Element dA for determining $\bar{x}$ and $\bar{y}$

$$
\begin{aligned}
\bar{x} &= \frac{\int_A x\,dA}{\int_A dA} \\[2mm]
&= \frac{\int_{x=0}^{b} \int_{y=0}^{(h/b)x} x\,dx\,dy}{\int_{x=0}^{b} \int_{y=0}^{(h/b)x} dx\,dy} \\[2mm]
&= \frac{\int_0^b x\left[\int_0^{(h/b)x} dy\right] dx}{\int_0^b \left[\int_0^{(h/b)x} dy\right] dx} \\[2mm]
&= \frac{\int_0^b \left(\frac{h}{b}\right) x^2\,dx}{\int_0^b \left(\frac{h}{b}\right) x\,dx} \\[2mm]
&= \frac{2}{3}b.
\end{aligned}
$$

The y coordinate of the centroid is

$$
\bar{y} = \frac{\int_A y\,dA}{\int_A dA} = \frac{\int_{x=0}^{b} \int_{y=0}^{(h/b)x} y\,dx\,dy}{\int_{x=0}^{b} \int_{y=0}^{(h/b)x} dx\,dy}
$$

$$
= \frac{1}{3}h,
$$

where the integrals have been evaluated in the same way as in the determination of $\bar{x}$.

Example C.2 Centroid of an Area by Integration
Determine the coordinates of the centroid of the semicircular area in Fig. C.4.

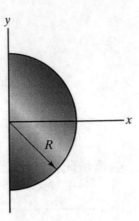

Fig. C.4

Solution
Using the element of area $dA = r\,dr\,d\theta$ in Fig. (a), the x coordinate of the centroid is

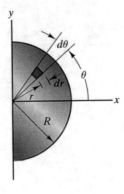

(a) Element dA for determining $\bar{x}$

$$\bar{x} = \frac{\int_A x \, dA}{\int_A dA} = \frac{\int_{r=0}^{R} \int_{\theta=-\pi/2}^{\pi/2} (r \cos \theta)(r \, dr \, d\theta)}{\int_{r=0}^{R} \int_{\theta=-\pi/2}^{\pi/2} r \, dr \, d\theta}$$

$$= \frac{\int_0^R r^2 \left[\int_{-\pi/2}^{\pi/2} \cos \theta \, d\theta \right] dr}{\int_0^R r \left[\int_{-\pi/2}^{\pi/2} d\theta \right] dr}$$

$$= \frac{\int_0^R 2r^2 \, dr}{\int_0^R \pi r \, dr}$$

$$= \frac{4R}{3\pi}.$$

Because the area is symmetric about the x axis, $\bar{y} = 0$.

Problems

For Problems C.1–C.6, use integration to determine the coordinates of the centroids of the areas shown.

C.1

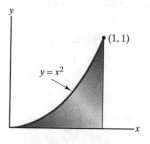

Problem C.1

C.2

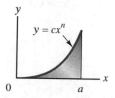

Problem C.2

C.3

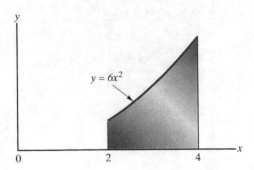

$y = 6x^2$

Problem C.3

C.4

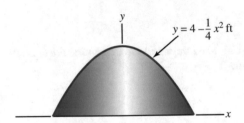

$y = 4 - \dfrac{1}{4} x^2 \text{ ft}$

Problem C.4

C.5

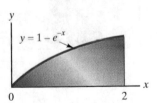

$y = 1 - e^{-x}$

Problem C.5

C.6

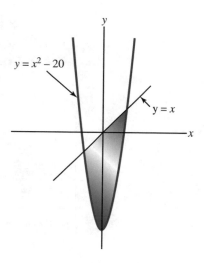

$y = x^2 - 20$

$y = x$

Problem C.6

C.7 Determine the coordinates of the centroid of the area by integrating as shown in Example C.2.

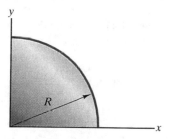

R

Problem C.7

C.8 Determine the coordinates of the centroid of the area shown in Problem C.7 by using an element of area in the form of a vertical "strip."

Composite Areas

An area consisting of a combination of simple parts is called a *composite area*. Figure C.5(a) shows a composite area consisting of a triangle, a rectangle, and a semicircle, labeled 1, 2, and 3, respectively. The x coordinate of the centroid of the composite area is

Fig. C.5 **(a)** Composite area composed of three parts. **(b)** Centroids of the parts

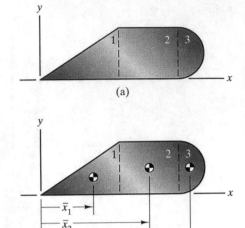

(a)

(b)

$$\bar{x} = \frac{\int_A x\ dA}{\int_A dA} = \frac{\int_{A_1} x\ dA + \int_{A_2} x\ dA + \int_{A_3} x\ dA}{\int_{A_1} dA + \int_{A_2} dA + \int_{A_3} dA}. \tag{C.2}$$

The x coordinates of the centroids of the parts are shown in Fig. C.5(b). From the definition of the x coordinate of the centroid of part 1,

$$\bar{x}_1 = \frac{\int_{A_1} x\,dA}{\int_{A_1} dA},$$

it is seen that

$$\int_{A_1} x\,dA = \bar{x}_1 A_1.$$

Using this equation and equivalent expressions for parts 2 and 3, Eq. (C.2) can be written as

$$\bar{x} = \frac{\bar{x}_1 A_1 + \bar{x}_2 A_2 + \bar{x}_3 A_3}{A_1 + A_2 + A_3}.$$

Thus the x coordinate of the centroid of the composite area can be expressed in terms of the x coordinates of the centroids of its parts. The coordinates of the centroid of a composite area with an arbitrary number of parts are

$$\bar{x} = \frac{\sum_i \bar{x}_i A_i}{\sum_i A_i}$$

$$\bar{y} = \frac{\sum_i \bar{y}_i A_i}{\sum_i A_i}.$$

(C.3)

As demonstrated in Example C.3, these equations can be used to determine the centroids of composite areas containing "holes," or cutouts, by treating the cutouts as negative areas.

Example C.3 Centroid of a Composite Area
Determine the coordinates of the centroid of the area in Fig. C.6.

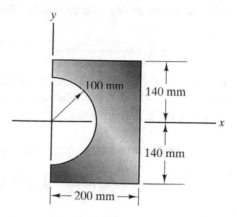

Fig. C.6

Solution
The area can be treated as a composite consisting of the rectangle without the semicircular cutout and the area of the cutout, which will be called parts 1 and 2, respectively [Fig. (a)].

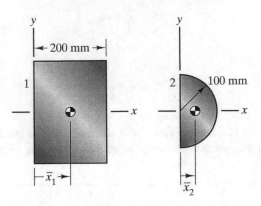

(a) Rectangle 1 and semicircular cutout 2

From Appendix D, the x coordinate of the centroid of the cutout is

$$\bar{x}_2 = \frac{4R}{3\pi} = \frac{4(100)}{3\pi} \text{ mm.}$$

The information for determining the x coordinate of the centroid of the composite area is summarized in Table C.1. Notice that the cutout is treated as a negative area. The x coordinate of the centroid is

$$\bar{x} = \frac{\bar{x}_1 A_1 + \bar{x}_2 A_2}{A_1 + A_2} = \frac{(100)\left[(200)(280)\right] - \left[(4)(100)/3\pi\right]\left[\frac{1}{2}\pi(100)^2\right]}{(200)(280) - \frac{1}{2}\pi(100)^2}$$

$$= 122.4 \text{ mm.}$$

Table C.1 Information for determining $\bar{x}$

	$\bar{x}_i$ (mm)	A_i (mm^2)	$\bar{x}_i A_i$ (mm^3)
Part 1 (rectangle)	100	$(200)(280)$	$(100)[(200)(280)]$
Part 2 (cutout)	$\frac{4(100)}{3\pi}$	$-\frac{1}{2}\pi(100)^2$	$-\frac{4(100)}{3\pi}\left[\frac{1}{2}\pi(100)^2\right]$

Because the area is symmetric about the x axis, $\bar{y} = 0$.

Problems

By treating the areas shown in Problems C.9–C.14 as composites and using the results in Appendix D, determine the coordinates of the centroids of the areas.

C.9

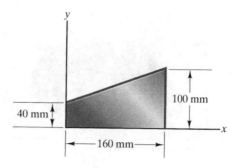

Problem C.9

C.10

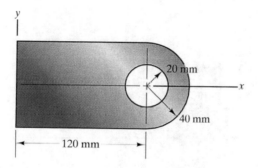

Problem C.10

C.11

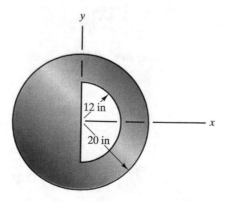

Problem C.11

C.12

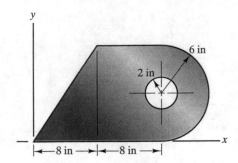

Problem C.12

C.13

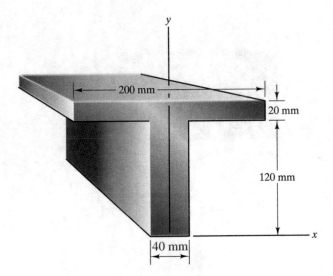

Problem C.13

C.14

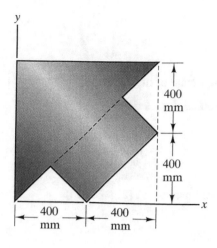

Problem C.14

Moments of Inertia of Areas

The moments of inertia of an area are integrals similar in form to those used to determine the centroid of the area. Consider an area A in the x-y plane [Fig. C.7(a)]. The moments and product of inertia of A are:

Fig. C.7 (a) Area A in the x-y plane. (b) Coordinates of dA

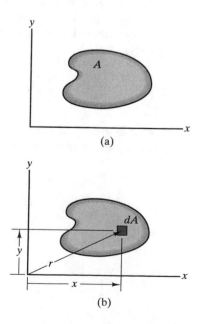

1. Moment of inertia about the x axis

$$I_x = \int_A y^2 \, dA, \tag{C.4}$$

where y is the y coordinate of the differential element of area dA [Fig. C.7(b)].

2. Moment of inertia about the y axis

$$I_y = \int_A x^2 \, dA, \tag{C.5}$$

where x is the x coordinate of dA [Fig. C.7(b)].

3. Product of inertia

$$I_{xy} = \int_A xy \, dA. \tag{C.6}$$

4. Polar moment of inertia

$$J_O = \int_A r^2 \, dA, \tag{C.7}$$

where r is the radial distance from the origin O to dA [Fig. C.7(b)].

The dimensions of the moments of inertia of an area are $(\text{length})^4$. The definitions of I_x, I_y and J_O imply that they have positive values for any area. The polar moment of inertia about the origin is equal to the sum of the moments of inertia about the x and y axes:

$$J_O = \int_A r^2 \, dA = \int_A \left(x^2 + y^2 \right) dA = I_y + I_x.$$

The definition of I_{xy} implies that if A is symmetric about either the x axis or the y axis, the product of inertia is zero.

Example C.4 Moments of Inertia of an Area by Integration
Determine the moments of inertia of the triangular area in Fig. C.8.

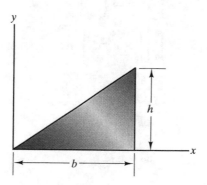

Fig. C.8

Solution
Use of Double Integration
The moments of inertia can be evaluated using the rectangular element
$dA = dx\,dy$ shown in Fig. (a).

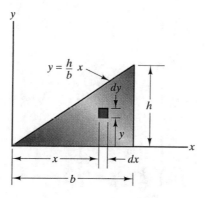

(a) Element dA for determining the moments of inertia

Moment of inertia about the x axis

$$I_x = \int_A y^2 dA = \int_{x=0}^{b} \int_{y=0}^{(h/b)x} y^2 \, dx \, dy$$

$$= \int_0^b \left[\int_0^{(h/b)x} y^2 dy \right] dx$$

$$= \int_0^b \frac{h^3}{3b^3} x^3 dx$$

$$= \frac{1}{12} bh^3.$$

Moment of inertia about the y axis

$$I_y = \int_A x^2 dA = \int_{x=0}^{b} \int_{y=0}^{(h/b)x} x^2 \, dx \, dy$$

$$= \int_0^b x^2 \left[\int_0^{(h/b)x} dy \right] dx$$

$$= \int_0^b \frac{h}{b} x^3 dx$$

$$= \frac{1}{4} hb^3.$$

Product of inertia

$$I_{xy} = \int_A xy \, dA = \int_{x=0}^{b} \int_{y=0}^{(h/b)x} xy \, dx \, dy$$

$$= \int_0^b x \left[\int_0^{(h/b)x} y \, dy \right] dx$$

$$= \int_0^b \frac{h^2}{2b^2} x^3 dx$$

$$= \frac{1}{8} b^2 h^2.$$

Polar moment of inertia

$$J_O = I_x + I_y = \frac{1}{12} bh^3 + \frac{1}{4} hb^3.$$

Use of a Strip Element

In some cases it is advantageous to determine moments of inertia using an element of area in the form of a strip, as was done in locating centroids of areas. Consider an element of area dA in the form of a vertical strip of width dx and height H that is at a position x [Fig. (b)]. The area $dA = H\,dx$, so the moment of inertia of the strip about the y axis is

$$\left(I_y\right)_{\text{strip}} = \int_A x^2\,dA$$
$$= Hx^2\,dx.$$

Using the element of area $dA' = dx\,dy$ shown in Fig. (c), the moment of inertia of the strip about the x axis is

$$\left(I_x\right)_{\text{strip}} = \int_A y^2\,dA'$$
$$= dx\int_0^H y^2\,dy$$
$$= dx\left[\frac{y^3}{3}\right]_0^H$$
$$= \frac{1}{3}H^3\,dx,$$

and the product of inertia of the strip is

$$\left(I_{xy}\right)_{\text{strip}} = \int_A xy\,dA'$$
$$= x\,dx\int_0^H y\,dy$$
$$= x\,dx\left[\frac{y^2}{2}\right]_0^H$$
$$= \frac{1}{2}H^2x\,dx.$$

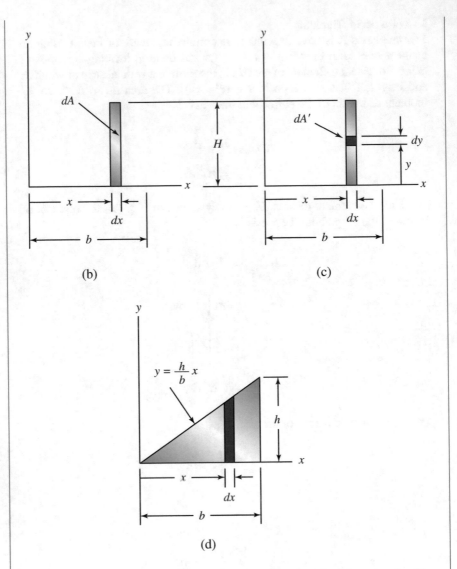

(b)

(c)

(d)

These results can be applied to the triangular area in this example. The height of the strip is $H = (h/b)x$. The moment of inertia about the y axis is obtained by integrating the moment of inertia of the strip across the width of the triangle:

$$I_y = \int_0^b \left(I_y\right)_{\text{strip}}$$

$$= \int_0^b Hx^2 \, dx$$

$$= \int_0^b \frac{h}{b} x^3 \, dx$$

$$= \frac{h}{b} \left[\frac{x^4}{4}\right]_0^b$$

$$= \frac{1}{4} hb^3.$$

In the same way, the moment of inertia about the x axis is

$$I_x = \int_0^b \left(I_x\right)_{\text{strip}}$$

$$= \int_0^b \frac{1}{3} H^3 \, dx$$

$$= \int_0^b \frac{h^3}{3b^3} x^3 \, dx$$

$$= \frac{1}{12} bh^3,$$

and the product of inertia is

$$I_{xy} = \int_0^b \left(I_{xy}\right)_{\text{strip}} dx$$

$$= \int_0^b \frac{1}{2} H^2 x \, dx$$

$$= \int_0^b \frac{h^2}{2b^2} x^3 \, dx$$

$$= \frac{1}{8} b^2 h^2.$$

Example C.5 Moments of Inertia of an Area by Integration
Determine the moments of inertia of the semicircular area in Fig. C.9.

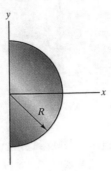

Fig. C.9

Solution
The moments of inertia can be evaluated using the element $dA = r\,dr\,d\theta$ shown in Fig. (a).

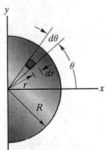

(a) Element dA for determining the moments of inertia

Moment of inertia about the x axis

$$I_x = \int_A y^2\,dA = \int_{r=0}^{R} \int_{\theta=-\pi/2}^{\pi/2} (r\,\sin\theta)^2 (r\,dr\,d\theta)$$

$$= \int_0^R r^3 \left[\int_{-\pi/2}^{\pi/2} \sin^2\theta\,d\theta \right] dr$$

$$= \int_0^R \frac{\pi}{2} r^3\,dr$$

$$= \frac{1}{8}\pi R^4.$$

Moment of inertia about the y axis

$$I_y = \int_A x^2 dA = \int_{r=0}^{R} \int_{\theta=-\pi/2}^{\pi/2} (r\ \cos\theta)^2 (r\, dr\, d\theta)$$

$$= \int_0^R r^3 \left[\int_{-\pi/2}^{\pi/2} \cos^2\theta\, d\theta \right] dr$$

$$= \int_0^R \frac{\pi}{2} r^3 dr$$

$$= \frac{1}{8}\pi R^4.$$

Product of inertia

Because the area is symmetric about the x axis, $I_{xy} = 0$.

Polar moment of inertia:

$$J_O = I_x + I_y$$

$$= \frac{1}{8}\pi R^4 + \frac{1}{8}\pi R^4$$

$$= \frac{1}{4}\pi R^4.$$

Problems

For Problems C.15–C.18, use integration to determine the moments of inertia I_x, I_y and I_{xy} for the areas shown.

C.15

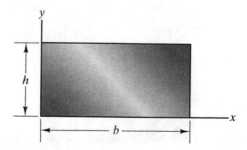

Problem C.15

C.16

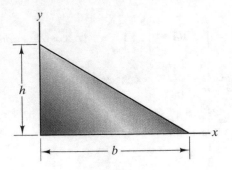

Problem C.16

C.17

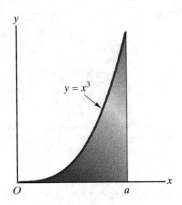

Problem C.17

C.18

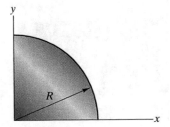

Problem C.18

Parallel Axis Theorems

Suppose that the moments of inertia of an area A are known in terms of a coordinate system $x'y'$ with its origin at the centroid of A and the objective is to determine the moments of inertia in terms of a parallel coordinate system xy [Fig. C.10(a)]. This can be accomplished using results known as parallel axis theorems.

Let the coordinates of the centroid of A in the xy coordinate system be (d_x, d_y), and let $d = \sqrt{d_x^2 + d_y^2}$ be the distance from the origin of the xy coordinate system to the centroid [Fig. C.10(b)]. Two preliminary results are needed to derive the parallel axis theorems: In terms of the $x'y'$ coordinate system, the coordinates of the centroid of A are

$$\bar{x}' = \frac{\int_A x'\, dA}{\int_A dA}, \quad \bar{y}' = \frac{\int_A y'\, dA}{\int_A dA}.$$

But the origin of the $x'y'$ coordinate system is located at the centroid of A, so $\bar{x}' = 0$ and $\bar{y}' = 0$. Therefore,

$$\int_A x'\, dA = 0, \qquad \int_A y'\, dA - 0. \tag{C.8}$$

Moment of inertia about the x axis In terms of the xy coordinate system, the moment of inertia of A about the x axis is

$$I_x = \int_A y^2\, dA, \tag{C.9}$$

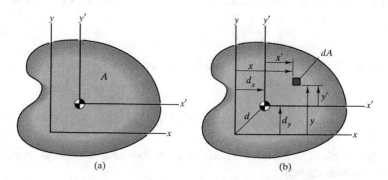

(a)　　　　　　　　　(b)

Fig. C.10 (a) Area A and the coordinate systems $x'y'$ and xy. (b) Coordinates of the differential element dA

where y is the coordinate of the element of area dA relative to the xy coordinate system. From Fig. C.10(b), it can be seen that $y = y' + d_y$, where y' is the coordinate of dA relative to the $x'y'$ coordinate system. Substituting this expression into Eq. (C.9) gives

$$I_x = \int_A (y' + d_y)^2 dA = \int_A (y')^2 dA + 2d_y \int_A y' dA + d_y^2 \int_A dA.$$

The first term on the right side of this equation is the moment of inertia of A about the x' axis. From Eq. (C.8), the second term on the right equals zero. Therefore,

$$I_x = I_{x'} + d_y^2 A. \tag{C.10}$$

This is a parallel axis theorem. It relates the moment of inertia of A about the x' axis through the centroid to the moment of inertia about the parallel axis x.

Moment of inertia about the y axis In terms of the xy coordinate system, the moment of inertia of A about the y axis is

$$I_y = \int_A x^2 dA = \int_A (x' + d_x)^2 dA$$
$$= \int_A (x')^2 dA + 2d_x \int_A x' dA + d_x^2 \int_A dA.$$

From Eq. (C.8), the second term on the right equals zero. Therefore, the parallel axis theorem that relates the moment of inertia of A about the y' axis through the centroid to the moment of inertia about the parallel axis y is

$$I_y = I_{y'} + d_x^2 A. \tag{C.11}$$

Product of inertia The parallel axis theorem for the product of inertia is

$$I_{xy} = I_{x'y'} + d_x d_y A. \tag{C.12}$$

Polar moment of inertia The parallel axis theorem for the polar moment of inertia is

$$J_O = J'_O + \left(d_x^2 + d_y^2\right) A = J'_O + d^2 A, \tag{C.13}$$

where d is the distance from the origin of the $x'y'$ coordinate system to the origin of the xy coordinate system.

Example C.6 Moment and Product of Inertia of a Composite Area
Determine I_x and I_{xy} for the area in Fig. C.11.

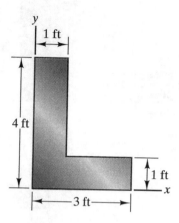

Fig. C.11

Solution

The moments of inertia can be determined by treating the area as a composite consisting of the rectangular parts 1 and 2 shown in Fig. (a). For each part, introduce a coordinate system $x'y'$ with its origin at the centroid of the part [Fig. (b)]. The moments of inertia of the rectangular parts in terms of these coordinate systems are given in Appendix D.

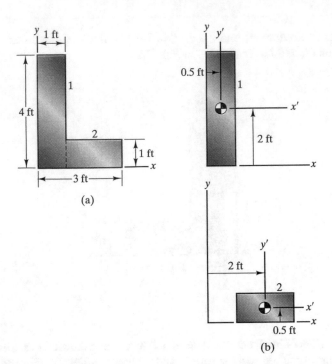

(a) Dividing the area into rectangles 1 and 2. (b) Parallel coordinate systems $x'y'$ with origins at the centroids of the parts

Equation (C.10) can be used to determine the moment of inertia of each part about the x axis (Table C.2).

Table C.2 Determining the moments of inertia of the parts about the x axis

	d_x (ft)	d_y (ft)	A (ft^2)	$I_{x'}$ (ft^4)	$I_x = I_{x'} + d_y^2 A$ (ft^4)
Part 1	0.5	2	$(1)(4)$	$\frac{1}{12}(1)(4)^3$	21.33
Part 2	2	0.5	$(2)(1)$	$\frac{1}{12}(2)(1)^3$	0.67

The moment of inertia of the composite area about the x axis is

$$I_x = (I_x)_1 + (I_x)_2 = 21.33 + 0.67 = 22.00 \text{ ft}^4.$$

Using Eq. (C.12), I_{xy} is determined for each part in Table C.3. The product of inertia of the composite area is

$$I_{xy} = (I_{xy})_1 + (I_{xy})_2 = 4 + 2 = 6 \text{ ft}^4.$$

Table C.3 Determining the products of inertia of the parts in terms of the xy coordinate system

	d_x (ft)	d_y (ft)	A (ft^2)	$I_{x'y'}$	$I_{xy} = I_{x''y''} + d_x d_y A$ (ft^4)
Part 1	0.5	2	(1)(4)	0	4
Part 2	2	0.5	(2)(1)	0	2

Example C.7 Moment of Inertia of a Composite Area

Determine the moment of inertia I_y for the area in Fig. C.12.

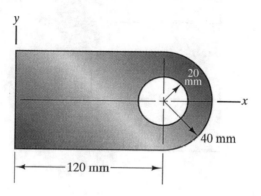

Fig. C.12

Solution

The area can be divided into a rectangle, a semicircle, and a circular cutout, labeled parts 1, 2, and 3 in Fig. (a). The moments of inertia of the parts in terms of the $x'y'$ coordinate systems and the location of the centroid of the semicircular part are given in Appendix D. In Table C.4, Eq. (C.10) is used to determine the moment of inertia of each part about the x axis.

Table C.4 Determining the moments of inertia of the parts

	d_x(mm)	A (mm^2)	$I_{y'}$ (mm^4)	$I_y = I_{y'} + d_x^2 A$ (mm^4)
Part 1	60	(120)(80)	$\frac{1}{12}(80)(120)^3$	4.608E7
Part 2	$120 + \frac{(4)(40)}{3\pi}$	$\frac{1}{2}\pi(40)^2$	$\left(\frac{\pi}{8} - \frac{8}{9\pi}\right)(40)^4$	4.744E7
Part 3	120	$\pi(20)^2$	$\frac{1}{4}\pi(20)^4$	1.822E7

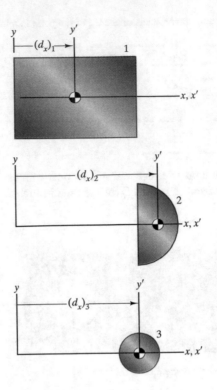

(a) Parts 1, 2, and 3

The moment of inertia of the composite area about the y axis is

$$I_y = (I_y)_1 + (I_y)_2 - (I_y)_3 = (4.608 + 4.744 - 1.822) \times 10^7$$
$$= 7.530\text{E}7 \text{ mm}^4.$$

Example C.8 Moment of Inertia of a Composite Area
The cross section of an I-beam is shown in Fig. C.13. Determine its moment of inertia about the x axis.

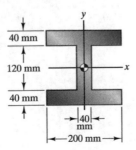

Fig. C.13

Solution

The area can be divided into the rectangular parts shown in Fig. (a). Introducing coordinate systems $x'y'$ with their origins at the centroids of the parts [Fig. (b)], Eq. (C.10) can be used to determine their moments of inertia about the x axis (Table C.5).

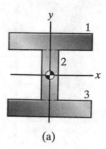

(a)

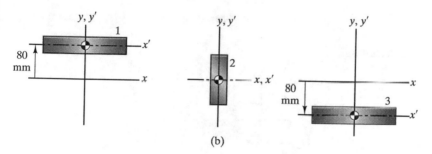

(b)

(a) Dividing the I-beam cross section into parts. (b) Parallel coordinate systems $x'y'$ with origins at the centroids of the parts

Table C.5 Determining the moments of inertia of the parts about the x axis

	d_y(mm)	A (mm²)	$I_{x'}$ (mm⁴)	$I_x = I_{x'} + d_y^2 A$ (mm⁴)
Part 1	80	(200)(40)	$\frac{1}{12}(200)(40)^3$	5.23E7
Part 2	0	(40)(120)	$\frac{1}{12}(40)(120)^3$	0.576E7
Part 3	-80	(200)(40)	$\frac{1}{12}(200)(40)^3$	5.23E7

The moment of inertia of the cross section is

$$I_x = (I_x)_1 + (I_x)_2 + (I_x)_3 = (5.23 + 0.576 + 5.23) \times 10^7 = 11.0\text{E7 mm}^4.$$

Problems

Solve Problems C.19–C.26 by using results in Appendix D and the parallel axis theorems.

C.19 Determine I_x.

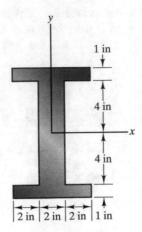

Problem C.19

C.20 Determine I_x.

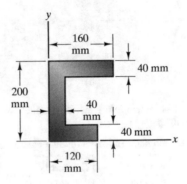

Problem C.20

C.21 Determine I_x.

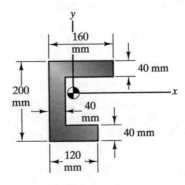

Problem C.21

C.22 Determine I_{xy}.

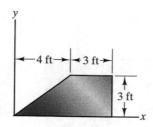

Problem C.22

C.23 Determine I_{xy}.

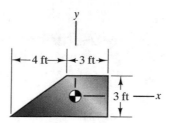

Problem C.23

C.24 Determine I_x and I_y.

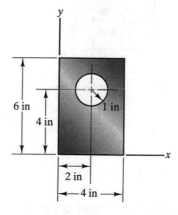

Problem C.24

C.25 Determine I_x and I_y.

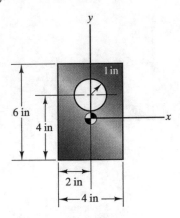

Problem C.25

C.26 Determine I_x and I_y.

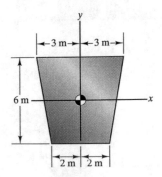

Problem C.26

Rotated and Principal Axes

In many engineering applications, it is necessary to determine moments of inertia of areas for various choices of the angular orientation of the coordinate system relative to the area. Determining the angular orientation of the coordinate system for which the value of a given moment of inertia is a maximum or minimum is also frequently necessary. These procedures are discussed in this section.

Rotated Axes

Consider an area A, a coordinate system xy, and a second coordinate system $x'y'$ that is rotated through an angle θ relative to the xy coordinate system [Fig. C.14(a)]. Suppose the moments of inertia of A are known in terms of the xy coordinate system. The objective is to determine the moments of inertia in terms of the $x'y'$ coordinate system.

In terms of the radial distance r to a differential element of area dA and the angle α in Fig. C.14(b), the coordinates of dA in the xy coordinate system are

$$x = r \cos \alpha, \tag{C.14}$$

$$y = r \, \sin \, \alpha. \tag{C.15}$$

The coordinates of dA in the $x'y'$ coordinate system are

$$x' = r \cos(\alpha - \theta) = r(\cos \alpha \, \cos \theta + \sin \alpha \, \sin \theta), \tag{C.16}$$

$$y' = r \sin(\alpha - \theta) = r(\sin \alpha \, \cos \theta - \cos \alpha \, \sin \theta). \tag{C.17}$$

Substituting Eqs. (C.14) and (C.15) into Eqs. (C.16) and (C.17) yields equations relating the coordinates of dA in the two coordinate systems:

$$x' = x \cos \theta + y \sin \theta, \tag{C.18}$$

$$y' = -x \, \sin \, \theta + y \, \cos \, \theta. \tag{C.19}$$

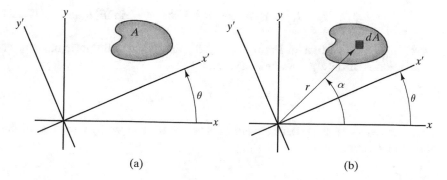

(a)   (b)

Fig. C.14 (a) The $x'y'$ coordinate system is rotated through an angle θ relative to the xy coordinate system. (b) Differential element of area dA

These expressions can be used to derive relations between the moments of inertia of A in terms of the xy and $x'y'$ coordinate systems.

Moment of inertia about the x' axis The moment of inertia of A about the x' axis is

$$
\begin{aligned}
I_{x'} &= \int_A (y')^2 dA = \int_A (-x\ \sin\ \theta + y\ \cos\ \theta)^2 dA \\
&= \cos^2\theta \int_A y^2 dA - 2\ \sin\ \theta\ \cos\ \theta \int_A xy\,dA + \sin^2\theta \int_A x^2 dA,
\end{aligned}
$$

which gives the result

$$
I_{x'} = I_x \cos^2\theta - 2I_{xy}\ \sin\ \theta\ \cos\ \theta + I_y\ \sin^2\theta. \tag{C.20}
$$

Moment of inertia about the y axis The moment of inertia of A about the y' axis is

$$
\begin{aligned}
I_{y'} &= \int_A (x')^2 dA = \int_A (x\ \cos\ \theta + y\ \sin\ \theta)^2 dA \\
&= \sin^2\theta \int_A y^2 dA + 2\ \sin\ \theta\ \cos\ \theta \int_A xy\,dA + \cos^2\theta \int_A x^2 dA,
\end{aligned}
$$

which gives the result

$$
I_{y'} = I_x\ \sin^2\theta + 2I_{xy}\ \sin\ \theta\ \cos\ \theta + I_y\ \cos^2\theta. \tag{C.21}
$$

Product of inertia In terms of the $x'y'$ coordinate system, the product of inertia of A is

$$
I_{x'y'} = (I_x - I_y)\ \sin\ \theta\ \cos\ \theta + (\cos^2\theta - \sin^2\theta)I_{xy}. \tag{C.22}
$$

Polar moment of inertia From Eqs. (C.20) and (C.21), the polar moment of inertia in terms of the $x'y'$ coordinate system is

$$
J_O' = I_{x'} + I_{y'} = I_x + I_y = J_O.
$$

Thus the value of the polar moment of inertia is unchanged by a rotation of the coordinate system.

Principal Axes

The moments of inertia of A in terms of the $x'y'$ coordinate system in Fig. C.14(a) depend on the angle θ. Consider the following question: For what values of θ is the moment of inertia $I_{x'}$ a maximum or minimum? To consider this question, it is convenient to use the identities

$$\sin 2\theta = 2 \sin \theta \, \cos \theta,$$
$$\cos 2\theta = \cos^2\theta - \sin^2\theta = 1 - 2\sin^2\theta = 2\cos^2\theta - 1,$$

to express Eqs. (C.20, C.21, and C.22) in the forms

$$I_{x'} = \frac{I_x + I_y}{2} + \frac{I_x - I_y}{2} \cos 2\theta - I_{xy} \, \sin 2\theta, \qquad (C.23)$$

$$I_{y'} = \frac{I_x + I_y}{2} - \frac{I_x - I_y}{2} \cos 2\theta + I_{xy} \, \sin 2\theta, \qquad (C.24)$$

$$I_{x'y'} = \frac{I_x - I_y}{2} \sin 2\theta + I_{xy} \, \cos 2\theta. \qquad (C.25)$$

Let a value of θ at which $I_{x'}$ is a maximum or minimum be denoted by θ_p. To determine θ_p, the derivative of Eq. (C.23) with respect to 2θ is equated to zero, obtaining the equation

$$\tan 2\theta_p = \frac{2I_{xy}}{I_y - I_x}. \qquad (C.26)$$

Equating the derivative of Eq. (C.23) with respect to 2θ to zero to determine a value of θ for which $I_{y'}$ is a maximum or minimum also leads to Eq. (C.26). The second derivatives of $I_{x'}$ and $I_{y'}$ with respect to 2θ are opposite in sign:

$$\frac{d^2 I_{x'}}{d(2\theta)^2} = -\frac{d^2 I_{y'}}{d(2\theta)^2},$$

which means that at angles θ_p for which $I_{x'}$ is a maximum and $I_{y'}$ is a minimum and at angles θ_p for which $I_{x'}$ is a minimum and $I_{y'}$ is a maximum.

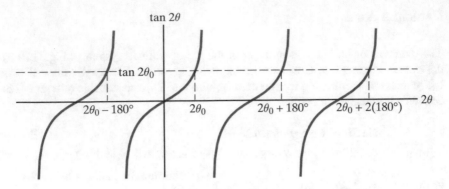

Fig. C.15 For a given value of $\tan 2\theta_0$, there are multiple roots $2\theta_0 + n(180°)$

A rotated coordinate system $x'y'$ that is oriented so that the derivative of Eq. (C.24) with respect to 2θ is equal to zero is called a set of *principal axes* of the area A. The corresponding moments of inertia $I_{x'}$ and $I_{y'}$ are called the *principal moments of inertia*. It can be shown that if x' and y' are principal axes, the product of inertia $I_{x'y'}$ equals zero. This is also a sufficient condition: if $I_{x'y'}$ equals zero, x' and y' are principal axes.

Once the orientation of the principal axes is determined by solving Eq. (C.26) for θ_p, the principal moments of inertia can be determined from Eqs. (C.23) and (C.24). Alternatively, by substituting Eq. (C.26) into Eqs. (C.23) and (C.24), it can be shown that the principal moments of inertia are given by

$$\text{principal moments of inertia} = \frac{I_x + I_y}{2} \pm \sqrt{\left(\frac{I_x - I_y}{2}\right)^2 + \left(I_{xy}\right)^2}. \quad \text{(C.27)}$$

Because the tangent is a periodic function, Eq. (C.26) does not yield a unique solution for the angle θ_p. However, it does determine the orientation of the principal axes within an arbitrary multiple of $90°$. Observe in Fig. C.15 that if $2\theta_0$ is a solution of Eq. (C.26), then $2\theta_0 + n(180°)$ is also a solution for any integer n. The resulting orientations of the $x'y'$ coordinate system are shown in Fig. C.16.

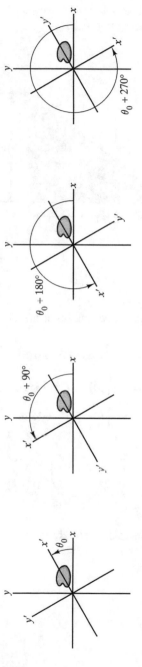

Fig. C.16 The orientation of the $x'y'$ coordinate system is determined only within a multiple of 90°

Example C.9 Determining Principal Axes

Determine a set of principal axes and the corresponding principal moments of inertia for the triangular area in Fig. C.17.

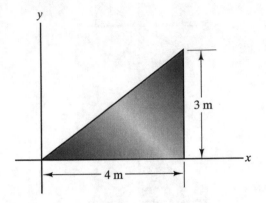

Fig. C.17

Solution

From Appendix D, the moments of inertia of the triangular area are

$$I_x = \frac{1}{12}(4)(3)^3 = 9\,\text{m}^4,$$

$$I_y = \frac{1}{4}(4)^3(3) = 48\,\text{m}^4,$$

$$I_{xy} = \frac{1}{8}(4)^2(3)^2 = 18\,\text{m}^4.$$

From Eq. (C.26)

$$\tan 2\theta_p = \frac{2I_{xy}}{I_y - I_x} = \frac{2(18)}{48 - 9} = 0.923,$$

the angle $\theta_p = 21.4°$. Principal axes corresponding to this value of θ_p are shown in Fig. (a).

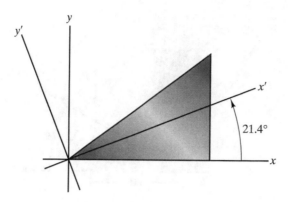

(a) Principal axes corresponding to $\theta_p = 21.4°$

Substituting $\theta_p = 21.4°$ into Eqs. (C.23) and (C.24) gives the values of the principal moments of inertia:

$$I_{x'} = \frac{I_x + I_y}{2} + \frac{I_x - I_y}{2} \cos 2\theta - I_{xy} \sin 2\theta$$

$$= \frac{9 + 48}{2} + \frac{9 - 48}{2} \cos\left[(2)(21.4°)\right] - (18) \sin\left[(2)(21.4°)\right] = 1.96\,\mathrm{m}^4,$$

$$I_{y'} = \frac{I_x + I_y}{2} - \frac{I_x - I_y}{2} \cos 2\theta + I_{xy} \sin 2\theta$$

$$= \frac{9 + 48}{2} - \frac{9 - 48}{2} \cos\left[(2)(21.4°)\right] + (18) \sin\left][(2)(21.4°)\right] = 55.0\,\mathrm{m}^4.$$

These results can also be obtained from Eq. (C.5).

The product of inertia corresponding to a set of principal axes is zero. In this example, substituting $\theta_p = 21.4°$ into Eq. (C.25) confirms that $I_{x'y'} = 0$.

Example C.10 Determining Principal Axes
The moments of inertia of the area in Fig. C.18 in terms of the xy coordinate system are $I_x = 22$ ft^4, $I_y = 10$ ft^4 and $I_{xy} = 6$ ft^4. **(a)** Determine $I_{x'}, I_{y'}$ and $I_{x'y'}$ for $\theta = 30°$. **(b)** Determine a set of principal axes and the corresponding principal moments of inertia.

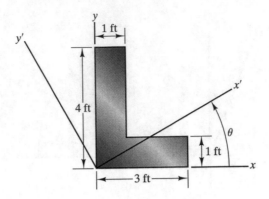

Fig. C.18

Solution
(a) Setting $\theta = 30°$ in Eqs. (C.23, C.24, and C.25) gives

$$
\begin{aligned}
I_{x'} &= \frac{I_x + I_y}{2} + \frac{I_x - I_y}{2} \cos\ 2\theta - I_{xy}\ \sin\ 2\theta \\
&= \frac{22 + 10}{2} + \frac{22 - 10}{2} \cos\left[(2)\,(30°)\right] - (6)\ \sin\left[(2)\,(30°)\right] = 13.8\ \text{ft}^4, \\
I_{y'} &= \frac{I_x + I_y}{2} - \frac{I_x - I_y}{2} \cos\ 2\theta + I_{xy} \sin\ 2\theta \\
&= \frac{22 + 10}{2} - \frac{22 - 10}{2} \cos\left[(2)\,(30°)\right] + (6)\ \sin\left[(2)\,(30°)\right] = 18.2\ \text{ft}^4, \\
I_{x'y'} &= \frac{I_x - I_y}{2} \sin\ 2\theta + I_{xy} \cos\ 2\theta \\
& -\frac{22 - 10}{2} \sin\left[(2)\,(30°)\right] + (6) \cos\left[(2)\,(30°)\right] = 8.2\ \text{ft}^4.
\end{aligned}
$$

(b) Substituting the moments of inertia in terms of the xy coordinate system into Eq. (C.26)

$$
\tan 2\theta_p = \frac{2I_{xy}}{I_y - I_x} = \frac{2(6)}{10 - 22} = -1,
$$

gives the result $\theta_p = -22.5°$. The principal axes corresponding to this value of θ_p are shown in Fig. (a).

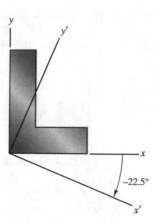

(a) Set of principal axes corresponding to $\theta_p = -22.5°$

Substituting $\theta_p = -22.5°$ into Eqs. (C.23) and (C.24) gives the principal moments of inertia:

$$I_{x'} = 24.5 \text{ ft}^4, \qquad I_{y'} = 7.5 \text{ ft}^4.$$

Problems

C.27 Determine $I_{x'}, I_{y'}$ and $I_{x'y'}$.

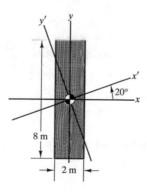

Problems C.27–28

C.28 Determine a set of principal axes and the corresponding principal moments of inertia.

C.29 The moments of inertia of the rectangular area in terms of the xy coordinate system shown are $I_x = 76.0$ m^4, $I_y = 14.7$ m^4 and $I_{xy} = 25.7$ m^4. Using these values, determine a set of principal axes and the corresponding principal moments of inertia.

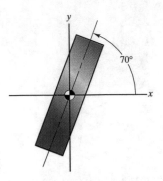

Problem C.29

C.30∗ Determine a set of principal axes and the corresponding principal moments of inertia.

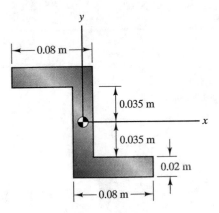

Problem C.30

C.31∗ Determine the moments of inertia $I_{x'}, I_{y'}$ and $I_{x'y'}$ if $\theta = 15°$.

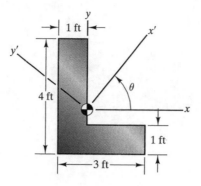

Problems C.31–32

C.32∗ Determine a set of principal axes and the corresponding principal moments of inertia.

C.33∗ Determine a set of principal axes and the corresponding principal moments of inertia.

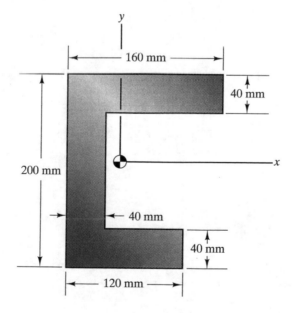

Problem C.33

C.34 Derive Eq. (C.22) for the product of inertia by using the same procedure used to derive Eqs. (C.20) and (C.21).

Appendix D: Properties of Areas

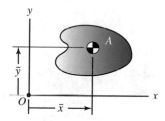

The coordinates of the centroid of the area A are

$$\bar{x} = \frac{\int_A x\,dA}{\int_A dA}, \qquad \bar{y} = \frac{\int_A y\,dA}{\int_A dA}.$$

The moment of inertia about the x axis I_x, the moment of inertia about the y axis I_y, and the product of inertia I_{xy} are

$$I_x = \int_A y^2\,dA, \qquad I_y = \int_A x^2\,dA, \qquad I_{xy} = \int_A xy\,dA.$$

The polar moment of inertia about O is

$$J_O = \int_A r^2\,dA = \int_A \left(x^2 + y^2\right) dA = I_x + I_y.$$

© Springer Nature Switzerland AG 2020
A. Bedford, K. M. Liechti, *Mechanics of Materials*,
https://doi.org/10.1007/978-3-030-22082-2

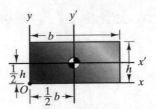

Rectangular area

Area $= bh$

$$I_x = \frac{1}{3}bh^3, \qquad I_y = \frac{1}{3}hb^3, \qquad I_{xy} = \frac{1}{4}b^2h^2$$

$$I_{x'} = \frac{1}{12}bh^3, \qquad I_{y'} = \frac{1}{12}hb^3, \qquad I_{x'y'} = 0$$

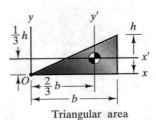

Triangular area

Area $= \frac{1}{2}bh$

$$I_x = \frac{1}{12}bh^3, \qquad I_y = \frac{1}{4}hb^3, \qquad I_{xy} = \frac{1}{8}b^2h^2$$

$$I_{x'} = \frac{1}{36}bh^3, \qquad I_{y'} = \frac{1}{36}hb^3, \qquad I_{x'y'} = \frac{1}{72}b^2h^2$$

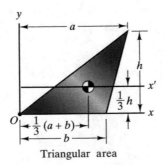

Triangular area

$$\text{Area} = \frac{1}{2}bh$$

$$I_x = \frac{1}{12}bh^3, \qquad I_{x'} = \frac{1}{36}bh^3$$

Circular area

$$\text{Area} = \pi R^2$$

$$I_{x'} = I_{y'} = \frac{1}{4}\pi R^4, \qquad I_{x'y'} = 0$$

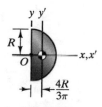

Semicircular area

$$\text{Area} = \frac{1}{2}\pi R^2$$

$$I_x = I_y = \frac{1}{8}\pi R^4, \qquad I_{xy} = 0$$

$$I_{x'} = \frac{1}{8}\pi R^4, \qquad I_{y'} = \left(\frac{\pi}{8} - \frac{8}{9\pi}\right)R^4, \qquad I_{x'y'} = 0$$

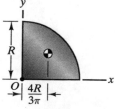

Quarter-circular area

$$\text{Area} = \frac{1}{4}\pi R^2$$

$$I_x = I_y = \frac{1}{16}\pi R^4, \qquad I_{xy} = \frac{1}{8}R^4$$

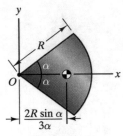

Circular sector

$$\text{Area} = \alpha R^2$$

$$I_x = \frac{1}{4}R^4\left(\alpha - \frac{1}{2}\sin\ 2\alpha\right), \qquad I_y = \frac{1}{4}R^4\left(\alpha + \frac{1}{2}\sin\ 2\alpha\right), \qquad I_{xy} = 0$$

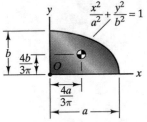

Quarter-elliptical area

$$\text{Area} = \frac{1}{4}\pi ab$$

$$I_x = \frac{1}{16}\pi ab^3, \qquad I_y = \frac{1}{16}\pi a^3 b, \qquad I_{xy} = \frac{1}{8}a^2 b^2$$

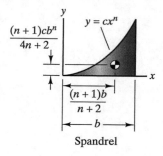

Spandrel

$$\text{Area} = \frac{cb^{n+1}}{n+1}$$

$$I_x = \frac{c^3 b^{3n+1}}{9n+3}, \qquad I_y = \frac{cb^{n+3}}{n+3}, \qquad I_{xy} = \frac{c^2 b^{2n+2}}{4n+4}$$

Appendix E: Structural Steel Shapes

W (Wide-Flange) Shapes

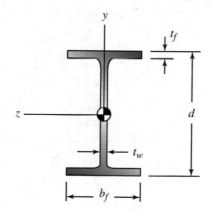

Fig. E.1

© Springer Nature Switzerland AG 2020
A. Bedford, K. M. Liechti, *Mechanics of Materials*,
https://doi.org/10.1007/978-3-030-22082-2

Metric units

Designation*	Cross-sectional area A, mm^2	Dimensions, mm				Moments of inertia and section moduli			
						Axis y		Axis z	
		d	b_f	t_f	t_w	I_y, 10^6 mm^4	S_y, 10^3 mm^3	I_z, 10^6 mm^4	S_z, 10^3 mm^3
W610X551	70200	711	347	69.1	38.6	484	2790	5570	15700
W610X372	47400	669	335	48.0	26.4	301	1800	3530	10600
W610X262	33300	641	327	34.0	19.1	199	1220	2370	7380
W610X174	22200	616	325	21.6	14.0	124	762	1470	4780
W610X125	15900	612	229	19.6	11.9	39.3	343	985	3220
W610X82	10500	599	178	12.8	10.0	12.1	136	565	1890
W460X260	33100	509	289	40.4	22.6	163	1130	1440	5640
W460X193	24700	489	283	30.5	17.0	116	818	1020	4190
W460X144	18400	472	283	22.1	13.6	83.6	591	727	3080
W460X106	13500	469	194	20.6	12.6	25.1	259	488	2080
W460X82	10500	460	191	16.0	9.9	18.7	196	371	1610
W460X52	6640	450	152	10.8	7.6	6.39	83.8	212	944
W310X500	63700	427	340	75.1	45.1	494	2910	1690	7910
W310X283	36000	365	322	44.1	26.9	245	1520	787	4310
W310X158	20100	327	310	25.1	15.5	125	808	388	2370
W310X86	11000	310	254	16.3	9.1	44.6	351	198	1280
W310X52	6670	318	167	13.2	7.6	10.2	122	119	747
W310X21	2680	303	101	5.7	5.1	0.981	19.5	36.9	244
W150X37.1	4750	162	154	11.6	8.1	7.1	92	22.3	275
W150X29.8	3800	157	153	9.3	6.6	5.53	72.3	17.3	220
W150X22.5	2870	152	152	6.6	5.8	3.88	51	12.2	160
W150X24	3060	160	102	10.3	6.6	1.84	36	13.4	168
W150X18	2290	153	102	7.1	5.8	1.25	24.5	9.18	120
W150X13	1620	148	100	4.9	4.3	0.825	16.5	6.16	83.3

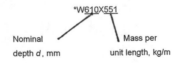

*W610X551

Nominal depth d, mm

Mass per unit length, kg/m

US Customary Units

| Designation[*] | Cross-sectional area A, in^2 | Dimensions, in. | | | | Moments of inertia and section Moduli | | | |
| | | | | | | Axis y | | Axis z | |
		d	b_f	t_f	t_w	I_y, in^4	S_y, in^3	I_z, in^4	S_z, in^3
W24X370	109.0	28.0	13.7	2.72	1.52	1160	170	13400	957
W24X279	82.0	26.7	13.3	2.09	1.16	823	124	9600	718
W24X192	56.3	25.5	13.0	1.46	0.81	530	81.8	6260	491
W24X131	38.5	24.5	12.9	0.96	0.61	340	53	4020	329
W24X94	27.7	24.3	9.1	0.88	0.52	109	24	2700	222
W24X55	16.3	23.6	7.0	0.51	0.40	29.1	8.3	1360	115
W18X175	51.3	20.0	11.4	1.59	0.89	391	68.8	3450	344
W18X130	38.2	19.3	11.2	1.20	0.67	278	49.9	2460	256
W18X86	25.3	18.4	11.1	0.77	0.48	175	31.6	1530	166
W18X65	19.1	18.4	7.6	0.75	0.45	54.8	14.4	1070	117
W18X50	14.7	18.0	7.5	0.57	0.36	40.1	10.7	800	88.9
W18X35	10.3	17.7	6.0	0.43	0.30	15.3	5.12	510	57.6
W12X336	98.8	16.8	13.4	2.96	1.78	1190	177	4060	483
W12X190	55.8	14.4	12.7	1.74	1.06	589	93	1890	263
W12X96	28.2	12.7	12.2	0.90	0.55	270	44.4	833	131
W12X53	15.6	12.1	10.0	0.58	0.35	95.8	19.2	425	70.6
W12X30	8.8	12.3	6.5	0.44	0.26	20.3	6.24	238	38.6
W12X14	4.2	11.9	4.0	0.23	0.20	2.36	1.19	88.6	14.9
W6X25	7.4	6.4	6.1	0.46	0.32	17.1	5.61	53.6	16.8
W6X20	5.9	6.2	6.0	0.37	0.26	13.3	4.41	41.5	13.4
W6X15	4.5	6.0	6.0	0.26	0.23	9.32	3.11	29.3	9.77
W6X16	4.7	6.3	4.0	0.41	0.26	4.43	2.2	32.1	10.2
W6X12	3.6	6.0	4.0	0.28	0.23	2.99	1.5	22.1	7.31
W6X9	2.7	5.9	3.9	0.22	0.17	2.2	1.11	16.4	5.56

*W12X96

Nominal depth, d, in.   Weight per unit length, lb/ft

S (American Standard) Shapes

Fig. E.2

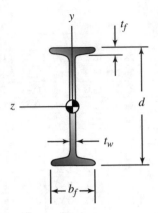

Metric units

		Dimensions, mm				Moments of inertia and section moduli			
						Axis y		Axis z	
Designation*	Cross-sectional area A, mm^2	d	b_f	t_f	t_w	I_y, 10^6 mm^4	S_y, 10^3 mm^3	I_z, 10^6 mm^4	S_z, 10^3 mm^3
S610X180	22900	622	204	20.3	27.7	34.6	338	1310	4220
S610X158	20100	622	200	15.7	27.7	32.0	320	1220	3930
S610X149	18900	610	184	18.9	22.1	19.7	215	992	3250
S610X134	17100	610	181	15.9	22.1	18.6	205	935	3070
S610X119	15100	610	178	12.7	22.1	17.5	197	875	2870
S510X143	18200	516	183	20.3	23.4	20.8	227	696	2700
S510X128	16300	516	179	16.8	23.4	19.4	216	655	2540
S510X112	14200	508	162	16.1	20.2	12.3	152	531	2090
S510X98.2	12500	508	159	12.8	20.2	11.4	144	495	1950
S460X104	13300	457	159	18.1	17.6	10.0	126	384	1680
S460X81.4	10400	457	152	11.7	17.6	8.63	113	334	1460
S380X74	9470	381	143	14.0	15.8	6.49	90.5	202	1060
S380X64	8130	381	140	10.4	15.8	5.94	85.0	186	974
S310X74	9450	305	139	17.4	16.7	6.48	93.2	126	829
S310X60.7	7690	305	133	11.7	16.7	5.61	84.1	113	739
S310X52	6610	305	129	10.9	13.8	4.10	63.5	95.0	624
S310X47.3	6010	305	127	8.9	13.8	3.88	61.1	90.4	593
S250X52	6630	254	126	15.1	12.5	3.45	55.0	61.1	481
S250X37.8	4810	254	118	7.9	12.5	2.80	47.3	51.3	404
S200X34	4360	203	106	11.2	10.8	1.78	33.5	26.9	265
S200X27.4	3480	203	102	6.9	10.8	1.53	30.2	23.9	235
S150X25.7	3260	152	90.6	11.8	9.1	0.953	21.0	10.9	143
S150X18.6	2360	152	84.6	5.9	9.1	0.750	17.7	9.2	120
S130X15	1890	127	76.3	5.4	8.3	0.497	13.0	5.1	80.4

US Customary Units

Designation[*]	Cross-sectional area A, in^2	Dimensions, in.				Moments of inertia and section moduli			
						Axis y		Axis z	
		d	b_f	t_f	t_w	I_y, in^4	S_y, in^3	I_z, in^4	S_z, in^3
S24X121	35.5	24.5	8.05	1.09	0.8	83	20.6	3160	258
S24X106	31.1	24.5	7.87	1.09	0.62	76.8	19.5	2940	240
S24X100	29.3	24	7.25	0.87	0.745	47.4	13.1	2380	199
S24X90	26.5	24	7.13	0.87	0.625	44.7	12.5	2250	187
S24X80	23.5	24	7	0.87	0.5	42	12	2100	175
S20X96	28.2	20.3	7.2	0.92	0.8	49.9	13.9	1670	165
S20X86	25.3	20.3	7.06	0.92	0.66	46.6	13.2	1570	155
S20X75	22	20	6.39	0.795	0.635	29.5	9.25	1280	128
S20X66	19.4	20	6.26	0.795	0.505	27.5	8.78	1190	119
S18X70	20.5	18	6.25	0.691	0.711	24	7.69	923	103
S18X54.7	16	18	6	0.691	0.461	20.7	6.91	801	89
S15X50	14.7	15	5.64	0.622	0.55	15.6	5.53	485	64.7
S15X42.9	12.6	15	5.5	0.622	0.411	14.3	5.19	446	59.4
S12X50	14.6	12	5.48	0.659	0.687	15.6	5.69	303	50.6
S12X40.8	11.9	12	5.25	0.659	0.462	13.5	5.13	270	45.1
S12X35	10.2	12	5.08	0.544	0.428	9.84	3.88	228	38.1
S12X31.8	9.31	12	5	0.544	0.35	9.33	3.73	217	36.2
S10X35	10.3	10	4.94	0.491	0.594	8.3	3.36	147	29.4
S10X25.4	7.45	10	4.66	0.491	0.311	6.73	2.89	123	24.6
S8X23	6.76	8	4.17	0.425	0.441	4.27	2.05	64.7	16.2
S8X18.4	5.4	8	4	0.425	0.271	3.69	1.84	57.5	14.4
S6X17.25	5.06	6	3.57	0.359	0.465	2.29	1.28	26.2	8.74
S6X12.5	3.66	6	3.33	0.359	0.232	1.8	1.08	22	7.34
S5X10	2.93	5	3	0.326	0.214	1.19	0.795	12.3	4.9

Nominal depth, d, in.　　Weight per unit length, lb/ft

C (American Standard Channel) Shapes

Fig. E.3

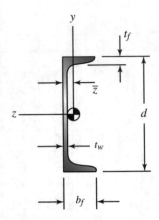

Metric units

Designation[*]	Cross-sectional area A, mm^2	Dimensions, mm				Moments of inertia and section moduli				Centroid position
		d	b_f	t_f	t_w	Axis y		Axis z		z, mm
						I_y, 10^6 mm^4	S_y, 10^3 mm^3	I_z, 10^6 mm^4	S_z, 10^3 mm^3	
C380X74	9480	381	94.4	16.5	18.2	4.57	61.7	168	882	20.3
C380X60	7580	381	89.4	16.5	13.2	3.82	54.8	145	761	19.8
C380X50.4	6420	381	86.4	16.5	10.2	3.36	50.6	131	688	20
C310X45	5690	305	80.5	12.7	13	2.13	33.6	67.4	442	17.1
C310X37	4730	305	77.4	12.7	9.8	1.85	30.7	60	394	17.1
C310X30.8	3920	305	74.7	12.7	7.2	1.61	28.2	53.7	353	17.7
C250X45	5690	254	77	11.1	17.1	1.63	27	43	339	16.5
C250X37	4740	254	73.3	11.1	13.4	1.39	24.1	37.9	299	15.7
C250X30	3790	254	69.6	11.1	9.6	1.16	21.5	32.8	258	15.4
C250X22.8	2890	254	66	11.1	6.1	0.944	18.9	28	220	16.1
C230X30	3790	229	67.3	10.5	11.4	1	19.1	25.3	222	14.8
C230X22	2840	229	63.1	10.5	7.2	0.796	16.5	21.2	186	14.9
C230X19.9	2540	229	61.8	10.5	5.9	0.728	15.6	19.9	174	15.3
C200X27.9	3550	203	64.2	9.9	12.4	0.821	16.5	18.3	180	14.4
C200X20.5	2600	203	59.5	9.9	7.7	0.632	13.9	15	148	14.1
C200X17.1	2180	203	57.4	9.9	5.6	0.545	12.7	13.5	133	14.5
C180X22	2790	178	58.4	9.3	10.6	0.568	12.7	11.3	127	13.5
C180X18.2	2320	178	55.7	9.3	8	0.483	11.4	10.1	113	13.3
C180X14.6	1850	178	53.1	9.3	5.3	0.398	10.1	8.84	99.5	13.7
C150X19.3	2460	152	54.8	8.7	11.1	0.436	10.5	7.21	94.7	13.1
C150X15.6	1980	152	51.7	8.7	8	0.358	9.19	6.29	82.6	12.7
C150X12.2	1540	152	48.8	8.7	5.1	0.286	8	5.44	71.3	13
C130X13	1700	127	47.9	8.1	8.3	0.26	7.27	3.7	58.3	12.1
C130X10.4	1270	127	44.5	8.1	4.8	0.196	6.09	3.11	49	12.3

US Customary Units

Designation[*]	Cross-sectional area A, in^2	Dimensions, in.				Moments of inertia and section moduli				Centroid position
						Axis y		Axis z		
		d	b_f	t_f	t_w	I_y, in^4	S_y, in^3	I_z, in^4	S_z, in^3	z, in.
C15X50	14.7	15	3.72	0.65	0.716	11	3.77	404	53.8	0.799
C15X40	11.8	15	3.52	0.65	0.52	9.17	3.34	348	46.5	0.778
C15X33.9	9.95	15	3.4	0.65	0.4	8.07	3.09	315	42	0.788
C12X30	8.81	12	3.17	0.501	0.51	5.12	2.05	162	27	0.674
C12X25	7.34	12	3.05	0.501	0.387	4.45	1.87	144	24	0.674
C12X20.7	6.08	12	2.94	0.501	0.282	3.86	1.72	129	21.5	0.698
C10X30	8.81	10	3.03	0.436	0.673	3.93	1.65	103	20.7	0.649
C10X25	7.34	10	2.89	0.436	0.526	3.34	1.47	91.1	18.2	0.617
C10X20	5.87	10	2.74	0.436	0.379	2.8	1.31	78.9	15.8	0.606
C10X15.3	4.48	10	2.6	0.436	0.24	2.27	1.15	67.3	13.5	0.634
C9X20	5.87	9	2.65	0.413	0.448	2.41	1.17	60.9	13.5	0.583
C9X15	4.41	9	2.49	0.413	0.285	1.91	1.01	51	11.3	0.586
C9X13.4	3.94	9	2.43	0.413	0.233	1.75	0.954	47.8	10.6	0.601
C8X18.75	5.51	8	2.53	0.39	0.487	1.97	1.01	43.9	11	0.565
C8X13.75	4.04	8	2.34	0.39	0.303	1.52	0.848	36.1	9.02	0.554
C8X11.5	3.37	8	2.26	0.39	0.22	1.31	0.775	32.5	8.14	0.572
C7X14.75	4.33	7	2.3	0.366	0.419	1.37	0.772	27.2	7.78	0.532
C7X12.25	3.6	7	2.19	0.366	0.314	1.16	0.696	24.2	6.92	0.525
C7X9.8	2.87	7	2.09	0.366	0.21	0.957	0.617	21.2	6.07	0.541
C6X13	3.81	6	2.16	0.343	0.437	1.05	0.638	17.3	5.78	0.514
C6X10.5	3.08	6	2.03	0.343	0.314	0.86	0.561	15.1	5.04	0.5
C6X8.2	2.39	6	1.92	0.343	0.2	0.687	0.488	13.1	4.35	0.512
C5X9	2.64	5	1.89	0.32	0.325	0.624	0.444	8.89	3.56	0.478
C5X6.7	1.97	5	1.75	0.32	0.19	0.47	0.372	7.48	2.99	0.484

*C15X50

Nominal depth, d, in.

Weight per unit length, lb/ft

L (Angle) Shapes

Fig. E.4

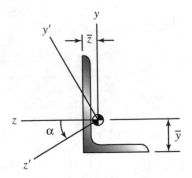

Metric units

Designation[*]	Weight per unit length	Cross-sectional area A	Moments of inertia and section moduli							
			Axis y		Axis z		Axis y'	Centroid position		
			I_y,	S_y,	I_z	S_z	$I_{y'}$	y	z	
	kN/m	mm^2	10^6 mm^4	10^3 mm^3	10^6 mm^4	10^3 mm^3	10^6 mm^4	mm	mm	tan α
L203X203X25.4	0.749	9730	37	258	37	258	15.258	59.9	59.9	1
L203X203X19	0.572	7430	29	199	29	199	11.829	57.5	57.5	1
L203X203X12.7	0.389	5050	20.2	137	20.2	137	8.161	55	55	1
L203X152X25.4	0.648	8420	16.1	146	33.6	247	8.839	67.2	41.9	0.543
L203X152X19	0.496	6450	12.8	113	26.4	191	6.855	64.8	39.6	0.551
L203X152X12.7	0.338	4390	9.03	78.3	18.4	131	4.781	62.4	37.2	0.558
L203X102X25.4	0.549	7130	4.85	64.5	29	230	3.265	77	26.5	0.247
L203X102X19	0.422	5480	3.9	50.3	22.9	178	2.557	74.6	24.1	0.258
L152X152X19	0.42	5460	11.7	109	11.7	109	4.849	44.9	44.9	1
L152X152X12.7	0.287	3720	8.29	75.4	8.29	75.4	3.348	42.5	42.5	1
L152X152X9.5	0.217	2830	6.4	57.6	6.4	57.6	2.581	41.2	41.2	1
L152X102X19	0.344	4480	3.61	48.6	10.2	102	2.129	52.6	27.3	0.428
L152X102X12.7	0.236	3060	2.61	34	7.24	70.9	1.495	50.2	24.9	0.44
L152X102X9.5	0.179	2330	2.04	26.2	5.61	54.2	1.159	49	23.7	0.446
L152X89X12.7	0.224	2910	1.77	26.1	6.91	69.2	1.084	52.6	21.1	0.344
L152X89X9.5	0.171	2220	1.39	20.1	5.35	53	0.836	51.4	19.8	0.35
L102X102X19	0.27	3510	3.19	46	3.19	46	1.376	32.2	32.2	1
L102X102X12.7	0.186	2420	2.31	32.3	2.31	32.3	0.958	29.9	29.9	1
L102X102X6.4	0.096	1250	1.26	17	1.26	17	0.510	27.4	27.4	1
L102X89X12.7	0.174	2260	1.58	24.8	2.22	31.6	0.757	31.6	25.3	0.75
L102X89X9.5	0.133	1730	1.24	19.2	1.74	24.4	0.586	30.4	24.1	0.755
L102X89X6.4	0.0902	1170	0.871	13.2	1.21	16.7	0.405	29.1	22.8	0.759
L102X76X12.7	0.162	2100	1.01	18.2	2.1	30.9	0.551	33.5	20.9	0.543
L102X76X6.4	0.084	1090	0.564	9.76	1.15	16.3	0.297	31	18.4	0.558

The moment of inertia about the z' axis can be determined from the relation $I_{y'} + I_{z'} = I_y + I_z$.

*L203X152X25.4

Vertical leg length, mm   Horizontal leg length, mm   Thickness, mm

U.S. Customary Units

Designation*	Weight per unit length lb/ft	Cross-sectional area A in^2	Moments of inertia and section moduli					Centroid position		
			Axis y		Axis z		Axis y'			
			I_y in^4	S_y in^3	I_z in^4	S_z in^3	$I_{y'}$ in^4	y in.	z in.	tan α
L8X8X1	51.3	15.1	89.1	15.8	89.1	15.8	36.75	2.36	2.36	1
L8X8X3/4	39.2	11.5	69.9	12.2	69.9	12.2	28.35	2.26	2.26	1
L8X8X1/2	26.7	7.84	48.8	8.36	48.8	8.36	19.82	2.17	2.17	1
L8X6X1	44.4	13.1	38.8	8.92	80.9	15.1	21.46	2.65	1.65	0.542
L8X6X3/4	34	9.99	30.8	6.92	63.5	11.7	16.62	2.55	1.56	0.55
L8X6X1/2	23.2	6.8	21.7	4.79	44.4	8.01	11.49	2.46	1.46	0.557
L8X4X1	37.6	11.1	11.6	3.94	69.7	14	7.91	3.03	1.04	0.247
L8X4X3/4	28.9	8.49	9.37	3.07	55	10.9	6.13	2.94	0.949	0.257
L6X6X3/4	28.8	8.46	28.1	6.64	28.1	6.64	11.58	1.77	1.77	1
L6X6X1/2	19.6	5.77	19.9	4.59	19.9	4.59	8.03	1.67	1.67	1
L6X6X3/8	14.9	4.38	15.4	3.51	15.4	3.51	6.20	1.62	1.62	1
L6X4X3/4	23.6	6.94	8.63	2.95	24.5	6.23	5.09	2.07	1.07	0.428
L6X4X1/2	16.2	4.75	6.22	2.06	17.3	4.31	3.55	1.98	0.981	0.441
L6X4X3/8	12.3	3.61	4.86	1.58	13.4	3.3	2.73	1.93	0.933	0.446
L6X3-1/2X1/2	15.4	4.52	4.24	1.59	16.6	4.23	2.58	2.07	0.829	0.343
L6X3-1/2X3/8	11.7	3.44	3.33	1.22	12.9	3.23	2.00	2.02	0.781	0.349
L4X4X3/4	18.5	5.43	7.62	2.79	7.62	2.79	3.25	1.27	1.27	1
L4X4X1/2	12.7	3.75	5.52	1.96	5.52	1.96	2.26	1.18	1.18	1
L4X4X1/4	6.58	1.93	3	1.03	3	1.03	1.18	1.08	1.08	1
L4X3-1/2X1/2	11.9	3.5	3.76	1.5	5.3	1.92	1.79	1.24	0.994	0.75
L4X3-1/2X3/8	9.1	2.68	2.96	1.16	4.15	1.48	1.39	1.2	0.947	0.755
L4X3-1/2X1/4	6.18	1.82	2.07	0.794	2.89	1.01	0.95	1.14	0.897	0.759
L4X3X1/2	11.1	3.25	2.4	1.1	5.02	1.87	1.30	1.32	0.822	0.543
L4X3X1/4	5.75	1.69	1.33	0.585	2.75	0.988	0.69	1.22	0.725	0.558

*L8X6X1

Vertical leg length, in.

Horizontal leg length, in.

Thickness, in.

The moment of inertia about the z' axis can be determined from the relation $I_{y'} + I_{z'} = I_y + I_z$

Appendix F: Deflections and Slopes of Prismatic Beams

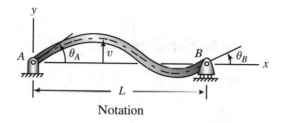

Notation

Simply Supported Beams

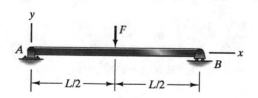

© Springer Nature Switzerland AG 2020
A. Bedford, K. M. Liechti, *Mechanics of Materials*,
https://doi.org/10.1007/978-3-030-22082-2

$$v = -\frac{F}{48EI}\left(3L^2 x - 4x^3\right), \qquad 0 \le x \le L/2$$

$$v' = \theta = -\frac{F}{16EI}\left(L^2 - 4x^2\right), \qquad 0 \le x \le L/2$$

$$v = -\frac{F}{48EI}\left(-L^3 + 9L^2 x - 12Lx^2 + 4x^3\right), \qquad L/2 \le x \le L$$

$$v' = \theta = -\frac{F}{16EI}\left(3L^2 - 8Lx + 4x^2\right), \qquad L/2 \le x \le L$$

$$\theta_A = -\theta_B = -\frac{FL^2}{16EI}.$$

$$v = -\frac{F}{48EI}\left(3L^2\langle x\rangle - 4\langle x\rangle^3 + 8\left\langle x - \frac{L}{2}\right\rangle^3\right).$$

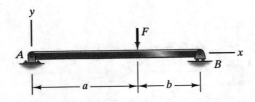

$$v = -\frac{Fb}{6EIL}\left[(a^2 + 2ab)x - x^3\right], \qquad 0 \le x \le a$$

$$v' = \theta = -\frac{Fb}{6EIL}\left(a^2 + 2ab - 3x^2\right), \qquad 0 \le x \le a$$

$$v = -\frac{Fa}{6EIL}\left[-La^2 + (a^2 + 2L^2)x - 3Lx^2 + x^3\right], \qquad a \le x \le L$$

$$v' = \theta = -\frac{Fa}{6EIL}\left[(a^2 + 2L^2) - 6Lx + 3x^2\right], \qquad a \le x \le L$$

$$\theta_A = -\frac{Fab(L+b)}{6EIL}, \qquad \theta_B = \frac{Fab(L+a)}{6EIL}.$$

$$v = -\frac{F}{6EIL}\left[b(a^2 + 2ab)\langle x\rangle - b\langle x\rangle^3 + L\langle x - a\rangle^3\right].$$

$$v = -\frac{M_0}{6EIL}\left(2L^2x - 3Lx^2 + x^3\right),$$

$$v' = \theta = -\frac{M_0}{6EIL}\left(2L^2 - 6Lx + 3x^2\right),$$

$$\theta_A = -\frac{M_0 L}{3EI}, \qquad \theta_B = \frac{M_0 L}{6EI}.$$

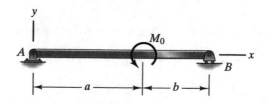

$$v = -\frac{M_0}{6EIL}\left[\left(6aL - 3a^2 - 2L^2\right)x - x^3\right], \qquad 0 \le x \le a$$

$$v' = \theta = -\frac{M_0}{6EIL}\left(6aL - 3a^2 - 2L^2 - 3x^2\right), \qquad 0 \le x \le a$$

$$v = -\frac{M_0}{6EIL}\left[3a^2L - \left(3a^2 + 2L^2\right)x + 3Lx^2 - x^3\right], \qquad a \le x \le L$$

$$v' = \theta = \frac{M_0}{6EIL}\left(3a^2 + 2L^2 - 6Lx + 3x^2\right), \qquad a \le x \le L$$

$$\theta_A = -\frac{M_0}{6EIL}\left(6aL - 3a^2 - 2L^2\right), \qquad \theta_B = \frac{M_0}{6LEI}\left(3a^2 - L^2\right),$$

$$v = -\frac{M_0}{6EIL}\left[\left(6aL - 3a^2 - 2L^2\right)\langle x\rangle - \langle x\rangle^3 + 3L\langle x - a\rangle^2\right].$$

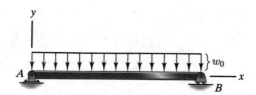

$$v = -\frac{w_0}{24EI}\left(L^3x - 2Lx^3 + x^4\right), \qquad v' = \theta = -\frac{w_0}{24EI}\left(L^3 - 6Lx^2 + 4x^3\right),$$

$$\theta_A = -\theta_B = -\frac{w_0 L^3}{24EI}.$$

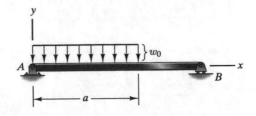

$$v = -\frac{w_0}{24EIL}\left[(a^4 - 4a^3L + 4a^2L^2)x + (2a^2 - 4aL)x^3 + Lx^4\right], \qquad 0 \le x \le a$$

$$v' = -\frac{w_0}{24EIL}\left[a^4 - 4a^3L + 4a^2L^2 + (6a^2 - 12aL)x^2 + 4Lx^3\right], \qquad 0 \le x \le a$$

$$v = -\frac{w_0a^2}{24EIL}\left[-a^2L + (4L^2 + a^2)x - 6Lx^2 + 2x^3\right], \qquad a \le x \le L$$

$$v' = \theta = -\frac{w_0a^2}{24EIL}\left(4L^2 + a^2 - 12Lx + 6x^2\right), \qquad a \le x \le L$$

$$\theta_A = -\frac{w_0a^2}{24EIL}(2L - a)^2, \qquad \theta_B = \frac{w_0a^2}{24EIL}(2L^2 - a^2).$$

$$v = -\frac{w_0}{24EIL}\left[(a^4 - 4a^3L + 4a^2L^2)\langle x\rangle + (2a^2 - 4aL)\langle x\rangle^3 + L\langle x\rangle^4 - L\langle x - a\rangle^4\right].$$

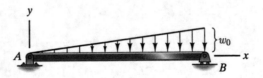

$$v = -\frac{w_0}{360LEI}\left(7L^4x - 10L^2x^3 + 3x^5\right),$$

$$v' = \theta = -\frac{w_0}{360LEI}\left(7L^4 - 30L^2x^2 + 15x^4\right),$$

$$\theta_A = -\frac{7w_0L^3}{360EI}, \qquad \theta_B = \frac{w_0L^3}{45EI}.$$

Cantilever Beams

$$v = -\frac{F}{6EI}\left(3Lx^2 - x^3\right), \qquad v' = \theta = -\frac{F}{2EI}\left(2Lx - x^2\right),$$

$$v_B = -\frac{FL^3}{3EI}, \qquad \theta_B = -\frac{FL^2}{2EI}.$$

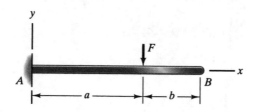

$$v = -\frac{F}{6EI}\left(3ax^2 - x^3\right), \qquad v' = \theta = -\frac{F}{2EI}\left(2ax - x^2\right), \qquad 0 \le x \le a$$

$$v = -\frac{Fa^2}{6EI}(3x - a), \qquad v' = \theta = -\frac{Fa^2}{2EI}, \qquad a \le x \le L$$

$$v_B = -\frac{Fa^2}{6EI}(3L - a), \qquad \theta_B = -\frac{Fa^2}{2EI}.$$

$$v = -\frac{F}{6EI}\left(3a\langle x\rangle^2 - \langle x\rangle^3 + \langle x - a\rangle^3\right).$$

$$v = -\frac{M_0 x^2}{2EI}, \qquad v' = \theta = -\frac{M_0 x}{EI},$$

$$v_B = -\frac{M_0 L^2}{2EI}, \qquad \theta_B = -\frac{M_0 L}{EI}.$$

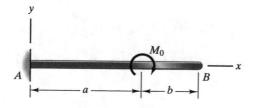

$$v = -\frac{M_0 x^2}{2EI}, \qquad v' = \theta = -\frac{M_0 x}{EI}, \qquad 0 \le x \le a$$

$$v = -\frac{M_0 a}{2EI}(2x - a), \qquad v' = \theta = -\frac{M_0 a}{EI}, \qquad a \le x \le L$$

$$v_B = -\frac{M_0 a}{2EI}(2L - a), \qquad \theta_B = -\frac{M_0 a}{EI}.$$

$$v = -\frac{M_0}{2EI}\left(\langle x \rangle^2 - \langle x - a \rangle^2\right).$$

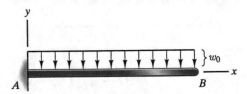

$$v = -\frac{w_0}{24EI}\left(6L^2 x^2 - 4Lx^3 + x^4\right), \qquad v' = \theta = -\frac{w_0}{6EI}\left(3L^2 x - 3Lx^2 + x^3\right),$$

$$v_B n = -\frac{w_0 L^4}{8EI}, \qquad \theta_B = -\frac{w_0 L^3}{6EI}.$$

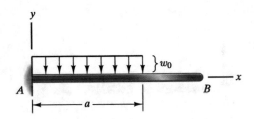

$$v = -\frac{w_0}{24EI}\left(6a^2 x^2 - 4ax^3 + x^4\right), \qquad 0 \le x \le a$$

$$v' = \theta = -\frac{w_0}{6EI}\left(3a^2 x - 3ax^2 + x^3\right), \qquad 0 \le x \le a$$

$$v = -\frac{w_0 a^3}{24EI}(4x - a), \qquad v' = \theta = -\frac{w_0 a^3}{6EI}, \qquad a \le x \le L$$

$$v_B = -\frac{w_0 a^3}{24EI}(4L - a), \qquad \theta_B = -\frac{w_0 a^3}{6EI}.$$

$$v = -\frac{w_0}{24EI}\left(6a^2 \langle x \rangle^2 - 4a\langle x \rangle^3 + \langle x \rangle^4 - \langle x - a \rangle^4\right).$$

$$v = -\frac{w_0}{120LEI}\left(20L^3x^2 - 10L^2x^3 + x^5\right),$$

$$v' = \theta = -\frac{w_0}{24LEI}\left(8L^3x - 6L^2x^2 + x^4\right),$$

$$v_B = -\frac{11w_0L^4}{120EI}, \qquad \theta_B = -\frac{w_0L^3}{8EI}.$$

$$v = -\frac{w_0}{120LEI}\left(10L^3x^2 - 10L^2x^3 + 5Lx^4 - x^5\right),$$

$$v' = \theta = -\frac{w_0}{24LEI}\left(4L^3x - 6L^2x^2 + 4Lx^3 - x^4\right),$$

$$v_B = -\frac{w_0L^4}{30EI}, \qquad \theta_B = -\frac{w_0L^3}{24EI}.$$

Appendix G: Answers to Even-Numbered Problems

Unless stated otherwise, x components are positive to the right, y components are positive upward, and counterclockwise moments are positive.

© Springer Nature Switzerland AG 2020
A. Bedford, K. M. Liechti, *Mechanics of Materials*,
https://doi.org/10.1007/978-3-030-22082-2

1.2	$A_x = 0, A_y = 2.5$ kN, $B = 27.5$ kN.
1.4	$A = 35$ lb, $B = 25$ lb.
1.6	$P = 0, V = -144$ kN, $M = -288$ kN-m.
1.8	$\mathbf{R} = 173\mathbf{i} + 100\mathbf{j}$ (N), $\mathbf{M} = -396\mathbf{k}$ (N-m).
1.10	$m = 52.3$ kg.
1.12	84.9 kN (T)
1.14	$F = 2.48$ kip.
1.16	Member CE, 4.72 kN (T). Member DE, 3.92 kN (C)
1.18	$AC : 150$ kN (T), $BC : 90.1$ kN (T), $BD : 225$ kN (C).
1.20	$C_x = 0, C_y = F, D_x = 0, D_y = -2F, E = F.$
1.22	3876 lb. (compression)
1.24	6000 lb. (tension)
1.26	1.70 kN (tension)
1.28	1.65 kN (compression)
1.30	$A_x = 15$ kN, $A_y = -8$ kN, $E_x = -15$ kN, $E_y = 14$ kN.
1.32	$\bar{x} = 667$ mm, $\bar{y} = 133$ mm.
1.34	$\bar{x} = 90.3$ mm, $\bar{y} = 59.4$ mm.
1.36	Distance = 3.28 ft.
1.38	**(a)** $A_x = 0, A_y = 80.7$ N, $B = 210$ N. **(b)** $w = 223$ N/m.
1.40	$A_x = 0, A_y = 159$ lb, $B = 169$ lb.
1.42	$R_x = 0, R_y = -913$ N, $M_R = 1970$ N-m clockwise
1.46	$A = c_0/2, B_x = 0, B_y = -c_0/2.$
2.2	**(a)** $A' = 0.00354$ m^2. **(b)** $\sigma_{av} = 1.20$E6 Pa, $\tau_{av} = 1.20$E6 Pa.
2.4	**(a)** $\sigma_{av} = 318$ psi. **(b)** $\sigma_{av} = -637$ psi.
2.6	$P = 8.18$ kN, $\tau_{av} = -168$ kPa.
2.8	$\sigma_{av} = 267$ psi, $|\tau_{av}| = 533$ psi.
2.10	$\sigma_{av} = 20,400$ psi.

2.12	$\sigma_{av} = 74.6x$ kPa. (x in meters)
2.14	$\sigma_{av} = 22.8$ psi, $|\tau_{av}| = 62.6$ psi.
2.16	$\sigma_{av} = 200$ kPa, $|\tau_{av}| = 83.3$ kPa.
2.18	$\sigma_{av} = 5.96$ MPa (864 psi).
2.20	(a) $\sigma_{av} = 9.42$ MPa. (b) $\sigma_{av} = 13.3$ MPa.
2.22	(a) $\tau_{av} = P/A$. (b) $\tau_{av} = P/(2A)$.
2.24	$\tau_{av} = 41.7$ MPa.
2.26	$\sigma_{av} = 102$ ksf.
2.28	$|\tau_{av}| = 3.60$ ksi.
2.30	$\sigma_{av} = 31.1$ MPa, $|\tau_{av}| = 9.17$ MPa.
2.32	$\sigma_{av} = -129$ psi, $|\tau_{av}| = 153$ psi.
2.34	$\tau_{av} = 18.75$ psi.
2.36	$\tau_{av} = 21.2$ MPa.
2.38	$\tau_{av} = F/(bt)$.
2.40	$\tau_{av} = 1.82$ psi.
2.42	$\sigma = 50$ kPa.
2.44	$\sigma = 2860$ psi.
2.46	(a) $70\mathbf{i} - 10\mathbf{j} - 5\mathbf{k}$ (kN). (b) $-70\mathbf{i} + 10\mathbf{j} + 5\mathbf{k}$ (kN).
2.48	$\sigma_{av} = 0.533$ MPa.
2.50	$\mathbf{t}_{av} = 1.955\mathbf{i} + 0.090\mathbf{j} + 1.060\mathbf{k}$ (MPa).
2.52	$a = 111$ lb/in.^{3}.
2.54	$dL' = 1.15dL$.
2.56	$L' = 6.24$ in.
2.58	$L' = 0.2113$ m.
2.60	$\varepsilon = 0.00625$.
2.62	$\delta = -3.60$ mm.
2.64	$\delta = 0.028$ in.

(continued)

2.66	$\varepsilon_{AB} = 0.1062.$
2.68	$\varepsilon = 0.00454.$
2.70	$\varepsilon = 0.0075.$
2.72	$\varepsilon_1 = 0.000333, \varepsilon_2 = -0.008333.$
2.74	$\gamma_{12} = 0.698.$
2.76	$\gamma_{12} = 0.0870.$
2.78	$\gamma_{12} = -0.234.$
2.80	$\varepsilon_x = 0.00296, \varepsilon_y = 0.00204, \varepsilon_z = 0.00204, \gamma_{xy} = -0.00370, \gamma_{yz} = 0, \gamma_{xz} = 0.$
2.82	$\sigma_{av} = 0, \tau_{av} = 398$ psi.
2.84	$\sigma_{av} = 320$ psi.
2.86	$\sigma_{av} = 200$ kPa, $|\tau_{av}| = 0.$
2.88	$\sigma_{av} = -750$ kPa.
2.90	$\sigma_{av} = -500$ kPa, $\tau_{av} = 267$ kPa.
2.92	$\tau_{av} = 92.9$ MPa.
2.94	$\tau_{av} = 2.32$ MPa.
2.96	$L' = 0.2008$ m.
2.98	$\delta = -0.028$ in.
2.100	$\varepsilon = 0.02.$
2.102	$\varepsilon = 0.476.$
3.2	(a) $\sigma_{AB} = 2.83$ MPa. (b) $\sigma_{BC} = 0.943$ MPa.
3.4	$\sigma = 568$ kPa.
3.6	$\sigma = 4.5$ ksi.
3.8	$\sigma_{BC} = -578$ MPa, $\sigma_{DG} = 1440$ MPa.
3.10	$\sigma = -10.3$ MPa.
3.12	$\sigma = 4.90$ ksi.
3.16	$\sigma_{BC} = 106$ MPa.
3.18	$F = 2170$ N.
3.20	$\sigma_\theta = 1.75$ ksi, $\tau_\theta = -4.82$ ksi.

3.22	$\theta = 50.2°$, $P = 61.0$ kN.
3.24	(a) $|\tau_{\theta}| = 0$. (b) $|\tau_{\theta}| = 18.2$ MPa. (c) $|\tau_{\theta}| = 27.9$ MPa.
3.26	$F = 5200$ lb.
3.28	$F = 173$ kN.
3.30	$\sigma_{AB} = -1900$ psi.
3.32	$\sigma_3 = -63.1$ MPa.
3.34	10.02 in.
3.36	$\sigma = -0.7$ GPa.
3.38	$E = 76.4$ GPa, $\nu = 0.351$.
3.40	57.1 kip
3.42	$\delta_{AB} = 0.107$ in, $\delta_{AC} = -0.0571$ in.
3.44	$x = b/(1 + E_{BC}L_{DG}/E_{DG}L_{BC})$.
3.46	0.01 in.
3.48	1.95° clockwise
3.50	$\delta_{BE} = 0.0411$ in.
3.52	$\delta_{AB} = 5.28$ mm, $\delta_{AC} = -3.96$ mm.
3.54	$\delta_{BC} = 0.845$ mm, $\delta_{BD} = 0.602$ mm.
3.56	-0.0199 mm.
3.60	$\sigma = 2F/(3A)$.
3.62	$\sigma_{rod} = -87.0$ MPa, $\sigma_{sleeve} = -16.0$ MPa.
3.64	$\sigma_{BC} = -7.16$ MPa.
3.66	$F_1 = 76.2$ kN.
3.68	$b = 0.0681$ in.
3.70	$\sigma_{AB} = -F\cos^2\theta/[A(1 + \cos^3\theta)]$, $\sigma_{AC} = -F/[A(1 + \cos^3\theta)]$.
3.72	310 kN
3.74	$h = 2.95$ mm.
3.76	$\sigma_{BC} = -72.1$ MPa, $\sigma_{DG} = 126$ MPa, $\sigma_{HI} = 390$ MPa.

(continued)

3.78	9.92 kN
3.80	23.5 kN
3.82	0.268 mm to the left, 0.247 mm upward
3.84	$\delta = 0.0116$ in.
3.86	$\delta = 0.392$ mm.
3.88	$\sigma_L = 30.6$ ksi.
3.90	$\delta = 0.127$ mm.
3.92	$\delta = 6.13$ mm.
3.94	$\delta = 0.0588$ m.
3.96	$\delta = 0.0682$ m.
3.98	$x = 1.26$ m, displacement $= 0.101$ mm.
3.100	$\sigma_A = 57.6$ ksi.
3.102	0.00137 mm
3.104	$\delta = -F/\pi E d\ \tan^2 \alpha.$
3.106	$\delta = 0.0131$ mm.
3.108	200.132 mm.
3.110	30.024 mm
3.112	$\sigma = 0.$
3.114	(a, b) $\delta = 0.0111$ in.
3.116	134 °F
3.118	$\sigma = -7470$ psi.
3.120	$\sigma_A = -16.8$ MPa, $\sigma_B = -67.2$ MPa.
3.122	$\sigma_{\text{sleeve}} = -557$ MPa.
3.124	16,000 lb. downward
3.126	0.369 in. to the right, 0.184 in. upward
3.128	1.20 mm to the left, 19.01 mm downward
3.130	$\sigma_{AB} = \sigma_{AD} = 46.3$ MPa, $\sigma_{AC} = 79.8$ MPa.
3.132	$\sigma_{AB} = -96.0$ MPa, $\sigma_{AC} = -109$ MPa.

3.134	0.916 mm
3.136	Either 2014-T6 or 7075-T6
3.138	ASTM-A514
3.140	3.70 in^2.
3.142	

Graph for 3.142: vertical axis $\sigma_{allow} V/FL$ (scale 0 to 12), horizontal axis θ, degrees (scale 0 to 90).

3.144	2014-T6 or 7075-T6
3.150	$A = 0.00912$ in^2.
3.152	$A_3 = 1850$ mm^2.
3.154	$\sigma_{AB} = 2.31$ MPa.
3.156	$\sigma_{AB} = 488$ MPa, $\sigma_{AC} = 345$ MPa.

(continued)

3.158	$\sigma = 1.12$ MPa.
3.160	$\sigma_\theta = 120.0$ ksi, $\lvert \tau_\theta \rvert = 69.3$ ksi.
3.162	$D' = 0.7495$ in.
3.164	$\varepsilon_{AB} = 0.000349$, $\varepsilon_{CD} = 0.000698$, $\varepsilon_{EF} = 0.001047$.
3.166	$\delta_{AB} = 0.292$ in, $\delta_{AC} = -0.137$ in.
3.168	A : 400E6 N (tension); B : 400E6 N (compression).
3.170	40 °C
3.172	$\sigma_{AB} = 81.7$ MPa, $\sigma_{AC} = -35.5$ MPa, $\sigma_{AD} = -82.6$ MPa.
4.2	$\beta = 89.0\,^\circ$.
4.4	$\beta = 89.9\,^\circ$.
4.6	$G = 75.8$ GPa.
4.8	$\sigma_\theta = 10.4$ MPa, $\lvert \tau_\theta \rvert = 6$ MPa.
4.10	$\tau = 16.2$ MPa.
4.12	**(a)** $\sigma_\theta = -17.3$ ksi, $\lvert \tau_\theta \rvert = 10$ ksi. **(b)** 20 ksi
4.14	$\tau = 3.97$ MPa, $\theta = 20.5^\circ$.
4.16	$J = 23.6$ in^4.
4.18	$\lvert \tau \rvert = 17.0$ MPa.
4.20	$\lvert \tau \rvert = 19.9$ MPa.
4.22	$\lvert \tau \rvert = 11.7$ MPa.
4.24	$\lvert \tau \rvert = 31.8$ MPa.
4.26	$R_o = 69.5$ mm.
4.28	$\phi = 3.96$ rad (227°).
4.30	$\phi = 0.000382$ rad (0.0219°)
4.32	**(a)** $\lvert \tau \rvert = 637$ psi. **(b)** 0.00110 rad (0.0629°).
4.34	$\lvert \tau_{AB} \rvert = 19.9$ ksi, $\lvert \tau_{BC} \rvert = 8.49$ ksi.

4.36	$|\tau_{AB}| = 30.9$ MPa, $|\tau_{BC}| = 61.8$ MPa, $|\tau_{CD}| = 124$ MPa.
4.38	$|\tau_{AB}| = 37.7$ MPa, $|\tau_{CD}| = 28.3$ MPa.
4.40	$r_C = 108$ mm.
4.42	$|T_O| = 13.7$ in-kip.
4.44	$|T_{AB}| = 1110$ N-m, $|T_{BC}| = 92.3$ N-m.
4.46	$\phi = 0.0107$ rad $(0.612°)$.
4.48	$|\phi_A| = 1.82°$, $|\phi_B| = 0.180°$.
4.50	656 MPa
4.52	$|\tau| = 21.9$ MPa.
4.54	$|\tau| = 17.9$ ksi.
4.56	$|\tau| = 2.83$ ksi.
4.58	$a = 0.251$ m^{-1}.
4.60	$T = 6.19$ N-m.
4.62	102 N-m
4.64	$|\tau| = 40.7$ MPa.
4.66	$c_0 = 7200$ in-lb/in., $|\tau| = 18.7$ ksi.
4.70	$|T_{left}| = 5c_0L/192, |T_{right}| = c_0L/64$.
4.72	$T = 6.49$ kN-m.
4.74	$\phi = 117°$.
4.76	$r_Y = 0.553$ in, $\phi = 74.6°$.
4.78	$T = 15.5$ kN-m.
4.80	$\phi = 142°$.
4.84	$\phi = 0.00937$ rad $(0.537°)$.
4.86	(a) $|\tau| = 27.8$ MPa. (b) $|\tau| = 19.4$ MPa.
4.88	$|\tau| = 30.4$ MPa.
4.90	$\phi = 0.975°$.

(continued)

4.92	$|\tau| = 5540$ psi.
4.94	$t = 0.0802$ in, $\phi = 1.95°$.
4.96	$\phi = 4.27°$.
4.98	$|\tau| = 68.5$ MPa.
4.100	(a) $|\tau| = 1591.5$ lb/in^2. (b) $|\tau| = 1590.6$ lb/in^2.
4.102	$|\tau_{max}| = 31.8$ MPa.
4.104	7075-T6 aluminum alloy
4.116	$G = 1.22$E7 psi.
4.118	$\gamma = 0.00346$.
4.120	$\phi = 0.000668$ rad (0.0383°).
4.122	$T = 1.99$ kN·m.
4.124	$\sigma_\theta = 20.5$ kPa, $|\tau_\theta| = 24.4$ kPa.
4.126	$|\tau_{AB}| = 8.13$ ksi, $|\tau_{BC}| = 4.06$ ksi.
4.128	$|\tau| = 325$ MPa.
5.2	$P_C = 0, V_C = -0.5$ kN, $M_C = 0.5$ kN·m.
5.4	$P_B = 0, V_B = 2$ kN, $M_B = -2$ kN·m.
5.6	(b) $A_x = 0, A_y = 0, B_y = 0$. (c) $P_C = 0, V_C = 0, M_C = -4$ kN·m.
5.8	(a) $P_A = 0, V_A = 4$ kN, $M_A = 4$ kN·m. (b) $P_A = 0, V_A = 2$ kN, $M_A = 3$ kN·m.
5.10	$P_A = 0, V_A = 16.7$ lb, $M_A = 575$ in-lb.
5.12	$P_A = 0, V_A = -475$ lb, $M_A = -1275$ ft-lb.
5.14	$P_A = 0, V_A = 4.8$ kN, $M_A = 13.6$ kN·m.
5.16	$P_C = 0, V_C = -3.7$ kN, $M_C = 14.1$ kN·m.
5.18	$P_B = 10.7$ kN, $V_B = -12$ kN, $M_B = 1.4$ kN·m.
5.20	$P_B = 90$ kN, $V_B = -40$ kN, $M_B = -20$ kN·m.

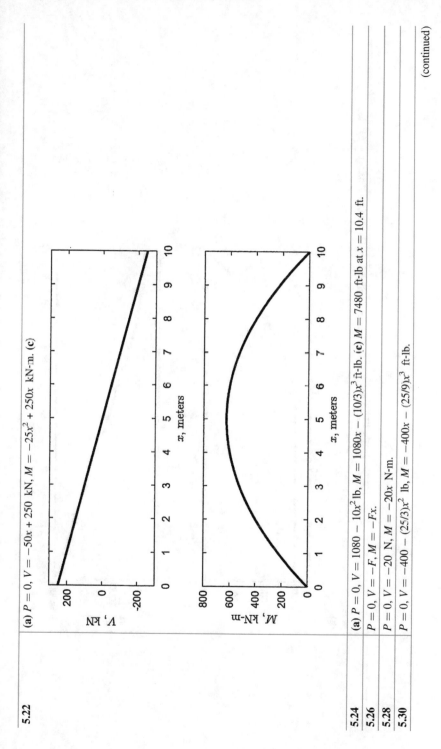

5.22 (a) $P = 0$, $V = -50x + 250$ kN, $M = -25x^2 + 250x$ kN·m. (c)

5.24 (a) $P = 0$, $V = 1080 - 10x^2$ lb, $M = 1080x - (10/3)x^3$ ft-lb. (c) $M = 7480$ ft-lb at $x = 10.4$ ft.

5.26 $P = 0$, $V = -F$, $M = -Fx$.

5.28 $P = 0$, $V = -20$ N, $M = -20x$ N·m.

5.30 $P = 0$, $V = -400 - (25/3)x^2$ lb, $M = -400x - (25/9)x^3$ ft-lb.

(continued)

5.32

(a) $0 < x < 6$ ft: $P = 0$, $V = 300$ lb, $M = 300x - 3000$ ft-lb. $6 < x < 12$ ft: $P = 0$, $V = 300 - (25/3)(x - 6)^2$ lb, $M = -3000 + 300x - (25/9)(x - 6)^3$ ft-lb.

(b)

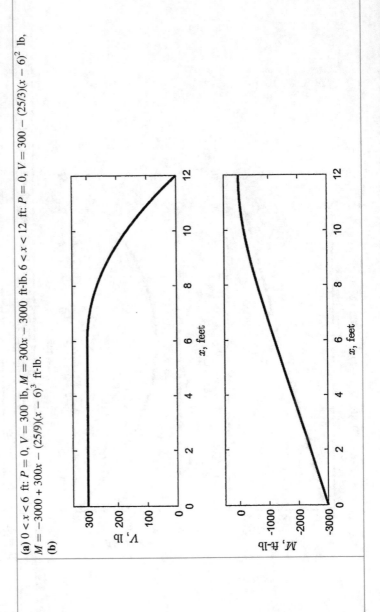

5.34 | $M = 578$ in-lb at $x = 9.33$ in.

5.36

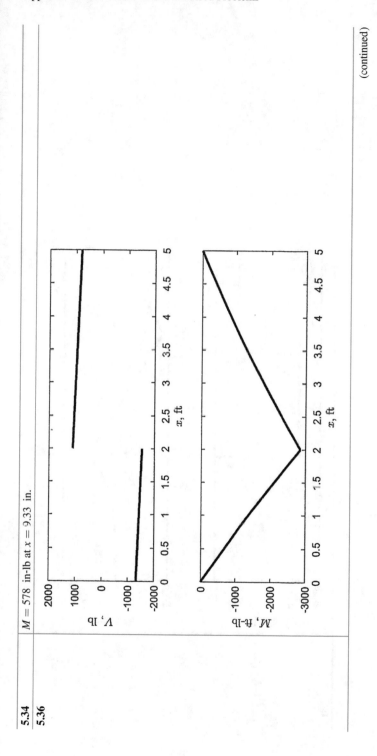

(continued)

5.38

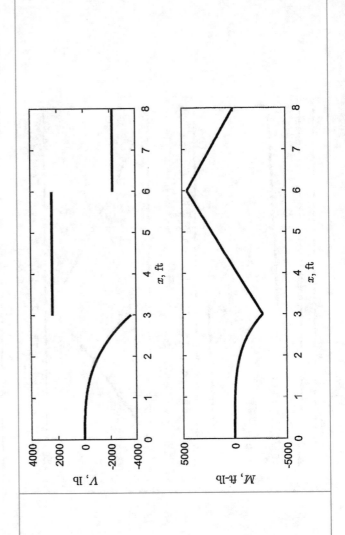

5.40

5.42 $V = -31.3 + 10x + 0.25x^2 - 0.2x^3$ kN.

5.44 $V = -w_0(x - x^2/2L)$, $M = -w_0(x^2/2 - x^3/6L)$.

5.46 $V = F$, $M = Fx$.

5.48 $V = w_0L/6 - w_0x^2/2L$, $M = (Lx - x^3/L)w_0/6$.

5.50 $V = w(L/2 - x)$, $M = -(w/2)(L^2/6 - Lx + x^2)$.

5.54 $0 < x < 2$ m: $P = 0$, $V = x$ kN, $M = x^2/2$ kN-m. $2 < x < 5$ m: $P = 0$,
$V = -4 + x$ kN, $M = 8 - 4x + x^2/2$ kN-m. $5 < x < 6$ m: $P = 0$,
$V = -6 + x$ kN, $M = 18 - 6x + x^2/2$ kN-m.

(continued)

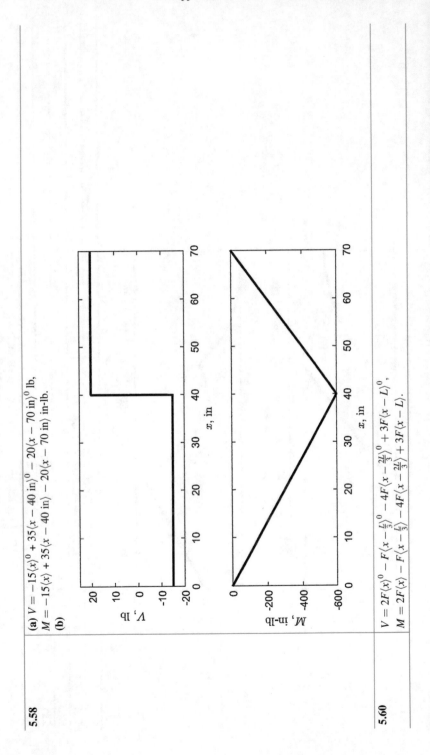

5.58

(a) $V = -15\langle x\rangle^0 + 35\langle x - 40\text{ in}\rangle^0 - 20\langle x - 70\text{ in}\rangle^0$ lb,

$M = -15\langle x\rangle + 35\langle x - 40\text{ in}\rangle - 20\langle x - 70\text{ in}\rangle$ in-lb.

(b)

5.60

$V = 2F\langle x\rangle^0 - F\langle x - \frac{L}{3}\rangle^0 - 4F\langle x - \frac{2L}{3}\rangle^0 + 3F\langle x - L\rangle^0$,

$M = 2F\langle x\rangle - F\langle x - \frac{L}{3}\rangle - 4F\langle x - \frac{2L}{3}\rangle + 3F\langle x - L\rangle$.

5.62

(a) $V = 40\langle x \rangle^0 - 6\langle x - 10 \text{ in}\rangle + 80\langle x - 30 \text{ in}\rangle^0$ lb,
$M = 40\langle x \rangle - 3\langle x - 10 \text{ in}\rangle^2 + 80\langle x - 30 \text{ in}\rangle$ in-lb.

(b)

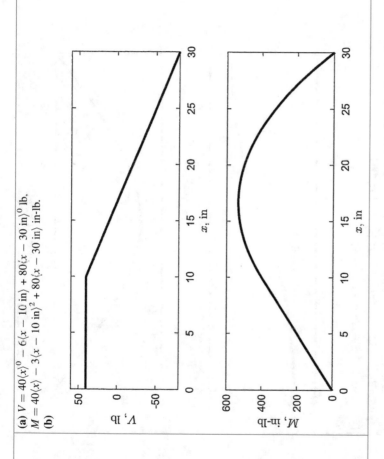

(continued)

5.64

(a) $V = 12\langle x \rangle^0 + 12\langle x - 12\,\text{m}\rangle^0 - \frac{1}{3}\langle x \rangle^2 + \frac{2}{3}\langle x - 6\,\text{m}\rangle^2$ kN,

$M = 12\langle x \rangle + 12\langle x - 12\,\text{m}\rangle - \frac{1}{9}\langle x \rangle^3 + \frac{2}{9}\langle x - 6\,\text{m}\rangle^3$ kN-m.

(b)

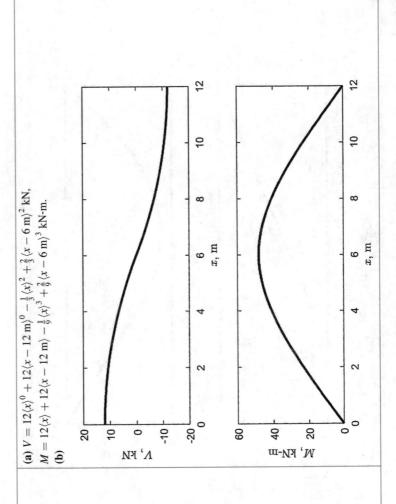

5.66

(a) $V = -12\langle x \rangle^{-1} + 8\langle x \rangle^{0} + 4\langle x - 6\,\text{m} \rangle^{0} - 6\langle x - 2\,\text{m} \rangle + 6\langle x - 4\,\text{m} \rangle$ kN,

$M = -12\langle x \rangle^{0} + 8\langle x \rangle + 4\langle x - 6\,\text{m} \rangle - 3\langle x - 2\,\text{m} \rangle^{2} + 3\langle x - 4\,\text{m} \rangle^{2}$ kN-m.

(b)

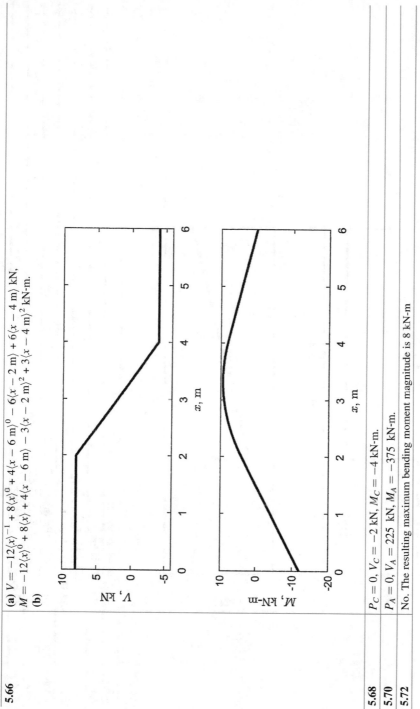

5.68 $P_C = 0$, $V_C = -2$ kN, $M_C = -4$ kN-m.

5.70 $P_A = 0$, $V_A = 225$ kN, $M_A = -375$ kN-m.

5.72 No. The resulting maximum bending moment magnitude is 8 kN-m

(continued)

5.74

6.2 (a) $\sigma_x = -9380$ psi. (b) 18, 800 psi.

6.4 $M = 492$ kN-m.

6.6 $M = 177,000$ in-lb.

6.8 (a) $M = 22.1$ N-m. (b) $\sigma_{\max} = 8.28$ MPa.

6.10 (a) $\sigma_x = 129$ MPa. (b) $\sigma_x = 78.6$ MPa.

6.12 (a) $M = 2700$ N-m at $x = 2.25$ m. (b) 11.8 MPa.

6.14 (a) $\sigma_x = 13.1$ ksi. (b) $\sigma_x = 6.75$ ksi.

6.16	$\sigma_x = 1.24$ GPa.
6.18	FS $= 1.87$.
6.20	$h = 3.90$ in.
6.22	(a) FS $= 2.01$. (b) FS $= 4.69$.
6.28	W150X29.8
6.30	W6X16
6.32	Steel, 1520 psi; aluminum, 876 psi
6.34	$H = 0.02$ m.
6.36	$H = 2.8$ in.
6.38	$E_B = 216$ GPa.
6.40	(a) $\sigma_x = 26.3$ MPa. (b) $\sigma_x = 13.8$ MPa.
6.42	(a) $w_0 = 1320$ lb/in. (b) $w_0 = 2580$ lb/in.
6.44	$\sigma_x = 3.02$ MPa.
6.46	$d = 0.0274$ m.
6.48	(a) 640,000 in-lb. (b) 960,000 in-lb.
6.50	$w_0 = 50, 300$ N/m.
6.52	$x = \sqrt{12}$ m, $d = 0.0293$ m.
6.54	$w_0 = 22, 700$ N/m, $x = 0.6$ m.
6.56	$w_0 = 116$ lb/in, $x = 18$ in.
6.60	$M_U = 11.0$ kN-m.
6.62	$\sigma_x = 31.3$ MPa.
6.64	$\sigma_x = -14$ MPa.
6.66	$|\sigma_x| = 9210$ psi.

(continued)

6.68	$|\sigma_x| = 21,400$ psi.
6.70	$|\sigma_x| = 2.08$ MPa.
6.72	$\sigma_x = -18,300$ psi.
6.74	$\sigma_x = -48.6$ MPa.
6.76	(a) $\tau_{av} = 4.44$ MPa.. (b) $\tau_{av} = 2.78$ MPa.
6.78	$\tau_{av} = 576$ kPa at $x = 0$, $y' = 0$ and at $x = 8$ m, $y' = 0$.
6.80	(a) $\tau_{av} = -2810$ psi. (b) $\tau_{av} = -1230$ psi.
6.84	$\tau_{av} = 6.86$ MPa.
6.86	$y' = 1$ in, $\tau_{av} = 4170$ psi.
6.88	$\tau_{av} = 11.4$ MPa.
6.90	$\tau_{av} = 6.31$ MPa.
6.92	$\tau_{av} = 5.35$ MPa.
6.94	3 kN/bolt
6.96	13.1 kN/bolt
6.98	577 N/bolt
6.100	$\tau = 30.9$ MPa.
6.102	$\tau = 2250$ psi.
6.104	$\tau = 138\zeta$ MPa (ζ in meters)

6.106

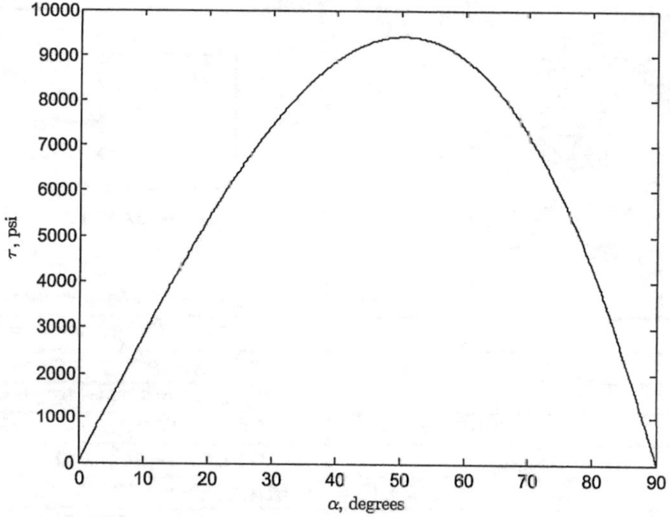

6.108	$\tau = 2.65\alpha + 9.46 \sin \alpha$ MPa.
6.110	$e = 0.312$ m.
6.112	$e = 0.0332$ m.
6.114	$e = 7.54$ in.
6.116	$e = 0.0637$ m.
6.118	$e = 3.12$ in.
6.120	$e = 0.0394$ m.
6.122	$M = 16,000$ in-lb.
6.124	6.84 MPa.
6.126	$w_0 = 13.3$ kN/m.

(continued)

6.128	$w_0 = 9.32$ kN/m.
6.130	$x = 2.86$ m, $d = 50.7$ mm.
6.132	$\sigma_x = -4.11$ MPa.
6.134	**(a)** $\tau_{av} = -19.9$ MPa. **(b)** $\tau_{av} = -11.1$ MPa.
6.136	$\tau_{av} = 223$ psi.
7.2	**(a)** $\sigma_y = -4$ MPa. **(b)** $|\tau| = 3.35$ MPa.
7.4	**(a)** $\sigma_x = 360$ psi, $\sigma_y = -480$ psi, $\sigma_z = 0$, $\tau_{xy} = 0$, $\tau_{xz} = -640$ psi, $\tau_{xz} = 0$, $\tau_{yz} = 0$.
	(b) $\sigma_{x'} = -150$ psi, $\sigma_{y'} = 29.6$ psi, $\tau_{x'y'} = -760$ psi.
7.6	$\sigma_{x'} = 25$ ksi, $\sigma_{y'} = -25$ ksi, $\tau_{x'y'} = 0$.
7.8	$\sigma_x = 64.00$ MPa, $\sigma_y = 85.00$ MPa, $\tau_{xy} = 0.00$ MPa.
7.10	$\sigma_x = 38.8$ MPa, $\sigma_y = -22.8$ MPa, $\tau_{xy} = 14.3$ MPa.
7.12	$\sigma = -2.23$ ksi, $|\tau| = 1.60$ ksi.
7.14	$\sigma_{(a)} = -56.6$ MPa, $\sigma_{(b)} = -50.0$ MPa, $\sigma_{(c)} = -33.2$ MPa.
7.16	$\sigma = -7.86$ MPa, $\tau = 13.50$ MPa.
7.18	$\tau_{x'y'} = 4.90$ MPa, $\theta = 19.5°$ or $\tau_{x'y'} = -4.90$ MPa, $\theta = 40.2°$.
7.20	$\tau_{xy} = -78.4$ psi, $\tau_{x'y'} = -114$ psi.
7.24	$\sigma_1 = \sigma_x, \sigma_2 = 0, \tau_{max} = |\sigma_x/2|$.

7.26 $\sigma_1 = 20$ MPa, $\sigma_2 = 10$ MPa, $\tau_{max} = 5$ MPa.

7.28 $\sigma_1 = 8.22$ ksi, $\sigma_2 = -10.22$ ksi, $\tau_{max} = 9.22$ ksi.

7.30 Absolute maximum shear stress $= 10$ MPa.

7.32 $\sigma_1 = 52.4$ MPa, $\sigma_2 = -32.4$ MPa, absolute maximum shear stress $= 42.4$ MPa.

7.36 No, the absolute maximum shear stress is 174 MPa

7.38 $\sigma_{x'} = 25$ ksi, $\sigma_{y'} = -25$ ksi, $\tau_{x'y'} = 0$.

(continued)

7.40	$\sigma_x = 64.0$ MPa, $\sigma_y = 85.0$ MPa, $\tau_{xy} = 0.00$ MPa.
7.42	$\theta = -20°$.
7.44	$\sigma = 5.8$ MPa, $|\tau| = 4.9$ MPa.
7.46	$\sigma = 177$ psi, $\tau = -237$ psi.
7.48	See the answer to Problem 5–2.24
7.50	See the answer to Problem 5–2.26
7.52	$\sigma_1 = 52.5$ MPa, $\sigma_2 = -32.5$ MPa, $\tau_{max} = 42.5$ MPa.
7.54	Absolute maximum shear stress = 6.6 ksi.
7.56	$\sigma_1 = 40.5$ ksi, $\sigma_2 = 0$, $\sigma_3 = -15.5$ ksi, $\tau_{abs} = 28.0$ ksi.
7.58	$\sigma_1 = 85.0$ MPa, $\sigma_2 = 65.0$ MPa, $\sigma_3 = 0$, $\tau_{abs} = 42.5$ MPa.
7.60	$\sigma_1 = \sigma_2 = 240$ MPa, $\sigma_3 = -120$ MPa, $\tau_{abs} = 180$ MPa.
7.62	$\sigma_1 = 409$ ksi, $\sigma_2 = 148$ ksi, $\sigma_3 = -257$ ksi, $\tau_{max} = 333$ ksi.
7.64	(a), (b) $\sigma_1 = 8.22$ ksi, $\sigma_2 = 0$, $\sigma_3 = -10.2$ ksi.
7.68	$\sigma_1 = 46.4$ MPa, $\sigma_2 = -7.42$ MPa.
7.70	$\sigma_1 = 1700$ psi, $\sigma_2 = -5890$ psi, $\tau_{max} = 3800$ psi.
7.72	$T = 1.38$ kN·m.
7.74	$\sigma_1 = 2860$ psi, $\sigma_2 = -318$ psi, $\tau_{max} = 1590$ psi.
7.76	$\sigma_x = 1.11$ MPa, $\sigma_y = 0$, $\tau_{xy} = 0$.
7.78	$F = 1235$ N.
7.80	$\sigma = 142$ ksi, $\tau_{abs} = 71.3$ ksi.
7.82	$\sigma_h = 250$ MPa.
7.84	$\sigma_h = 15$ MPa, $\tau_{abs} = 7.65$ MPa.
7.86	$t = 6.02$ mm.
7.92	$\sigma = 7.10$ MPa, $|\tau| = 1.57$ MPa.
7.94	$\mathbf{t} = 2.67\mathbf{i} - 0.332\mathbf{j} + 3.00\mathbf{k}$ (ksi).
7.96	$\sigma = 6.86$ GPa, $|\tau| = 2.15$ GPa.
7.98	$|\tau| = 10.0$ MPa.
7.100	$\sigma = \sigma_1$, $|\tau| = 0$.

7.102	$\sigma = -353$ psi, $|\tau| = 44.5$ psi.
7.106	$\theta = -20.0°$.
7.108	$\sigma = -353$ psi, $\tau = -44.5$ psi.
7.110	Absolute maximum shear stress $= 6.54$ ksi.
7.112	$\sigma = 3530$ psi, $|\tau| = 1290$ psi.
7.114	$\tau_{abs} = 250$ MPa.
7.116	$\sigma_x = 13.5$ MPa, $\sigma_y = 0$, $\tau_{xy} = 4.77$ MPa.
7.118	$\sigma_x = 375$ psi, $\sigma_y = 0$, $\tau_{xy} = 0$.
7.120	2.11E5 Pa.
8.2	$\varepsilon_x = -0.004$, $\varepsilon_z = 0.0015$, $\gamma_{xz} = 0.00244$.
8.4	$\gamma_{xy} = 0.000997$.
8.6	$\theta = -20.0°$.
8.8	$\varepsilon_x = 0.00640$, $\varepsilon_y = 0.00850$, $\gamma_{xy} = 0$.
8.10	$\gamma_{x'y'} = 0.0098$, $\theta = 19.5°$ or $\gamma_{x'y'} = -0.0098$, $\theta = 40.2°$.
8.12	$PQ = 0.997$ mm.
8.14	$\gamma_{xy} = -0.00360$.
8.16	$\varepsilon_x = -0.00201$, $\varepsilon_y = 0.00811$, $\gamma_{xy} = -0.00250$.
8.18	1.57626 rad $(90.313°)$
8.20	$\varepsilon_a = 0.00232$, $\varepsilon_b = 0.00203$, $\varepsilon_c = 0.00475$.
8.22	$\varepsilon_x = 0.003$, $\varepsilon_y = -0.001$, $\gamma_{xy} = 0$.
8.24	$\varepsilon_d = -0.00605$.

(continued)

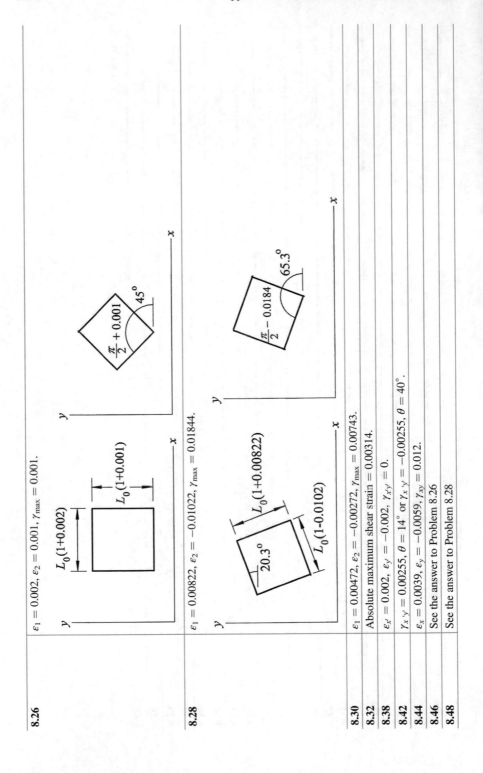

8.26 $\varepsilon_1 = 0.002$, $\varepsilon_2 = 0.001$, $\gamma_{max} = 0.001$.

8.28 $\varepsilon_1 = 0.00822$, $\varepsilon_2 = -0.01022$, $\gamma_{max} = 0.01844$.

8.30 $\varepsilon_1 = 0.00472$, $\varepsilon_2 = -0.00272$, $\gamma_{max} = 0.00743$.

8.32   Absolute maximum shear strain $= 0.00314$.

8.38 $\varepsilon_{x'} = 0.002$, $\varepsilon_{y'} = -0.002$, $\gamma_{xy} = 0$.

8.42 $\gamma_{x'y'} = 0.00255$, $\theta = 14°$ or $\gamma_{x'y'} = -0.00255$, $\theta = 40°$.

8.44 $\varepsilon_x = 0.0039$, $\varepsilon_y = -0.0059$, $\gamma_{xy} = 0.012$.

8.46   See the answer to Problem 8.26

8.48   See the answer to Problem 8.28

8.50	$\varepsilon_1 = 0.00368, \varepsilon_2 = -0.00608, \gamma_{max} = 0.00977.$
8.52	$\sigma_x = 42.5$ MPa, $\sigma_y = -37.2$ MPa, $\sigma_z = -26.4$ MPa, $\tau_{xy} = 19.4$ MPa, $\tau_{yz} = -12.9$ MPa, $\tau_{xz} = 10.8$ MPa.
8.54	The required condition is that $\sigma_x + \sigma_y = 0$.
8.56	$\sigma_x = 63.8$ ksi, $\sigma_y = 83.3$ ksi, $\tau_{xy} = -93.6$ ksi.
8.58	(a) $\lambda = 46.2$ GPa, $\mu = 30.8$ GPa. (b) $K = 66.7$ GPa.
8.60	$\sigma_x = 34.2$ ksi, $\sigma_y = -33.5$ ksi, $\sigma_z = -44.8$ ksi, $\tau_{xy} = 56.4$ ksi, $\tau_{yz} = 0, \tau_{xz} = -56.4$ksi.
8.62	(a) $\varepsilon_x = \sigma_x/E, \varepsilon_y = \varepsilon_z = -\nu\sigma_x/E$, other strain components equal zero. (b) Volume $= (1 + \sigma_x/E)(1 - \nu\sigma_x/E)^2 LA.$
8.64	$\sigma_x = -55.5$ MPa, $\sigma_y = 104$ MPa, $\tau_{xy} = -94.3$ MPa.
8.66	$\sigma_x = 413$ MPa, $\sigma_y = 444$ MPa, $\tau_{xy} = -74.5$ MPa.
8.68	$\varepsilon_{x'} = 0.002, \varepsilon_{y'} = -0.002, \gamma_{x'y'} = 0.$
8.70	$\varepsilon_x = 0.00360, \varepsilon_y = 0.00320, \gamma_{xy} = -0.00358.$
8.72	$\varepsilon_x = -0.00160, \varepsilon_y = 0.00100, \gamma_{xy} = 0.00201.$
8.74	Absolute maximum shear strain $= 0.00559.$
8.76	$\sigma_x = -253$ MPa, $\sigma_y = 809$ MPa, $\tau_{xy} = -234$ MPa.
9.2	$|v| = 3.03$ mm.
9.4	$w = 19.3$ lb/in.
9.6	Displacement magnitude is 10.4 mm, slope is zero
9.8	$v = -0.558$ in, $v' = -0.0106$ rad.
9.10	$|v| = 28.2$ mm.
9.22	$\sigma_x = 5.04$ MPa.
9.24	$|v| = 3.02$ mm at $x = 1.63$ m.
9.26	$M_0 = 98.2$ kN-m.
9.28	$x = 36.4$ in.
9.30	$A_x = 0, A_y = -3M_0/(2L), M_A = M_0/2$ clockwise, $B = 3M_0/(2L).$
9.32	$A_x = 0, A_y = 9w_0 L/40, M_A = 7w_0 L^2/40$ counterclockwise, $B_y = 11w_0 L/40.$

(continued)

9.34	$F_0 = 12, 200$ lb, $M_0 = 294, 000$ in-lb.
9.36	$x = 30.4$ in and 114 in.
9.38	$w_0 = 38.1$ kN/m.
9.40	$v = -w_0 L^4/(768EI)$.
9.42	$v = 27.1$ mm.
9.44	$A = 2295$ lb, $M_A = 91, 800$ in-lb counterclockwise.
9.46	$v = -(w_0/384EI)(16x^4 - 38Lx^3 + 29L^2x^2 - 8L^3x + L^4)$.
9.48	$v = (M_0/24LEI)[4\langle x \rangle^3 - 12L\langle x - L/2 \rangle^2 - 4\langle x - L \rangle^3 - L^2\langle x \rangle]$.
9.50	$v = -10.1$ mm.
9.52	$v = -20.0$ mm.
9.54	$F = 3.83$ kN.
9.56	$v_B = -M_0 L^2/2EI$.
9.58	$v_B = -w_0 L^4/8EI$.
9.60	$v_B = -6.22$ mm.
9.62	$v = -(Fx^2/6EI)(3L - x) + M_0 x^2/2EI$.
9.64	$v = -(w_0 x/24EI)(L^3 - 2Lx^2 + x^3) + (Fx/48EI)(3L^2 - 4x^2)$.
9.66	$A = C = 1.5$ kN, $B = 5$ kN.
9.68	$v = -0.225$ in.
9.70	$v = -w_0 L^4/(384EI)$.
9.72	$B = 2.71$ kN, $C = 1.71$ kN.
9.76	$A_x = 0$, $A_y = 11F/16$, $M_A = 3LF/16$ counterclockwise, $B_y = 5F/16$.
9.78	$v = -0.007L^3 F/(EI)$.
9.80	$|M_0| = 72, 700$ in-lb.

9.82	$v = -66.6$ mm.
10.2	$P = 965$ kip.
10.4	$P = 202$ kN.
10.6	$P = 806$ kN.
10.10	$R = 21.5$ mm.
10.12	$F = 80.4$ kip.
10.14	$F = 3.02$ kN.
10.16	$m = 250$ kg.
10.18	$b = 3.10$ in, $F = 5.66$ kip.
10.20	$R_{AB} = 28.0$ mm, $R_{CD} = 0.798$ mm.
10.22	$F = 368$ N.
10.24	$F = 3.18$ kN.
10.26	$P = 1.25$ MN.
10.28	$P = 31.7$ kip.
10.30	$P = 588$ kN.
10.32	$P = 7.20$ MN. It bends in the x-y plane
10.34	$P = 1.66$ MN. It bends in the x-y plane

(continued)

10.36

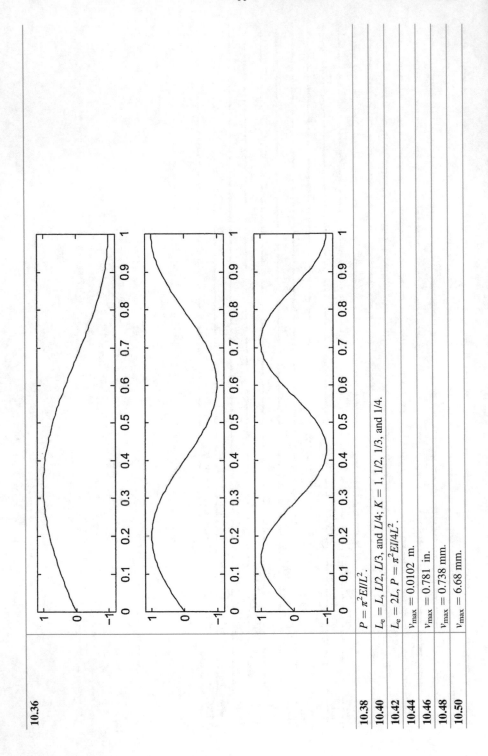

10.38 $P = \pi^2 EI/L^2$.

10.40 $L_e = L$, $L/2$, $L/3$, and $L/4$; $K = 1$, $1/2$, $1/3$, and $1/4$.

10.42 $L_e = 2L$, $P = \pi^2 EI/4L^2$.

10.44 $v_{max} = 0.0102$ m.

10.46 $v_{max} = 0.781$ in.

10.48 $v_{max} = 0.738$ mm.

10.50 $v_{max} = 6.68$ mm.

10.52	$P = 264$ kN.
10.56	$P = 6.70$ MN.
10.58	$R_i = 15.5$ mm.
10.60	$h/b = 2.86$.
10.62	$\sigma_{max} = 161$ MPa.
11.2	**(a)** $U = 0.00167$ J (joules, or N-m). **(b)** $W = 0.00167$ J.
11.4	**(a)** $U = 386$ in-lb. **(b)** $W = 386$ in-lb.
11.6	$\theta = 0.00321$ rad (0.184°) counterclockwise
11.8	$F = 98.6$ kN.
11.10	$v = \left(1 + 2\sqrt{2}\right) FL_0/(EA)$.
11.12	$v = 0.833$ in.
11.14	$v = FL^3/12EI$.
11.16	$u = 1.24$ MJ/m³.
11.20	$v = 0.0977$ in.
11.22	$\theta_B = w_0 L^3/6EI$.
11.24	$v_B = M_0 L^2/2EI$.
11.26	$v = FL^3/12EI$.
11.28	$3M_0/2L$ upward.
11.30	$v = 0.257$ in.
11.32	$v = 11.1$ mm.
12.2	$W = 6$ in.
12.4	Spacing is satisfactory
12.6	$C = 1.35$.
12.8	$d_2 = 1.0$ in.
12.10	Naval brass, hard
12.12	$a = 0.0855$ in.
12.14	10.2 MW.

(continued)

12.16	$W = 500$ kN.
12.18	Case (a).
12.20	$M = 50.7$ kN-m.
12.22	$a = 0.137$ m, $C = 1.4$.
12.24	$M = 23.7$ kN-m.
12.26	It will fail
12.28	It will suffice
12.30	No
12.32	15% increase
12.34	$FS_T = 1.20$.
12.36	$R = \left[\frac{10}{\pi \sigma_Y} \sqrt{M^2 + T^2} \right]^{1/3}$.
12.38	$FS_M = 13.1$.
12.40	$\tau_{xy} = 120$ MPa or $\tau_{xy} = -120$ MPa.
12.42	3320 cycles
12.44	Damage is 0.036, 0.376, and 0.520. 9.561E3 cycles
12.46	15.2 years
12.48	33, 800 cycles.
12.50	5.34 years
12.52	$\sigma = 6.76$ ksi.
12.54	(a) $\sigma = 17.4$ ksi. (b) $\sigma = 35.0$ ksi. (c) $\sigma = 47$ ksi.
12.56	$h = 0.5$ m.
12.58	Yes, the plate will fail
12.60	Yes, it will
12.62	$a = 0.221$ in.
12.64	$c_0 = 28.6$ mm. It is satisfied.
12.66	936 kilocycles.
12.68	$W = 2040$ lb.

12.70	(a) Yes. (b) Yes. (c) 650 kilocycles.
C.2	$\bar{x} = a(n+1)/(n+2)$, $\bar{y} = ca^n(n+1)/(4n+2)$.
C.4	$\bar{x} = 0$, $\bar{y} = 1.6$ ft.
C.6	$\bar{x} = 0.5$, $\bar{y} = -7.6$.
C.8	$\bar{x} = \bar{y} = 4R/(3\pi)$.
C.10	$\bar{x} = 70.9$ mm, $\bar{y} = 0$.
C.12	$\bar{x} = 12.0$ in, $\bar{y} = 5.49$ in.
C.14	$\bar{x} = 344$ mm, $\bar{y} = 456$ mm.
C.16	$I_x = \frac{1}{12}bh^3$, $I_y = \frac{1}{12}hb^3$, $I_{xy} = \frac{1}{24}b^2h^2$.
C.18	$I_x = I_y = \frac{1}{16}\pi R^4$, $I_{xy} = \frac{1}{8}R^4$.
C.20	$I_x = 2.65\text{E}8$ mm^4.
C.22	$I_{xy} = 92.3$ ft^4.
C.24	$I_x = 237$ in^4, $I_y = 115$ in^4, $I_{xy} = 119$ in^4.
C.26	$I_x = 88.8$ m^4, $I_y = 65$ m^4.
C.28	$\theta_p = 0$, $I_x = 85.33$ m^4, $I_y = 5.33$ m^4.
C.30	$\theta_p = 35.6°$, $I_x = 1.02\text{E-}5$ m^4, $I_y = 1.12\text{E-}6$ m^4.
C.32	$\theta_p = 26.6°$, $I_x = 10$ ft^4, $I_y = 2.5$ ft^4.

Index

A

Absolute maximum shear stress
  determination of, 520
  explanation of, 522, 527, 557
  as failure criterion, 845, 891
Aging, 196
Allowable stress
  in axially loaded bars, 195
  beam design issues and, 383–391
  design issues related to, 491
  and failure criteria, 283
  in pressure vessels, 579–588
  in torsion, 282, 285
American standard channel (C) structural
    steel shapes, 964–965
American standard (S) structural steel shapes,
    960–962
Angle (L) structural steel shapes, 965–967
Angle of twist, 219
Areas
  centroids, 907
  moments and products of inertia of, 921
  properties of, 953
Average stress
  in axially loaded bars, 40, 41, 89
  determined on given plane within an
    object, 42, 593
  examples determining, 42
  in pin, 41
Axial force
  in beams, 301–307
  material behavior and, 125, 126
  prismatic bars subjected to, 99–112,
    124
Axial strain, 124–127

Axially loaded bars
  average stress in, 40
  Castigliano's second theorem and, 807–810
  design issues and, 195–197, 205
  distributed loads and nonprismatic,
    164–174
  material behavior and, 125
  statically indeterminate problems in,
    145–157
  strain energy of, 791
  strains in prismatic, 125, 202
  stress concentrations and, 825, 831
  stress in prismatic, 40, 99–104
  thermal strains in, 181–188, 204

B

Bars
  average stress in axially loaded, 40
  distributed loads, 167
  with gradually varying cross sections, 164
  subjected to combined loads, 566–575
  torsion of prismatic circular, 218–230, 290
  See also Axially loaded bars
Beams
  assymmetric cross sections, 424
  axial force, shear force, and bending
    moment in, 301–307, 322, 357
  Castigliano's second theorem and,
    806, 813
  composite, 395–407, 492, 500
  deflections of, 322, 671–726, 969–975
    (*see also* Deflections of beams)
  design issues related to, 383–391, 491
  elastic-perfectly plastic, 410–420, 493, 495

Printed in the United States
By Bookmasters